T0145227

The series "Advances in Intelligent Systems and Computing" contains publications on theory, applications, and design methods of Intelligent Systems and Intelligent Computing. Virtually all disciplines such as engineering, natural sciences, computer and information science, ICT, economics, business, e-commerce, environment, healthcare, life science are covered. The list of topics spans all the areas of modern intelligent systems and computing such as: computational intelligence, soft computing including neural networks, fuzzy systems, evolutionary computing and the fusion of these paradigms, social intelligence, ambient intelligence, computational neuroscience, artificial life, virtual worlds and society, cognitive science and systems, Perception and Vision, DNA and immune based systems, self-organizing and adaptive systems, e-Learning and teaching, human-centered and human-centric computing, recommender systems, intelligent control, robotics and mechatronics including human-machine teaming, knowledge-based paradigms, learning paradigms, machine ethics, intelligent data analysis, knowledge management, intelligent agents, intelligent decision making and support, intelligent network security, trust management, interactive entertainment, Web intelligence and multimedia.

The publications within "Advances in Intelligent Systems and Computing" are primarily proceedings of important conferences, symposia and congresses. They cover significant recent developments in the field, both of a foundational and applicable character. An important characteristic feature of the series is the short publication time and world-wide distribution. This permits a rapid and broad dissemination of research results.

** Indexing: The books of this series are submitted to ISI Proceedings, EI-Compendex, DBLP, SCOPUS, Google Scholar and Springerlink **

More information about this series at http://www.springer.com/series/11156

Advances in Intelligent Systems and Computing

Volume 834

Series Editor

Janusz Kacprzyk, Systems Research Institute, Polish Academy of Sciences, Warsaw, Poland

Advisory Editors

Nikhil R. Pal, Indian Statistical Institute, Kolkata, India

Rafael Bello Perez, Faculty of Mathematics, Physics and Computing, Universidad Central de Las Villas, Santa Clara, Cuba

Emilio S. Corchado, University of Salamanca, Salamanca, Spain

Hani Hagras, Electronic Engineering, University of Essex, Colchester, UK

László T. Kóczy, Department of Automation, Széchenyi István University, Gyor, Hungary

Vladik Kreinovich, Department of Computer Science, University of Texas at El Paso, El Paso, TX, USA

Chin-Teng Lin, Department of Electrical Engineering, National Chiao Tung University, Hsinchu, Taiwan

Jie Lu, Faculty of Engineering and Information Technology, University of Technology Sydney, Sydney, NSW, Australia

Patricia Melin, Graduate Program of Computer Science, Tijuana Institute of Technology, Tijuana, Mexico

Nadia Nedjah, Department of Electronics Engineering, University of Rio de Janeiro, Rio de Janeiro, Brazil

Ngoc Thanh Nguyen, Faculty of Computer Science and Management, Wrocław University of Technology, Wrocław, Poland

Jun Wang, Department of Mechanical and Automation Engineering, The Chinese University of Hong Kong, Shatin, Hong Kong

Jeng-Shyang Pan · Jerry Chun-Wei Lin ·
Bixia Sui · Shih-Pang Tseng
Editors

Genetic and Evolutionary Computing

Proceedings of the Twelfth International
Conference on Genetic and Evolutionary
Computing, December 14–17, Changzhou,
Jiangsu, China

 Springer

Editors
Jeng-Shyang Pan
Fujian Provincial Key Laboratory of Big
Data Mining and Applications
Fujian University of Technology
Fuzhou, Fujian, China

Bixia Sui
Changzhou College of Information
Technology
Changzhou, Jiangsu, China

Jerry Chun-Wei Lin
Department of Computing, Mathematics,
and Physics
Western Norway University of Applied
Sciences
Bergen, Norway

Shih-Pang Tseng
College of Smart Living and Management
Tajen University
Pingtung, Taiwan

ISSN 2194-5357   ISSN 2194-5365 (electronic)
Advances in Intelligent Systems and Computing
ISBN 978-981-13-5840-1   ISBN 978-981-13-5841-8 (eBook)
https://doi.org/10.1007/978-981-13-5841-8

Library of Congress Control Number: 2019930370

This Springer imprint is published by the registered company Springer Nature Singapore Pte Ltd.
The registered company address is: 152 Beach Road, #21-01/04 Gateway East, Singapore 189721,
Singapore

Conference Organization

Honorary Chairs

Yong Zhou, Changzhou College of Information Technology, China
Jeng-Shyang Pan, Fujian University of Technology, China
Jhing-Fa Wang, Tajen University, Taiwan
Xin Tong, Fujian University of Technology, China

Advisory Committee Chairs

Tzung-Pei Hong, National University of Kaohsiung, Taiwan
Bixia Sui, Changzhou College of Information Technology, China

General Chairs

Jerry Chun-Wei Lin, Western Norway University of Applied Sciences, Norway
Pyke Tin, University of Computer Studies, Myanmar
Vaclav Snasel, VSB-Technical University of Ostrava, Czech Republic

Program Chairs

I-Hsin Ting, National University of Kaohsiung, Taiwan
Lyuchao Liao, Fujian University of Technology, China
Philippe Fournier-Viger, Harbin Institute of Technology (Shenzhen), China

Shih-Pang Tseng, Tajen University, Taiwan
Shu-Chuan Chu, Flinders University, Australia

Local Organization Chair

Jianhui Zhao, Changzhou College of Information Technology, China

Electronic Media Chairs

Shi-Jian Liu, Fujian University of Technology, China
Thi-Thi Zin, University of Miyazaki, Japan

Publication Chair

Chien-Ming Chen, Harbin Institute of Technology (Shenzhen), China

Finance Chair

Jui-Fang Chang, National Kaohsiung University of Sciences and Technology,
Taiwan

Program Committee Members

An-chao Tsai, Tajen University, Taiwan
Ashish Sureka, Ashoka University, India
Bay Vo, Ho Chi Minh City University of Technology, Vietnam
Bo Zhang, Changzhou Vocational Institute of Mechatronic Technology, China
Chao Huang, Southeast University, China
Chien-Ming Chen, Harbin Institute of Technology (Shenzhen), China
Chuan Zhu, Changzhou College of Information Technology, China
Chun-Hao Chen, Tamkang University, Taiwan
Gang Sun, Nanjing College of Information Technology, China
Hong Zhu, Changzhou Vocational Institute of Textile and Garment, China
Hui Wang, Nanchang Institute of Technology, China
Hsun-Hui Huang, Tajen University, Taiwan

Hua Lou, Changzhou College of Information Technology, China
I-Hsin Ting, National University of Kaohsiung, Taiwan
Jean Lai, Hong Kong Baptist University, Hong Kong
Jerry Chun-Wei Lin, Western Norway University of Applied Sciences, Norway
Ji Zhang, University of Southern Queensland, Australia
Jimmy Ming-Tai Wu, Shandong University of Science and Technology, China
Kawuu W. Lin, National Kaohsiung University of Sciences and Technology, Taiwan
Lan Cao, Changzhou College of Information Technology, China
Ming-Jin Li, Huaian Vocational College of Information Technology, China
Ming-Wei Liu, Tianjin Bohai Vocational Technical College, China
Pei-Wei Tsai, Swinburne University of Technology, Australia
Qian Yin, Changzhou College of Information Technology, China
Shouyun Chu, Changzhou Vocational Institute of Light Industry, China
Thi-Thi Zin, University of Miyazaki, Japan
Tsu-Yang Wu, Fujian University of Technology, China
Tzung-Pei Hong, National University of Kaohsiung, Taiwan
Wei-Min Chen, Yili Vocational and Technical College, China
Wen-Quan Zeng, Guangdong University of Science and Technology, China
Wensheng Gan, Harbin Institute of Technology (Shenzhen), China
Xiang-Yang Liu, Nanjing University, China
Xiao-Hua Zou, Changzhou College of Information Technology, China
Xiao-Jun Fan, Shanghai University, China
Xiao-Jun Wu, Jiangnan University, China
Xiao-Yan Tang, Changzhou College of Information Technology, China
Xin-Chang Zhang, Nanjing University of Information Science and Technology, China
Xin-Hua Li, Shandong Polytechnic, China
Yi-Chun Lin, Tajen University, Taiwan
Yong Zhou, Changzhou Vocational Institute of Engineering, China
Yong-Hua Xie, Nanjing Institute of Industry Technology, China
Yuh-Chung Lin, Tajen University, Taiwan
Yun-Fei Wu, Changzhou College of Information Technology, China
Yu-Rong Ren, Changzhou University, China
Zai-Wu Gong, Nanjing University of Information Science and Technology, China
Ze-Zhong Xu, Changzhou Institute of Technology, China
Zhong-Ping Qian, Changzhou College of Information Technology, China

Preface

This volume composes the Proceedings of the Twelfth International Conference on Genetic and Evolutionary Computing (ICGEC 2018), which was hosted by Changzhou College of Information Technology and was held in Changzhou, Jiangsu, China on December 14–17, 2018. ICGEC 2018 was technically cosponsored by Changzhou College of Information Technology (China), Fujian Provincial Key Lab of Big Data Mining and Applications (Fujian University of Technology, China), National Demonstration Center for Experimental Electronic Information and Electrical Technology Education (Fujian University of Technology, China), Tajen University (Taiwan), National University of Kaohsiung (Taiwan), Shandong University of Science and Technology (China), Western Norway University of Applied Sciences (Norway), and Springer. It aimed to bring together researchers, engineers, and policymakers to discuss the related techniques, to exchange research ideas, and to make friends. Eighty-one excellent papers were accepted for the final proceeding. Six plenary talks were kindly offered by: Prof. James, Jhing-Fa Wang (President of Tajen University, Taiwan, IEEE Fellow), Prof. Zhigeng Pan (Hangzhou Normal University, China), Prof. Xiudong Peng (Chengzhou College of Information Technology, China), Prof. Jiuyong Li (University of South Australia, Australia), Prof. Philippe Fournier-Viger (Harbin Institute of Technology (Shenzhen), China), and Prof. Peter Peng (University of Calgary, Canada). We would like to thank the authors for their tremendous contributions. We would also like to express our sincere appreciation to the reviewers, Program Committee members, and the Local Committee members for making this conference successful.

Fuzhou, China   Jeng-Shyang Pan
Bergen, Norway   Jerry Chun-Wei Lin
Changzhou, China   Bixia Sui
Pingtung, Taiwan   Shih-Pang Tseng
December 2018

Contents

Part I
Nature Inspired Constrained Optimization

A New Advantage Sharing Inspired Particle Swarm Optimization Algorithm

Lingping Kong and Václav Snášel

Abstract Particle swarm optimization algorithm is a widely used computational method for optimizing a problem. This algorithm has been applied to many applications due to its easy implementation and few particles required. However, there is a big problem with the PSO algorithm, all the virtual particles converged to a point which may or may not be the optimum. In the paper, we propose an improved version of PSO by introducing the idea of advantage sharing and pre-learning walk mode. The advantage sharing means that the good particles share their advantage attributes to the evolving ones. The pre-learning walk mode notices one particle if it should continue to move or not which uses the feedback of the last movement. Two more algorithms are simulated as the comparison methods to test Benchmark function. The experimental results show that our proposed scheme can converge to a better optimum than the comparison algorithms.

Keywords Particle swarm optimization · Advantage sharing · Benchmark function

1 Introduction

A collective behavior of natural or artificial is the concept of Swarm intelligence (SI) [1]. The agents follow very simple rules, and although there is no centralized control structure guiding how members should behave, communications between such agents converge to the global behavior. Swarm Intelligence-based techniques can be used in a number of applications. Swarm Intelligence-based algorithms include Cat Swarm Optimization [2], Genetic algorithm (GA), Artificial Bee Colony Optimization [3],

L. Kong · V. Snášel (✉)
Faculty of Electrical Engineering and Computer Science, VSB-Technical
University of Ostrava, Ostrava, Czech Republic
e-mail: vaclav.snasel@vsb.cz

L. Kong
e-mail: konglingping2007@163.com

© Springer Nature Singapore Pte Ltd. 2019
J.-S. Pan et al. (eds.), *Genetic and Evolutionary Computing*,
Advances in Intelligent Systems and Computing 834,
https://doi.org/10.1007/978-981-13-5841-8_1

Particle swarm optimization (PSO) [4], Shuffle frog leaping algorithm (SFLA) [5], and so on.

PSO is originally proposed by Kennedy and Eberhart for simulating social behavior, then this algorithm was simplified and was observed to be performing optimization. A number of candidate particles represent problem solutions are simulated and iteratively improved toward the best. The most common type of implementation defines the updating mode with two equations, velocity updating, and position updating. There are many variants of PSO [6], some researchers seek ways to simplify the algorithm, some versions improve the updating mode by adding extra features to it [7], some applications combine PSO with other algorithms attributes to solve problems [8].

Ahmad Nickabadi is one of the authors who studied the parameter decision of updating equation. The author analyzed and discussed three main groups of choosing inertia weight value, constant, time-varying, and adaptive inertia weights. Except for that, the author proposed a new approach with adaptive inertia weight parameter, the populations situation can be reflected based on the feedback parameter which uses the success rate of the particles [9]. Moradi [10] combines the genetic algorithm and particle swarm optimization for location decision use on a distribution system. The author uses the combination strategy to minimize network power losses and improve the voltage stability, and a good performance is demonstrated by carrying out this strategy on 33 and 69 bus systems. A hybrid algorithm used a fuzzy adaptive particle swarm optimization (FAPSO) and Nelder–Mead simplex search is proposed by Taher Niknam [11]. This hybrid algorithm completes the local search through the Nelder–Mead process and accomplishes the global search by FAPSO. The author verified the hybrid algorithm by testing it on two typical systems consisting of 13 and 40 thermal units. Cheng [12] proposed a social learning particle swarm optimization (SLPSO). This algorithm adopts a dimension-dependent parameter to relief the parameter settings and it also uses demonstrators in the current swarm to guide the particle evolving, which unlike the classical PSO variants. The experimental results show the algorithm performs well on low-dimensional and high-dimensional problems.

The rest of paper is organized as follows: Sect. 2 introduces the disadvantage of PSO algorithm and presents the idea of advantage sharing, then proposes the AS-PSO algorithm. Section 3 gives the results of comparison experiments tested on Benchmark function. Section 4 concludes the paper.

2 Advantage Sharing Particle Swarm Optimization

The particle swarm optimization (PSO) is a computational method and also a population-based stochastic algorithm. PSO optimizes a problem by updating candidate solutions iteratively, which also means that there are no selection and random walk during the running process. The candidate members are active based on its local best-known location over the individuals position and velocity, and their movements are influenced by the best one known location in searching space. Even PSO can search very large spaces of candidate solutions, but it does not guarantee an optimum solution. In other words, the PSO algorithm may converge to a local optimum. So this paper proposes an improved version of PSO with the idea of advantage sharing and pre-learning walk, called AS-PSO.

In the advantage sharing, one individual is guided toward several the best-known positions instead of one best position, the common properties of those better positions could be the characteristics of the optimum location in a big probability, and advantage sharing decreases the occurrence of trapping into a local optimum. Before the population updating, each particle will be set up a movement flag, which labels the moving direction of its last step, and before each particles move, there is also a maximum steps threshold holds the particles moving times. The movement flag works as the pioneer wizard, it tells the particles whether the last movement gets better or not. If one particle moves in a direction and gets improvement, then it might get more improvement after moving along this direction. AS-PSO algorithm consists of a set of virtual particle population. During the searching process, every particle will evolve based on the basic PSO operation. The different points are in three parts, first before an operation, each particle will be given a movement flag, if the previous movement is good, then continue to move until it reaches the biggest moving threshold. Second, if the first step searching does not improve its current status, then goes to the advantage sharing operation. Third, adding a small proportional random walk operation. The process of AS-PSO in detail is introduced as follows:

2.1 Steps of the AS-PSO

Step 0. Setting-up phase. Define a problem space with D dimensions, the virtual population P with N particles. The evolving iteration times is t. define the final solution variable as g_{best}, and $threshold_1$ is the maximum step number that one individual can move, $threshold_2$ is the probability of random walk for a particle. Set $MC = 0$ (Moving Count, MC).

Step 1. Initialize population $P = \{p_1, p_2, \ldots, p_N\}$, and each particle has its random position and velocity within the search space. For particle j, its position can be labeled as $X_j = \{x_1, x_2, \ldots, x_D\}$, the velocity is $V_j = \{v_1, v_2, \ldots, v_D\}$. Other than that, each particle has p_h value, it stores its historical best-known ever position.

Step 2. Compute the evaluation value (V) for the population, store the one with best evaluation value particle as the g_{best}. Sort the population based on the evaluation value in a ascending order.

Step 3. Take a particle into **Updating mode** (Introduced in next section), and update its searching location. Do the same updating process for all the particles.

Step 4. Evaluate the new locations of particles, and compare the current evaluation value with its p_h value. If the current location is better then replace the p_h value.

Step 5. Loop to *step* 2 until a condition is met, either the loop index reaches a maximum value or the g_{best} is satisfied with the need.

2.2 Updating Mode

Suppose particle j, P_j is under updating mode, its $X_j = \{x_{j,1}, x_{j,2}, \ldots, x_{j,D}\}$ and $V_j = \{v_{j,1}, v_{j,2}, \ldots, v_{j,D}\}$. Copy $\widetilde{p}_j = p_j$.

Step 1. Set $MC = MC + 1$, if MC is smaller than $threshold_1$, change the velocity and position of $\widetilde{p}_j$ based on Eqs. 1 and 2. Otherwise, exit the updating mode. in the equation w, c_1 and c_2 are positive constants, r_1 and r_2 are two random numbers in the range $[0,1]$, symbol $i \in [1, 2, \ldots, D]$.

$$v_{j,i}(t + 1) = w \times v(t) + c_1 \times r_1 \times (x_{p_h,i} - x_{j,i}) + c_2 \times r_2 \times (x_{g_{best},i} - x_{j,i})$$
(1)

$$x_{j,i}(t + 1) = x_{j,i}(t) + v_{j,i}(t + 1)$$
(2)

Step 2. Compare the $\widetilde{p}_j$ to the p_j, if $\widetilde{p}_j$ is better, then replace p_j with $\widetilde{p}_j$ and go back to Step 1. Otherwise, check the MC value, if MC equals to 0, go to step 3, If not, exit the updating mode.

Step 3. Generate a random number ran. if ran is smaller than $threshold_2$. Use Eq. 3 to update p_j. Otherwise, use Eq. 4. In the Eq. 3 m is a integer, which controls the number of good particles $gp = \{gp_1, gp_2, \ldots, gp_m\}$ used in advantage sharing process. In the Eq. 4, $x_{boundary}$ is a searching space related value.

$$gp_i = r + j, \quad r \in [0, N - j]$$
(3)
$$x_{j,i} = \left(\sum_{k=1}^{m} x_{gp_k,i}\right) \div m$$

$$x_{j,i}(t + 1) = rand \times x_{boundary}, \quad rand \in [0, 1]$$
(4)

2.3 Pseudocode of AS-PSO

The pseudocode of AS-PSO is showed as

3 Experiment Results

Three different algorithms are simulated in our experiment: *PSO*, *SLPSO*, our scheme. Rosenbrock (5), Rastrigin (6) and Sphere (7) this three benchmark functions are used to test their performance. The benchmark function $f(x)$ is shown as follows:

Algorithm 1 Pseudocode of AS-PSO

1: **function** AS- PSO($P = \{p_1, p_2, \ldots, p_N\}, g_{best}$)
2: Input $P, threshold_1, threshold_2, D, c_1, c_2, x_{boundary}$
3: Output: g_{best}
4:     **for** $i = 0$ to N **do**
5:         Initialize p_i, Compute V(p_i)
6:         Store p_h
7:     **end for**
8:     Store g_{best}
9:     **for** $t = 1$ to iteration times **do**
10:         **for** $j = 0$ to N **do**
11:             $\widetilde{p_j} = p_j$
12:             $MC = 0$
13:             **for** MC to $threshold_1, MC+ = 1$ **do**
14:                 **for** $i = 0$ to D, do $\widetilde{p_j}$ **do**
15:                     $v_{j,i}(t + 1) = v(t) + c_1 \times r_1 \times (x_{p_h,i} - x_{j,i}) + c_2 \times r_2 \times (x_{g_{best},i} - x_{j,i})$
16:                     $x_{j,i}(t + 1) = x_{j,i}(t) + v_{j,i}(t + 1)$
17:                 **end for**
18:                 **if** V(p_j)is worse than V($\widetilde{p_j}$) **then**
19:                     $p_j = \widetilde{p_j}$
20:                 **end if**
21:             **end for**
22:             **if** $ran \leq threshold_2$,do p_j **then**
23:                 Sort(P)
24:                 **if** $MC = 0$ **then**
25:                     **for** $i = 0$ to m **do**
26:                         $gp_i = r + j, \quad r \in [0, N - j]$
27:                         $x_{j,i}+ = (\sum_{k=1}^{m} x_{gp_k,i}) \div m$
28:                     **end for**
29:                 **else**
30:                     **for** $i = 0$ to m **do**
31:                         $x_{j,i}(t + 1) = rand \times x_{boundary}, \quad rand \in [0, 1]$
32:                     **end for**
33:                 **end if**
34:             **end if**
35:         **end for**
36:         Store p_h for P, Store g_{best}
37:     **end for**
38:     Output g_{best}
39: **end function**

$$f_1(x) = f(x_1, x_2, \ldots, x_d) = \sum_{i=1}^{d}[100 \times (x_{2i-1}^2 - x_{2i})^2 + (x_{2i} - 1)^2] \quad (5)$$

Table 1 Symbols description

Symbols	Description	Values
N	The number of population	16
D	The number of dimension	3
m	The good particles used in advantage sharing	2
c_1, c_2	Acculate coefficent	1.49445
$x_{boundary}$	The boundary value of space	$[-100, 100]$
$threshold_1$	The steps value in Updating model process	2
$threshold_2$	The probability of random walk and advantage sharing	0.5
iteration	The iteration times used in three algorithm	100

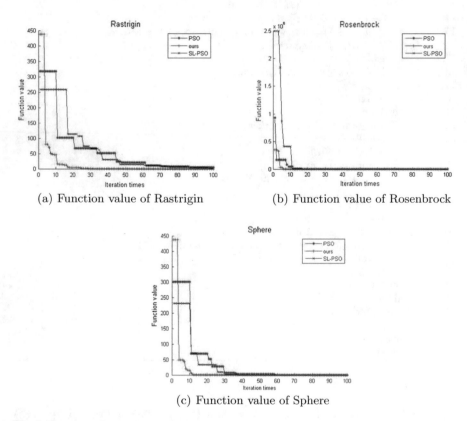

(a) Function value of Rastrigin   (b) Function value of Rosenbrock

(c) Function value of Sphere

Fig. 1 Function value convergence curves

$$f_2(x) = f(x_1, x_2, \ldots, x_d) = Ad + \sum_{i=1}^{d}\left[x_i^2 - A \times \cos(2\pi x_i)\right] \qquad (6)$$

Table 2 Comparison results in Function value

BenchMark	Methods		
	PSO	SLPSO	AS-PSO
$f_1(x)$	4248.89485	0.53217	0.20247
$f_2(x)$	6.35623	2.55794	0.000000
$f_3(x)$	0.13459	0.00036	0.000000

Table 3 The g_{best} positions in Rastrigrin Function

Values	Methods		
	PSO	SLPSO	AS-PSO
x_1	1.057379	−0.065119	7.132366×10^{-8}
x_2	−0.0499616	0.0495942	-3.333312×10^{-8}
x_3	2.014951	0.080029	2.703721×10^{-8}

$$f_3(x) = f(x_1, x_2, \ldots, x_d) = \sum_{i=1}^{d} x_i^2 \tag{7}$$

The parameters used in experiment are showed in Table 1.

The convergence curves of three benchmark function in three algorithms are showed in Fig. 1, the x-coordinate is the iteration times, and the y-coordinate is the function value. The AS-PSO has the best convergence speed, followed by SLPSO, and finally PSO.

Table 2 lists the final benchmark function value, from the table we can tell that AS-PSO shows a better performance, the benchmark function value are around 0.000000, which is close to the minimum value of Rastrigin and Sphere. Table 3 lists the final position x_1, x_2, x_3 of g_{best} in Rastrigin function, its position axes should be (0.0, 0.0, 0.0) in minimum value. From the table, it tells AS-PSO is better than the other comparison algorithms.

4 Conclusion

In this paper, we propose a variant of particle swarm optimization with the idea of advantage sharing and pre-learning walk. One particle evolves its position based on several other excellent individuals instead of one best-known member, which is Advantage sharing. It is right in a big probability to go more steps along the benefit obtained way. Furthermore, this algorithm also adds a small proportional random walk operation for avoiding converging local optimum. In the end, the original PSO and a recently proposed improved PSO are simulated. The experimental results show that AS-PSO has a good convergence speed and the final results are better than the comparison algorithms.

References

1. Derrac, J., Salvador, G., Daniel, M., Francisco, H.: A practical tutorial on the use of nonparametric statistical tests as a methodology for comparing evolutionary and swarm intelligence algorithms. Swarm Evolut. Comput. **1**(1), 3–18 (2011)
2. Shu-Chuan, C., Pei-Wei, T., Jeng-Shyang, P.: Cat swarm optimization. In: Pacific Rim International Conference on Artificial Intelligence. Springer, Berlin, Heidelberg (2006)
3. Thi-Kien, D., Tien-Szu, P., Trong-The, N., Shu-Chuan, C.: A compact articial bee colony optimization for topology control scheme in wireless sensor networks. J. Inf. Hiding Multimed. Signal Process. **6**(2), 297–310 (2015)
4. Wuling, R., Cuiwen, Z.: A localization algorithm based On SFLA and PSO for wireless sensor network. Inf. Technol. J. **12**(3), 502–505 (2013)
5. Kaur, P., Shikha, M.: Resource provisioning and work flow scheduling in clouds using augmented Shuffled Frog Leaping Algorithm. J. Parallel Distrib. Comput. **101**, 41–50 (2017)
6. Ishaque, K., Zainal, S., Muhammad, A., Saad, M.: An improved particle swarm optimization (PSO) based MPPT for PV with reduced steady-state oscillation. IEEE Trans. Power Electron. **27**(8), 3627–3638 (2012)
7. Kennedy, J.: Particle swarm optimization. In: Encyclopedia of machine learning, pp. 760–766. Springer, Boston, MA (2011)
8. Marinakis, Y., Magdalene, M.: A hybrid genetic particle swarm optimization algorithm for the vehicle routing problem. Expert Syst. Appl. **37**(2), 1446–1455 (2010)
9. Nickabadi, A., Mohammad, M.E., Reza, S.: A novel particle swarm optimization algorithm with adaptive inertia weight. Appl. Soft Comput. **11**(4), 3658–3670 (2011)
10. Moradi, M.H., Abedini, M.: A combination of genetic algorithm and particle swarm optimization for optimal DG location and sizing in distribution systems. Int. J. Electrical Power Energy Syst. **34**(1), 66–74 (2012)
11. Niknam, T.: A new fuzzy adaptive hybrid particle swarm optimization algorithm for non-linear, non-smooth and non-convex economic dispatch problem. Appl. Energy **87**(1), 327–339 (2010)
12. Cheng, R., Yaochu, J.: A social learning particle swarm optimization algorithm for scalable optimization. Inf. Sci. **291**, 43–60 (2015)

Density Peak Clustering Based on Firefly Algorithm

Jiayuan Wang, Tanghuai Fan, Zhifeng Xie, Xi Zhang and Jia Zhao

Abstract In density peak clustering the choice of cut-off distance is not theoretically supported, and to address this concern, we propose density clustering based on firefly algorithm. The certainty between data is determined on the basis of density estimation entropy. The cut-off distance corresponding to the minimum entropy is found by iterative optimization of FA, and then substituted into the standard density clustering algorithm. Simulation experiments are conducted on eight artificial datasets. Compared with the standard density peak clustering, our method can choose the cut-off distance in a self-adaptive manner on different datasets, which improves the clustering effect.

Keywords Density peak clustering · Cut-off distance · Density estimation entropy · Firefly algorithm

1 Introduction

Clustering is a process of clustering dataset samples into clusters on the basis of similarity. As a result, samples within the same cluster have a higher similarity, and those across different clusters are low in similarity [1–3]. Clustering is able to extract

J. Wang · T. Fan (✉) · Z. Xie · X. Zhang · J. Zhao
School of Information Engineering, Nanchang Institute of Technology, Nanchang 330099, China
e-mail: fantanghuai@163.com

J. Wang
e-mail: wangjiayuan1201@163.com

Z. Xie
e-mail: xiezhifeng_nit@163.com

X. Zhang
e-mail: zhangxi95@vip.qq.com

J. Zhao
e-mail: zhaojia925@163.com

© Springer Nature Singapore Pte Ltd. 2019
J.-S. Pan et al. (eds.), *Genetic and Evolutionary Computing*,
Advances in Intelligent Systems and Computing 834,
https://doi.org/10.1007/978-981-13-5841-8_2

11

hidden pattern and rules from a large amount of data and considered an important means to identify knowledge from the data. At present, clustering has found extensive applications in data analysis [4] and engineering system [5].

Many clustering algorithms have emerged to deal with different types of datasets. Conventional clustering algorithms include partitioning-based clustering [6], hierarchical clustering method [7], density-based clustering method [8] and network-based clustering method [9]. To improve the clustering efficiency and reduce complexity, Rodriguez et al. [10] described clustering by fast search and find of density peaks (DPC) in 2014. This algorithm can rapidly find the density peak points for datasets of any shape and efficiently perform the assignment of data points and removal of outliers. Therefore, DPC is fit for clustering analysis over massive data. DPC is based on the concept of cut-off distance d_c, used as a density measure for the samples. The choice of d_c has a large impact on the clustering effect. However, d_c is generally determined subjectively with the current DPC, and to do this, the empirical value first needs to be obtained from extensive prior experiments for different datasets, which is definitely a major drawback for DPC.

Firefly Algorithm (FA) [11–14] is a new optimization technique based on swarm intelligence [15], proposed by Yang Xin-she in 2008. In this algorithm, each firefly is randomly distributed in the solution space and has a fitness value assigned by optimization corresponding to the intensity of the emitted light. The firefly determines the direction of its movement by comparing the intensity of the emitted light, and the distance of movement is determined by relative attractiveness. After the population evolves for some generations, most fireflies will be attracted to the firefly with the highest intensity of the emitted light, and the optimization task is completed. FA is structurally simple, needs fewer parameters and applies to many optimization problems.

Here, FA is introduced to address the drawbacks of DPC, and density peak clustering based on firefly algorithm (FADPA) is proposed as a modified approach. This novel algorithm uses density estimation entropy as the objective function of d_c, which is optimized iteratively by FA. Cut-off distance d_c is determined self-adaptively for different datasets, and then the clustering is performed by using standard DPC.

2 Relevant Studies

2.1 Density Peak Clustering

DPC can automatically identify the cluster center of the dataset samples and achieve high-efficiency clustering for datasets of any shape. Its basic working principle is as follows. An ideal clustering center has two basic features: (1) the local density of the cluster center is higher than that of its neighbors; (2) the distance between different cluster centers is relatively large. To find the cluster centers that meet the above criteria, the local density ρ_i for sample i and the distance δ_i from sample i

to sample j that has a local density greater than sample i and is nearest to it. It is defined as follows:

$$\rho_i = \sum_{j \neq i} \chi(d_{ij} - d_c), \tag{1}$$

where d_{ij} is the Euclidean distance between sample i and sample j; d_c is the cut-off distance. When $x < 0$, $\chi(x) = 1$; otherwise $\chi(x) = 0$.

$$\delta_i = \min_{j:\rho_j > \rho_i} (d_{ij}) \tag{2}$$

For sample i with the largest local density ρ_i, $\delta_i = \max(d_{ij})$.

It is known from formula (1) that the local density ρ of the sample is a discrete value and greatly influenced by the cut-off distance d_c. To reduce the influence of d_c on the local density ρ of the sample, literature [16] uses a Gaussian kernel function to compute the local density of the sample.

$$\rho_i = \sum_{j \neq i} \exp\left(-\left(\frac{d_{ij}}{d_c}\right)^2\right) \tag{3}$$

DPC constructs the decision graph of relative distance δ about local density ρ. Samples with large δ and ρ are the density peak points. However, when the size of each cluster is small, the difference between ρ and δ is insignificant for the sample. As a result, the samples are sparse and the density peak points are unclear. So it is very difficult to find the density peak points by comparing ρ or δ alone. To solve this problem, ρ and δ are normalized [17]. The cluster centers are found by using γ decision graph, and are defined as follows:

$$\gamma_i = \rho_i \cdot \delta_i \tag{4}$$

Apparently, sample points with large γ can be chosen as the centers. For the remaining sample j, DPC categorizes it into the cluster where the sample with a larger local density than j and nearest to j is located. In this way, the remaining sample is assigned.

2.2 Firefly Algorithm

FA is an optimization process based on swarm intelligence, which mimics the behavior of fireflies in nature attracting the opposite gender by the emitted light. To reduce the algorithm complexity, the gender of the fireflies is neglected. It is generally believed that the fireflies emitting less bright light will be attracted to those emitting

brighter light. The movement distance is determined by the relative attractiveness
between the fireflies. In the standard FA, relative attractiveness between the fireflies
is given by

$$\beta = \beta_0 e^{-\gamma r_{ij}^2}, \tag{5}$$

where r_{ij} is the Euclidean distance between firefly i and firefly j; β_0 is the attractiveness of the firefly at the distance $r_{ij} = 0$, a constant; γ is the light absorption
parameter, usually set to 1. For any two fireflies, their Euclidean distance is given by

$$r_{ij} = \|x_i - x_j\| = \sqrt{\sum_{d=1}^{D} (x_{id} - x_{jd})^2}, \tag{6}$$

where D is the problem dimension.

Hence the movement of firefly i to firefly j can be defined by

$$x_{id}(t+1) = x_{id}(t) + \beta_0 e^{-\gamma r_{ij}^2} (x_{jd}(t) - x_{id}(t)) + \alpha(t)\varepsilon, \tag{7}$$

where x_{id} and x_{jd} are the d-th dimension of firefly i and j, respectively. Parameter
$\varepsilon = (rand() - 1/2)$, and $rand()$ is a random function distributed uniformly in the
range [0, 1]. $\alpha(t)$ is the step factor, with the value range [0, 1]. t is the number of
iterations.

Through the above, the position of firefly is translated into an optimization problem. The brightness of the firefly is the function value of the optimization problem.
The position of the firefly is constantly updated through iterative optimization of the
firefly population, until the optimal solution is found to the solution.

3 Density Peak Clustering Based on Firefly Algorithm (FADPC)

DPC defines different d_c for different datasets and achieves a good clustering effect.
However, the choice of d_c is not supported by theory, and is usually done according
to the general principle of d_c ensuring that the mean number of neghbours for each
data point accounts for about 1–2% of the total data points. In this study, the number
of neiighbors is the number of data points with $d_{ij} < d_c$. This principle is only an
empirical one drawn from several datasets, and its universality remains to be verified.

FADPC aims for a more reasonably selected d_c by constructing an objective
function to solve d_c by FA. Thus appropriate d_c values are obtained self-adaptively
for different datasets, and the clustering is made more accurate.

3.1 Density Estimation Entropy

By the definition of DPC, the sample points are ranked in accordance with the local density ρ. The δ values are computed sequentially, and the samples with high δ and ρ are identified as clusters. Moreover, the samples are ranked and categorized in a decreasing order of ρ. If any sample is miscategorized, the samples following it will be also miscategorized. Therefore, the objective function is designed to make all ρ values of the samples to be uniformly distributed in a decreasing order as much as possible.

Information entropy [18] is a measure of system uncertainty and has found applications in clustering algorithm in recent years. Consider a dataset $D = \{x_1, x_2, \ldots, x_n\}$ containing n samples in space. Let the value of density function of each sample be $\varphi_i = \sum_{j=1}^{n} e^{\left(\frac{\|x_i - x_j\|}{\delta}\right)^2}$, then the density estimation entropy is defined as

$$H = -\sum_{i=1}^{n} \frac{\varphi_i}{Z} \log\left(\frac{\varphi_i}{Z}\right), \tag{8}$$

where $Z = \sum_{i=1}^{n} \varphi_i$ is the normalization factor.

By comparing formula (8) and formula (3), it can be found that δ in the value of density function for each sample has the same meaning as the cut-off distance d_c. Optimizing δ is in essence to optimize the cut-off distance d_c. If the entire dataset is considered as a system, the best clustering effect can be achieved when the entire system is most stable and the relationship between the data has the highest certainty.

To obtain δ corresponding to the minimum density estimation entropy H, optimization is performed by FA. The density estimation entropy H is taken as the objective function to be solved. Each firefly particle represents a δ value, and FA is implemented to obtain a δ that makes H value minimum for the clustering. This approach can overcome the drawback of manual parameter configuration and improve the accuracy of clustering.

3.2 Steps of FADPC

Formula (8) is used as the objective function for FA, which is then implemented to optimize δ. The optimized result is substituted as d_c into DPC for the clustering. The implementation steps of the algorithm are shown below

Step 1: Choose dataset sample $A_{m \times n}$ (m is the number of data points, and n is the dimension). Calculate the distance matrix $D_{m \times m}$ (Euclidean distance between any two data points) for the dataset.

Step 2: Initialize the population and the population size is N. Each firefly $x_i(i = 1, 2, \ldots, n)$ represents a δ value, and the search range is $(D_{\min}, D_{\max})$. The number of iterations is T.

Step 3: Assess the population and calculate the value of objective function $H(x_i)$ corresponding to each firefly. Record δ_{best} that makes H minimum.

Step 4: Start the iteration using the full attraction model. The value of objective function is compared between the fireflies successively. If $H(x_i) < H(x_j)$, then firefly j moves towards firefly i according to formula (7); otherwise, it does not move.

Step 5: Reassess the population and update δ_{best}.

Step 6: Determine whether the stopping criteria are met. If yes, then stop the algorithm and record δ_{best}; otherwise, return to step 3.

Step 7: Substitute δ_{best} as d_c into the DPC algorithm and do the clustering.

4 Simulation Experiment

To verify the FADPC, it is compared against DPC by simulation experiments on eight artificial datasets [19]. The population size N = 30, T = 300. The experiment is conducted in Win10 64bit environment and using Matlab R2016b software. Clustering quality is evaluated based on RI [20] and NMI [21], and the experimental results are analyzed.

All artificial datasets are 2D datasets of different shapes, including circular, mixed and continuous, and they can test the performance of the clustering algorithm comprehensively. The basic features of artificial datasets are shown in Table 1.

Both two algorithms are run for 10 times, and the results with the best clustering effect are selected. The clustering quality is assessed by RI and NMI. Table 2 is the comparison of RI and NMI on the artificial datasets between the two algorithms.

As shown in Table 2, the clustering accuracy of FADPC is much higher than that of DPC. Among 8 artificial datasets, Flame and Spiral have the smallest data size and simple shape. Both algorithms achieve the best clustering effect on these two

Table 1 Basic feature of artificial datasets

Dataset	Instances	Dimensions	Clusters
Flame	240	2	2
Jain	373	2	2
Aggregation	788	2	7
Pathbased	300	2	3
Spiral	312	2	3
Compound	399	2	6
R15	600	2	15
D31	3100	2	31

Table 2 Comparison of RI and NMI on the artificial datasets between the two algorithms

Dataset	RI		NMI	
	DPC	FADPC	DPC	FADPC
Flame	1	1	1	1
Jain	0.5179	0.8122	0.0967	0.5784
Aggregation	0.8922	0.9470	0.8354	0.9166
Pathbased	0.6920	0.7509	0.4201	0.5530
Spiral	1	1	1	1
Compound	0.8589	0.9093	0.8136	0.7971
R15	0.9889	0.9991	0.9695	0.9942
D31	0.9333	0.9880	0.8202	0.9354

datasets, with RI and NMI of 1. On the other six artificial datasets RI of FADPC is higher than that of DPC, and NMI of FADPC is higher than that of DPC on five artificial datasets. The greatest difference between the two algorithms is observed on Jain. RI is 0.8122 for FADPC versus 0.5179 for DPC, and NMI is 0.5784 for FADPC versus 0.0967 for DPC. On R15 and D31, DPC achieves a good clustering effect, though FADPC is even better. RI of DPC is 0.9889 and 0.9333 on these two algorithms, respectively, and that of FADPC is 0.9991 and 0.9880, respectively; NMI of DPC is 0.9695 and 0.8202, respectively, and that of FADPC is 0.9942 and 0.9354, respectively. On Compound alone is NMI of FADPC (0.7971) is lower than that of DPC (0.8136), though the difference is not of statistical significance.

5 Conclusion

This study describes density peak clustering based on FA (FADPC) and introduces density estimation entropy to determine the choice of cut-off distance d_c. In this way, the choice of d_c in DPC no longer depends on empirical and subjective factors and after this modification; FADPC can apply to the clustering of unknown datasets. Next, simulation experiments are performed on eight artificial datasets. The results indicate that the clustering accuracy of FADPC is higher than that of DPC.

Acknowledgements This research was supported by the National Natural Science Foundation of China under Grant (Nos. 61663029, 51669014, 61563036), The Science Fund for Distinguished Young Scholars of Jiangxi Province under Grant (No. 2018ACB21029), National Undergraduate Training Programs for Innovation and Entrepreneurship under Grant (No. 201711319001) and the project of Nanchang Institute of Technology's graduate student innovation program under Grant (Nos. YJSCX2017023, YJSCX20180023).

References

1. Xu, R.: Survey of Clustering Algorithms. IEEE Press (2005)
2. Frey, B.J., Dueck, D.: Clustering by passing messages between data points. Science **315**(5814), 972–976 (2007)
3. Saxena, A., Prasad, M., Gupta, A., et al.: A review of clustering techniques and developments. Neurocomputing **267**, 664–681 (2017)
4. Karami, A., Guerrero-Zapata, M.: A fuzzy anomaly detection system based on hybrid PSO-Kmeans algorithm in content-centric networks. Neurocomputing **149**, 1253–1269 (2015)
5. Shen, J., Hao, X., Liang, Z., et al.: Real-time superpixel segmentation by dbscan clustering algorithm. IEEE Trans. Image Process. **25**(12), 5933–5942 (2016)
6. Yu, Q., Luo, Y., Chen, C., et al.: Outlier-eliminated k -means clustering algorithm based on differential privacy preservation. Appl. Intel. **45**(4), 1179–1191 (2016)
7. Goldberger, J., Tassa, T.: A hierarchical clustering algorithm based on the Hungarian method. Pattern Recogn. Lett. **29**(11), 1632–1638 (2008)
8. Viswanath, P., Pinkesh, R.: l-DBSCAN: a fast hybrid density based clustering method. In: International Conference on Pattern Recognition, pp. 912–915. IEEE Computer Society (2006)
9. Huang, J., Hong, Y., Zhao, Z., et al.: An energy-efficient multi-hop routing protocol based on grid clustering for wireless sensor networks. Clust. Comput. **20**(3), 1–13 (2017)
10. Rodriguez, A., Laio, A.: Clustering by fast search and find of density peaks. Mach. Learn. Sci. **344**(6191), 1492–1496 (2014)
11. Yang, X.S.: Nature-Inspired Metaheuristic Algorithms. Luniver Press (2008)
12. Lv, Li, Zhao, Jia: The Firefly algorithm with Gaussian disturbance and local search. J. Signal Process. Syst. **90**(8–9), 1123–1131 (2018)
13. Lv, L., Zhao, J., Wang, J., et al.: Multi-objective firefly algorithm based on compensation factor and elite learning. Future Gener. Comput. Syst. (2018). https://doi.org/10.1016/j.future.2018.07.047
14. Zhang, X., Zhao J., Li, P., et al.: Soft subspace clustering algorithm based on improved Firefly algorithm. J. Nanchang Inst. Technol.**37**(4), 61–67 (2018)
15. Sun, H., Xie, H., Zhao, J.: Global optical guided artificial bee colony algorithm based on sinusoidal selection probability model. J. Nanchang Inst. Technol. **37**(4), 61–67 (2018)
16. Fahad, A., Alshatri, N., Tari, Z., et al.: A survey of clustering algorithms for big data: taxonomy and empirical analysis. IEEE Trans. Emerg. Top. Comput. **2**(3), 267–279 (2014)
17. Vinh, N.X., Epps, J., Bailey, J.: Information theoretic measures for clusterings comparison: is a correction for chance necessary? In: International Conference on Machine Learning, pp. 1073–1080. ACM (2009)
18. Scrucca, L., Fop, M., Murphy, T.B., et al.: mclust 5: Clustering, Classification and Density Estimation Using Gaussian Finite Mixture Models. R. J. **8**(1), 289–317 (2016)
19. Franti, P., Virmajoki, O., Hautamaki, V.: Fast Agglomerative Clustering Using a k-Nearest Neighbor Graph. IEEE Trans. Pattern Anal. Mach. Intell. **28**(11), 1875–1881 (2006)
20. Rand, W.: Objective criteria for the evaluation of clustering methods. J. Am. Stat. Assoc. **66**(336), 846–850 (1971)
21. Strehl, A., Ghosh, J.: Cluster ensembles: a knowledge reuse framework for combining multiple partitions. J. Mach. Learn. Res. **3**(3), 583–617 (2003)

A Multi-population QUasi-Affine TRansformation Evolution Algorithm for Global Optimization

Nengxian Liu, Jeng-Shyang Pan, Xiangwen Liao and Guolong Chen

Abstract In this paper, we propose a new Multi-Population QUasi-Affine TRansformation Evolution (MP-QUATRE) algorithm for global optimization. The proposed MP-QUATRE algorithm divides the population into three sub-populations with a sort strategy to maintain population diversities, and each sub-population adopts a different mutation scheme to make a good balance between exploration and exploitation capability. In the experiments, we compare the proposed algorithm with DE algorithm and QUATRE algorithm on CEC2013 test suite for real-parameter optimization. The experimental results indicate that the proposed MP-QUATRE algorithm has a better performance than the competing algorithms.

Keywords QUATRE algorithm · Differential evolution · Multi-population · Global optimization

1 Introduction

In the recent few decades, many swarm-based intelligence algorithms, such as particle swarm optimization (PSO) [1], Ant Colony Optimization (ACO) [2], Differential Evolution (DE) [3], Cat Swarm Optimization (CSO) [4], Ebb-Tide-Fish (ETF) algorithm [5], and QUasi-Affine TRansformation Evolution (QUATRE) algorithm [6], have been developed and applied to various practical problems.

The QUATRE algorithm was proposed by Meng et al. in 2016 to conquer representational/positional bias of DE algorithm. Related works of QUATRE algorithm can be found in literature [6–12]. The QUATRE is a simple and powerful algorithm, which has many advantages and has been applied to hand gesture segmentation [8].

N. Liu · J.-S. Pan (✉) · X. Liao · G. Chen
College of Mathematics and Computer Science, Fuzhou University, Fuzhou, China
e-mail: jspan@cc.kuas.edu.tw

J.-S. Pan
Fujian Provincial Key Lab of Big Data Mining and Applications,
Fujian University of Technology, Fuzhou, China

© Springer Nature Singapore Pte Ltd. 2019
J.-S. Pan et al. (eds.), *Genetic and Evolutionary Computing*,
Advances in Intelligent Systems and Computing 834,
https://doi.org/10.1007/978-981-13-5841-8_3

However, it has some weaknesses as the DE algorithm such as it will be premature convergence, and will search stagnation and may be easily trapped into local optima. Population diversities play important role in reducing these weaknesses. Partitioning population into several sub-populations to maintain population diversities and to enhance the performance of algorithms, such as PSO, CSO and DE, can be found in previous literatures [13–15]. On the other hand, different mutation schemes in QUATRE algorithm have different performance over different optimization problems [8] as each mutation scheme has different exploration and exploitation ability. And the sort strategy has been used to improve the performance of QUATRE algorithm [10]. Therefore, in this paper, in order to improve the performance of QUATRE algorithm, we propose a novel multi-population QUATRE algorithm with sort strategy and each sup-population has different mutation schemes. The proposed algorithm is different from the S-QUATRE mentioned in [10], which divides the population into the better and the worse groups and only evolves the individuals in the worse group.

The rest of the paper is composed as follows: the QUATRE algorithm is briefly reviewed in Sect. 2. Our proposed Multi-Population QUasi-Affine TRansformation Evolution (MP-QUATRE) algorithm is given in Sect. 3. The experimental analysis of MP-QUATRE algorithm under CEC2013 test suite for real parameter optimization is presented, and the proposed algorithm is compared with the DE and QUATRE algorithms in Sect. 4. Finally, the conclusion is given in Sect. 5.

2   QUasi-Affine TRansformation Evolutionary Algorithm

QUATRE is a swarm-based intelligence algorithm, individuals in which evolve according to Eq. 1. $\mathbf{X} = [X_{1,G}, X_{2,G}, \ldots, X_{i,G}, \ldots, X_{ps,G}]^T$ denotes the individual population matrix with ps different individuals. $X_{i,G} = [x_{i1}, x_{i1}, \ldots, x_{i1}, \ldots x_{iD}]$ denotes the location of ith individual of the Gth generation, which is the ith row vector of the matrix $\mathbf{X}$, and each individual $X_{i,G}$ is a candidate solution for an optimization problem, and D denotes the dimension number of objective function, where i $\in \{1, 2, \ldots, ps\}$. $\mathbf{B} = [B_{1,G}, B_{2,G}, \ldots, B_{i,G}, \ldots, B_{ps,G}]^T$ denotes the donor matrix, and Table 1 lists several different calculation schemes [7] for it. "$\otimes$" denotes component-wise multiplication of the elements in each matrix, same as ".*" operation in MATLAB software.

$$\mathbf{X} \leftarrow \mathbf{M} \otimes \mathbf{X} + \overline{\mathbf{M}} \otimes \mathbf{B} \tag{1}$$

$\mathbf{M}$ is an evolution matrix and $\overline{\mathbf{M}}$ is a binary inverted matrix of $\mathbf{M}$, and their elements are either 0 or 1. The binary invert operation means to invert the values of the matrix. The corresponding values of zero elements in matrix $\mathbf{M}$ are ones in $\overline{\mathbf{M}}$, while the corresponding values of one element in matrix $\mathbf{M}$ are zeros in $\overline{\mathbf{M}}$. An example of binary inverse operation is given in Eq. 2.

Table 1 The four schemes for calculating matrix **B** in QUATRE algorithm

No	QUATRE variants	Equation
1	QUATRE/best/1	$\mathbf{B} = \mathbf{X}_{gbest,G} + F \cdot (\mathbf{X}_{r1,G} - \mathbf{X}_{r2,G})$
2	QUATRE/rand/1	$\mathbf{B} = \mathbf{X}_{r0,G} + F \cdot (\mathbf{X}_{r1,G} - \mathbf{X}_{r2,G})$
3	QUATRE/target/1	$\mathbf{B} = \mathbf{X} + F \cdot (\mathbf{X}_{r1,G} - \mathbf{X}_{r2,G})$
4	QUATRE/target-to-best/1	$\mathbf{B} = \mathbf{X} + F \cdot (\mathbf{X}_{gbest,G} - \mathbf{X}) + F \cdot (\mathbf{X}_{r1,G} - \mathbf{X}_{r2,G})$

$$\mathbf{M} = \begin{bmatrix} 1 & & & \\ 1 & 1 & & \\ & & \ddots & \\ 1 & 1 & \ldots & 1 \end{bmatrix}, \overline{\mathbf{M}} = \begin{bmatrix} 0 & 1 & 1 & 1 \\ 0 & 0 & 1 & 1 \\ & & \ldots & 1 \\ 0 & 0 & \ldots & 0 \end{bmatrix} \tag{2}$$

Evolution matrix M is transformed from an initial matrix $\mathbf{M}_{ini} \cdot \mathbf{M}_{ini}$ is initialized by a lower triangular matrix with the elements set to ones. Transforming from $\mathbf{M}_{ini}$ to **M** contains two steps. In the first step, we randomly permute every element in each row vector of $\mathbf{M}_{ini}$. In the second step, we randomly permute the row vectors with all elements of each row vector unchanged. Equation 3 gives an example of the transformation with ps = D. When the population size ps is larger than the dimension number of optimization problem, matrix $\mathbf{M}_{ini}$ needs to be extended according to ps. Equation 4 gives an example of ps = 2D + 2. More generally, when ps%D = k, the first k rows of the D × D lower triangular matrix are included in $\mathbf{M}_{ini}$, and **M** is adaptively changed according to $\mathbf{M}_{ini}$ [8].

In Table 1, $\mathbf{X}_{gbest,G} = \left[\mathbf{X}_{gbest,G}, \mathbf{X}_{gbest,G}, \ldots, \mathbf{X}_{gbest,G}\right]^{T}$ is a row-vector-duplicated matrix with each row vector equaling to the global best individual $\mathbf{X}_{gbest,G}$ of the Gth generation. $\mathbf{X}_{ri,G}$, $i \in \{0, 1, 2\}$ denotes a set of random matrices which are generated by randomly permutating the sequence of row vectors in the population matrix **X** of the Gth generation.

$$M_{ini} = \begin{bmatrix} 1 & & & \\ 1 & 1 & & \\ & & \ddots & \\ 1 & 1 & \ldots & 1 \end{bmatrix} \sim \begin{bmatrix} & & 1 & \\ & \ldots & & \\ 1 & 1 & \ldots & 1 \\ & 1 & & 1 \end{bmatrix}^{T} = M \tag{3}$$

$$
M_{\text{ini}} = \begin{bmatrix} 1 & & & & \\ 1 & 1 & & & \\ & \cdots & & & \\ 1 & 1 & \ldots & 1 & \\ 1 & & & & \\ 1 & 1 & & & \\ & \cdots & & & \\ 1 & 1 & \ldots & 1 & \\ & \vdots & & & \\ 1 & & & & \\ 1 & 1 & & & \end{bmatrix} \sim \begin{bmatrix} 1 & & \ldots & 1 \\ & & \ldots & 1 \\ & \cdots & & \\ 1 & 1 & & \\ & 1 & & \\ 1 & 1 & \ldots & 1 \\ & & 1 & \\ 1 & & & 1 \\ & \vdots & & \\ 1 & 1 & \ldots & 1 \\ & 1 & & \end{bmatrix} = M \tag{4}
$$

3 Multi-population QUasi-Affine TRansformation Evolution (MP-QUATRE) Algorithm

In this section, we present a new Multi-Population QUasi-Affine TRansformation Evolution (MP-QUATRE) algorithm to enhance the performance of the QUATRE algorithm. In MP-QUATRE, we first sort the ps individuals in the population **X** after initialization according to the fitness values of the objective function. Then the individuals in the sorted sequence are divided into three sub-populations, say pop_{better}, pop_{middle}, and pop_{worse}, respectively. The pop_{better} contains the top $\lfloor \frac{ps}{3} \rfloor$ ($\lfloor \rfloor$ is a floor operator) individuals that had better fitness values, and the pop_{middle} contains the middle $\lfloor \frac{ps}{3} \rfloor$ individuals that had middle fitness values, and the pop_{worse} contains the rest individuals that had worse fitness values. And each sub-population evolves using a different mutation scheme to make a balance between exploration and exploitation, and to enhance the convergence capability. The pop_{better} evolves using mutation scheme "QUATRE/best/1", which has good convergence capability by employing the best individual found so far and this scheme exploits excellent information around best individual. The pop_{middle} evolves using mutation scheme "QUATRE/rand/1", which is robust and the most commonly used scheme, and it is an appropriate scheme for exploration to maintain population diversities. The pop_{worse} evolves using mutation scheme "QUATRE/target-to-best/1", which has good exploration and convergence ability due to the individuals in this sub-population guided by the best individual and two randomly selected individuals. After the evolution step, greedy selection is made to ensure that the individual with a better fitness will survive to the next generation. Figure 1 gives an illustration of the MP-QUATRE algorithm and the pseudo code of MP-QUATRE algorithm is shown in Algorithm 1.

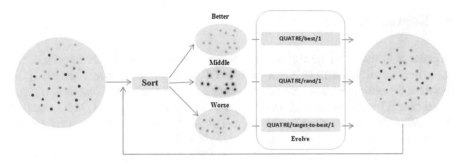

Fig. 1 Illustration of MP-QUATRE algorithm

Algorithm 1. Pseudo code of MP-QUATRE Algorithm

Initialization:
    Initialize the searching space V, dimension D, the benchmark function f(X),
    population size ps, individual population **X**, and calculate fitness values of all individuals.
Iteration:
1: **while** exeTime < MaxIteration|!stopCriterion **do**
2:   Sort individuals in the population and then partition them into pop_{better}, pop_{middle}, and pop_{worse} according to sorted sequence
3:   Generate evolution matrices $\mathbf{M}_{better}$, $\mathbf{M}_{middle}$, and $\mathbf{M}_{worse}$ according to Eq. 3.
4:   Calculate donor matrix $\mathbf{B}_{better}$ according to scheme "QUATRE/best/1"
     Calculate donor matrix $\mathbf{B}_{middle}$ according to scheme "QUATRE/rand/1"
     Calculate donor matrix $\mathbf{B}_{worse}$ according to scheme "QUATRE/target-to-best/1"
5:   Generate matrices $\overline{\mathbf{M}}_{better}$, $\overline{\mathbf{M}}_{middle}$, and $\overline{\mathbf{M}}_{worse}$ according to Eq.2.
6:   Evolve individuals in each sub-population according to Eq.1.
7:   **for** i = 1:ps **do**
8:     **if** $f(X_i)$ is more optimal than $f(X_{pbest,i})$ **then**
9:       $X_{pbest,i} \leftarrow X_i$
10:    **end if**
11: **end for**
12: $\mathbf{X} = \mathbf{X}_{pbest}$
13: $X_{gbest} = opt\{\mathbf{X}_{pbest}\}$.
14: **end while**
Output:
    The global optima X_{gbest}, $f(X_{pbest})$.

4   Experimental Analysis

In order to evaluate the performance of the proposed MP-QUATRE algorithm, we make the comparison with DE algorithm and QUATRE algorithm over CEC2013 [16] test suite for real-parameter optimization, which includes unimodal functions (f_1–f_5), multi-modal functions (f_6–f_{20}) and composition functions (f_{21}–f_{28}). More detailed descriptions of functions can be found in [16]. All test functions' search ranges are $[-100, 100]^D$ and they are shifted to the same global best location, $O\{o_1, o_2, \ldots, o_d\}$.

The parameter settings of the MP-QUATRE and QUATRE algorithm are ps = 100, F = 0.7, D = 30, nfe = ps × D × 1000 (nfe denotes the number of function evaluation). The parameter settings of the DE algorithm are ps = 100, F = 0.7, Cr = 0.1, D = 30, nfe = ps × D × 1000. We run each algorithm on each benchmark function 50 times independently. The best, mean, and standard deviation of the function error are collected in Table 2. The simulation results of some benchmark functions are shown in Fig. 2.

From Table 2, we can see that, for the best value, the DE algorithm finds four minimum values of CEC2013 benchmark functions. The QUATRE algorithm finds 11 minimum values of CEC2013 benchmark functions. Our proposed MP-QUATRE algorithm has an overall better performance than the contrasted algorithms, and it finds 16 minimum values of CEC2013 benchmark functions in comparison with DE and QUATRE algorithms, especially for multi-modal functions. For the mean, our proposed MP-QUATRE algorithm also has an overall better performance than the contrasted algorithms. For the standard deviation, the DE algorithm has better performance than QUATRE and MP-QUATRE algorithms, and MP-QUATRE algorithm has better performance than QUATRE algorithm. On the whole, our proposed MP-QUATRE algorithm has a better performance than all other contrasted algorithms.

5   Conclusion

In this paper, a new MP-QUATRE algorithm is proposed for optimization problems. The MP-QUATRE algorithm divides the population into three sub-populations with sort strategy, and each sub-population employs a different mutation scheme to balance the exploration and exploitation capability. The proposed algorithm is verified under CEC2013 test suite for real-parameter optimization. The experimental results demonstrate that the proposed MP-QUATRE algorithm has a better performance than the DE algorithm and QUATRE algorithm. In future works, we plan to find a parameter adaptive scheme for MP-QUATRE algorithm.

Table 2 Comparison results of best, mean and standard deviation of 50-run fitness error among DE, QUATRE and MP-QUATRE algorithms under CEC2013 test suite. The best results of the comparisons are emphasized in **BOLDFACE** and the tier results are in *ITALIC* fonts

30D	DE/best/1/bin			QUATRE/best/1			MP-QUATRE		
No	Best	Mean	Std	Best	Mean	Std	Best	Mean	Std
1	2.27E−13	2.27E−13	**0.00E+00**	*0.00E+00*	**2.2737E−14**	6.8905E−14	*0.0000E+00*	1.4552E−13	1.1025E−13
2	1.49E+07	2.69E+07	7.16E−06	6.58E+04	3.2526E+05	1.7428E+05	**5.2506E+04**	**1.4902E+05**	**7.0589E+04**
3	1.89E+08	6.71E+08	2.54E+08	**6.43E−02**	1.1505E+06	3.0040E+06	6.3978E−01	**7.9880E+05**	**2.2499E+06**
4	2.38E+04	3.73E+04	6.37E+03	**4.13E+00**	**2.1215E+01**	**1.5656E+01**	9.5577E+00	3.6332E+01	1.8646E+01
5	1.14E−13	1.36E−13	4.59E−14	**0.00E+00**	**1.0914E−13**	**2.2504E−14**	1.1369E−13	1.2506E−13	3.4452E−14
6	1.62E+01	2.46E+01	1.16E+01	2.99E−04	7.3092E+00	1.0870E+01	**1.2641E−07**	**4.4369E+00**	**9.7019E+00**
7	4.53E+01	5.85E+01	**7.79E+00**	**1.14E+00**	**2.1349E+01**	1.7591E+01	1.1747E+00	2.1391E+01	1.2265E+01
8	2.08E+01	2.09E+01	**4.77E−02**	2.08E+01	2.1004E+01	5.8828E−02	**2.0601E+01**	**2.0892E+01**	9.1526E−02
9	2.31E+01	2.91E+01	**1.90E+00**	**5.81E+01**	**1.6079E+01**	5.5873E+00	1.2308E+01	1.8765E+01	3.6367E+00
10	6.78E+00	1.55E+01	4.32E+00	**0.00E+00**	**2.3154E−02**	**1.5496E−02**	7.3960E−03	3.7256E−02	2.2052E−02
11	**5.68E−14**	**6.96E−01**	**1.01E+00**	1.39E+01	2.6411E+01	8.3326E+00	1.1940E+01	2.0695E+01	5.2725E+00
12	1.15E+02	1.51E+02	**1.53E+01**	3.94E+01	7.5733E+01	1.8919E+01	**2.0894E+01**	**6.0852E+01**	1.9002E+01
13	1.26E+02	1.69E+02	**1.32E+01**	5.22E+01	1.1252E+02	3.2964E+01	**1.9149E+01**	**9.5689E+01**	2.6836E+01

(continued)

Table 2 (continued)

30D	DE/best/1/bin			QUATRE/best/1			MP-QUATRE		
14	**1.30E+00**	**2.66E+01**	**5.01E−01**	1.01E+02	8.1082E+02	2.6067E+02	3.5809E−01	4.7348E+02	2.2837E+02
15	5.12E+03	6.13E+03	**3.94E+02**	3.35E+03	5.1268E+03	7.7891E+02	**2.2109E+03**	**3.8174E+03**	6.9016E+02
16	1.53E+00	2.31E+00	**3.26E−01**	1.32E+00	2.4246E+00	4.5226E−01	**4.5301E−01**	**1.0366E+00**	3.8000E−01
17	2.60E+01	**3.12E+01**	**9.13E−01**	2.14E+01	5.5165E+01	1.1812E+01	**8.1234E+00**	3.5413E+01	6.2249E+00
18	1.98E+02	2.20E+02	**1.14E+01**	1.08E+02	1.6029E+02	2.5169E+01	**4.8507E+01**	**7.9656E+01**	1.5823E+01
19	2.79E+00	3.87E+00	**3.92E−01**	1.69E+00	3.6567E+00	7.7921E−01	**1.0719E+00**	**1.9876E+00**	6.2029E−01
20	1.21E+01	1.28E+01	**2.56E−01**	1.03E+01	1.2062E+01	6.4629E−01	**9.9562E+00**	**1.1164E+01**	6.0585E−01
21	*2.00E+02*	**2.90E+02**	7.68E−01	*2.00E+02*	3.1932E+02	8.3201E+01	*2.0000E+02*	*2.9897E+02*	**7.6236E+01**
22	**1.17E+02**	**2.50E+02**	**1.83E+02**	4.56E+02	8.3787E+02	2.5055E+02	1.7574E+02	5.4529E+02	2.5622E+02
23	5.52E+03	6.51E+03	**4.16E+02**	3.72E+03	5.3213E+03	8.4120E+02	**2.2909E+03**	**3.8158E+03**	7.5643E+02
24	2.57E+02	2.72E+02	**6.35E+00**	**2.11E+02**	**2.3769E+02**	1.1610E+01	2.1228E+02	2.4756E+02	1.2892E+01
25	2.79E+02	2.89E+02	**4.63E+00**	**2.42E+02**	**2.5758E+02**	8.0331E+00	2.4481E+02	2.6480E+02	1.0985E+01
26	2.01E+02	**2.02E+02**	**5.02E−01**	2.00E+02	2.4558E+02	6.4221E+01	**2.0001E+02**	2.6574E+02	7.2447E+01
27	9.49E+02	1.05E+03	**4.10E+01**	5.56E+02	**6.9637E+02**	9.3585E+01	**5.3018E+02**	7.6499E+02	9.8916E+01
28	3.00E+02	3.24E+02	1.68E+02	**1.00E+02**	3.7948E+02	2.8873E+02	3.0000E+02	**3.2023E+02**	**1.4306E+02**
Win	3	6	20	9	9	3	14	14	13	5
Draw	1	0	0	2	0	0	2	0	0
Total	4	6	20	11	9	3	16	13	5

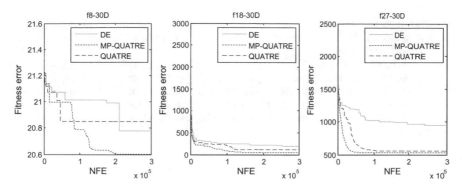

Fig. 2 Simulation of functions f_8, f_{18}, and f_{27} with 30D

References

1. Kennedy, J., Eberhart, R.: Particle swarm optimization. In: Proceedings of IEEE International Conference on Neural Networks, vol. 4, pp. 1942–1948. IEEE (1995)
2. Wang, K., Liu, Y.Q., et al.: Improved particle swarm optimization algorithm based on gaussian-grid search method. J. Inf. Hiding Multimed. Signal Process. **9**(4), 1031–1037 (2018)
3. Dorigo, M., Maniezzo, V., Colorni, A.: Ant system: optimization by a colony of cooperating agents. IEEE Trans. Syst. Man Cybern. Part B Cybern. **26**(1), 29–41 (1996)
4. Storn, R., Price, K.: Differential evolution-a simple and efficient heuristic for global optimization over continuous spaces. J. Glob. Optim. **11**(4), 341–359 (1997)
5. Chu, S.C., Tsai, P.W., Pan, J.S.: Cat swarm optimization. In: The 9th Pacific Rim International Conference on Artificial Intelligence (PRICAI), pp. 854–858 (2006)
6. Meng, Z., Pan, J.S., Alelaiwi, A.: A new meta-heuristic ebb-tide-fish inspired algorithm for traffic navigation. Telecommun. Syst. **62**(2), 1–13 (2016)
7. Meng, Z., Pan, J.S., Xu, H.: QUasi-Affine TRansformation Evolutionary (QUATRE) algorithm: a cooperative swarm based algorithm for global optimization. Knowl.-Based Syst. **109**, 104–121 (2016)
8. Meng, Z., Pan, J.S.: QUasi-Affine TRansformation Evolutionary (QUATRE) algorithm: the framework analysis for global optimization and application in hand gesture segmentation. In: 2016 IEEE 13th International Conference on Signal Processing, pp. 1832–1837 (2016)
9. Meng, Z., Pan, J.S.: Monkey king evolution: a new memetic evolutionary algorithm and its application in vehicle fuel consumption optimization. Knowl.-Based Syst. **97**, 144–157 (2016)
10. Meng Z, Pan J.S.: QUasi-Affine TRansformation Evolutionary (QUATRE) algorithm: a parameter-reduced differential evolution algorithm for optimization problems. In: 2016 IEEE Congress on Evolutionary Computation (CEC), pp. 4082–4089. IEEE (2016)
11. Pan, J.S., Meng, Z., Chu, S., Roddick, J.F.: QUATRE algorithm with sort strategy for global optimization in comparison with DE and PSO variants. In: The Euro-China Conference on Intelligent Data Analysis and Applications, pp. 314–323 (2017)
12. Meng, Z., Pan, J.S., Li, X.: The quasi-affine transformation evolution (QUATRE) algorithm: an overview. In: The Euro-China Conference on Intelligent Data Analysis and Applications, pp. 324–333 (2017)
13. Meng, Z., Pan, J.S.: QUasi-Affine TRansformation Evolution with External ARchive (QUATRE-EAR): an enhanced structure for differential evolution. Knowl.-Based Syst. **155**, 35–53 (2018)
14. Chang, J.F., Chu, S.C., Roddick, J.F., Pan, J.S.: A parallel particle swarm optimization algorithm with communication strategies. J. Inf. Sci. Eng. **21**(4), 809–818 (2005)

15. Tsai, P.W., Pan, J.S., Chen, S.M., Liao, B.Y., Hao, S.P.: Parallel cat swarm optimization. In Proceedings of the Seventh International Conference on Machine Learning and Cybernetics, pp. 3328–3333 (2008)
16. Cui, L.Z., Li, G.H., Lin, Q.Z., et al.: Adaptive differential evolution algorithm with novel mutation strategies in multiple sub-populations. Comput. Oper. Res. **67**, 155–173 (2016)
17. Liang, J.J., et al.: Problem definitions and evaluation criteria for the CEC 2013 special session on real-parameter optimization. In: Computational Intelligence Laboratory, Technical Report 201212. Zhengzhou University, Zhengzhou, China and Nanyang Technological University, Singapore (2013)

Particle Swarm Optimization Based Parallel Input Time Test Suite Construction

Yunlong Sheng, Chang'an Wei and Shouda Jiang

Abstract Testing of Real-Time Embedded Systems (RTESs) under input timing constraints is a critical issue, especially for parallel input factors. Test suites which can cover more possibilities of input time and discover more defects under input timing constraints are worthy of study. In this paper, Parallel Input Time Test Suites (*PITTSs*) are proposed to improve the coverage of input time combinations. *PITTSs* not only cover all the neighbor input time point combinations of each factor, but also cover all the input time point combinations between each two factors of the same input. Particle swarm optimization based the *PITTS* construction algorithm is presented and benchmarks with different configurations are conducted to evaluate the algorithm' performance. A real world RTES is tested with a *PITTS* as application and we have reason to believe that *PITTSs* are effective and efficient for testing RTESs under input timing constraints of parallel input factors.

Keywords Real-time embedded systems · Parallel input ·
Time test suite construction · Particle swarm optimization

1 Introduction

Real-time embedded systems (RTESs) have been extensively implemented in various fields, especially for business, industry and safety critical applications [1]. When RTESs are used to safety critical fields, such as aviation and spaceflight, failure would pose a significant security risk to the personnel and environment related and where the verification of the function of their software is of the utmost importance

Y. Sheng · C. Wei (✉) · S. Jiang
Automatic Test and Control Institute, Harbin Institute of Technology, Harbin, China
e-mail: weichangan@hit.edu.cn

Y. Sheng
e-mail: 13B901008@hit.edu.cn

S. Jiang
e-mail: jsd@hit.edu.cn

© Springer Nature Singapore Pte Ltd. 2019
J.-S. Pan et al. (eds.), *Genetic and Evolutionary Computing*,
Advances in Intelligent Systems and Computing 834,
https://doi.org/10.1007/978-981-13-5841-8_4

[2]. Software function testing is an area of software testing, which is related to the ability of testing software to perform functional testing in a specific time under given environmental conditions. [3].

RTES testing is a huge challenge since RTESs run in a physical environment that may consist of a large number of sensors and actuators, and various input timing constraints are placed on them by the real-time behavior of the external world to which they are interfaced [1]. A timing constraint is typically a time interval which limits the input time interval between two neighbor inputs strictly [4]. Once any time interval is violated, RTESs may produce unexpected or severe results.

Timing constraints in the input space increase the difficulty of testing [5]. If there is not any existing of input timing constraints, an identical input sequence from input space can only result in a unique expected consequence or behavior. However, in the presence of input timing constraints, an identical input sequence may lead to diverse consequences or behaviors based on different input time intervals between each two neighbor inputs in the input sequence. Usually, RTESs have multiple parallel input factors. The correlation of input time combinations among factors needs to be taken into account. To test these aspects completely, it had better configure different time intervals between the same neighbor inputs. It is obvious that a parallel input sequence may be changed into plenty of unique sequences, when each input is assigned a specific time. The existence of timing constraints makes test suite construction more complex, as an extra dimension is added to test suite construction.

The conventional method to test the systems under test (SUTs) with timing constraints is random testing [1]. For inputs each with a specific input time interval, random testing selects a time point for each input depended on its related interval randomly. Test cases constructed by random testing may not cover all the neighbor input time point combinations of each factor and all the time point combinations between each two factors of the same input compared to combinatorial testing (CT) [6] which is a depth testing method [7]. A test suite which covers as many related input time point combinations as possible can contribute to finding more defects caused by input timing.

In this paper, Parallel Input Time Test Suites (*PITTSs*) are proposed to improve the coverage of correlation of input time. *PITTSs* not only cover all the neighbor input time point combinations of each factor, but also cover all the input time point combinations between each two factors of the same input. Particle swarm optimization based the *PITTS* construction algorithm is presented and benchmarks with different configurations are conducted to evaluate the algorithm's performance. A real world RTES, called missile signal monitoring system, is tested with a *PITTS* as application.

2 Background

Input timing constraints describe the input factors which are subject to time limits. Each input factor usually has an independent input relationship. Testers simply need to consider the input timing of input factors. Thus, input time can be selected in a timed sequence according to given input timing constraints. In the case of parallel input factors, multiple input factors are input in a timed sequence concurrently, such as three factors f_1, f_2, and f_3 which are restricted to be executed in time intervals [0, 1], [2, 5], and [7, 10] respectively. There may exist timed correlation among factors or among inputs of a single factor.

Thus, we propose *PITTSs*, to improve the coverage of correlation of input time. *PITTSs* not only cover all the neighbor input time point combinations of each factor, but also cover all the input time point combinations between each two factors of the same input. *PITTSs* improve the coverage of input time points of RTESs.

Definition 1 Consider that A is a parallel input time test suite, denoted by $PITTS(n; F, I_1, I_2, \ldots, I_m)$, where n is the number of matrices in the *PITTS*, F is the number of factors and $I_1, I_2, \ldots, I_m$ are m optional input time point sets with $|I_1| = |I_2| = \ldots = |I_m| = I$. A satisfies the following two conditions:

(1) $\forall$ two factors f_i, f_j $(i \neq j)$, and $\forall a \in I_l, b \in I_l$, there is at least a matrix $M_{m \times F}$ in A, such that $M[l; i] = a$ and $M[l; j] = b$;

(2) $\forall f_i$, and $\forall a \in I_j, b \in I_{j+1}$, there is at least a matrix $M_{m \times F}$ in A, such that $M[j; i] = a$ and $M[j + 1; i] = b$.

Consider three factors f_1, f_2 and f_3 which are restricted to be executed in three intervals [0, 1], [2, 5], and [7, 10] respectively. Two boundary time points in each interval are selected and the *PITTS* is shown in Table 1.

Table 1 A *PITTS* example of three factors

Test case 1				Test case 2				Test case 3				Test case 4			
no.	f_1	f_2	f_3	no.	f_1	f_2	f_3	no.	f_1	f_2	f_3	no.	f_1	f_2	f_3
1	0	1	0	1	1	0	0	1	1	1	1	1	0	0	1
2	5	5	5	2	2	5	2	2	5	2	2	2	2	2	5
3	10	10	7	3	7	7	7	3	7	10	10	3	10	7	10

3   Test Suite Generation

As finding a minimum test suite which only satisfies the first condition in Definition 1 is NP-complete [8], to construct minimum *PITTSs* is also NP-complete. In this paper, Particle Swarm Optimization (PSO) is used to construct *PITTSs* with as small sizes as possible for the simplicity of its algorithm structure over other optimization methods.

3.1   Particle Swarm Optimization

Particle swarm optimization is an optimizing method proposed by Kennedy according to the foraging behavior of bird groups in 1995 [9]. During the optimization searching process, the population flies in the direction with the maximum fitness factor. Particle position $X_i^t = (x_{i1}^t, x_{i2}^t, \cdots, x_{ik}^t)$ represents a solution to the problem. Particle velocity $V_i^t = (v_{i1}^t, v_{i2}^t, \cdots, v_{ik}^t)$ represents the degree of adjustment. Each particle records the optimal position in the solution domain, which is the best solution up to now, called *pBest*. Population records the optimal position in the solution domain, which is the best solution up to now, called *gBest*. As a result of the values of the factors are discrete, we employ the discrete form of particle swarm optimization used in test suite generation [10, 11]. The update rules of positions and velocities are as follows:

$$v_{ij}^{t+1} = \omega v_{ij}^t + c_1 r_1 (pBest_{ij}^t - x_{ij}^t) + c_2 r_2 (gBest_j^t - x_{ij}^t), \tag{1}$$

$$x_{ij}^{t+1} = x_{ij}^t + v_{ij}^{t+1}, \tag{2}$$

where t is the number of iterations, (c_1, c_2) are two acceleration coefficients, ω is the inertia weight from 0 to 1, and (r_1, r_2) are two random values from 0 to 1.

3.2   Parallel Input Time Test Suite Construction Algorithm

The parallel input time test suite construction algorithm is shown as Algorithm 1.
     The algorithm is illustrated from the following three aspects.

(1) Fitness factors

Fitness factors are the criteria for selecting best particles. To record the input time point combinations of each two factors of the same input, such as f_i and f_j with $f_i \neq f_j$, $H_{f_i f_j}^{I_u} = \{(a, b) | a, b \in [0, |I_u| - 1]\}$ is used, where $u \in [1, m]$. To record the neighbor

---

Algorithm 1 parallel input time test suite construction algorithm

---

Require: the number of factors F, input time sets $I_1, I_2, ..., I_m$, swarm size P and
    iteration time R

Ensure: parallel input time test suite T

 1: initialize fitness factor structures;

 2: **while** TRUE **do**

 3:     initialize an empty matrix $t_{m \times F}$;

 4:     **for** m **do**

 5:         initialize the population with a population size of P;

 6:         **for** R iterations **do**

 7:             update the population;

 8:         **end for**

 9:         $t_{m \times F}$ appends $gBest$;

10:     **end for**

11:     **if** the fitness factor of $t_{m \times F}$ is greater than 0 **then**

12:         T appends $t_{m \times F}$;

13:         update the fitness factor structures;

14:         **if** fitness factor structures are empty **then**

15:             break while;

16:         **end if**

17:     **end if**

18: **end while**

---

input time point combinations of the same factor, such as f_i, $H_{f_i}^{I_v I_{v+1}} = \{(a, b) | a \in [0, |I_v| - 1], b \in [0, |I_{v+1}| - 1]\}$ is used, where $v \in [1, m - 1]$. $H_{f_i f_j}^{I_u}$ and $H_{f_i}^{I_v I_{v+1}}$ are constructed in line 1. Fitness factors of particles are calculated in line 5 and line 7. A fitness factor is the sum of the number of input time point combinations between each two factors of the same input recorded in $H_{f_i f_j}^{I_u}$ and the number of two neighbor input time point combinations of each factor recorded in $H_{f_i}^{I_v I_{v+1}}$. Only if the fitness factors of subset test suite $t_{m \times F}$ is greater than 0 in line 11, it can be appended into T in line 12. Then, the input time point combinations between each two factors and the neighbor input time point combinations of each factor covered in $t_{m \times F}$ are removed from $H_{f_i f_j}^{I_u}$ and $H_{f_i}^{I_v I_{v+1}}$ in line 13. The algorithm will finish when $H_{f_i f_j}^{I_u}$ and $H_{f_i}^{I_v I_{v+1}}$ are empty in lines from 14 to 16.

(2) Population initialization

The position and velocity of each particle in the population are initialized by generating random spaces. Each subset $t_{m \times F}$ has m lines, and for each line of it R iterations are repeated in lines from 4 to 10. For the input time point combination generation, the position of each particle is in the form of a F-dimensional vector, $X_i^0 = (x_{i1}^0, x_{i2}^0, \cdots, x_{ik}^0)$, where each dimension x_{ij}^0 is a random input time point from the corresponding time set $[0, |I_r| - 1]$ $(1 \leq r \leq m)$. The velocity of each particle is also in the form of a k-dimensional vector, $V_i^0 = (v_{i1}^0, v_{i2}^0, \cdots, v_{ik}^0)$, where each dimension v_{ij}^0 is a random input time point between $-(|I_r| - 1)/2$ and $(|I_r| - 1)/2$.

(3) Population update

The position and velocity of each particle are updated in line 7 in accordance with Eqs. 1 and 2. After every update, if x_{ij}^{t+1} exceeds the range $[0, |I_r| - 1]$, then x_{ij}^{t+1} is replaced with the previous value x_{ij}^t. If v_{ij}^{t+1} exceeds the range $-(|I_r| - 1)/2$, $(|I_r| - 1)/2$, then v_{ij}^{t+1} is replaced with a random value between $-(|I_r| - 1)/2$ and $(|I_r| - 1)/2$. $pBest$ of each particle and $gBest$ of the population are also updated.

4 Experiments and Applications

In this section, we adopt a benchmark with eight SUTs which have different configurations. Configurations of eight SUTs and their *PITTSs* are shown in Table 2. They have different factors, number of input time point and optional time points in each input. The number of input time point combinations between each two factors and the number of neighbor input time point combinations of each factor are listed respectively. As this algorithm is non-deterministic, we performed 50 independent runs per SUT for a statistical analysis. The size of particles is 160, the iteration number is 20, $\omega = 0.3$ and $c_1 = c_2 = 1.375$ which are referenced in [10]. The units of the generated time are seconds. In the table, we can see that PSO constructs as small sizes of *PITTSs* as possible while spending a reasonable amount of time. In case of the test suite sizes, they are feasible and practical to perform.

A missile signal monitoring system is used as a RTES with parallel input timing constraints. The missile signal monitoring system independently monitors direct

Table 2 *PITTSs* constructed by PSO

No.	SUTs			PSO			
	$F = m = I$	$|H_{f_if_j}^{I_u}|$	$|H_{f_i}^{I_vI_{v+1}}|$	Time		Size	
				Best	Average	Best	Average
1	3	81	54	0.2866	0.3449	10	11.7333
2	4	384	192	1.0044	1.3093	22	24.1
3	5	1250	500	2.8217	3.2699	38	40.3667
4	6	3240	1080	6.5101	7.4731	59	61.7333
5	7	7203	2058	13.8973	15.6619	85	89.1667
6	8	14336	3584	27.4272	32.6284	117	121.0333
7	9	26244	5832	51.9760	56.6438	154	159.9333
8	10	45000	9000	91.7392	98.3428	200	204.2333

voltage signals, alternating voltage signals and serial port signals, etc. This system monitors five minutes each time and stores the monitoring results into data files. We just test the function of direct voltage signal monitoring in this paper. The signal monitoring system monitors four direct voltage signal channels simultaneously for five minutes and stores the monitoring into a data file. As the monitoring system monitors direct voltage signals in a timed sequence and records monitoring results uniformly, *PITTSs* are applicable.

As the monitored missile always sends direct voltage signals every 5 s or so, we divide five minutes into 60 five-second intervals and in each interval a direct voltage signal is input. Thus, 60 inputs will be performed in each channel. In each five-second interval, we select five integer time points as optional input time. A $PITTS(46; 4, I_1, I_2, \ldots, I_{60})$ with $|I_1| = |I_2| = \ldots = |I_{60}| = 5$ can be constructed in 62.8213 s. Because of the limited space, the generated *PITTSs* are not listed.

We perform the *PITTS* with 46 feasible test cases. In each test 60 different direct voltage values are performed to input in the four channels. Thus, four direct voltage value groups need to exist in the monitoring record data file, and in each of them sixty different values exist. The practice test results conform to the expected results. The *PITTS* verifies the function of direct voltage signal monitoring and discovers no faults.

5   Conclusions

In this paper, parallel input time test suites which are appropriate for parallel input timing constraints are presented. Particle swarm optimization based on the *PITTS* construction algorithm is also presented and benchmarks with different configurations are conducted to evaluate the algorithm's performance. A real world RTES under input timing constraints is tested. *PITTSs* which are high-level and detailed test cases satisfy SUTs' high-level functional requirements. We have reason to believe that *PITTSs* are effective and efficient and will play an important role in improving the quality and reliability of RTESs.

Acknowledgements The research work presented in this paper is supported by National Defense Basic Scientific Research Project of China (xx2016xxxxx). The authors also gratefully acknowledge the helpful comments and suggestions of the reviewers, which have improved the presentation.

References

1. Arcuri, A., Iqbal, M.Z., Briand, L.: Black-box system testing of real-time embedded systems using random and search-based testing. In: 22nd IFIP WG 6.1 International Conference on Testing Software and Systems, pp. 95–110. Springer, Berlin (2010)
2. Bell, R.: Introduction and Revision of IEC 61508. Adv. Syst. Saf. **42**, 273–291 (2011)

3. Ferrer, J., Kruse, P.M., Peter, M., Chicano, F., Alba, E.: Search based algorithms for test sequence generation in functional testing. Inf. Soft. Technol. **58**, 419–432 (2015)
4. En-Nouaary, A., Hamou-Lhadj A.: A boundary checking technique for testing real-time systems modeled as timed input output automata. In: Eighth International Conference on Quality Software, pp. 209–215. IEEE Computer Society, Washington DC (2008)
5. Krichen, M.: A formal framework for black-box conformance testing of distributed real-time systems. Int. J. Crit. Comput. Based Syst. **3**, 26–43 (2012)
6. Nie, C., Wu, H., Niu, X., Kuo, F., Leung, H., Colbourn, C.: Combinatorial testing, random testing, and adaptive random testing for detecting interaction triggered failures. Inf. Softw. Technol. **62**, 198–213 (2015)
7. Wei, C.A., Sheng, Y.L., Jiang, S.D.: Combinatorial test suites generation method based on fuzzy genetic algorithm. J. Inf. Hiding Multimed. Signal Process. **6**, 968–976 (2015)
8. Lei Y., Tai K.C.: In-parameter-order: a test generation strategy for pairwise testing. In: Proceedings of 3rd IEEE International Symposium on High-Assurance Systems Engineering, pp. 254–261. IEEE Computer Society, Washington DC (1998)
9. Kennedy, J., Eberhart, R.: Particle swarm optimization. In: ICNN'95—International Conference on Neural Networks, pp. 1942–1948. IEEE, Washington DC (1995)
10. Ahmed, B.S., Zamli, K.Z., Lim, C.P.: Application of particle swarm optimization to uniform and variable strength covering array construction. Appl. Soft Comput. **12**, 1330–1347 (2012)
11. Sheng, Y.L., Sun, C., Jiang, S.D., Wei, C.A.: Extended covering arrays for sequence coverage. Symmetry **10** 146–1–146–26 (2018)

Improved Whale Optimization Algorithm and Its Application to UCAV Path Planning Problem

Jeng-Shyang Pan, Jenn-Long Liu and En-Jui Liu

Abstract This study proposes an improved whale optimization algorithm (WOA), termed improved WOA, by proposing a new judgment criterion for selecting the process of encircling prey or searching for prey in the WOA. The new judgment criterion is a self-tuning parameter that is based on the quality of agent's fitness instead of a random value used in the original WOA. The agent with higher fitness, i.e., superior agent, updates its position towards the best agent found so far. On the contrary, the agent with lower fitness, i.e., inferior agent, updates its position toward a reference agent which is selected randomly from the population. The performance of the proposed WOA is examined by testing six benchmark functions on low, medium, and high dimensions. Furthermore, the proposed WOA is applied to the path planning of unmanned combat aerial vehicle (UCAV). The computed results of flight path and optimal cost obtained using the improved WOA will be compared with those obtained using the original WOA.

Keywords Improved whale optimization algorithm · Exploitation · Exploration · Path planning · Unmanned combat aerial vehicle

J.-S. Pan
College of Information Science and Engineering, Fujian University of Technology,
Fuzhou 350118, Fujian, China
e-mail: jengshyangpan@fjut.edu.cn

J.-L. Liu (✉)
Department of Information Management, I-Shou University, Kaohsiung 84001, Taiwan
e-mail: jlliu@isu.edu.tw

E.-J. Liu
Department of Power Mechanical Engineering, National Tsing Hua University,
Hsinchu 30013, Taiwan
e-mail: adw2579@gmail.com

© Springer Nature Singapore Pte Ltd. 2019   37
J.-S. Pan et al. (eds.), *Genetic and Evolutionary Computing*,
Advances in Intelligent Systems and Computing 834,
https://doi.org/10.1007/978-981-13-5841-8_5

1 Introduction

In recent years, many metaheuristic algorithms have been widely applied as powerful packages to solve optimization problems in the engineering fields. There are many well-known population-based metaheuristic optimization algorithms, such as Particle Swarm Optimization (PSO) [1], Ant Colony System (ACS) [2], Artificial Bee Colony (ABC) [3], Cuckoo Search (CS) [4], and Whale Optimization Algorithm (WOA) [5]. Among the aforementioned metaheuristic optimization algorithms, WOA, proposed by Mirjalili and Lewis [5], is a recently developed optimization algorithm. Basically, the algorithm is inspired by the foraging behavior of humpback whales. The whales in a group hunt school of small fishes by shrinking and encircling around them to herd them to close to the sea surface and also generating bubbles along a helix-shaped movement to perform a spiral bubble-net attack [5–7]. The processes of shrinking encircling and spiral updating represent exploitation phase, and the encircling with a random search for prey represent the exploration phase. WOA has designed a judgment criterion for selecting the process of shrinking encircling mechanism or search for prey. Due to the simplicity of WOA in implementation and only two main internal parameters to be adjusted, the algorithm has shown competition with other metaheuristic algorithms from the testing results of different benchmark functions and engineering design problems [5].

Similar to other basic metaheuristic algorithms, the main problems faced by WOA are slow convergence and premature convergence. The original WOA can efficiently solve low-dimensional and unimodal optimization problems, however, the performance of the computation algorithm decreases when solving high-dimensional and multimodal optimization problems. Therefore, a number of variants of WOA were presented in the literature. To enhance the convergence speed and exploitation mechanism, Mafarja and Mirjalili proposed a hybrid algorithm by incorporating Simulated Annealing (SA) into WOA to perform a local search around the best-found solution [6]. Considering that the Lévy flight trajectory [8], usually applied in CS algorithm, is helpful for increasing the diversity of the population against premature convergence and enhancing the ability of escaping local optima, Ling et al. proposed a Lévy flight Whale Optimization Algorithm (LWOA) to obtain a better tradeoff between exploration and exploitation in the WOA [9]. They validated that the use of a Lévy flight trajectory can achieve fast convergence and premature convergence avoidance for the original WOA. Since the updated solution is mostly depending on the current best candidate solution, Hu et al. [10] proposed the original WOA with inertia weights, which is similar to the modified PSO algorithm [11], to obtain an improved whale optimization algorithm. In 2018, Kaur and Arora introduced a Chaotic Whale Optimization Algorithm (CWOA) with chaotic sequence from 10 chaotic maps to replace the critical parameter (p shown in [5]) instead of 50–50% probability that whale either follows the shrinking encircling or logarithmic path during optimization in the original WOA [12]. Oliva et al. [13] showed that the chaotic variable benefits to shift the local and global searching ability of WOA. The basic WOA defines a judgment criterion ($|\vec{A}| > 1$ or $|\vec{A}| < 1$ as shown in [5]) to transit the process for selecting the process of shrinking encircling prey (when $|\vec{A}| < 1$) or search for prey (when $|\vec{A}| > 1$) over the course of iterations. However, the authors found that the judgment criterion ($|\vec{A}|$) designed in the original WOA strongly depends on the random distribution only and could cause inefficiency in searching the optimal solution and convergence speed. Therefore, this study introduces a new self-tuning judgment criterion based on the value of agent's fitness instead of random distribution to enhance the convergence speed and optimal solution achievement of the original WOA. The improved method has the advantages of accelerating the convergence speed and preserving the robustness of the original WOA. The performance of the proposed WOA, termed improved WOA, is assessed by computing six benchmark functions on low, medium, and high dimensions. Furthermore, the improved WOA is applied to the path planning problem for unmanned combat aerial vehicle (UCAV).

The rest of this paper is organized as follows. Section 2 presents the UCAV path planning problem. The basic of the original WOA algorithm and the proposed improved WOA are presented in Sect. 3. In Sect. 4, the experimental results are presented and results are analyzed. Finally, in Sect. 5, conclusions are given.

2   UCAV Path Planning Problem

UCAV is a type of UAV that is manipulated by the base control personnel to perform reconnaissance, surveillance, communications, navigation, deception, target detection, rescue, and attack missions by the uses of a remote control technology and automatic flight control system. The planned path can ensure UCAV to arrive at the destination along the specified optimal path with a minimum probability of being assaulted by enemy and less fuel consumption. However, the path planning of UCAV is an NP-hard optimization problem. The earliest path planning research began in the 1960s. Since then some available algorithms had been proposed, such as A* search algorithm and dynamic programming algorithm. The objective function of path planning generally is the minimization for the distance of flight path, fuel consumption, and possibility of exposure to threat sources, such as radar, missile, tank, artillery, or gun threats.

As shown in Fig. 1, a UCAV will encounter counterattacks from ground-based air defense weapons if it flies into the regions of threat sources. After collecting the data related to terrain information, enemy positions, and settings of threats, a UCAV is programmed to fly from the starting point (marked "S") to the target point (marked "T") according to the global optimal path instead of the local optimal one. As depicted in Fig. 1, the shortest path between points "S" and "T" is a straight line. However, it could be an infeasible solution since it usually passes through threat sources.

Since the path planning belongs to an NP hard problem, metaheuristic algorithms can solve such a problem better. Some metaheuristic algorithms have been applied to the UAV/UCAV path planning, such as ACO [14], PSO [15], FSA [16], and ABC [17]. In the above research literature, most of researchers made modifications for the basic algorithm to enhance the performance of search ability of the algorithm first,

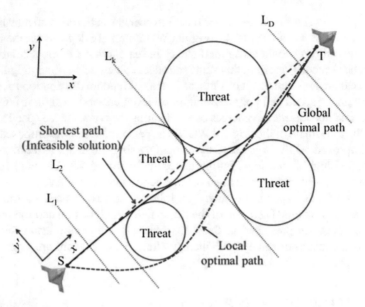

Fig. 1 Demonstration of UCAV path planning

and then applied the improved version to the path planning problems. Considering that there is no WOA applying to the complex UCAV path planning problem so far, this study proposes an improved WOA. After that, it applies to the path planning problems.

3   Whale Optimization Algorithm

A mentioned in Sect. 2, a newly purposed WOA is inspired by the special hunting behavior of humpback whales. The behavior is modeled by WOA as below.

3.1   Stage of Encircling Prey

Humpback whales hunt a school of krill or small fishes with two mechanisms: encircling around them or creating bubble-nets. First, they observe and memorize the locations of the prey, and then encircle around them. When encircling the prey, whales update their location towards the best location obtained so far. Therefore, the model in the stage of encircling prey can be formulated as follows [5]:

$$\vec{X}(t+1) = \vec{X}^*(t) - \vec{A} \cdot \vec{D}; \quad \vec{D} = \left| \vec{C} \cdot \vec{X}^*(t) - \vec{X}(t) \right| \tag{1}$$

where $\vec{A}$ and $\vec{D}$ are coefficient vectors; t is the index of current iteration; $\vec{X}^*(t)$ represents the position vector of the best solution found so far, and $\vec{X}(t)$ represents position vector; ($\cdot$) denotes an element-by-element multiplication and | | denotes the absolute value. The vectors $\vec{A}$ and $\vec{C}$ are given based on random functions as follows:

$$\vec{A} = 2\vec{a} \cdot \vec{r} - \vec{a}; \quad \vec{C} = 2 \cdot \vec{r} \tag{2}$$

Since the coefficient $\vec{r}$ is a random vector in [0, 1], then the ranges of vectors $\vec{A}$ and $\vec{C}$ are in the intervals $[-\vec{a}, \vec{a}]$ and [0, 2], respectively. The value of $\vec{a}$ is designed to decrease linearly from 2 to 0 from the start to the end of iteration.

3.2 Stage of Bubble-Net Attacking (Exploitation Phase)

In the stage of bubble-net attacking, whales swim around the prey within the shrinking circle as well as simultaneously move along a spiral-shaped path to form distinctive bubbles along a 9-shaped path [5]. The two types of behavior in the stage of bubble-net attacking include shrinking encircling mechanism and spiral updating position.

Shrinking Encircling Mechanism. As mentioned above, the value of $\vec{a}$ decreases linearly from 2 to 0. Since the range of vectors $\vec{A}$ is in $[-\vec{a}, \vec{a}]$, so the value of $\vec{A}$ decreases over the course of iterations. Therefore, the shrinking encircling mechanism can be achieved by using Eq. (1).

Spiral Updating Position. In this phase, each whale updates its position according to a spiral-shaped path. The spiral-shaped equation between the positions of whale and prey can be expressed as follows:

$$\vec{X}(t+1) = \vec{D}' \cdot e^{bl} \cdot \cos(2\pi\, l) + \vec{X}^*(t); \quad \vec{D}' = \left| \vec{X}^*(t) - \vec{X}(t) \right| \tag{3}$$

where b is a constant for defining the logarithmic shape and l is a random number in $[-1, 1]$; $\vec{D}'$ represents the distance between positions of prey (i.e., the best solution found so far, denoted by $\vec{X}^*(t)$) and whale (denoted by $\vec{X}(t)$).

To model this simultaneous behavior in the stage of bubble-net attacking, Mirjalili and Lewis assume that there is a probability of selection, denoted by *prob*, choosing either to carry out the shrinking encircling mechanism or the spiral-shaped movement for updating the position of whale during iterations. The simultaneous behavior in the stage of bubble-net attacking is modeled as follows:

$$\vec{X}(t+1) = \begin{bmatrix} \vec{X}^*(t) - \vec{A} \times \vec{D} & \text{if } p < prob \\ \vec{D}' \cdot e^{bl} \cdot \cos(2\pi\, l) + \vec{X}^*(t) & \text{otherwise} \end{bmatrix} \tag{4}$$

where p is a random number between [0, 1].

3.3   Stage of Search for Prey (Exploration Phase)

The vector $\vec{A}$ shown in Eq. (2) also can be applied to the stage of the search for prey when whales explore the possible position of prey. The location of a whale is updated by selecting a whale randomly from the population instead of the best one.

$$\vec{X}(t+1) = \vec{X}_{rand}(t) - \vec{A} \cdot \vec{D}; \quad \vec{D} = \left[\vec{C} \cdot \vec{X}_{rand}(t) - \vec{X}(t) \right] \tag{5}$$

where $\vec{X}_{rand}$ represents a position vector of whale selected by randomness. The value of $\vec{A}$, represented by $|\vec{A}|$ exists in two cases: greater than 1 or less than 1. When $|\vec{A}| > 1$, the search agent will move far away from a reference whale and the search emphasizes exploration to allow the WOA performing a global search. Yet, when $|\vec{A}| < 1$, the search agent will update the position according to the current best solution and the search emphasizes exploitation to allow the WOA performing a local search.

3.4   Improved WOA

As seen in the previous Sect. 3.3, the original WOA depends on the value of $\vec{A}$ to determine that the search agent either performs the process of shrinking encircling prey (when $|\vec{A}| < 1$) or searches for prey (when $|\vec{A}| > 1$) over the course of iterations. However, the vector $\vec{A}$ is given just based on random function only and does not have any information of solution about the search agent. Thus, the search agent may move towards wrong directions resulting in decreasing the convergence speed and getting stuck in local optima. Accordingly, this study proposes a self-tuning judgment criterion (R) which is designed based on the agent's fitness (*fit*) to replace the original judgment criterion ($|\vec{A}|$) used in the basic WOA. The proposed self-tuning judgment criterion is formulated as follows:

$$R = (fit_{best} - fit) / (fit_{best} - fit_{worst}) \tag{6}$$

where fit_{best} and fit_{worst} represent the best and worst values of agent's fitness for the entire population so far. When the value of R for a search agent is larger than 0.5, i.e., the value of agent's fitness is smaller than the mean of the population, the agent belongs to an inferior one. We update the position of the search agent according to a randomly chosen search agent by using Eq. (5). On the contrary when the value of R is smaller than 0.5, i.e., the value of agent's fitness is larger than the mean of the population, the agent belongs to a superior one. We update the position of the search agent towards the best search agent by using Eq. (1). Therefore, the improved updating rules for search agent either performs the process of shrinking encircling prey or searches for prey is as follows:

$$\vec{X}(t+1) = \begin{bmatrix} \vec{X}^*(t) - \vec{A} \cdot \left[\vec{C} \cdot \vec{X}^*(t) - \vec{X}(t) \right] & \text{if } R < 0.5 \\ \vec{X}_{rand}(t) - \vec{A} \cdot \left[\vec{C} \cdot \vec{X}_{rand}(t) - \vec{X}(t) \right] & otherwise \end{bmatrix} \quad (7)$$

4 Results and Discussion

4.1 Experiments of Benchmark Functions

In this work, six benchmark functions for minimization were tested to evaluate the proposed improved WOA. The six benchmark functions were Sphere, Rosenbrock, Griewank, Rastrigin, Ackley, and Schwefel 2.26 [5, 12]. The optimal solutions for functions 1–5 all equal zero, whereas the function 6 in D-dimensional space approximates $-418.9829 \times D$. In this work, the population size of whales was 30, and the dimensions for the six scalable benchmark functions were 30, 50, and 100. In addition, 30 independent runs were performed for each case.

Table 1 compares the best mean solutions (Mean), standard deviation (Std), and best (Best) solutions obtained using the original and improved WOAs for the cases

Table 1 Solutions obtained using the WOAs for the six benchmark functions in 50 dimensions

Function	Solution	Original WOA	Improved WOA	Known optimum
$f_1(\vec{x})$	Mean	2.05409930e−071	1.62274858e−088	0
	(Std)	(1.03859405e−070)	(7.84066369e−088)	
	Best	9.59586816e–088	1.13888443e−099	
$f_2(\vec{x})$	Mean	4.82279404e+001	4.77371066e+001	0
	(Std)	(3.13200219e−001)	(5.06004034e−001)	
	Best	4.73412706e+001	4.67434796e+001	
$f_3(\vec{x})$	Mean	3.70074342e−018	0.00000000e+000	0
	(Std)	(1.99291132e−017)	(0.00000000e+000)	
	Best	0.00000000e+000	0.00000000e+000	
$f_4(\vec{x})$	Mean	3.78956126e−015	0.00000000e+000	0
	(Std)	(2.04074119e−014)	(0.00000000e+000)	
	Best	0.00000000e+000	0.00000000e+000	
$f_5(\vec{x})$	Mean	4.67773968e−015	3.25665421e−015	0
	(Std)	(2.58370205e−015)	(2.30850464e−015)	
	Best	8.88178420e−016	8.88178420e−016	
$f_6(\vec{x})$	Mean	−1.75031578e+004	−1.91221191e+004	−20949.145
	(Std)	(3.08483191e+003)	(2.29871602e+003)	
	Best	−2.09473180e+004	−2.09491444e+004	

44   J.-S. Pan et al.

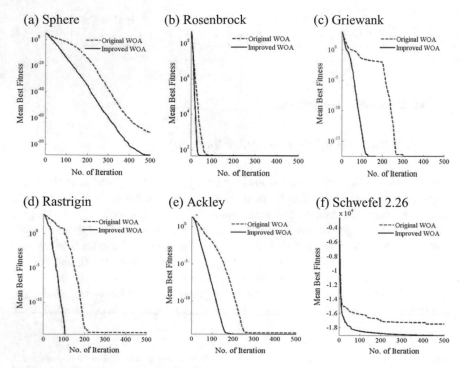

Fig. 2 The comparison of solutions for the six objective functions in 50-dimensional space

of 50 dimensions. From Table 1, the solutions obtained using the improved WOA
were significantly better than those obtained using the original WOA. Namely, the
present improved WOA outperforms the original WOA in terms of the mean, standard
deviations, and best solution. The table clearly indicates that the presented WOA can
achieve optimal solutions effectively. Also, the solutions that were evaluated by the
improved WOA were better than those obtained using the original WOA for the
small- and large-scale problems in 30 and 100 dimensions. Experimental results
revealed that the proposed WOA significantly outperforms the original WOA for all
function evaluations.

Figure 2 displays the convergence histories of mean solutions that were evaluated
using the two WOAs for the cases of D = 50. Clearly, the improved WOA outperforms the original WOA in terms of both convergence speed and optimal solutions.
The presented WOA converged to a near optimum very rapidly. Noted that the convergence histories of solutions were plotted using a logarithmic y-axis to reveal the
variations among the convergence speed and optimal solutions. From the computational results, the proposed WOA can effectively find out the global optima, and it
is a promising algorithm for solving optimization problems.

4.2   Simulations of UCAV Path Planning

Two flight scenarios of path planning of UCAV were simulated. The UCAV flies from the starting point (x_s, y_s) to the target point (x_t, y_t). The positions and ranges of enemy threat sources were denoted as (x_{obs}, y_{obs}), and r_{obs}, respectively. The parameter settings of scenarios 1 and 2 were listed in Table 2 as below.

The objective function (f) was set to the lowest cost of the flight path which was computed along the starting point to the destination, that is, to minimize the objective function. The objective function was expressed as follows:

$$\text{minization } f = \lambda \times f_{threat} + (1 - \lambda) \times f_{fuel} \tag{8}$$

$$f_{threat} = \int_s^t \gamma_{threat} \, dl; \quad f_{fuel} = \int_s^t \gamma_{fuel} \, dl \tag{9}$$

From Eq. (8), the objective function of UCAV was expressed as the tradeoff of the threat cost and the cost of fuel consumption. In this paper, $\lambda = 0.5$ was used. Therefore, when the simulated flight paths are located outside the threat source areas (i.e., the flight path is a feasible solution), the value of (f_{threat}) in Eq. (9) will equal zero. Assume the UCAV flies at a constant speed, then its fuel consumption rate will be a constant value, and the objective function can be simplified as the minimum of total flight length along the feasible solution path when $\gamma_{fuel} = 2$. Moreover, the flight paths were smoothed by using B-spline fitting for the computed optimal points.

As shown in Fig. 3a, b, both two algorithms can effectively find out feasible paths. Namely, the simulation results showed that both two WOAs can get the solutions with locating outside the threat regions. As displayed in Fig. 3a, b, the optimal paths obtained using the improved WOA were better than those obtained using the original WOA. Table 3 lists the comparisons of mean, standard deviation, and bet solutions obtained using the two WOAs. Clearly, the optimal solution obtained using the improved WOA was the best. The straight-line distances from the starting point to the target point, i.e., the shortest distance between the two points, for scenarios 1 and 2 were 134.3503 and 141.4213, respectively. Clearly, the distances of optimal path were slightly larger than those of straight-line ones.

Table 2 Parameter settings for scenarios 1 and 2

Scenario 1	Scenario 2
$x_s = 5; y_s = 5; x_t = 100; y_t = 100$	$x_s = 0; y_s = 0; x_t = 100; y_t = 100$
$x_{obs} = [10, 10, 20, 40, 40, 50, 75, 80]$	$x_{obs} = [30, 50, 65, 0, 50, 75, 100, 50]$
$y_{obs} = [30, 50, 80, 15, 50, 70, 70, 40]$	$y_{obs} = [20, 15, 55, 24, 80, 90, 70, 36]$
$r_{obs} = [14, 10, 20, 12, 15, 12, 14, 12]$	$r_{obs} = [10, 21, 10, 17, 13, 10, 20, 11]$

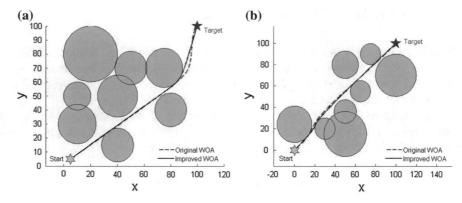

Fig. 3 The comparison of flight paths evaluated by the WOAs for scenarios **a** 1 and **b** 2

Table 3 The comparison of solutions for the two scenarios of UCAV path planning

Scenario	Solution	Original WOA	Improved WOA	Straight-line distance (infeasible path)
1	Mean	163.211	156.496	134.3503
	(Std)	(15.7875)	(17.8122)	
	Best	141.416	139.978	
2	Mean	162.46	156.572	141.4213
	(Std)	(23.9354)	(16.5456)	
	Best	142.13	141.915	

5 Conclusion

This study proposed a new self-tuning judgment criterion which was designed based on the agent's fitness to enhance the performance of the original WOA. The enhanced version was named improved WOA herein. Six scalable benchmark functions were tested to evaluate the performance of the improved WOA. Comparison of the solutions revealed that the presented WOA had an excellent convergence speed, as evidenced by the plots of convergence history and performed well in terms of the mean, standard deviation, and best solutions. Moreover, the improved WOA was applied to simulate the path planning of UCAV. Two scenarios of path planning problem were simulated by using the original and improved WOAs. From the simulation results, the proposed improved WOA can effectively achieve the best paths that avoid locating inside the regions of prescribed threat sources.

References

1. Kennedy, J., Eberhart, R.C.: Particle swarm optimization. In: IEEE International Conference on Neural Networks, vol. 4, pp. 1942–1948. IEEE Press, Perth, Australia, New York (1995). https://doi.org/10.1109/icnn.1995.488968

2. Dorigo, M., Gambardella, L.M.: Ant colony system: a cooperative learning approach to the traveling salesman problem. IEEE Trans. Evol. Comput. **1**(1), 53–66 (1997)
3. Karaboga, D., Basturk, B.: On the performance of artificial bee colony (ABC) algorithm. Appl. Soft Comput. **8**(1), 687–697 (2008)
4. Yang, X.S., Deb, S.: Engineering optimisation by cuckoo search. Int. J. Math. Model. Numer. Optim. **1**(4), 330–343 (2010)
5. Mirjalili, S., Lewis, A.: The whale optimization algorithm. Adv. Eng. Softw. **95**, 51–67 (2016)
6. Mafarja, M.M., Mirjalili, S.: Hybrid whale optimization algorithm with simulated annealing for feature selection. Neurocomputing **260**, 302–312 (2017)
7. Sayed, G.I., Darwish, A., Hassanien, A.E., Pan, J.S.: Breast cancer diagnosis approach based on meta-heuristic optimization algorithm inspired by the bubble-net hunting strategy of whales. In: Pan, J.S., Lin, J.W., Wang, C.H., Jiang, X. (eds.) Proceedings of the Tenth International Conference on Genetic and Evolutionary Computing, Advances in Intelligent Systems and Computing, vol. 536, pp. 306–313. Springer, Heidelberg (2016)
8. Kamaruzaman, A.F., Zain, A.M., Yusuf, S.M., Udin, A.: Lévy flight algorithm for optimization problems-a literature review. Appl. Mech. Mater. **421**, 496–501 (2013)
9. Ling, Y., Zhou, Y., Luo, Q.: Lévy flight trajectory-based whale optimization algorithm for global optimization. IEEE Access **5**, 6168–6186 (2017). https://doi.org/10.1109/ACCESS.2017.2695498
10. Hu, H., Bai, Y., Xu, T.: Improved whale optimization algorithms based on inertia weights and theirs applications. Int. J. Circuits Syst. Signal Process. **11**, 12–26 (2017)
11. Shi, Y., Eberhart, R.C.: A modified particle swarm optimizer. In: IEEE International Conference on Evolutionary. IEEE Press, Anchorage, AK, New York (1998). https://doi.org/10.1109/icec.1998.699146
12. Kaur, G., Arora, S.: Chaotic whale optimization algorithm. J. Comput. Des. Eng. **5**, 275–284 (2018)
13. Oliva, D., Aziz, M.A.E., Hassanien, A.E.: Parameter estimation of photovoltaic cells using an improved chaotic whale optimization algorithm. Appl. Energy **200**, 141–154 (2017)
14. Zhao, Q., Zhen, Z., Gao, C., Ding, R.: Path planning of UAVs formation based on improved ant colony optimization algorithm. In: Guidance, Navigation and Control Conference, pp. 1549–1552. IEEE Press Yantai, China, New York (2014). https://doi.org/10.1109/cgncc.2014.7007423
15. Zhang, Y., Wu, L., Wang, S.: UCAV path planning by fitness-scaling adaptive chaotic particle swarm optimization. Math. Probl. Eng. **2013**, 1–9 (2013). https://doi.org/10.1155/2013/705238
16. Ma, Q., Lei, X.: Application of artificial fish school algorithm in UCAV path planning. In: 2010 IEEE Fifth International Conference on Bio-Inspired Computing: Theories and Applications, pp. 555–559 Changsha, China (2010). https://doi.org/10.1109/bicta.2010.5645185
17. Li, B., Gong, L.G., Yang, W.L.: An improved artificial bee colony algorithm based on balance-evolution strategy for unmanned combat aerial vehicle path planning. Sci. World J. **2014**, 1–10 (2014). https://doi.org/10.1155/2014/232704

Part II
Recent Advances on Evolutionary
Optimization Technologies

Effects of Centrality and Heterogeneity on Evolutionary Games

Xin Ge, Hui Li and Lili Li

Abstract In the evolutionary games based on the heterogeneous populations, recent research has shown that the degree of players in the network plays an important role and often determine the level of cooperation. Yet, the individual influence described by centralities remains inadequate in quantifying the effect of promoting cooperation. In this work we have comprehensively investigated how the representative centrality metrics impact the fate of cooperation on different levels of heterogeneous populations. Simulation results show that on the whole, centrality characteristic is efficient to facilitate cooperation in social dilemmas except the Clustering, and Degree is neither the sole nor the best one. Meanwhile, there is an optimal level of heterogeneity that maximizes the cooperators regardless of the influence of centralities.

Keywords Prisoner's dilemma games · Network reciprocity · Cooperation · Centrality metric

1 Introduction

How the large-scale cooperation emerges, sustains and evolves has become a challenging subject for many different fields from biology, economic to social sciences. The evolutionary game theory [1, 2] provides a convenient and fundamental framework for the elucidation of this domain [3]. In the well-known Prisoner's Dilemma(PD) games, one of seminal works solve this dilemma is performed by Nowak [4] who proposed direct and spatial reciprocity or referred as network reciprocity. Meanwhile, compelling evidence has been accumulated that a plethora of biological, social, and technological real-world networks of contacts are mostly het-

X. Ge (✉) · H. Li
College of Information Science and Technology, Dalian Maritime University, Dalian, China
e-mail: ge_xin@dlmu.edu.cn

L. Li
College of Marine Electrical Engineering, Dalian Maritime University, Dalian, China

School of Mathematics, Liaoning Normal University, Dalian 116029, China

© Springer Nature Singapore Pte Ltd. 2019
J.-S. Pan et al. (eds.), *Genetic and Evolutionary Computing*,
Advances in Intelligent Systems and Computing 834,
https://doi.org/10.1007/978-981-13-5841-8_6

erogeneous [5, 6]. That is to say, the roles of different nodes may be significantly impact the evolutionary dynamics in real-world structured populations. To quantitatively measure the roles of different nodes, numerous centrality metrics have been proposed, among which most studies have focused on node degree [7]. However, the networks with identical degree distribution perhaps exhibit great difference in regarding of individual importance. Then, one natural question is: to what extent and in what manner the centralities affect evolutionary behavior and cooperation level?

To achieve a comprehensive insight of this issue, we have performed a systematic study and simulations based on PD games, taking into account seven most representative metrics of individual centralities on prototypical model for heterogeneous structure of populations. Here, we are interested in the distinctive role of different centralities on promoting cooperation, especially which centralities are efficient metric identifying cooperative hubs. We incorporate centralities via preferentially assigning strategy among populations according to individual centrality scores. As we will show, our work reveals a nontrivial role of the population structure indicated by heterogeneity and centrality in the evolution of cooperation.

2 Methods

2.1 Network Model

The structured heterogeneous populations can be modeled through mapping players to nodes of heterogeneous network. The Barabasi–Albert (*BA*) model provides the best known model leading to overall scale-free degree distributions $d(k) = k^{-\gamma}$. In this study, we adopt the single-scale version implemented by Eppstein David, etc. [8] (*ED* model) because it reflects the real-world connectivity particular the social relationship such as acquaintances. Besides strength of heterogeneity, another crucial difference between these two models is that *BA* model does not contain node of $k = 1$, leading to all the nodes have identical Coreness. While Coreness is one vital metric to measure the influence of individuals, which has attracted intensive attention and has been as a baseline in the comparison of different centralities [9]. To avoid stochastic effect, we have checked larger network size up to $N = 10^5$ and smaller one, down to values of $N = 100$ We found that populations size $N > 1500$ is able to remain results unchanged thus in the present study, the simulations were performed for the setup of $k = 2000$ and the average degree $\langle k \rangle = 2 \sim 8$.

2.2 Game Model

In the PD game, we use payoff matrix $R = 2$ and $P = 1$ and correspondingly $0 \le S \le 1$ and $2 \le T \le 3$. The studied region in the T $-$ S plane we employed was sampled in steps of 0.1, thus encompassing $11 \times 11 = 121$ parameter combinations. The evolutionary process with synchronous update procedure comprises discrete elementary steps (game round) where the whole population plays simultaneously. The traditional process (referred as benchmark simulation) consists of two steps, initialization and update. Initially, a portion of individuals $\rho_C(0)$ (here 0 indicates the first time step) randomly chosen are assigned cooperation strategy and others defection strategy. As to strategy updating, player x with strategy s_x imitates the strategy s_y of another player y, chosen randomly from the neighborhood of x. Player x takes over the strategy s_y with a probability determined by Fermi rule [10]

$$\mathcal{P}\left(s_y^t \to s_x^{t+1}\right) = \frac{1}{1 + e^{-[(p_y^t - p_x^t)/k]}},$$

where k quantifies the uncertainty related to the strategy changing process, here the selected value of $k = 0.1$ is a traditional and frequently employed choice.

Since stimulating specific individuals to cooperate is a feasible and applicable approaches in the purpose of promoting cooperation. All the players are ranked according to a centrality metric then select top-I ones as initial cooperators, namely $\rho_C(0) = I/N$. We employ the centrality metrics Degree (Deg), Hindex (Hin), Coreness (Cor), Clustering (Clu), Closeness (Clo), Eigenvector (Eig), and Betweenness (Bet) measuring influential populations. It is vital to investigate the correlation between any two centrality metrics used in the present study. As detailed in Fig. 1, we computed the Pearson correlation coefficients φ between any two centrality metrics in networks of $\langle k \rangle = 4$. The outcomes indicate that strong linear correlations do exist between certain centrality metrics. It is natural to expect they maybe have akin impact on evolutionary dynamic. Note that for Degree and Eigenvector $\varphi = 1$, which means we will get the same set of nodes according to sorted scores of these two metrics, thus in the next section, the Eigenvector is omitted in the results.

For the population size we have found that, $N = 2000$ and times well over 5000 steps warrants a correct convergence and steady outcomes, in agreement with many other works in this field, like for example in [11]. To measure the cooperation level, we use a quantitative measure $\mathcal{C}$ for the overall asymptotic cooperation in the PD game, given by averaging the $\rho_c(\infty)$ over the corresponding region in the T-S plane. Moreover, to assure suitable accuracy and the stochastic effect all the final results are obtained via averaging 100 independent runs for each set of parameters.

3 Results and Discussion

3.1 Effect of Heterogeneity

To assess the fundamental evolution of PD game, we begin by presenting the result of benchmark dynamic in Figs. 2 and 3 which illustrate the impact of average degree $\langle k \rangle$

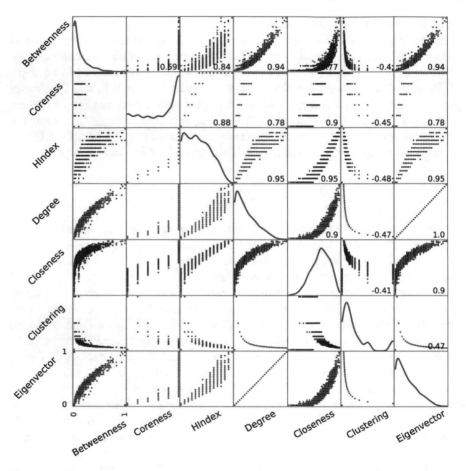

Fig. 1 Correlation matrix and distributions of centralities on network of $\langle k \rangle = 4$. The φ of any two centrality metrics are shown on upper triangular insets

and initial proportion of cooperators $\rho_C(0)$ on the outcome of the $T - S$ space panel. It is clearly evidenced that heterogeneity drastically boosts cooperation compared with the evolution underlying random and well-mixed population that inevitably evolves to full defection. Despite the heterogeneity generally play a positive role in driving cooperation, we need scrutinize the results in terms of the correlation between heterogeneity and final evolutionary state. In the *ED* model, the level of heterogeneity γ is proportional to $\langle k \rangle$, therefore each column of Fig. 2 (e.g., panel a, d, g, j) records the final abundance of cooperators with the heterogeneity varying from lower level ($\gamma = 1.8$) to higher level ($\gamma = 2.6$). According to naive intuition, cooperation should gradually thrive with the average degree increasing due to the consensus that interconnectivity between networks does promote cooperation by means of enhanced reciprocity, yet we show that the influence of heterogeneity is not monotonous toward cooperation level, $\langle k \rangle = 2$ being the exception from $\langle k \rangle = 4, 6, 8$. To suppress the

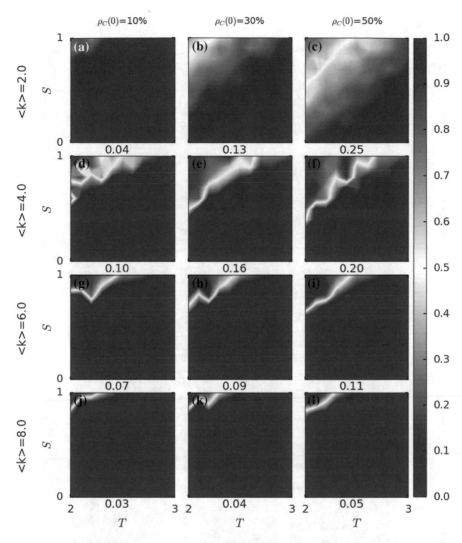

Fig. 2 Asymptotic density of cooperators C on different heterogenous structured populations, $\langle k \rangle = 2, 4, 6, 8$

bias of stochastic disturbance and fluctuation, we further inspect the conditions of smaller resolution of $\langle k \rangle$ and the result is given in Fig. 3. It shows that irrespective of initial fraction of cooperators the mean $\rho_C(0)$ on $T - S$ panel reaches the maximum at $\langle k \rangle = 2.7$ (see panel j, k, i). This result suggests that the cooperation level does not monotonously depend on the connection density of underlying network (e.g., the $\rho_C(0) = 0.13, 0.19, 0.18, 0.27, 0.26$ corresponding to $\langle k \rangle = 2.0, 2.3, 2.7, 2.4$ when $\rho_C(0) = 0.3$), in other words, existing an intermediate heterogencity optimally sustaining cooperation. It implies that if a population structure evolves to become

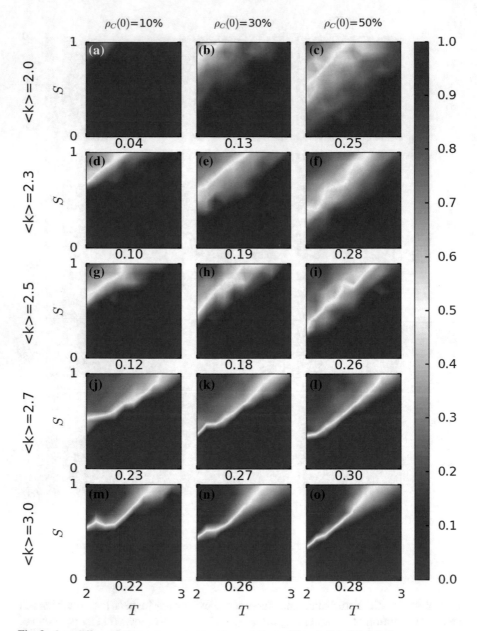

Fig. 3 $\mathcal{C}$ on different heterogenous structured populations, $\langle k \rangle = 2, 2.3, 2.5, 2.7, 3$

increasingly or decreasingly heterogeneous beyond the threshold, the population will eventually reach a structure under which cooperation becomes less viable as shown in the panel a of Fig. 3. It can be predicted that the cooperators will thoroughly die out on $T - S$ panel when degree exceeds one threshold that closes to the average degree of well-mixed structure. Moreover, this result is robust to variations of $\rho_C(0)$, and thus indicates a high degree of universality. Previous research has focused on the mechanism of network reciprocity, while the existence of threshold lacks interpretation [12]. We deduce it stems from the fact that those individuals with more connections have more opportunity to participate in the majority of the interactions. In this case, they accumulate large fitness based on payoffs and determine the outcome of evolution. However, if the overall connection is too sparse (lower degree) or too dense(higher degree), it is difficult for independent formation of cooperative patterns on each individual network due to the asymmetric strategy flow.

3.2 Effect of Centrality

In this section, we proceed exploring the evolution of cooperation considering nonrandom strategy assignment, one applicable and feasible approaches to foster cooperation. The results under varying heterogeneity $\langle k \rangle = 2, 4, 8, 2.7$ are depicted separately in Figs. 4, 5, 6 and 7 and , corresponding to the first, second, and fourth rows of Fig. 2 and the fourth row of Fig. 3, respectively. Notably, we have assessed the conditions where $\langle k \rangle = 2{-}3$ at interval 0.1 and $\langle k \rangle = 3{-}8$ at interval 1, while only present the representative results corresponding to the benchmark result in Figs. 2 and 3.

In most cases, preferential initialization elevates the final level of cooperation, which happens irrespectively of the initial fraction of cooperators and, more importantly, of the $\langle k \rangle$. For instance between panels m, n, o in Fig. 2 and panels j, k, l in Fig. 6, the mean level of cooperation rises from 0.1, 0.16 and 0.2 to 0.18, 0.22 and 0.57 when initially designate cooperators based on Coreness metric. Aside from aforementioned result, another considerable difference between random initiation in benchmark and centrality preferential initiation is that the optimal level of heterogeneity facilitating cooperation shifts from $\langle k \rangle = 2.7{-}\langle k \rangle = 2$. Accordingly, a natural question is that whether $\langle k \rangle = 2$ is the most optimal level amongst all available values of $\langle k \rangle$, not limited to this study. In fact, we can consider the extreme condition of minimum value $\langle k \rangle = 0$, where each node is isolated and has no connection to others. Due to the absence of direct links between individuals, we can infer that each cooperators can successfully resist the invasion of defectors and remain their initial strategy. In this case, the final outcome is completely determined by the initial fraction of cooperators. For instance, if $\rho_C(0) = 0.5$, the T−S space will be dominated by white color (denotes 0.5 in color bar), regardless of values of T and S. Apparently, $\langle k \rangle = 0$ is not the optimal value promoting cooperation and we can predict the optimal value is between 1 and 2. These results, consequently reveal that the optimal level of heterogeneity indeed exists not only in benchmark evolution,

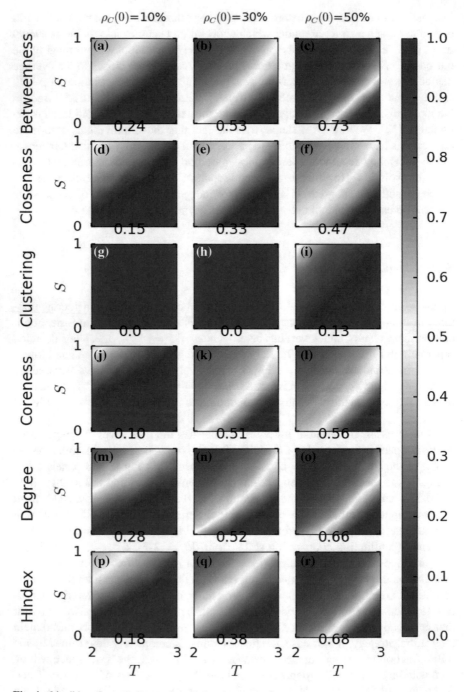

Fig. 4 $\mathcal{C}$ in $\langle k \rangle = 2$, centrality preferential assignment of strategy

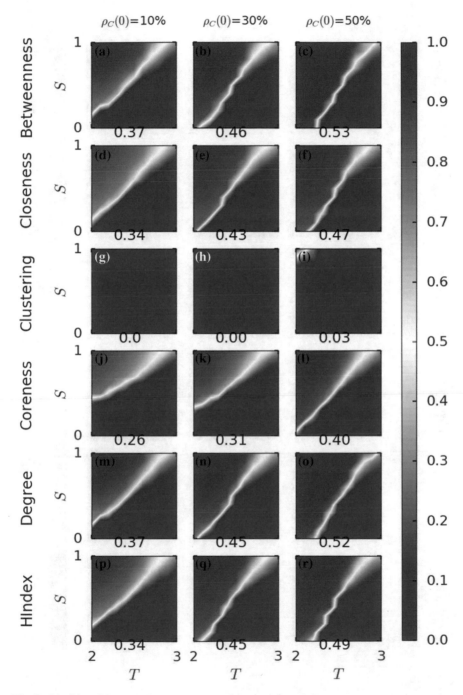

Fig. 5 C in $\langle k \rangle = 2.7$, centrality preferential assignment of strategy

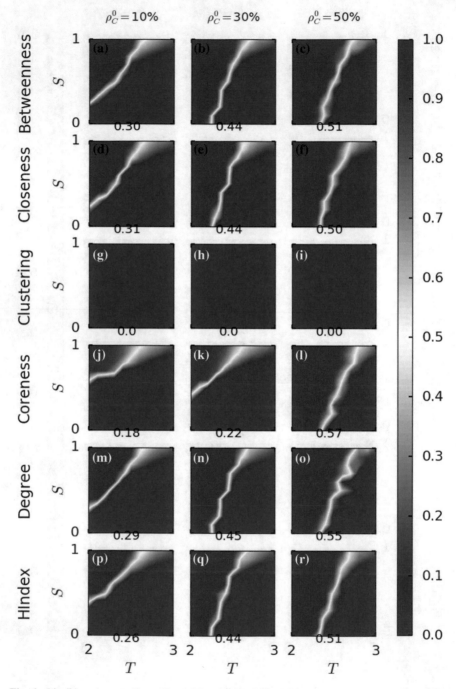

Fig. 6 $\mathcal{C}$ in $\langle k \rangle = 4$, centrality preferential assignment of strategy

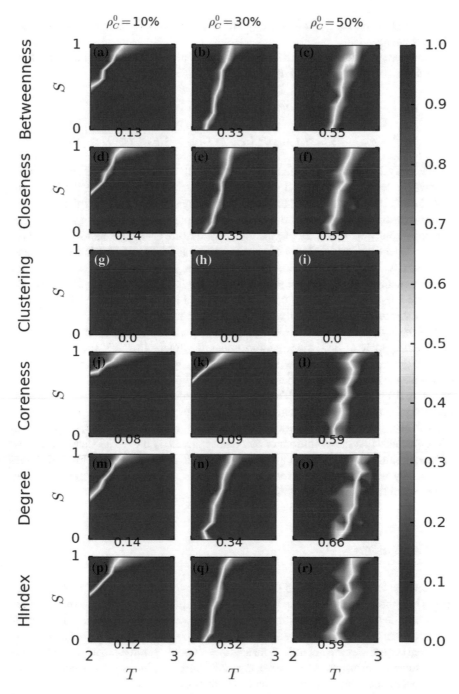

Fig. 7 C in $\langle k \rangle = 8$, centrality preferential assignment of strategy

but also in the version of preferentially initiation, and more importantly the optimal value is not identical.

To systematically compare the impact of different centrality metric, we rank these centralities according to C and dependency to $\rho_C(0)$ as following:

$$\begin{cases} C_{Deg} > C_{Bet} > C_{Hin} > C_{Clo} > C_{Cor} > C_{Clu} & \rho_C(0) = 0.1 \\ C_{Bet} > C_{Deg} > C_{Cor} > C_{Hin} > C_{Clo} > C_{Clu} & \rho_C(0) = 0.3 \\ C_{Bet} > C_{Hin} > C_{Deg} > C_{Cor} > C_{Clo} > C_{Clu} & \rho_C(0) = 0.5 \end{cases}$$

Basically, the centralities can be roughly classified into three classes according to relative effectiveness of promoting cooperation. It can be observed that the Betweenness, Degree, and Hindex as the first class noticeably and stably enhance the resilience of cooperation compared with other centralities, particularly in dense and stronger heterogeneous population. In the second class, despite Closeness and Coreness also foster cooperation, their influence is not as intense as the centralities in the first class, occasionally illustrating weak effect (see Fig. 4d, e, j and Fig. 6j, k). Lastly, the third class consisting of only Cluster always lead to the full-defection outcome, plays the opposite role, even worse than random selection of initial cooperators. This implies that Cluster thoroughly lost influence in the angle of helping cooperators.

Unlike in the benchmark cases depicted in Figs. 2 and 3, the relative advantages of centrality cannot hold stabilization under the fluctuation of initial fraction of cooperators $\rho_C(0)$, e.g., as shown in Fig. 4, in the case of $\langle k \rangle = 2$, Degree yields the best outcome if $\rho_C(0) = 0.1$ (panel m), whereas Betweenness becomes the optimal one if $\rho_C(0) = 0.5$ (panel c). This phenomenon means again that the influence of initial fraction of cooperators cannot be negligible, especially as decreasing $\rho_C(0)$, the gap of C between different centralities extends. As illustrated, these results confirm the hypothesis that the evolutionary outcome is driven by the initial state of the most influential individuals in the population [13]. This result reveal the limitation that a centrality which is optimal for one condition in the evolutionary game is often sub-optimal for a different condition.

4 Conclusion

We have systematically studied the evolution of cooperation on structured population where the heterogeneity can be continuously adjusted through controlling model parameter degree. Besides optimal heterogeneity, we find striking effect in promoting cooperation by deliberately initializing cooperators via ranking players, while different centralities are not able to measure evolutionary influence in general conditions, someone even counterintuitive, e.g., the Clustering index plays a negative role in cooperative evolution. On one hand, Hindex, Coreness, and Degree centralities are better indicators to locate important individuals as initial cooperators because of their prominent effect and low computational complexity. On the other hand, Degree is not the only, and in certain conditions not the most effective centrality facilitating cooperation. In our study, we directly apply the centralities to find the initial set of cooperators, which could be inefficient since their influences may be largely overlapped [14]. Therefore, comprehensively identify a set of influential nodes emerges as a significant cooperators may be an alternative approach.

Acknowledgements This paper is supported by Natural Science Foundation of Liaoning Province (No. 20170540097), Fundamental Research Funds for the Central Universities (No. 3132018127) and High-level Talents Innovation of Dalian City, China (grant number 2016RQ049).

References

1. Refer Smith, J.M. Evolution and the Theory of Games; Cambridge university press (1982)
2. Nowak, M.A.: Evolutionary dynamics: exploring the equations of life. Best Seller **82**(03) (2006)
3. Trivers, R.L.: The evolution of reciprocal altruism. Q. Rev. Biol. pp. 35–57 (1971)
4. Nowak, M.A.: Five rules for the evolution of cooperation. Science **314**, 1560–1563 (2006)
5. Amaral, L.A.N., Scala, A., Barthelemy, M., Stanley, H.E.: Classes of small-world networks. Proc. Natl. Acad. Sci. **97**, 11149–11152 (2000)
6. Dorogovtsev, S.N.; Mendes, J.F. Evolution of networks: From biological nets to the Internet and WWW; OUP Oxford, 2013
7. Santos, F.C., Santos, M.D., Pacheco, J.M.: Social diversity promotes the emergence of cooperation in public goods games. Nature **454**, 213–216 (2008)
8. Eppstein, D., Joseph, W.: A steady state model for graph power laws. arXiv preprint cs/0204001 (2002)
9. Liu, Y., Tang, M., Zhou, T., Do, Y.: Core-like groups result in invalidation of identifying super-spreader by k-shell decomposition. Sci. Rep. **5**, 9602 (2015)
10. SzabÓ, G., Toke, A.C.: Evolutionary prisoner's dilemma game on a square lattice. Phys. Rev. E, **58**(1), 69–73 (1997)
11. Wang, Z., Wang, L., Perc, M.: Degree mixing in multilayer networks impedes the evolution of cooperation. Phys. Rev. E **89**(5), 052813 (2014)
12. Cimini, G., Sánchez, A.: How evolution affects network reciprocity in Prisoner's Dilemma (2014). arXiv:1403.3043
13. Nowak, M.A., May, R.M.: Evolutionary games and spatial chaos. Nature **359**, 826–829 (1992)
14. Iyer, S., Killingback, T.: Evolution of Cooperation in Social Dilemmas on Complex Networks. PLoS Comput. Biol. **12**, e1004779 (2016)

An Improvement for View Synthesis Optimization Algorithm

Chang Liu, Kebin Jia and Pengyu Liu

Abstract View synthesis optimization (VSO) allows 3D video system to improve the quality of synthesized views (SVs). Based on the latest segment-based VSO scheme, we find that not all of the intra mode and skip interval are necessary to be considered. Therefore, in this paper, an early determination of intra mode and optimal skip interval based on VSO scheme in 3D-HEVC is presented. First, we discuss a measure to select the best intra mode. Then, we decide the optimal skip interval based on statistical analysis. Experimental results indicate that the proposed algorithm achieves a reduction of 21.711% with negligible loss in rate-distortion performance compared with the original 3D-HEVC encoder, and the encoding time has also reduced to 7.869% compared with the others.

Keywords VSO · SVDC · Intra mode · Skip interval · 3D-HEVC

1 Introduction

In recent years, 3D video gains a great impact in the market of cinema and other entertainment facilities. With the development of High Efficiency Video Coding (HEVC) [1], 3D-HEVC was proposed and completed in February in 2015 [2]. In order to satisfy people watching 3D video without aided equipment, 3D displayer

C. Liu · K. Jia (✉) · P. Liu
Faculty of Information Technology, Beijing University of Technology, 100124 Beijing, China
e-mail: kebinj@bjut.edu.cn

C. Liu
Beijing Laboratory of Advanced Information Networks, 100124 Beijing, China

K. Jia
Beijing Advanced Innovation Center for Future Internet Technology,
Beijing University of Technology, 100124 Beijing, China

P. Liu
Beijing Key Laboratory of Computational Intelligence and Intelligent System,
Beijing University of Technology, 100124 Beijing, China

© Springer Nature Singapore Pte Ltd. 2019
J.-S. Pan et al. (eds.), *Genetic and Evolutionary Computing*,
Advances in Intelligent Systems and Computing 834,
https://doi.org/10.1007/978-981-13-5841-8_7

employs layered depth video (LDV) [3], which is a sparse representation of MVD, or multiview video plus depth (MVD) format [4]. Specifically, MVD associates a small number of videos and depth map which is taken from different views. Thus, after encoding video and depth map, additional intermediate views can be synthesized with help from Depth Image Based Rendering (DIBR) [5]. However, it will make massive information and data to be stored and compressed. Therefore, an efficient 3D video coding tool in 3D-HEVC is needed to solve the above problem. This coding tool is called view synthesis optimization (VSO). The distortion measure D_{VSO} consists of the distortion in the virtual view and depth map, denoted as

$$D_{VSO} = (W_s \times D_s + W_d \times D_d)/(W_d + W_s) \tag{1}$$

where the distortion of virtual view and depth map are expressed by D_s and D_d, respectively; W_s and W_d are the weighting factors. However, according to previous studies, it was found that 3D-HEVC adopted VSO may increase the computational complexity of the encoder [6]. Due to this problem, there has been a growing interest in reducing the complexity of VSO [7–16].

On the one hand, several works use the approximate method to reduce the complexity. The work in [7] proposed a virtual view distortion function based on the features of texture images and depth maps to accurately estimate the distortion of virtual views. In order to decrease the virtual view distortion in intra coding modes, Zhang et al. and Silva et al. proposed an algorithm in [8, 9], respectively. In addition, based on the analysis of the virtual view distortion (VVD), Zheng et al. [10] proposed a fine virtual view distortion estimation method. A fast depth map encoding framework based on the Lagrange optimization adjustment at Coding Unit (CU) level was proposed in [11]. With any strategy adopted, however, the distortion calculated by these algorithms [7–11] may not represent the accurate synthesized view distortion due to the approximate values.

On the other hand, in order to calculate the exact synthesized view distortion, a model based on error spread was proposed by Gao et al. [12]. Wang and Yu [13] proposed a simple model, which was calculated pixel by pixel and similar to View Synthesis Distortion (VSD), to estimate synthesized view distortion [14]. Yuan et al. [15] and Ma et al. [16] also constructed a polynomial model and a novel zero-synthesized view difference (ZSVD) model in the VSO scheme for depth compression, respectively. However, it still needs to consume a large amount of computing resource and then affects 3D video application, so further studies are still essential for complexity reduction.

The rest of this paper is organized as follows. Section 2 introduces the previous studies about segment-based VSO in 3D-HEVC. In Sect. 3, an early determination of intra mode and optimal skip interval based on VSO scheme in 3D-HEVC is presented. Section 4 provides simulation results and the corresponding discussion and analysis. And Sect. 5 draws a conclusion.

2 Related Works

In this section, an adaptive segment-based skipping method for VSO in 3D-HEVC is reviewed [6].

The proposed method in [6] marks the zero distortion area based on the coded depth and texture information. Then the synthesized view distortion is calculated flexibly for each non-zero distortion area which means the original line-based calculation is replaced by the flexible segment-based calculation. Experimental results show the proposed method can reduce the depth map encoding time meanwhile maintaining the reconstructed quality of the original depth map and the synthesized view.

The texture smoothness criterion in [6] is defined by a threshold, T as follows:

$$|l_i - l_{i+1}| \leq T \tag{2}$$

where l_i and l_{i+1} is the pair of pixel along the horizontal line. If the pixel pair satisfies above criterion, the position error may not cause rendering error. According to the already encoded smoothing blocks in the texture frame, the threshold can be obtained from Eq. (3), as follows:

$$T = \frac{\sum_{j=0}^{N_h} \sum_{i=0}^{N_w-1} |l_{i,j} - l_{i+1,j}|}{IntraXXBlockNum(N_w - 1)N_h} \tag{3}$$

where N_h and N_w denote the size of each block, and IntraXXBlockNum denotes the number of blocks coded using one of the 35 intra modes.

3 Selection of Intra Mode and Optimal Skip Interval

It is well known that the decision of intra mode adopts the method called "select the best from all of the choices". The IntraXXBlockNum in [6] also adopts the same method. However, it will bring huge encoding complexity. Therefore, an early determination of intra mode and optimal skip interval based on VSO scheme in 3D-HEVC is proposed in this paper.

3.1 Intra Mode Selection

In 3D-HEVC intra prediction, there are 33 kinds of directional modes as shown in Fig. 1. For different video sequences, the intra mode of *IntraXXBlockNum* is also different. For instance, the motion vectors (MV) display of basketball and BQ square are illustrated in Fig. 2. In these two figures, the blue and red regions indicate the

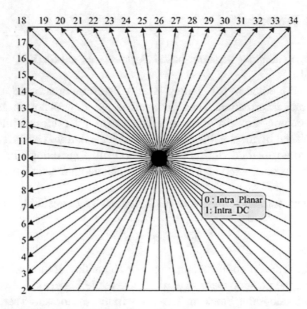

Fig. 1 33 directional modes in HEVC intra prediction

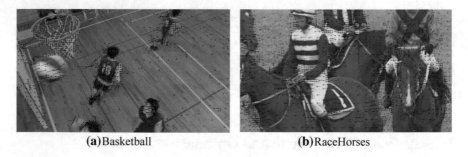

(a)Basketball   (b)RaceHorses

Fig. 2 The motion vectors display of basketball and racehorses

moving part in this video sequence, and the arrows are motion vectors. Based on the moving direction, the intra mode of *IntraXXBlockNum* can be decided.

Take the Balloons sequence as an example. The main process of the intra mode decision could be summarized as the following three steps:

Step 1: By observing the moving direction of the object in the video sequence, five most probable modes (MPMs) are selected as candidates. Figure 3 shows four inconsistent frames of the Balloons sequence. Because of the moving direction is up and down, vertical may be the best intra mode.

Step 2: In order to precisely predict the main moving direction, we can make use of edge detection in depth map to decide the final intra mode. In addition, an image gradient [17] can reflect the change of direction. And the image gradient can be obtained from Eqs. (4) and (5), as follows:

Fig. 3 Four inconsistent frames of balloons sequence

$$\text{magn}(\nabla f) = \sqrt{\left(\frac{df}{dx}\right)^2 + \left(\frac{df}{dy}\right)^2} = \sqrt{M_x^2 + M_y^2} \tag{4}$$

$$\text{dir}(\nabla f) = tan^{-1}\left(\frac{M_y}{M_x}\right) \tag{5}$$

where $\frac{df}{dx}$ is the gradient in the y direction and $\frac{df}{dy}$ is the gradient in the x direction. Gradient direction is calculated from Eq. (6), as follows:

$$\theta = \arctan\left(\frac{df}{dy}, \frac{df}{dx}\right) \tag{6}$$

The horizontal and vertical gradient of each pixel in current depth map block can be obtained from Eqs. (7) and (8), as follows:

$$G_x = \frac{1}{N^2} \sum_{i=1}^{N} \sum_{j=1}^{N} \begin{bmatrix} 1 & 0 \\ 0 & -1 \end{bmatrix} \times I(i, j) \tag{7}$$

$$G_y = \frac{1}{N^2} \sum_{i=1}^{N} \sum_{j=1}^{N} \begin{bmatrix} 0 & 1 \\ -1 & 0 \end{bmatrix} \times I(i, j) \tag{8}$$

where $I(i, j)$ is the depth map block, and (i, j) represents the position of pixel. And final gradient of the current depth map block can be decided form Eq. (9), as follows:

$$\theta_{current} = \arctan(G_x, G_y) \tag{9}$$

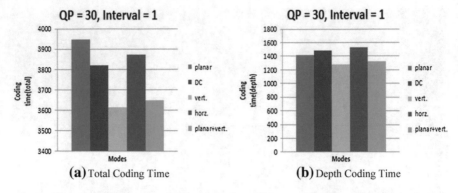

(a) Total Coding Time   (b) Depth Coding Time

Fig. 4 Coding time of balloons sequence under the QP of (30, 39) and interval equals to 1

where $\theta_{current}$ is the gradient angle. And the closest mode to it is vertical.

Step 3: In order to prove previous conclusion, the optimal intra mode can be selected from five candidates based on the coding result. Figure 4 is the coding result.

3.2 The Optimal Segment-Wise Interval Selection

There are several different skipping intervals to be tested. Based on the best intra mode, the optimal interval can also be determined. It can be seen from Fig. 5 that 1 is the best interval.

Fig. 5 Coding time of balloons sequence under the QP of (30, 39)

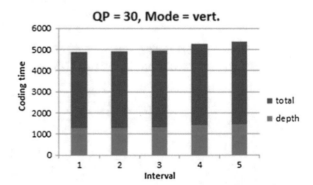

4 Experimental Results

To assess the efficiency of our proposed algorithm, this paper compares the proposed algorithm with the original 3D-HEVC encoder and the method in [16]. All the experiments are conducted on the 3D-HEVC Test Model (HTM16.0). Several sequences with two resolutions of 1920 × 1088 and 1024 × 768 were encoded. Specifically, the sequences include Balloons (1024 × 768), Kendo (1024 × 768), Newspaper (1024 × 768), Poznan_Hall2 (19201088) and Poznan Street (1920 × 1088). The experimental platform configuration is as follows: Intel (R) Core (TM) i5-4570 CPU @ 3.20 GHz and 6.00 GB RAM. In this paper, Quantization Parameter (QP) pairs for texture video and depth map (25, 34), (30, 39), (35, 42), (40, 45) have been adopted.

Table 1 shows the comparison of coding time among our proposed algorithm, original 3D-HEVC encoder and the method in [16]. In Table 1, ΔT_1 (%) represents the ratio of method in [16] to the original 3D-HEVC encoder in total coding time (sum of the total coding time of 4 QP values). And ΔT_2 (%) represents the ratio of the proposed method to the original 3D-HEVC encoder in total coding time (sum of the total coding time of 4 QP values). From Table 1, it can be found that the method in [16] can obtain 7.869% complexity reduction in average compared with original algorithm in HTM16.0, while the proposed method can obtain 21.711% complexity reduction in average compared with original algorithm in HTM16.0. It indicates that the proposed method is more effective than the method in [16] in reducing the coding complexity.

Table 1 Total coding time comparison

Sequence	QP	Total coding time (s)			ΔT_1 (%)	ΔT_2 (%)
		HTM16.0	Method in [16]	Proposed method		
Balloons	(25, 34)	5944.789	5624.960	4675.520		
	(30, 39)	4656.752	4397.461	3672.464		
	(35, 42)	4155.648	3674.273	3323.980	92.891	78.623
	(40, 45)	3722.879	3469.579	2857.633		
Kendo	(25, 34)	6019.188	5820.771	4619.810		
	(30, 39)	4828.753	4491.696	3738.671		
	(35, 42)	4331.341	3902.228	3275.018	92.062	75.570
	(40, 45)	4077.307	3513.383	2918.311		
Newspaper	(25, 34)	5270.937	5045.364	4183.430		
	(30, 39)	4457.943	3917.372	3437.376		
	(35, 42)	4041.508	3608.569	3021.150	89.739	76.800
	(40, 45)	3690.404	3097.791	2767.955		

(continued)

Table 1 (continued)

Sequence	QP	Total coding time (s)			ΔT_1 (%)	ΔT_2 (%)
		HTM16.0	Method in [16]	Proposed method		
Poznan_Hall2	(25, 34)	10483.147	9818.830	8812.998		
	(30, 39)	8551.049	8121.586	6960.763		
	(35, 42)	7918.812	7300.502	6255.602	92.966	80.681
	(40, 45)	7730.831	7003.200	5954.036		
PoznanStreet	(25, 34)	12317.722	12043.637	10382.116		
	(30, 39)	9933.270	9224.121	8023.070		
	(35, 42)	9218.814	8882.081	7101.003	92.999	79.772
	(40, 45)	8528.731	7048.226	6401.315		
Average					92.131	78.289

Table 2 BDBR comparison of the proposed algorithm compared with method in [16]

Sequence	Proposed algorithm compared to HTM16.0 (%)		
	Video PSNR/video bitrate	video PSNR/total bitrate	Synth PSNR/total bitrate
Balloons	0.0	−0.6	4.6
Kendo	0.0	−0.3	1.2
Newspaper	1.0	2.7	0.0
Poznan_Hall2	0.0	−0.8	4.3
PoznanStreet	0.1	−0.2	1.7
Average	0.22	0.16	2.36

Table 2 presents the BDBR performance of the proposed algorithm compared to the method in [16]. In Table 2, the "video PSNR/video bitrate" indicates the BDBR coding results for the coded texture views. The "video PSNR/total bitrate" indicates the BDBR coding results for the coded views, which include depth views. And the "synth PSNR/total bitrate" indicates the BDBR coding results for synthesized views. It can be seen from "video PSNR/video bitrate", "video PSNR/total bitrate", and "synth PSNR/total bitrate" that the ΔBDBR increases only 0.22%, 0.16%, and 2.36%, respectively. Therefore, the proposed algorithm can reduce the computational complexity without significant ΔBDBR loss.

5 Conclusion

In this paper, an efficient algorithm for reducing the computational complexity of 3D-HEVC encoder has been conducted. The previous work mainly focused on the problem of high computational complexity in calculating synthesized view distortion change (SVDC) for depth coding. However, the previous method still adopts full traversal mode when choosing intra mode and skip interval, which will greatly increase the coding complexity. In response to the problem, in this work, I put forward a method based on the VSO scheme to decide the intra mode and optimal segment-wise interval in advance for depth map. We decide intra mode and optimal segment-wise interval mainly through the subjective view of the video, the use of gradient operator for edge detection and statistical methods. This method can obtain 21.711% complexity reduction compared with the original method while the method proposed by Ma can obtain 7.869% complexity reduction compared with the original method.

Acknowledgements This work is supported by the Project for the National Natural Science Foundation of China (Grant No. 61672064), the Beijing Natural Science Foundation (Grant No. KZ201610005007), the China Postdoctoral Science Foundation (Grant No. 2015M580029), and the Beijing Postdoctoral Research Foundation (Grant No. 2015ZZ-23).

References

1. Sullivan, G.J., et al.: Overview of the high efficiency video coding (HEVC) standard. IEEE Trans. Circuits Syst. Video Technol. **22**(12), 1649–1688 (2012)
2. Sullivan, G.J., et al.: Standardized extensions of high efficiency video coding (HEVC). IEEE J. Sel. Top. Signal Process. **7**(6), 1001–1016 (2013)
3. Li, Z., et al.: A novel representation and compression method in layered depth video. In: Proceedings of the SPIE/COS Photonics Asia (2016)
4. Zhang, Q., Wu, Q., Wang, X.: Early skip mode decision for three-dimensional high efficiency video coding using spatial and interview correlations. J. Electron. Imaging **23**(5), 053017 (2014)
5. Guo, L., et al.: Hole-filling map-based coding unit size decision for dependent views in three-dimensional high-efficiency video coding. J. Electron. Imaging **25**(3), 033020 (2016)
6. Dou, H., et al.: Segment-based view synthesis optimization scheme in 3D-HEVC. J. Vis. Commun. Image Represent. **42**, 104–111 (2017)
7. Yang, C., et al.: Virtual view distortion estimation for depth map coding. In: Proceedings of the Visual Communication and Image Processing (VCIP), pp. 1–4 (2015)
8. Zhang, Y., et al.: Efficient multiview depth coding optimization based on allowable depth distortion in view synthesis. IEEE Trans. Image Process. **23**(11), 4879–4892 (2014)
9. De Silva, D., Fernando, W., Arachchi, H.: A new mode selection technique for coding depth maps of 3D video. In: Proceedings of the IEEE International Conference on Acoustics, Speech and Signal Processing (ICASSP), pp. 686–689 (2010)
10. Zheng, Z., et al.: Fine virtual view distortion estimation method for depth map coding. IEEE Signal Process. Lett. 1–1 (2017)
11. Yang, C., et al.: Depth map coding based on virtual quality. In: Proceedings of the IEEE International Conference on Acoustic, Speech and Signal Processing (ICASSP), pp. 1367–1371 (2016)
12. Gao, M., et al.: Estimation of end-to-end distortion of virtual for error-resilient depth map coding. In: Proceedings IEEE Processing International Conference on Image Processing, pp. 1816–1820 (2013)
13. Wang, L., Yu, L.: Rate-distortion optimization for depth map coding with distortion estimation of synthesized view. In: International Symposium on Proceedings of the Circuits and Systems (ISCAS), pp. 17–20 (2013)

14. Yang, M., et al.: A novel method of minimizing view synthesis distortion based on its non-monotonicity in 3D video. IEEE Trans. Image Process. **26**(11), 5122–5137 (2017)
15. Yuan, H., Kwong, S., Liu, J., Sun, J.: A novel distortion model and lagrangian multiplier for depth maps coding. IEEE Trans. Circuits Syst. Video Technol. **24**(3), 443–451 (2014)
16. Ma, S.W., Wang, S.Q., Gao, W.: Low complexity adaptive view synthesis optimization in HEVC based 3D video coding. IEEE Trans. Multimed. **16**(1), 266–271 (2014)
17. Canny, J.: A computational approach to edge detection. IEEE Trans. Pattern Anal. Mach. Intell. **8**(6), 679–698 (1986)

External Hierarchical Archive Based Differential Evolution

Zhenyu Meng, Jeng-Shyang Pan and Xiaoqing Li

Abstract Evolutionary Algorithms (EAs) have become much popular in tackling kinds of complex optimization problems nowadays, and Differential Evolution (DE) is one of the most popular EAs for real-parameter numerical optimization problems. Here in this paper, we mainly focus on an external hierarchical archive based DE algorithm. The external hierarchical archive in the mutation strategy of DE algorithm can further improve the diversity of trial vectors and the depth information extracted from the hierarchical archive can achieve a better perception of the landscape of objective function, both of which consequently help this new DE variant secure an overall better optimization performance. Commonly used benchmark functions are employed here in verifying the overall performance and experiment results show that the new algorithm is competitive with other state-of-the-art DE variants.

Keywords Depth information · Differential evolution · Evolutionary algorithm · Hierarchical archive

1 Introduction

There are many different optimization demands arising from different areas, e.g., computer science, engineering, and mechanics, etc., nowadays. The common approaches to tackle these optimization problems usually begin with designing the related objectives that can model the problem or the system. Usually, many of these objectives are of black-box characteristic or are of noises, therefore, some deterministic algorithm, e.g. Newton's Method, Quasi-Newton Methods etc., failed to solve these problems, and stochastic optimization algorithms come into effect. There are many stochastic optimization algorithms proposed in the literature these decades, and our focus in this paper is mainly on the optimization algorithms in evolutionary

Z. Meng · J.-S. Pan (✉) · X. Li
Fujian Key Lab of Big Data Mining and Applications, Fujian University
of Technology, Fuzhou, China
e-mail: jengshyangpan@gmail.com

© Springer Nature Singapore Pte Ltd. 2019
J.-S. Pan et al. (eds.), *Genetic and Evolutionary Computing*,
Advances in Intelligent Systems and Computing 834,
https://doi.org/10.1007/978-981-13-5841-8_8

computation domain. Evolutionary Algorithms (EAs), such as Particle Swarm Optimization (PSO) [1], Differential Evolution (DE) [2], are famous branches for tough optimization problems. However, there are still some weakness within them, for example, the PSO model has "two steps forward, one step back" weakness [3], and the DE model has both mutation weakness and crossover weakness [4–10].

Meng and Pan proposed an ebb-tide-fish algorithm to tackle the slow convergence weakness, a manifestation of the "two step forward, one step back" weakness, of PSO model in lower dimensional optimization [11, 12] by employing both the global best particle and turbulence particles of the population. As we know, the PSO model based evolutionary algorithm performed relative worse than the DE model based ones, and the difference operation was verified to be very useful on optimization problems. Therefore, an improved ebb-tide-fish algorithm incorporating difference operation [13, 14] was proposed to enhance the optimization performance of the former ebb-tide-fish algorithm, and this algorithm implemented an combination of PSO and DE algorithm. However, the weaknesses existing in DE model also hampered the further development of DE variants or DE model based EAs. These weaknesses of DE mainly lay in trial vector generation strategy and control parameter adaptation schemes.

As we know, the canonical DE algorithm originated from Genetic Anneal Algorithm, a combined Genetic Algorithm (GA) and Simulated Annealing (SA), therefore, the operations such as mutation, crossover and selection in GA were inherited into DE. There were mainly three different crossover schemes in GA, 1-point crossover, 2-point crossover and N-point crossover. The dependence on parameter separation of these three crossover schemes resulted in bias which is called positional bias in GA. In DE algorithm, there were usually two different crossover schemes mentioned in literature, one is exponential crossover and the other is binomial crossover. The exponential crossover actually implemented a combined 1-point crossover and 2-point crossover, therefore, positional weakness still existed in this crossover scheme. The other scheme, the binomial crossover, implemented an equal treatment of each parameters, and therefore, it was asserted tackled the positional bias, nevertheless, bias still existed in this crossover scheme from high dimensional perspective of view.

Meng et al. proposed several papers to illustrate this bias [4, 7, 8, 15], and then proposed a QUATRE structure to achieve a better evolution. Two focuses were mainly paid on in these papers: one is the selection probability of each potential candidate, and the other is the selection probability of the number of parameters inherited from donor vector [8]. These papers both tackled the selection bias with fixed crossover rate and dynamic changed crossover rate respectively. Furthermore, some tricks were also involved into the QUATRE structure [16–20], such as pair-wise competition strategy, sort strategy, etc., to enhance the overall performance of it. Besides the crossover scheme, mutation strategy also played important role in the overall optimization performance, and there were seven commonly used mutation strategies mentioned in literature, and these mutation equations are shown in Table 1. All the seven mutation strategies except for the sixth one usually combined the binomial crossover scheme to form new trial vector generation strategies while the sixth mutation strategy was also

Table 1 The generation of mutant vector $V_{i,G}$ in several different DE mutation strategies

No.	$DE/x/y$	Equation
1	DE/rand/1	$V_{i,G} = X_{r0,G} + F * (X_{r_1,G} - X_{r_2,G})$
2	DE/best/1	$V_{i,G} = X_{gbest,G} + F * (X_{r_1,G} - X_{r_2,G})$
3	DE/rand/2	$V_{i,G} = X_{r0,G} + F * (X_{r_1,G} - X_{r_2,G}) + F * (X_{r_3,G} - X_{r_4,G})$
4	DE/best/2	$V_{i,G} = X_{gbest,G} + F * (X_{r_1,G} - X_{r_2,G}) + F * (X_{r_3,G} - X_{r_4,G})$
5	DE/target-to-best/1	$V_{i,G} = X_{i,G} + F * (X_{gbest,G} - X_{i,G}) + F * (X_{r_1,G} - X_{r_2,G})$
6	DE/target-to-rand/1	$V_{i,G} = X_{i,G} + F * (X_{r0,G} - X_{i,G}) + F * (X_{r_1,G} - X_{r_2,G})$
7	DE/target-to-$p$best/1	$V_{i,G} = X_{i,G} + F * (X_{best,G}^p - X_{i,G}) + F * (X_{r_1,G} - X_{r_2,G})$

a unique trial vector generation strategy without combining any crossover scheme. Furthermore, all the seven mutation strategies except the last one didn't employ an external archive while the last mutation strategy employed an external archive to enhance the diversity of trial vectors. Empirically, the last mutation strategy performed very well in recent competitions, nevertheless, there still exists a certain weakness in this mutation strategy.

In this paper, a new external hierarchical archive with depth information was introduced into the mutation strategy of the DE algorithm, and the novel hierarchical archive enhanced the diversity of trial vectors as well as obtaining a better perception of the landscape of the objectives by the extracted depth information of the hierarchical archive. The rest of the paper is organized as follows: Sect. 2 presents a detailed description of the novel algorithm. Section 3 presents the experiment analysis on commonly used benchmark functions for optimization algorithm evaluation. Finally, Sect. 4 presents the conclusion.

2 The Hierarchical Archive in Differential Evolution

As is mentioned above, most of the mutation strategy do not involved external archive, however, the later proposed DE/target-to-$p$best/1/ mutation strategy [21] employing an external archive secured first ranks at recent competitions [21–23]. This is a good direction for the development of DE variants, and here in this section, we further extends the external archive into a hierarchical archive of the mutation strategy in DE algorithm. The hierarchical archive contains two levels of storage, the first-level storage has a maximum size equalling to population size ps, the second-level storage has a maximum size equalling to $r^{har} \cdot ps$. The function of these two storages are also different in the hierarchical archive, the first level storage has the same function as the external archive in JADE [21], and the second level storage records the past populations of solutions. The detailed hierarchical archive based mutation strategy in the novel DE algorithm is presented in Eq. 1

$$V_{i,G} = X_{i,G} + F \cdot (X_{best,G}^p - X_{i,G}) + F_1 \cdot (X_{r_1,G} - \widetilde{X}_{r_2,G}) + F_2 \cdot (X_{r_1,G} - \widetilde{X}_{r_3,G})$$
$$(1)$$

where $X_{i,G}$ denotes the target vector of the ith individual, $X_{best,G}^p$ denotes a randomly selected vector from the top $100p\%$ superior individuals, $X_{r_1,G}$ denotes a randomly selected vector from the population, $\widetilde{X}_{r_2,G}$ denotes a randomly selected vector from union $P \cup S_1$ where P denotes the set of solution in the current population and S_1 denotes the set of solutions in the first level of storage. $\widetilde{X}_{r_3,G}$ also denotes a randomly selected vector from another union $P \cup S_2$ where P also denotes the set of solution in the current population and S_2 denotes the set of solutions in the second level of storage. All the indices r_1, r_2, r_3 obeys random selection with restriction [5, 9]. Control parameters F, F_1 and F_2 are all scale factors of the difference pairs, and they are dynamically changed during the evolution. Generally, we set $F_1 = 0.9 \cdot F$ and $F_2 = 0.7 \cdot F$ as the default settings. From the mutation equation, we can see that one more difference pair is employed in the donor vector generation strategy. Because the extra difference pair is calculated by the difference between the current population and several past generations, this can be considered as depth information during the evolution. Furthermore, the extra difference pair not only enhances the diversity of trial vector but also obtains a better perception of the landscape of objectives.

For the scale factor F in the novel algorithm, it obeys Cauchy distribution, $F \sim C(\mu_F, \sigma_F)$, μ_F is the location parameter with its initial value equalling to 0.5, σ_F is the scale parameter which is set a constant value during the evolution, $\sigma_F = 0.1$. For the control parameter Cr, it obeys Normal distribution, $Cr \sim N(\mu_{Cr}, \sigma_{Cr})$, μ_{Cr} and σ_{Cr} are the mean and standard deviation of the distribution. μ_{Cr} is dynamically changed during the evolution and σ_{Cr} is fixed constant during the evolution, the initial value of μ_{Cr} is 0.8, and σ_{Cr} equals to 0.1 during the evolution. The adaptation scheme of both μ_F and μ_{Cr} are given in Eqs. 2 and 3 respectively

$$\begin{cases} w_k = \dfrac{\Delta f_j}{\sum_{k=1}^{|S_F|} \Delta f_j} \\ \Delta f_j = f(X_{j,G}) - f(U_{j,G}) \\ mean_{WL}(S_F) = \dfrac{\sum_{k=1}^{|S_F|} w_k \cdot S_{F,k}^2}{\sum_{k=1}^{|S_F|} w_k \cdot S_{F,k}} \\ \mu_{F,G+1} = \begin{cases} mean_{WL}(S_F), & if\ S_F \neq \varnothing \\ \mu_{F,G}, & otherwise \end{cases} \end{cases} \tag{2}$$

$$\begin{cases} w_k = \dfrac{\Delta f_j}{\sum_{k=1}^{|S_{Cr}|} \Delta f_j} \\ \Delta f_j = f(X_{j,G}) - f(U_{j,G}) \\ mean_{WL}(S_{Cr}) = \dfrac{\sum_{k=1}^{|S_{Cr}|} w_k \cdot S_{Cr,k}^2}{\sum_{k=1}^{|S_{Cr}|} w_k \cdot S_{Cr,k}} \\ \mu_{Cr,k,G+1} = \begin{cases} mean_{WL}(S_{Cr}), & if\ S_{Cr} \neq \varnothing \\ \mu_{Cr,k,G}, & otherwise \end{cases} \\ Cr_i = \begin{cases} 0, & if\ \mu_{Cr,r_i} = 0; \\ randn_i(\mu_{Cr,r_i}, 0.1) & otherwise \end{cases} \end{cases} \tag{3}$$

where all the symbols are the same as the ones in LSHADE algorithm. Further-more, the linear population size reduction scheme is also the same as the LSHADE algorithm with the equation shown in Eq. 4

$$ps_G = \begin{cases} round[\frac{ps_{min}-ps_{ini}}{nfe_{max}} \cdot nfe + ps_{ini}], & if\ G > 1 \\ ps_{ini}, & otherwise \end{cases} \tag{4}$$

where ps_{min}, ps_{ini}, nfe and nfe_{max} denotes the minimum population size, initial population size, current number of function evaluation and maximum number of function evaluation respectively.

3 Experiment Analysis of the Algorithm

In this part, the unimodal benchmark functions in CEC2013 are selected out as the test-bed for algorithm evaluation. All the experiments are conducted on 10-D and 30-D optimization of the benchmark respectively. The maximum function evalu-ations is a commonly used setting: $nfe_{max} = 10000 \cdot D$, 51 runs are conducted on each function, and all these results of mean/std (mean and standard deviation) of the total 51 runs are summarized in Table 2. "Better performance", "Similar

Table 2 Mean/std of $\Delta f = f - f*$ comparisons among several famous DE variants including JADE, LSHADE, jSO, LPALMDE and the new proposed hierarchical DE algorithm are presented here. The overall performances of these DE variants are evaluated under Wilcoxon's signed rank test with $\alpha = 0.05$. Symbols ">", "=" and "<" denoting "Better performance", "Similar performance" and "Worse performance" are concluded by the above measurement. Values that smaller than 1.0E−010 are considered as zeros herein

D = 10	JADE	LSHADE	jSO	LPALMDE	Proposed DE algorithm
fa_1	0/0(=)	0/0(=)	0/0(=)	0/0(=)	0/0
fa_2	0/0(=)	0/0(=)	0/0(=)	0/0(=)	0/0
fa_3	3.7651E+ 001/7.7285E+001(<)	1.9091E− 001/9.7404E−001(<)	1.3992E− 003/9.9919E−003(>)	4.1975E− 003/1.6957E−002(<)	3.7632E− 003/2.3185E−002
fa_4	0/0(=)	0/0(=)	0/0(=)	0/0(=)	0/0
fa_5	0/0(=)	0/0(=)	0/0(=)	0/0(=)	0/0
D = 30	JADE	LSHADE	jSO	LPALMDE	Proposed DE algorithm
fa_1	0/0(=)	0/0(=)	0/0(=)	0/0(=)	0/0
fa_2	7.0462E+ 003/5.8745E+003(<)	0/0(=)	3.7115E− 010/7.7174E−010(<)	3.5191E− 010/1.8040E−009(<)	0/0
fa_3	3.1464E+ 005/1.0965E+006(<)	9.7559E− 001/2.7597E+000(<)	2.3793E− 009/1.5886E−008(<)	7.3958E− 003/3.6215E−002(<)	0/0
fa_4	3.5640E+ 003/1.0960E+004(<)	0/0(=)	0/0(=)	0/0(=)	0/0
fa_5	0/0(=)	0/0(=)	0/0(=)	0/0(=)	0/0

performance" and "Worse performance" are denoted by ">", "=" and "<" respectively, and all these evaluations are concluded by the measurement of Wilcoxon's signed rank test with significant level $\alpha = 0.05$.

From the table we can see that the novel hierarchical archive based DE algorithm can obtain four better performances and six similar performances out of the ten cases in comparison with JADE algorithm; it also reveals two better performances eight similar performances out of the ten cases in comparison with LSHADE algorithm; it also reveals two better performances seven similar performances in comparison with jSO algorithm; it also reveals three better performance seven similar performances in comparison with LPALMDE algorithm. To summarize, the novel hierarchical archive based DE algorithm outperforms the other contrasted state-of-the-art DE variants.

4 Conclusion

Evolutionary algorithms for the tackling of tough optimization problems became more and more popular nowadays, and DE variants outperformed other EAs in recent competitions on real-parameter single objective optimization. Here in the paper, an external hierarchical archive based DE algorithm was proposed, the hierarchical archive further improved the diversity of trial vectors and the depth information extracted from the hierarchical archive helped to get a better perception of the landscape of objectives. Both of the two aspects helped the novel algorithm obtain an overall better performance on real-parameter single-objective numerical optimization. This new algorithm was verified on many commonly used benchmarks and experiment results showed that it was competitive with other state-of-the-art DE variants.

References

1. Kennedy, J., Eberhart, R.: Particle swarm optimization. Proc. IEEE Int. Conf. Neural Netw. **4**(2), 1942–1948 (1995)
2. Storn, R., Price, K.: Differential Evolution A Simple and Efficient Adaptive Scheme for Global Optimization Over Continuous Spaces. International Computer Science Institute, CA, Berkeley (1995)
3. van den Bergh, F., Engelbrecht, A.P.: A cooperative approach to particle swarm optimization. IEEE Trans. Evol. Comput. **8**(3), 225–239 (2004)
4. Meng, Z., Pan, J.S.: Quasi-affine transformation evolutionary (QUATRE) algorithm: A parameter-reduced differential evolution algorithm for optimization problems. In: 2016 IEEE Congress on Evolutionary Computation (CEC), pp. 4082–4089
5. Price, K., Storn, R.M., Lampinen, J.A.: Differential evolution: a practical approach to global optimization. Springer Science & Business Media (2006)
6. Meng, Z., Pan, J.-S., Kong, L.: Parameters with adaptive learning mechanism (PALM) for the enhancement of differential evolution. Knowl. Based Syst. **141**, 92–112 (2018)

7. Meng, Z., Pan, J.-S., Xu, H.: QUasi-Affine TRansformation Evolutionary (QUATRE) algorithm: A cooperative swarm based algorithm for global optimization. Knowl. Based Syst. **109**, 104–121 (2016)
8. Meng, Z., Pan, J.-S.: QUasi-Affine TRansformation Evolution with External ARchive (QUATRE-EAR): An enhanced structure for differential evolution. Knowl. Based Syst. **155**, 35–53 (2018)
9. Pan, J.S., Meng, Z., Xu, H., et al.: A Matrix-Based Implementation of DE Algorithm: The Compensation and Deficiency, International Conference on Industrial Engineering and Other Applications of Applied Intelligent Systems, pp. 72–81. Springer, Cham (2017)
10. Meng, Z., Pan, J.-S., Zheng, W.-M.: Differential evolution utilizing a handful top superior individuals with bionic bi-population structure for the enhancement of optimization performance, Enterprise Information Systems. https://doi.org/10.1080/17517575.2018.1491064
11. Meng, Z., Pan, J.-S.: A Simple and Accurate Global Optimizer for Continuous Spaces Optimization. Genetic and Evolutionary Computing. Springer International Publishing, pp. 121–129 (2015)
12. Zhenyu, M., Pan, J.-S., Abdulhameed, A.: A new meta-heuristic ebb-tide-fish-inspired algorithm for traffic navigation. Telecommun. Syst. 1–13 (2015)
13. Pan, J.S., Meng, Z., Chu, S.C., et al.: Monkey King Evolution: an enhanced ebb-tide-fish algorithm for global optimization and its application in vehicle navigation under wireless sensor network environment. Telecommun. Syst. **65**(3), 351–364 (2017)
14. Meng, Z., Pan, J.-S.: Monkey King Evolution: A new memetic evolutionary algorithm and its application in vehicle fuel consumption optimization. Knowl. Based Syst. **97**, 144–157 (2016)
15. Pan, J.-S., Meng, Z., Xu, H., et al.: QUasi-Affine TRansformation Evolution (QUATRE) algorithm: A new simple and accurate structure for global optimization. In: International Conference on Industrial, Engineering and Other Applications of Applied Intelligent Systems. Springer International Publishing, pp. 657–667 (2016)
16. Meng, Z., Pan, J.-S.: A Competitive QUasi-Affine TRansformation Evolutionary (C-QUATRE) Algorithm for global optimization. In: 2016 IEEE International Conference on Systems, Man, and Cybernetics (SMC), pp. 1644–1649. IEEE (2016)
17. Pan, J.-S., Zhenyu, M., Chu, S.-C., Roddick, J.F.: QUATRE algorithm with sort strategy for global optimization in comparison with DE and PSO variants. In: The Euro-China Conference on Intelligent Data Analysis and Applications, pp. 314–323. Springer, Cham (2017)
18. Zhenyu, M., Pan, J.-S., Li, X.: The QUasi-Affine TRansformation Evolution (QUATRE) algorithm: an overview. In: The Euro-China Conference on Intelligent Data Analysis and Applications, pp. 324–333. Springer, Cham (2017)
19. Zhenyu, M., Pan, J.-S., Li, X.: Transfer knowledge based evolution of an external population for differential evolution. In: International Conference on Smart Vehicular Technology, Transportation, Communication and Applications, pp. 222–229. Springer, Cham (2017)
20. Meng, Z., Pan, J.-S.: QUasi-Affine TRansformation Evolutionary (QUATRE) algorithm: The framework analysis for global optimization and application in hand gesture segmentation. In: 2016 IEEE 13th International Conference on Signal Processing (ICSP), pp. 1832–1837
21. Zhang, J., Sanderson, A.C.: JADE: adaptive differential evolution with optional external archive. IEEE Trans. Evolut. Comput. **13**(5), 945–958
22. Tanabe, R., Fukunaga, A.S.: Improving the search performance of shade using linear population size reduction. In: 2014 IEEE Congress on Evolutionary Computation (CEC), pp. 1658–1665, July 2014
23. Janez, B., Maucec, M.S., Boskovic, B.: Single objective real-parameter optimization: algorithm jSO. In: 2017 IEEE Congress on Evolutionary Computation (CEC), pp. 1311–1318. IEEE (2017)

A GIS-Based Optimization of ACO in UAV Network

Weifeng Sun, Yuanxun Xing, Guangqun Ma and Shumiao Yu

Abstract UAVs can carry out rescue work in the disaster area. The ant colony algorithm is used to search and rescue work in the disaster area based on GIS. Considering regional priority, an algorithm named priority-PAACO is proposed. Simulations and analysis show the effective of this algorithm.

Keywords UAV · GIS · Priority-PAACO · Parameter adjustment

1 Introduction

UAVs are often more suitable for tasks that are "dull, dirty or dangerous" than manned aircraft, and more commonly for civilian use [1]. Ant colony algorithm (ACO) is an effective swarm intelligence algorithm to solve the problem of path planning. The ant colony algorithm introduces parameters such as α and β to adjust the possibility to choose the next target. This paper optimizes the ACO to guide the search and rescue path of the drone disaster area. The use of optimized ant colony algorithm to guide UAVs to find ways should also take into account the special needs of the search and rescue scenarios in the disaster area. For example, in some areas, the disaster situation is more serious and needs to be rescued as soon as possible. Therefore, this paper introduces an optimized ant colony algorithm to adaptively adjust parameters with regional priorities.

W. Sun (✉) · Y. Xing · G. Ma · S. Yu
School of Software, Dalian University of Technology, Dalian 116620, China
e-mail: wfsun@dlut.edu.cn

Y. Xing
e-mail: xingyuanxun2017@163.com

G. Ma
e-mail: guangqun.ma@foxmail.com

S. Yu
e-mail: yusm1995@outlook.com

© Springer Nature Singapore Pte Ltd. 2019
J.-S. Pan et al. (eds.), *Genetic and Evolutionary Computing*,
Advances in Intelligent Systems and Computing 834,
https://doi.org/10.1007/978-981-13-5841-8_9

Cyber-GIS [2] is a global geographic information system, which collects geographical information from around the world. Cyber-GIS could combine the collected geographic topologies and topological relationships of coordinates to generate a visual geographic graph and store, update, manipulate, and analyze the graph.

2 Related Work

The researchers find that the parameters have great influence on the overall performance of ACO. So we adjusted parameters and proposed SAACO [3] which works in cloud computing. Since the SAACO algorithm did not consider user requirements, QSACO [4] was proposed to combine user QoS requirements. Tang et al. also proposed an algorithm called Cloud task scheduling based on load balancing ant colony optimization [5], which illustrates the superior performance of the ant colony algorithm in task scheduling in another perspective. This paper considers the difference between the urgency of rescue in the disaster area and the value of the ant colony algorithm parameters in different application scenarios. An optimized ant colony algorithm with adaptive adjustment of parameters with regional priority is proposed in the scenario.

3 Algorithm Design

Using UAVs to search and rescue scene of the disaster area is abstracted. Therefore, the areas that need to be searched and rescued are designed into a three-dimensional coordinate topology map. Each coordinate in the topological map represents an area that needs to be searched and rescued. There are two main types of UAVs in the model: one type is the temporary base station responsible for regional information collection. The other type is used to patrol each city, which is the carrier of information. In each area to be searched and rescued, a temporary UAV is hovering as a temporary base station, responsible for the collection and sorting of various disaster information in the area. Its height will be adjusted according to the geographical information of the area collected from the GIS system (for example, whether the area is a basin or a plateau, whether it is in a mountain area that is difficult to search). The temporary base station sets a priority level for each disaster-stricken area to represent the level of search and rescue. It organizes the collected information and sends the information to the UAV fleet in charge of carrying information that is cruising to the area. However, the focus of this paper is to design a UAV cruise path scheduling algorithm to ensure that the UAV can obtain a better search and rescue route based on the priority given by the temporary base station and the path length from the area. Combined with the special requirements of the UAV search and rescue scenario, an improved ant colony algorithm is designed.

The main improvements of the algorithm are as follows: one is to improve the value selection of the important parameters α and β of the traditional ant colony algorithm; the other is to set the corresponding priority, which refers to the level of disaster. Therefore, the UAVs can preferentially search and rescue the disaster area with higher priority while ensuring that the search and rescue route is as short as possible.

3.1 Improvement of the Values of Parameters α and β

In the traditional ACO, the value of the parameter α indicates the degree of importance of the amount of information left on each node. The value of β indicates the degree to which heuristic information is valued. The research results of the traditional ant colony algorithm also show that when α is 1 and β is 5, better experimental results can be obtained. That is to say, the values of α and β are actually empirical values, which have a good effect on the TSP problem in the traditional sense. However, in the scenario proposed in this paper, the path finding process of the UAVs should not only consider the length of the distance but also the priority of the disaster area. Then, the empirical values of the parameters α and β are no longer desirable. In the previous research work, we proposed a parameter adjustment method for refining the interval idea. The pseudo code of the specific method is shown in Fig. 1 (taking the adjustment scheme of the α parameter as an example, β is similar).

PSO-based algorithm is another way to adjust parameters. This algorithm is proposed in our earlier paper which is named QSACO [4]. The pseudo code of the algorithm is shown in Fig. 2.

Pseudo Code of Refine Interval Algorithm
1. Algorithm start
2. While step length bigger than the threshold we set:
3. Specify a range and step length for the parameter α
4. Find a smaller range of α which can reach a better result
5. Remake the step length
6. End while
7. Get a certain value of α and finish algorithm

Fig. 1 Thinning interval parameter adjustment method pseudo code

---

Pseudo Code of PSO-based Algorithm

1. Algorithm start
2. Initialize parameters α and β
3. Get current best position of particle according to the FITNESS
4. While not reach the number of iteration:
5. Calculate velocity and direction for each particle according to the current best particle
6. Adjust other particles towards current best particle
7. Bring each value of particles into FITNESS
8. Return new current best position of particle.
9. End while
10. Get accurate results and finish algorithm

Fig. 2 Parameter adjustment method based on particle swarm pseudo code

3.2 Primitive Ant Colony Algorithm

For the search and rescue scenes in the disaster area of the UAVs, the different disaster areas have different priorities. Therefore, it is necessary to design an algorithm that meets the needs of a particular scenario.

In the traditional ant colony algorithm, the parameter η_{ij} is the visibility from city i to city j, also known as heuristic information. In the TSP problem, generally take $\eta_{ij} = 1/d_{ij}$, from which d_{ij} represents the distance between i and j. η is an important parameter indicator for guiding ants to choose the next city in the ant colony algorithm. In the TSP problem, the standard for selecting the next city is only the distance between the two cities. In the scenario of this article, in the path finding plan for UAV search and rescue, the UAV chooses the next disaster area to consider not only the inter-area distance but also the priority of each affected area. Therefore, this paper introduces a new heuristic information algorithm, the formula is as follows:

$$\eta_{ij} = (1-p) * \left(\frac{d_{ava}}{d_{ij}} \right) + p * \left(\frac{W_{ava}}{W_j} \right) \tag{1}$$

From which, i indicates the number of the current ant area, j indicates the area number of the ant to go next time, p is the adjusting parameter, and $p \in [0, 1]$. d_{ij} indicates the distance between the area i and area j, d_{ava} represents the expectation of the sum of the distances of any two regions. The calculation method is as follows:

$$d_{ava} = \frac{1}{\binom{n}{2}} \sum_{i',j'} d_{i'j'} \tag{2}$$

W_j indicates the priority of the area j to be reached. W_{ava} represents all priority expectations. The calculation method is as follows:

$$W_{ava} = \frac{1}{n} \sum_{j'} W_{j'} \tag{3}$$

The benefit of such heuristic information setting is that it can take into account the dual indicators of distance and regional priority. In the formula, the adjustment parameter p introduced can set different values according to the area size and the topology structure. The advantage is that the distance and the priority proportion of the priority can be flexibly adjusted, and the search and rescue path finding of the drone is dynamically guided.

Combining priority and parameter optimization, we propose a Parameter Adjustment Ant Colony Optimization with priority (Priority-PAACO).

4   Experimental Results and Analysis

This paper uses Matlab R2017a as an experimental tool. The topologies with input sizes of 31 and 128 are selected. The experimental results are shown in Fig. 3.

In the following experiments, the parameter α of the traditional ant colony algorithm is 1 and the value of β is 5, both of which are empirical values. In the thinning interval method, the initial value interval of the parameter α is set to [0, 10]. The initial value interval of β is set to [1, 20], and the threshold is set to 0.5. The number of particle group iteration termination is set to 20. Priority-PAACO leads to the large number of cycles of the algorithm.

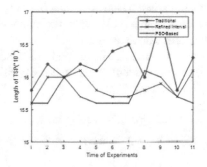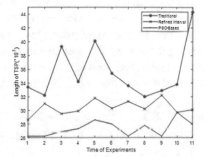

Fig. 3   Three algorithm results under 31,128 scale

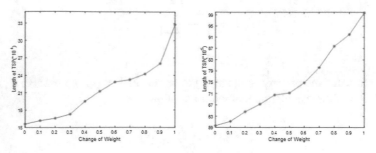

Fig. 4 The shortest path length corresponding to each value of p (scale 31,128)

It can be seen that in the case of a small input scale (scale 31), all three methods can find relatively good results. However, compared with the running time, the traditional ant colony algorithm performs shorter time. The parameter optimization algorithm based on particle swarm optimization algorithm takes a little longer in execution time.

In the case where the input scale is centered (the size is 128). The parameter optimization algorithm based on particle swarm optimization also shows excellent routing methods and the best path length from this algorithm is the shortest. But in general, the differences between the three algorithms are still not obvious. Based on the above three experimental results, we find that the larger the input scale, the stronger the particle cluster-based parameter optimization method is for finding the optimal path, and the variance is smaller.

Second, we analyze the results of the ant colony algorithm that introduces weights to find a reasonable value of the parameter p in the case of different input sizes.

In the experimental scenarios of 31, 128 input scales, the step size of p is set to 0.1, and the value of p is adjusted for each experiment. 50 experiments are performed on the same value of p, and the average path length is recorded. The experimental results obtained are shown in Fig. 4 when the input scale is 31 and 128.

When the input scale is small, changing the value of the parameter p, it can be found that the proportion of the impact on the shortest path is the smallest when p is 0.3, and the path length will increase very rapidly when p is greater than 0.3, that is, the growth rate is larger. After many experiments, at this scale, when the value of p is 0.3, the requirements of priority and path length can be balanced.

Taking input scale 31 as an example, this paper obtains a 31-area path map under a non-special experiment (as shown in Fig. 5) and the order of access to priority cities (the higher the priority value, the more the first access):

$$29 \rightarrow 30 \rightarrow 31 \rightarrow 27 \rightarrow 28 \rightarrow 26 \rightarrow 25 \rightarrow 24 \rightarrow 20 \rightarrow 21 \rightarrow 22 \rightarrow 3 \rightarrow 18$$
$$\rightarrow 17 \rightarrow 19 \rightarrow 23 \rightarrow 11 \rightarrow 12 \rightarrow 14 \rightarrow 13 \rightarrow 7 \rightarrow 2 \rightarrow 4 \rightarrow 5 \rightarrow 6 \rightarrow 16 \rightarrow 8$$
$$\rightarrow 9 \rightarrow 10 \rightarrow 15 \rightarrow 1$$

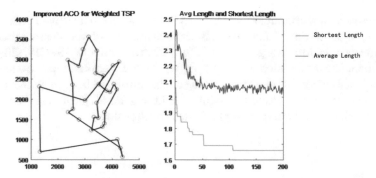

Fig. 5 The input scale is 31, the path map and the path change trend when p = 0.3

It can be seen that cities with higher priority are indeed preferentially visited to a certain extent. Because the scale is small, the topology is sparse, and the distance between adjacent regions is large, the parameter p is more likely to be allocated to the heuristic information portion determined by the path length, and less to the heuristic information portion determined by the priority. The experimental results of scale 128 show that the growth rate is the lowest at p = 0.5. Therefore, under the medium-scale input, the value of the parameter p is 0.5. According to the analysis, for the distribution with the topological density, the path length and priority factors should be considered equally, and the path optimization can be maximized. For the dense topology of large-scale input, the distance difference between any two points is not very large. In this case, the proportion of the priority part should be satisfied as much as possible. After experimental results and theoretical analysis, it is most reasonable to set p to 0.8.

In summary, the parameter adaptive adjustment algorithm with priority in this paper has significant contributions in guiding the search and rescue path of the UAV disaster area. Selecting a smaller parameter p to increase the path distance in the scenario of small-scale sparse topology is significantly effective to the weighting effect on the overall path planning. In the scenario of large-scale dense topology, selecting a larger parameter p and increasing the weight of the priority can obtain better results. Moreover, the experimental data in large-scale topology shows that the parameter adaptive adjustment method combined with particle swarm optimization has at least 17% improvement in performance compared with the traditional ant colony algorithm. The algorithm proposed in this paper does have the effect of improving performance.

5 Conclusion and Future Work

In this paper, two adjustment schemes in ACO are proposed. The parameter adjustment algorithm based on particle swarm optimization algorithm is superior to the other two in performance, but the problem of long time consuming still exists. In future research, it will combine with Cyber-GIS, to form a global network of disaster-affected geographic information database. Moreover, we will consider more complex experimental scenarios, use geographic information data imported by GIS, fully combine geographic information to set up complex three-dimensional search and rescue scenarios, and expand the search and rescue algorithm of the UAV groups into a three-dimensional search and rescue algorithm. In addition, more categories of UAVs will be expanded to give them more sophisticated functionality.

References

1. Yao, P., Wang, H.: Dynamic adaptive ant lion optimizer applied to route planning for unmanned aerial vehicle. Soft. Comput. **21**(18), 5475–5488 (2017)
2. Wang, S., Anselin, L., Bhaduri, B., et al.: CyberGIS software: a synthetic review and integration roadmap. Int. J. Geogr. Inf. Sci. **27**(11), 2122–2145 (2013)
3. Sun, W., Ji, Z., Sun, J., et al.: SAACO: a self adaptive ant colony optimization in cloudcomputing. In: 2015 IEEE Fifth International Conference on Big Data and Cloud Computing (BDCloud), pp. 148–153. IEEE (2015)
4. Sun, W., Xing, Y., Zhou, C., et al.: QSACO: A QoS-Based self-adapted ant colony optimization. In: 2017 5th IEEE International Conference on Mobile Cloud Computing, Services and Engineering (Mobile Cloud), pp. 157–160. IEEE (2017)
5. Tang, Z., Qi, L., Cheng, Z., et al.: An energy-efficient task scheduling algorithm in DVFS-enabled cloud environment. J. Grid Comput. **14**(1), 55–74 (2016)

Part III
Software Development and Reliability Systems

An Intelligent Temperature Measurement and Control System for the Steel Mill

Yuhang Huang, Yeqing Wang and Gefei Yu

Abstract To improve the measurement precision and the drift of temperature controlling, an intelligent temperature measurement and control system based on microcontroller has been designed. This system is suited for temperature controlling of smelting special steel furnace in the steel mill. It also has functions of the real-time detection, display, and over temperature alarm. The fitting curve and analysis results were obtained based on the actual measured data. Experimental results show that the error between the actual measured data and the fitting calculation data is less than 0.1 °C, indicating that the system has good performance.

Keywords Thermocouple · Temperature control · Data fitting

1 Introduction

The furnace temperature control during the smelting of special steel in the steel mill included process temperature control and end temperature control. If the temperature cannot be controlled, the temperature of molten steel will fluctuate greatly. It was very difficult to control the temperature of the process. In the previous control system, the point-by-point measurement method was adopted, and the control method was monotonous. The temperature fluctuation was large and the precision was low, so the traditional method was not suitable for smelting special steel in the steel mill [1].

In recent years, many approaches based on the microcontroller have been proposed. Ding et al. [2] indicated the microcontroller was accurate and reliable for temperature detection. It was also inexpensive, convenient, and flexible to use from the aspect of system maintenance. In the system based on microcontroller, it collected the actual temperature data at every moment and stored data in the microcontroller.

Y. Huang · Y. Wang (✉)
Changzhou College of Information Technology, Changzhou 213164, China
e-mail: yeqingwang2014@hotmail.com

G. Yu
Changzhou Institute of Mechatronic Technology, Changzhou 213000, China

© Springer Nature Singapore Pte Ltd. 2019
J.-S. Pan et al. (eds.), *Genetic and Evolutionary Computing*,
Advances in Intelligent Systems and Computing 834,
https://doi.org/10.1007/978-981-13-5841-8_10

The comparison between the stored data and the fitted curve data can be performed in the system. If the stored data is less than that of fitting curve, the heater will start. The measurement area will be heated continuously, and then it is treated with a constant temperature. The process temperature control was the key to the end temperature control. Based on this analysis, our team developed a temperature measurement system based on thermocouple for steel furnace. The thermocouple was suitable for the measurement of the special steel smelting temperature in an ultra-high-temperature environment and realized precise control of the furnace temperature [3].

2 System Circuit Design

2.1 Design Idea

A set of temperature measurement and control system is developed based on single-chip microcomputer. The temperature in the measured area can be accurately measured using the point temperature distribution measurement method, and a number of average temperatures can also be obtained. The measured temperature at different stages of the process is compared with the curve fitting data of the ideal steelmaking process and stored in the microcontroller. If the measured temperature is lower than the range of the stored curve, the measured area will continuously heat the heater, and then maintain a constant temperature [4]. The accuracy of the temperature sensor is 1 °C, and the temperature measurement range is 0–1800 °C.

The temperature drift of the thermocouple is very low, which is below 1 μV/°C, and the stability is good [5]. The electronic solid-state relay (SSR) is a load with a switching time up to 500 μs, and it is controlled by chopping with fast response. So, it is suitable for precise debugging of heating mode. The temperature changes are collected when the variation is large, and the pulse duty ratio of electronic SSR is automatically sent in. Different pulse duty ratios are used to adjust the heating time and heating frequency to ensure the accuracy of the control temperature [6].

2.2 The Principle of Temperature Measurement

Thermocouple is based on thermoelectric effect for a wide range of applications in the field of ultra-high-temperature measurement. It is composed of A and B with two different conductors (or semiconductors) in the closed loop. If the two junction temperatures of T and T_0 are different, the electromotive force (emf) in the circuit will be produced. The phenomenon is called the thermoelectric effect. The thermoelectric emf consists of two parts, one of them is the contact electromotive force and another is the temperature difference electromotive force [7].

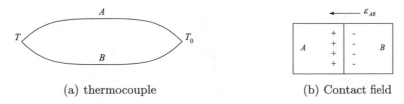

(a) thermocouple   (b) Contact field

Fig. 1 Contact electromotive force

(1) Contact electromotive force is also known as the Peltier electromotive force caused by the different free electron density $N_A(T)$ and $N_B(T)$. If $N_A(T) > N_B(T)$, the electrons are diffused from A to B, and the diffused electron current is formed. The potential in A is positive while it is negative in B. The internal static electric field is formed on both sides of the contact between A and B. The drift electron current caused by the internal static electric field is in the opposite direction of the diffused electron current. When the diffused electron current and the drift electron current are balanced, it is shown in Eq. (1) [8].

$$\varepsilon_{AB} = \frac{kT}{e} \ln \frac{N_A(T)}{N_B(T)} .$$ (1)

where e is the unit charge, k is Boltzmann constant, and the direction of the electromotive force is $B \to A$ as shown in Fig. 1.

(2) Temperature difference electromotive force is known as Thomson electromotive force and is shown in Fig. 2. It is caused by different temperatures T and T_0 in two ends of a conductor. If $T > T_0$, the free electron densities are different and the diffused electron current is formed caused by diffusing movement of free electron. The potential at high-temperature end is positive while that at low temperature end is negative. The internal static electric field is formed when the diffused electron current and the drift electron current are balanced as shown in Eq. (2) [9].

$$\varepsilon_A(T, T_0) = \frac{kT}{e} \ln N_A(T) \Big|_{T_0}^{T} - \frac{k}{e} \int_{T_0}^{T} \ln N_A(\tau)d\tau .$$ (2)

The direction of the electromotive force is $T_0 \to T$.

(3) The closed loop is composed of conductors A and B. The two contact points are, respectively, in temperatures of T and T_0 is shown in Fig. 3. The electromotive force in thermocouple loop is indicated in (3) and its direction is shown in Fig. 3.

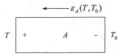

Fig. 2 Temperature difference electromotive force

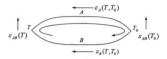

Fig. 3 Thermocouple loop

$$E_{AB}(T, T_0) = \varepsilon_{AB}(T) - \varepsilon_A(T, T_0) - \varepsilon_{AB}(T_0) + \varepsilon_B(T, T_0) . \qquad (3)$$

The total electromotive force in thermocouple loop is obtained from (1)–(3).

$$E_{AB}(T) = \frac{kT}{e} \int_{T_0}^{T} \ln \frac{N_A(\tau)}{N_B(\tau)} d\tau . \qquad (4)$$

2.3 The Block Diagram of System Circuit

This device includes a CPU module (STM32F407), that is, used for controlling the work of the whole furnace temperature. The controlling process includes the collected data, that is, from thermocouple 1 reflects the size of the furnace temperature after A/D conversion. The voltage signal of temperature from thermocouple 2 reflects the temperature in the heating process. The pulse signal of the solid-state relay is output controlled for adjusting the working status of electromagnetic controller and achieving the goal of the heater heating and temperature controlling. The outputted current signal 4–20 mA is for analog display instrument. The alarm signal reflecting over-temperature is transmitted to the driving circuit. At the same time, there is a kind of interface circuits such as RS232, high-speed USB interface, keyboard, and display interface circuit, etc. [10]. The entire unit block diagram includes IC1 microcontroller. Two sets of electric sampling circuit include thermocouple 1, preamplifier IC2, A/D converter IC3, thermocouple 2, preamplifier IC15, A/D converter IC14, voltage–current module IC4, voltage conversion module IC6, alarm driving circuit, USB quick interface circuit, and it consists of chip IC9 and the surrounding components of RS232 interface circuit, etc. The block diagram is shown in Fig. 4.

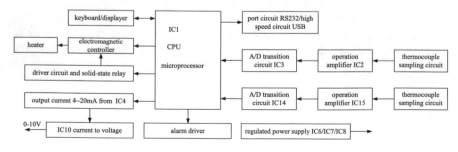

Fig. 4 System black diagram

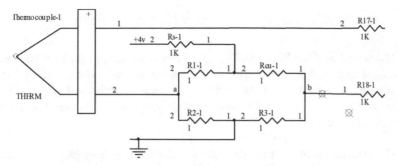

Fig. 5 The temperature sensor of thermocouple and the compensation circuit

2.4 The Sensors and Compensation Circuit

The selected sensor is Platinum Rhodium 10-Platinum thermocouple sensor and its measured temperature range of 0–1800 °C. The bridge compensation method is adopted as cold end compensation, and it compensates the fluctuations of the thermoelectric emf caused by temperature change in cold end through the emf produced from unbalanced bridge. The unbalanced bridge consists of resistors of four-bridge arms R_{1-1}, R_{2-1}, R_{3-1}, R_{cu-1}, and regulated power supply in bridge road as shown in Fig. 5. The resistor of R_{cu-1} and the cold end of thermocouple are in 0 °C and $R_{1-1} = R_{2-1} = R_{3-1} = 1\Omega$. The voltage of the power supply in bridge road is 4V by the regulated power supply. The resistor of R_{s-1} is a limit of flow resistance and the resistance is different due to the different categories of the thermocouple. The four-bridge arm resistances satisfy $R_{1-1} = R_{2-1} = R_{3-1} = R_{cu-1}$ when the bridge is balanced and it is without output in two endpoints of a and b. When the temperature of cold end is over 20 °C, the resistor of R_{cu-1} increases as the temperature rises while the thermoelectric emf decreases with the increase of the cold end temperature. U_{ab} is equal to the reduction of the thermoelectric emf. So, the output voltage remains the same after superposition U_{ab} with thermoelectric emf for achieving the purpose of the cold end compensation.

2.5 The Single Chip Microcomputer Control Circuit

The single-chip microcomputer circuit of IC1 (STM32F407RDT6) is 32 bits with advanced Cortex-M4 kernel. There are a variety of peripheral interfaces for cameras, high-speed USB, temperature sensors, and communications. It has a strong ability of floating-point arithmetic, high running speed, enhanced DSP processing instruction, more storage space, and a flexible FSMC external memory interface. The on-chip flash memory is 1 M and the embedded SRAM is 196 K. This circuit has the ultra-low power and the current consumption is only 38.6 mA while the frequency is 168 MHz [11].

2.6   Important Points of the Circuit Hardware Design

(1) The circuit of IC2 / IC15 (SL28617) is used as sensor preamplifier because the temperature drift is large and the output voltage is small. The amplifier circuit is of high common mode rejection ratio, high input impedance, high gain but low noise. The watchdog circuit, software trap, and other parts ensure the reliable work of the circuit.

(2) The connection lines of compensation wires are shielding processing, respectively, to avoid the distributed signal. The non-balance bridge compensates temperature drift of the thermocouple to prevent the changes in the thermoelectric potential due to temperature changes at the cold end.

(3) The multilayered board structure is used in EMC circuit design and it adopts complete GND layer. The wide line or laying copper in the power supply is to reduce the impedance of the power supply system. The ground wires of analog signal and the digital signal, high-frequency signal, and low-frequency signal are handled separately. The connection lines are as short as possible. They should avoid right-angle turns, sharp corners, and bifurcations. The distributed lines should be avoided just under the module [12].

3   Fitting Theory Method and the Result Analysis

In order to improve the efficiency of the system in programming, the temperature range must be found. In the temperature range, the relationship between the temperature of the thermocouple sensor and the thermoelectric emf is an approximate linear function. It is very beneficial for computing and programming.

The Least Square Method is used for curve fitting and is used to determine the ratio parameters of the thermocouple temperature measurement and the thermoelectric potential. The temperature ranges are over the whole thermocouple measuring range. The space of the fitting point is $1\,°C$ and the maximum fitting error is not more than $0.1\,°C$.

3.1   The Basic Theory of the Fitting

It supposes the relationship between the thermocouple temperature T and the thermoelectric emf E is $T = T(E)$. For T_n, the corresponding thermoelectric emf is $E_n, (n = 0, 1, \ldots)$. A set of data can be obtained $(E_0, T_0), (E_1, T_1), \ldots, (E_n, T_n)$ in the temperature range of -50–$1800\,°C$. The function $T = T(E)$ is not a single linear function and it can be used m-order polynomial as fitting function. The Least Square Method can use to fit curve for the $n + 1$ nodes [13].

It supposes

$$T = T(E) = f_m(E) = a_0 + a_1 E + a_2 E^2 + \cdots + a_m E^m, m < n - 1 . \quad (5)$$

Here, $a_j (j = 0, 1, 2, \ldots, m)$, for $E_i (i = 0, 1, 2, \ldots, n)$, the sum of the error square is

$$\delta = \sum_{i=0}^{n} \left[f_m(E_i) - T_i \right]^2 = \sum_{i=0}^{n} \left[a_0 + a_1 E_i + a_2 E_i^2 + \cdots + a_m E_i^m - T_i \right]^2 . \quad (6)$$

The partial derivative of the sum of the error square for a_j is zero.
$\frac{\partial \delta}{\partial a_j} = 0, (j = 0, 1, 2, \ldots, m)$ then

$$\frac{\partial \delta}{\partial a_0} = 2 \sum_{i=0}^{n} \left[a_0 + a_1 E_i + a_2 E_i^2 + \cdots + a_m E_i^m - T_i \right] = 0 .$$

$$\frac{\partial \delta}{\partial a_1} = 2 \sum_{i=0}^{n} \left[a_0 + a_1 E_i + a_2 E_i^2 + \cdots + a_m E_i^m - T_i \right] E_i = 0 . \quad (7)$$

$$\cdots$$

$$\frac{\partial \delta}{\partial a_j} = 2 \sum_{i=0}^{n} \left[a_0 + a_1 E_i + a_2 E_i^2 + \cdots + a_m E_i^m - T_i \right] E_i^m = 0 .$$

The matrix form is as

$$\begin{bmatrix} \sum_{i=0}^{n} 1 & \sum_{i=0}^{n} E_i & \cdots & \sum_{i=0}^{n} E_i^m \\ \sum_{i=0}^{n} E_i & \sum_{i=0}^{n} E_i^2 & \cdots & \sum_{i=0}^{n} E_i^{m+1} \\ \cdots & \cdots & \cdots & \cdots \\ \sum_{i=0}^{n} E_i^m & \sum_{i=0}^{n} E_i^{m+1} & \cdots & \sum_{i=0}^{n} E_i^{2m} \end{bmatrix} \begin{bmatrix} a_0 \\ a_i \\ \cdots \\ a_m \end{bmatrix} = \begin{bmatrix} \sum_{i=0}^{n} T_i \\ \sum_{i=0}^{n} E_i T_i \\ \cdots \\ \sum_{i=0}^{n} E_i^m T_i \end{bmatrix} . \quad (8)$$

Equation (8) is used to replace (5), and the relationship between the thermocouple temperature T and the thermoelectric emf could be obtained.

3.2 Analysis of the Calculation Result

The polynomial calculation of the indexing table for Platinum Rhodium 10-Platinum thermocouple and its result based on the criterion of IEG584-1 and GB3772-83 are

Table 1 Calculation of the indexing table for thermocouple and its results

Test temperature range /°C	Corresponding calculation formula /uV	Value of i	Calculation results
−50–629	$E = \sum_{i=0}^{4} a_i T^i$	0	0.000000
		1	5.394467
		2	1.241976×10^{-2}
		3	-2.234813×10^{-5}
		4	2.835217×10^{-8}
630–1064	$E = \sum_{i=0}^{2} b_i T^i$	0	-2.882337×10^2
		1	8.227554
		2	1.635382×10^{-3}
1065–1665	$E = \sum_{i=0}^{3} c_i (T^*)^i,$ $T^* = \frac{T-1365}{300}$	0	1.3843437×10^4
		1	3.6378679×10^3
		2	-5.0271203
		3	-4.2450448×10^1
1666–1767	$E = \sum_{i=0}^{3} d_i (T^{**})^i,$ $T^{**} = \frac{T-1715}{50}$	0	1.8113086×10^4
		1	5.6785365×10^2
		2	-1.2112390×10^1
		3	-2.7117587

shown in Table 1. The fitting results are shown in Table 2 [14]. The maximum error is less than 0.1 °C.

4   Conclusion

This paper proposed a design scheme, and gave a solution according to the problems encountered in reality. This scheme was an intelligent design of the circuit system based on microcontroller STM32F407, which was used for temperature measurement and control in a steel furnace. The temperature measurement range was from 0 to 1800 °C. The measurement accuracy was 1 °C, and the maximum error of the test result was less than 0.1 °C. According to the performance, this furnace temperature measurement and control system has reached the advanced level of similar products at home and abroad. The experiment results show that it has a wide range of practical value.

Table 2 Fitting results

Test temperature/$^\circ C$	Corresponding calculation formula	Coefficient c	Value of c	Piecewise node maximum error value/$^\circ C$
−50–159	$T = \sum_{i=1}^{4} c_i E^i$	$c1$	1.752109×10^2	8.814323×10^{-2}
		$c2$	-8.140778×10	
		$c3$	1005997×10^2	
		$c4$	-1.142815×10^2	
160–374	$T = \sum_{i=1}^{5} c_i E^i$	$c1$	1.7113011×10^3	5.588478×10^{-2}
		$c2$	-2.850047×10	
		$c3$	3.051468	
		$c4$	1.744674	
		$c5$	1.401330×10^{-1}	
375–687	$T = \sum_{i=1}^{4} c_i E^i$	$c1$	1.548615×10^2	8.347254×10^{-2}
		$c2$	-1.574068×10	
		$c3$	2.211235	
		$c4$	-1.447084×10^{-1}	
688–850	$T = \sum_{i=1}^{3} c_i E^i$	$c1$	1.334814×10^{-3}	8.188712×10^{-2}
		$c2$	-4.372191	
		$c3$	1.4000015×10^{-1}	
851–999	$T = \sum_{i=1}^{5} c_i E^i$	$c1$	8.061312×10	3.100488×10^{-2}
		$c2$	1.285153×10	
		$c3$	-1.512701	
		$c4$	6.279972×10^{-2}	
		$c5$	-1.179356×10^{-2}	
1000–1224	$T = \sum_{i=1}^{4} c_i E^i$	$c1$	9.722175×10	5.587614×10^{-2}
		$c2$	8.318394×10	
		$c3$	-1.512908	
		$c4$	9.617005	
1225–1632	$T = \sum_{i=1}^{3} c_i E^i$	$c1$	1.282868×10^2	5.319723×10^{-2}
		$c2$	-3.390247	
		$c3$	8.1678712×10^{-2}	
1633–1696	$T = \sum_{i=1}^{4} c_i E^i$	$c1$	1.182817×10^2	2.370372×10^{-2}
		$c2$	-8.76886×10^{-1}	
		$c3$	-1.506577×10^{-1}	
		$c4$	1.211935×10^{-2}	
1670–1767	$T = \sum_{i=1}^{4} c_i E^i$	$c1$	6.934563×10	1.877535×10^{-1}
		$c2$	5.308651	
		$c3$	-2.204555×10^{-1}	
		$c4$	4.066083×10^{-3}	

Acknowledgements This paper was supported by the project of Electronic Information Engineering Major in Jiangsu Province (Grant No. 2017GGZY02).

References

1. Lu, L., Gao, Y.: The temperature controlling system based on S3C44B0 (In Chinese). Inf. Microcomput. **5–2**, 113–115 (2006)
2. Ding, Q., Liu, J., Zeng, X., Li, W.: Optimization of casting temperature control system controlled by single chip microcomputer. Spec. Cast. Nonferrous Alloys **38**(03), 254–256 (2018)
3. Li, X., Yao, G.: The temperature controlling technology of high precision. Radio Technol. **1**, 35–36 (2005)
4. Wang, M.: Temperature controlling instrument of heat treatment. Electric Drive Autom. **3**, 21–24 (2007)
5. Chen, J.: The intelligent instrument design of temperature controlling system in resistance furnace and its application. Master Dissertation, Jiangsu University (2007)
6. Wan, X., Liu, B.: Idea of realizing in solid relay instead of ac contactor based on single chip microcomputer. Digit. Technol. Appl. **4**, 24–25 (2013)
7. Chen, Y., Jiang, H., Zhao, W., et al.: Fabrication and calibration of Pt-10%Rh/Pt thin film thermocouples. Measurement **48**, 248–251 (2014)
8. Gutmacher, D., Foelmli, C., Vollenweider, W., et al.: Comparison of fas sensor technologies for fire gas detection. Proced. Eng. **25**, 1121–1124 (2011)
9. Zhao, J., Feng, H., Xu, Z., et al.: Real-time automatic small target detection using saliency extraction and morphological theory. Opt. Laser Technol. **47**, 268–277 (2013)
10. Huang, Y., Zhu, W., Jiang, H., J, X.: Design of the circuit in a portable tracing instrument based on GPS technology. Int. J. Appl. Innov. Eng. Manag. **1**(2), 109–113 (2012)
11. Huang, Y., Zhu, W., Jiang, X.: A circuit design and its experimental analysis for electromagnetic flowmeter in measurement of sewage. Sens. Transducers **164**(2), 36–43 (2014)
12. Huang, Y., Mo, S., Tang, L.: Design of fishing boat tracing instrument compatible with more navigation system. Fish. Mod. **2**(2), 296–297 (2013)
13. Chen, J.: The common method of fitting for thermocouple temperature calculation formula of fitting. J. Univ. Petrol. **13**(5), 73–77 (1989)
14. Shen, W., Zhang, S.: Thermocouple indexing manual. Stand. Lab. Instrum. Mach. Ind. 59–60 (1983)

Influence of SVC and STATCOM Access on Voltage Stability of Doubly Fed Wind Turbine

Peiqiang Li, Zhengbang Xia, Lu-Qi Jie, Chuan Lin, Peidong Sun
and Pengcheng Cao

Abstract Wind power is random and intermittent, and large-scale wind power access will have a greater impact on the safe and stable operation of the power system. In order to study the influence of reactive power compensation device on the voltage stability of the doubly fed fan, this paper builds a system of IEEE14 node for the doubly fed asynchronous wind farm with SVC and STATCOM modules in MATLAB/PSAT and uses SVC in time domain simulation. It is applied to the wind power generation system separately from STATCOM, and analyzes the voltage stability under sudden changes in wind speed, three-phase short-circuit fault, and load sudden increase. The simulation results show that both SVC and STATCOM can improve the voltage stability of wind farm after grid disturbance, and STATCOM has better compensation stability than SVC.

Keywords SVC · STATCOM · Voltage stability · Doubly fed fan

P. Li (✉) · Z. Xia · L.-Q. Jie · C. Lin · P. Sun · P. Cao
Fujian University of Technology, Fuzhou 350118, Fujian, China
e-mail: 596905210@qq.com

Z. Xia
e-mail: 2423081432@qq.com

L.-Q. Jie
e-mail: 410759161@qq.com

C. Lin
e-mail: 492456598@qq.com

P. Sun
e-mail: 815420060@qq.com

P. Cao
e-mail: cpc670259572@sina.cn

© Springer Nature Singapore Pte Ltd. 2019
J.-S. Pan et al. (eds.), *Genetic and Evolutionary Computing*,
Advances in Intelligent Systems and Computing 834,
https://doi.org/10.1007/978-981-13-5841-8_11

1   Introduction

In recent years, wind energy as a renewable and clean energy has made wind power development rapid. The doubly fed asynchronous wind turbine (DFIG) is the earliest variable-speed constant-frequency wind turbine with the advantages of variable-speed operation and high wind energy conversion efficiency, but it also has the problems of general fans [1, 2]. When the fan delivers active power to the grid, it will also absorb reactive power from the grid. If a fault occurs in the grid, the grid cannot provide enough reactive power, which will lead to voltage instability or even instability. How to improve the voltage stability of asynchronous fans after they are connected to the grid has caused widespread concern [3, 4].

With the development of power electronics technology, a reactive power compensation device (FACTS) has appeared, which makes reactive power regulation to the power grid, and can improve transient stability in the power grid. The installation of a reactive power compensation device in a network containing a wind farm can stabilize the voltage and improve the operating performance of the system [5–9]. When a three-phase short circuit occurs in a wind farm, access to FACTS can mitigate the impact of the wind farm on the grid. In order to reduce the damage of the system to the fault, it is necessary to study the transient voltage stability of the wind turbine when the wind farm is connected to the grid, and take the most effective measures to ensure the stable operation of the wind farm and the access grid during the fault [10–15]. The above literature only studies the role of reactive power compensation devices from some disturbances and does not study from multiple disturbances. In this paper, the IEEE14 node is connected to the doubly fed asynchronous fan in MAT-LAB/PSAT as an example. Time-domain simulation was performed by connecting SVC and STATCOM respectively to verify the effect of FACTS device on improving system voltage stability in the event of sudden wind speed change, three-phase short circuit and load sudden increase disturbance. The simulation and analysis of the voltage stability of the two are carried out and compared.

2   Double-Fed Asynchronous Wind Turbine Model

Wind power systems use wind turbines to convert wind energy into mechanical energy, and generators and their inverter control systems convert mechanical energy into electrical energy. The pitch angle reference value of the pitch control output is responsible for the pitch angle control of the wind turbine, and the electromagnetic power reference value of the MPPT control output is responsible for the electromagnetic power control of the double-fed fan converter. Two mutually decoupled PWM converters control the modules, the rotor side converter controls the active power of the unit and the stator side reactive power, and the grid side converter controls the DC bus voltage.

2.1 Fan Model

The wind energy utilization coefficient equation is as shown in Eqs. (1) and (2).

$$c_p = 0.22(\frac{116}{\lambda_i} - 0.4\theta_p - 5)e^{-\frac{12.5}{\lambda_i}}, \tag{1}$$

$$\frac{1}{\lambda_i} = \frac{1}{\lambda + 0.08\theta_p} - \frac{0.035}{\theta_p^3 + 1}. \tag{2}$$

The power captured by the wind turbine from the wind can be expressed as

$$P_w = \frac{\rho}{2}A_r c_p(\lambda, \theta_p)v_w^3, \tag{3}$$

where ρ is the air density (1.225 kg/m^3 at 15° sea-level average pressure); λ is the tip-speed ratio; $A_r = \pi r^2$ is the blade swept area (m^2) of the wind turbine, R is blade radius (m); v_w represents wind speed (m/s); and θ_p is pitch angle.

2.2 DFIG Model

The relationship between the speed of the doubly fed generator and the current frequency of the stator and rotor windings is given as

$$f_1 = f \pm f_2, \tag{4}$$

where f_1 is the stator current frequency, f is the frequency of the synchronous generator output voltage, and f_2 is the rotor current frequency. When it is integrated into the power grid, if the difference between the synchronous speed of the generator and the speed of the generator rotor itself is constant when f_1 is 50 Hz, the stator winding potential frequency of the asynchronous generator will remain at 50 Hz.

Electrical torque depends on the q-axis component of the rotor current as given below:

$$T_e \approx -\frac{x_m V i_{qr}}{\omega_b(x_s + x_m)}. \tag{5}$$

3 FACTS Device

3.1 SVC Model

Static reactive power compensation device (SVC) can be equivalent to a dynamic reactive power supply in the system, which can quickly emit or absorb reactive power and keep the node voltage stable. There are three main forms of SVC: thyristor-controlled reactor (TCR), thyristor-switched capacitor (TSC), and a hybrid device (TCR–TSC). The TCR–TSC-type SVC is generally composed of n TSC branches and one TCR branch in parallel. It combines the advantages of both forms, which can reduce the risk of resonance and meet the dynamic reactive demand to some extent. The structure of the SVC controller used is shown in Fig. 1.

The model assumes a total reactance b_{svc} and the differential equation is

$$b_{svc} = \left(K_r \left(V_{ref} + V_{POD} - V \right) - b_{svc} \right) / T_r. \tag{6}$$

Then, the reactive power injected by the SVC at the node is

$$Q = b_{svc} V^2. \tag{7}$$

3.2 STATCOM Model

As a dynamic reactive power compensation device, STATCOM consists of a voltage source inverter and a DC power supply. It can be equivalent to an AC voltage source whose amplitude and phase can be controlled. According to the reactive power and active power commands input by the system, if the inverter terminal voltage is higher than the AC voltage at the connection point, the STATCOM will generate reactive power. The purpose of reactive power has a good effect on suppressing wind farm

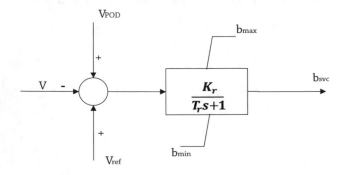

Fig. 1 SVC controller structure

Fig. 2 STATCOM
controller structure

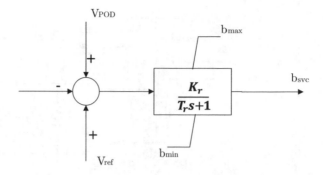

voltage fluctuation caused by disturbance, damping system oscillation, and improving system transient stability. The structure of the STATCOM controller used is shown in Fig. 2.

The model assumes a total current I_{SH} and the differential equation is given as

$$I_{sh} = \left(K_r\left(V_{ref} + V_{POD} - V\right) - I_{SH}\right)/T_r. \tag{8}$$

Then, the reactive power injected by the STATCOM at the node is

$$Q = I_{SH}V^2. \tag{9}$$

4   Case Analysis

According to the above theoretical analysis, this paper adopts the IEEE14 node system model based on MATLAB/PSAT. Its network diagram is shown in Fig. 3. At node 8, the generator is replaced by a double-fed fan of the same capacity. According to different research objects, different reactive compensation devices are added at the grid connection of the fan. In order to better analyze the effect of voltage stability when the FACTS device is connected to the wind farm and the performance difference between STATCOM and SVC, time domain simulation analysis will be carried out from three disturbance scenarios.

4.1   Voltage Stability Assessment When the Wind Speed Changes Suddenly

Disturbance of random winds in wind farms is an important aspect of wind farm voltage fluctuations. Consider the case where the random wind speed is more serious. The method of Wiweb distribution is used to describe the wind speed, and the parame-

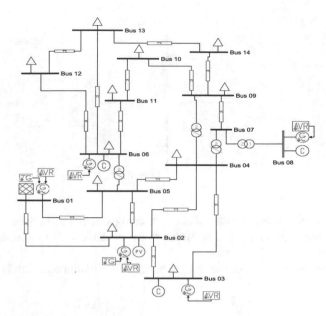

Fig. 3 IEEE14 node system

ters are changed so that the wind speed varies between 11 and 17 m/s. Sudden changes in wind speed cause a sudden change in the mechanical torque Tm, the rotational speed begins to increase, and the reactive power absorbed by the fan also increases, causing voltage instability. The installation of FACTS provides the reactive power consumed by the wind turbine and helps to increase voltage stability.

The voltage amplitude curves of the node voltages 7 and 8 before and after the installation of FACTS are shown in Fig. 4. The voltage values of node 7 and node 8 are 1.04 and 1.09, respectively. When the wind farm does not set reactive power compensation, it only relies on generator excitation. The voltage of the wind turbine unit voltage drops and oscillates up and down, and cannot recover for a long time. After the wind farm is installed with reactive compensation, the voltage at node 7 and node 8 is improved. When the SVC device is connected, the voltage amplitudes of the nodes 7 and 8 are slightly increased, but they continue to fluctuate up and down. The STATCOM device will add reactive power to the system. The bus voltage of node 7 rises to a stable value of 1.035 after 0.5s, and the bus voltage of node 8 rises to a stable value of 1.038 after 0.2s. This indicates that STATCOM has a significant effect on the recovery of wind farm node voltages compared to SVC.

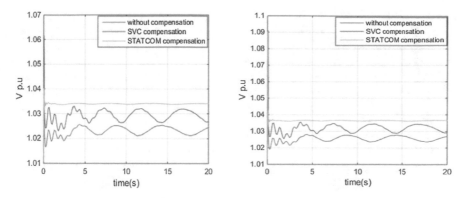

Fig. 4 Without compensation, SVC compensation, STATCOM compensation voltage

4.2 When Three-Phase Short Circuit Occurs in the Line

When a short circuit fault occurs in the power system line, the torque imbalance is caused. This will cause the electromagnetic power injected into the grid by the generator to decrease, and the amount of reactive power increased. If the power grid cannot provide sufficient dynamic reactive power support, the asynchronous power generator system is based on the voltage of each node which will inevitably fall to varying degrees.

The simulation setup node 5 has a three-phase short-circuit fault, and the fault is cleared after 0.2s. The comparison between SVC and STATCOM can be seen in Fig. 5. The short circuit has the most serious influence on the voltage, and the voltage drop is 0.9. When the fault occurs, when the SVC compensation device is connected, the voltage of node 7 increases by 0.05 and the voltage of node 8 increases by 0.05. Node 7 voltage is stable from 4.3s to 1.06 and node 8 voltage is stable at about 4s to 1.09. When STATCOM is connected, node 7 voltage is increased by 0.25, node 8 voltage is increased by 0.27, and final node 7 voltage is about 4.5s to 1.06. When the operation is stable, node 8 voltage at about 2s to 1.09 stable operating voltage level is higher than the addition of SVC. Both SVC and STATCOM can quickly compensate the reactive power required by the wind turbine to restore the fan outlet voltage to the rated level, so that the wind farm can operate stably.

4.3 Voltage Stability Assessment During Load Surge

After the simulation was set for 1s, the load on the whole network is suddenly increased to 1.1 times. The voltage amplitude curves of the node voltages 7 and 8 before and after the installation of FACTS are shown in Figs. 6. When a fault occurs, the load increases instantaneously and the voltage drops. The voltages of the nodes 7

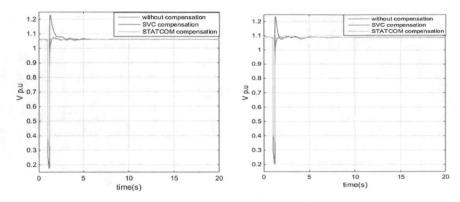

Fig. 5 Without compensation, SVC compensation, STATCOM compensation voltage

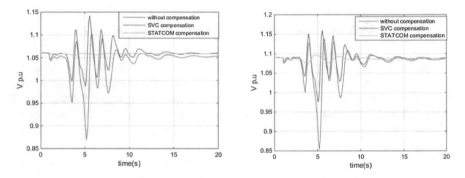

Fig. 6 Without compensation, SVC compensation, STATCOM compensation voltage

and 8 fall and fluctuate, and the stable value can be restored for a long time, but cannot be restored to the initial value. When the SVC compensation device is connected, when the fault occurs, the voltage fluctuates up and down from the initial value, and the swing amplitude is significantly reduced compared with the unmounted FACTS and returns to a stable initial value in about 17s. When the STATCOM is connected, the voltage fluctuation is very small and the voltage returns to a stable initial value around 9s, which is obviously improved compared with SVC.

5   Conclusion

After sudden changes in wind speed, short-circuit faults in the power grid, and sudden increase in load, the voltage will have different degrees of influence. Avant-garde measures should be taken to prevent voltage instability. Installation of FACTS has a significant effect on the improvement of the voltage level of the whole system. Both SVC and STATCOM can generate reactive power stable wind motor outlet voltage and grid voltage to improve the transient voltage stability of the system. In the event of a grid failure, STATCOM's ability to stabilize voltage is better than SVC, STATCOM reacts more quickly, and wind turbines return to normal operating voltage in less time.

Acknowledgements In this paper, the research was supported by Fujian Provincial Natural Science Foundation Projects (2018J01622).

References

1. Ming, W.: Analysis of the Influence of Wind and Light Storage Access on Power System Stability. Taiyuan University of Technology, Taiyuan (2016)
2. Yizhang, Y.: The Effect of Doubly-Fed Fan Access on the Stability of Small Disturbances in Power Systems. North China Electric Power University, Beijing (2016)
3. Jianbin, Y.: Simulation Study on Dynamic Characteristics of Large Wind Farms Connected to the Power Grid. Kunming University of Science and Technology, Kunming (2016)
4. Yi, W., Xiaorong, Z., Shuqiang, Z.: Modeling and Simulation of Wind Power Generation Systems. China Water Resources and Hydropower Press, Beijing (2015)
5. Xueling, Z., Yang, Z., Kun, G., et al.: Research on reactive power compensation of wind farms. Power Syst. Prot. Control **37**(16), 68–76 (2009)
6. Yuqi, Z., Dongdong, L., Zichao, L.: Reactive power coordination control strategy of doubly-fed induction wind turbine under fault. J. Electr. Power Syst. Autom. **7**(28), 12–18 (2016)
7. Zhiyue, Y., Fengting, L.: Simulation of dynamic reactive power compensation scheme for grid-connected wind farms. New Energy Wind Power **38**(11), 34–38 (2011)
8. Xiaofu, X., Hanyu, Z., Jinxin, O.: Simulation study on transient output characteristics of SVC doubly-fed wind turbines. Power Syst. Prot. Control **39**(19), 89–93 (2011)
9. Wei, Z.: Research on Reactive Power Compensation of Wind Farm Based on SVC and STATCOM. Xinjiang University, Xinjiang (2008)
10. Yongning, C., Hongliang, G., Weisheng, W., et al.: SVC and pitch angle control improve transient voltage stability of asynchronous wind farm. Autom. Electr. Power Syst. **31**(3), 95–100 (2007)
11. Wanpeng, S.: Research on STATCOM in Voltage Stability of Wind Power Grid-Connected System. Nanjing University of Posts and Telecommunications, Nanjing (2013)
12. Min, P.: Study on the Voltage Stability of Power System Considering the Influence of Static Synchronous Compensator. Guangxi University, Guangxi (2008)
13. Lixia, L., Shuwei, Y., Yanmin, Y.: Research and analysis of static var compensator in voltage stability. East China Electr. Power **8**(34), 14–16 (2006)
14. Xiao-Fu, X., Han-Xi, Z., Jin-Xin, O.: Simulation analysis of transient output characteristics of SVC doubly-fed wind turbines. Power Syst. Prot. Control **19**(39), 89–99 (2011)
15. Hong, Z., Wanzhen, T., Chuncheng, S., et al.: Comparative analysis of STATCOM and SVC in power system operation. Jilin Electr. Power **37**(3), 24–27 (2009)

Influence of Wind Farm Access on System Small Signal Stability

Peiqiang Li, Lu-Qi Jie, Zheng-Bang Xia and Cai-Jie Weng

Abstract With the increase of wind power installed capacity, research on the dynamic behavior and stabilization mechanism of the power system has had a certain impact. Frequent small disturbances will affect system stability and cause huge economic losses. In this context, based on the concept and analysis method of small signal stability, three different types of wind turbines are adopted, the eigenvalue analysis method in PSAT is used, the wind turbines have been studied in different types, different wind farm access distances and under different conditions of wind power penetration rate on the influence of the small signal stability when accessing WSCC 3-machine, 9-bus system. The result showed that the wind turbine type, the access distance, and the wind power penetration rate had different degrees of impact on the system.

Keywords Power system · Wind power integration · Wind turbines · Small signal stability · PSAT · Damping characteristic

1 Introduction

With the deteriorating environment and the increasing consumption of traditional energy, new energy is developing rapidly and the wind power is the most mature power generation technology in renewable energy [1]. China's wind power installed

P. Li (✉) · L.-Q. Jie · Z.-B. Xia · C.-J. Weng
School of Information Science and Engineering, Fujian University of Technology, Fuzhou 350118, China
e-mail: lpqcs@yahoo.com.cn

L.-Q. Jie
e-mail: 410759161@qq.com

Z.-B. Xia
e-mail: 2423081432@qq.com

C.-J. Weng
e-mail: caijieweng@gmail.com

© Springer Nature Singapore Pte Ltd. 2019
J.-S. Pan et al. (eds.), *Genetic and Evolutionary Computing*,
Advances in Intelligent Systems and Computing 834,
https://doi.org/10.1007/978-981-13-5841-8_12

capacity was 19.5 GW in 2017; The cumulative installed capacity was 188.232 GW [2] by the end of the year, ranking first in the world. Wind power technology has been vigorously developed due to the short construction period, low environmental requirements, abundant reserves, and high utilization rate [3]. However, wind power has been characterized by strong randomicity and uncertainty, and large-scale integration of wind power will cause certain interference to the system. It is widely concerned to study the small signal stability of wind power systems connected to power systems.

In recent years, many research work has been done at home and abroad for the problem of small signal stability of wind power integration. References [4–6] analyzed the small signal stability of wind power systems including Squirrel Cage Induction Generator (SCIG) to power systems. References [4–8] analyzed the small signal stability of wind power systems including Doubly Fed Induction Generator (DFIG) to power systems. References [5, 7, 9, 10] analyzed the small signal stability of wind power systems including Direct-drive Permanent Magnet Synchronous Generator (DDPMSG) to power systems. Most of the above documents are one or two types of wind turbines for comparison and research, and there are few detailed comparisons of three types of fans. Based on the research of the above literature, based on the MATLAB toolbox Power System Analysis Toolbox (PSAT), this paper studies the effects of different types of the wind turbines, the wind farm access distance, and the wind power penetration rate.

2 Mathematical Models of Wind Turbine

The mathematical models of the wind turbine in this paper are from reference [11]. There are three main types of wind farms used: SCIG, DFIG, DDPMSG. The following assumptions are made to make the mathematical models of the wind turbine discussed applicable to dynamic simulation of power systems: (1) neglecting magnetic saturation; (2) sinusoidal distribution of magnetic flux; (3) no loss except copper loss; (4) sinusoidal distribution of stator voltage and current at fundamental frequency [12].

2.1 Mathematical Model of SCIG

In the synchronous rotating coordinate, dq coordinate system, the stator voltage equation of generator, the rotor voltage equation, the stator flux linkage equation, the rotor flux linkage equation, the electromagnetic torque equation are as follows:

$$u_{ds} = d\psi_{ds}/dt - \omega_1\psi_{qs} + R_s i_{ds}, u_{qs} = d\psi_{qs}/dt + \omega_1\psi_{ds} + R_s i_{qs} \qquad (1)$$

$$0 = d\psi_{dr}/dt + R_r i_{dr} - s\omega_1 \psi_{qr},\, 0 = d\psi_{qr}/dt + R_r i_{qr} - s\omega_1 \psi_{dr} \qquad (2)$$

$$\psi_{ds} = (L_s + L_m)i_{ds} + L_m i_{dr},\, \psi_{qs} = (L_s + L_m)i_{qs} + L_m i_{qr} \qquad (3)$$

$$\psi_{dr} = L_m i_{ds} + (L_r + L_m)i_{dr},\, \psi_{qr} = L_m i_{qs} + (L_r + L_m)i_{qr} \qquad (4)$$

$$T_e = E'_d i_{ds} + E'_q i_{qs} \qquad (5)$$

In formulas (1)–(5), ω_1 is synchronous rotating speed. $u_{ds}, u_{qs}, \psi_{ds}, \psi_{qs}, i_{ds}, i_{qs}$ are stator voltage, flux linkage, and current d axis and q axis component, respectively; L_s is stator inductor; L_r is rotor inductor; L_m is magnetic inductance.

2.2 Mathematical Model of DFIG

In the synchronous rotating coordinate, dq coordinate system, the stator voltage equation of generator, the rotor voltage equation, the stator flux linkage equation, the rotor flux linkage equation, the electromagnetic torque equation are as follows:

$$u_{ds} = R_s i_{ds} - \omega_1 \psi_{qs} + d\psi_{ds}/dt,\, u_{qs} = R_s i_{qs} + \omega_1 \psi_{ds} + d\psi_{qs}/dt \qquad (6)$$

$$u_{dr} = R_r i_{dr} - \omega_s \psi_{qr} + d\psi_{dr}/dt,\, u_{qr} = R_r i_{qr} + \omega_s \psi_{dr} + d\psi_{qr}/dt \qquad (7)$$

$$\psi_{ds} = L_{ss} i_{ds} + L_m i_{dr},\, \psi_{qs} = L_{ss} i_{qs} + L_m i_{qr},\, L_{ss} = L_s + L_m \qquad (8)$$

$$\psi_{dr} = L_{rr} i_{dr} + L_m i_{ds},\, \psi_{qr} = L_{rr} i_{qr} + L_m i_{qs},\, L_{rr} = L_r + L_m \qquad (9)$$

$$T_e = p_n L_m (i_{ds} i_{qr} - i_{qs} i_{dr}) \qquad (10)$$

In formulas (6)–(10), R_s is stator resistance; R_r is rotor resistance; $u_{dr}, u_{qr}, i_{dr}, i_{qr}, \psi_{dr}, \psi_{qs}$ are generator rotor voltage, current, flux d axis and q axis component, respectively; ω_s is shift frequency, $\omega_s = s\omega_1$; s is transfer rate; L_s is stator leakage reactance; L_r is rotor leakage reactance; L_m is mutual inductance between stator and rotor; T_e is electromagnetic torque; p_n is number of pole pairs.

2.3 Mathematical Model of DDPMSG

In the synchronous rotating coordinate, dq coordinate system, the stator voltage equation of generator, the stator flux linkage equation, the electromagnetic torque equation, the electromagnetic torque equation are as follows:

$$u_{ds} = R_s i_{ds} + d\psi_{ds}/dt - \omega_1 \psi_{qs}, u_{qs} = R_s i_{qs} + d\psi_{qs}/dt + \omega_1 \psi_{ds} \tag{11}$$

$$\psi_{ds} = L_d i_{ds} + \psi_f, \psi_{qs} = L_q i_{qs} \tag{12}$$

$$T_e = 1.5 p_n (\psi_{ds} i_{qs} - \psi_{qs} i_{ds}) \tag{13}$$

In formula (11)–(13), L_d, L_q are d axis and q axis synchronous inductor, regarded as constant. ω_f is rotor flux, regarded as constant.

3 Small Signal Stability Analysis Method

Small signal stability refers to the ability of the system to automatically recover to the initial operating state after spontaneous disturbance or nonperiodic loss of synchronization after a small disturbance [13]. The following describes the small signal stability analysis method. The system used for small signal stability analysis method is a differential algebraic equation (DAE), as follows:

$$d_x/d_t = f(x, y), 0 = g(x, y) \tag{14}$$

In the formula, x represents the state variable describing the dynamic characteristics of the system in the differential equations; y represents the input vector of the system in the algebraic equations.

According to the basic idea of Lyapunov's first method, the stability of the linear system can be used to study the stability of the actual nonlinear power system. The Stable-Operation-Point $(x_{(0)}, y_{(0)})$ is linearized, then

$$\begin{bmatrix} d\Delta x/dt \\ 0 \end{bmatrix} = \begin{bmatrix} \tilde{A} & \tilde{B} \\ \tilde{C} & \tilde{D} \end{bmatrix} \begin{bmatrix} \Delta x \\ \Delta y \end{bmatrix} \tag{15}$$

The operational vectors Δy are removed in the upper, then:

$$d\Delta x/dt = A\Delta x \tag{16}$$

In the formula, A is called $n \times n$-dimensional coefficient matrix or state matrix to evaluate the stability of a small interference system.

For the conjugate eigenvalue $\lambda = \sigma \pm j\omega$, corresponding to the generator rotor motion equation, it is called the low-frequency oscillation mode. The real part reflects the nature of the attenuation, and the imaginary part reflects the frequency. The frequency of oscillation (Hz) and the damping ratio are defined as

$$f = \omega/2\pi \qquad (17)$$

$$\xi = -\sigma/\sqrt{\sigma^2 + \omega^2} \qquad (18)$$

The damping ratio is an indicator to measure the small signal stability of the power system. When the damping ratio is negative, the system small disturbance is unstable. Conversely, the system will reach a new steady-state equilibrium point, and the larger the damping ratio, the faster the system oscillation decays, the fewer the number of oscillations.

According to the small disturbance rotor angle oscillation mode, the oscillation characteristics can be divided into a local oscillation mode and an inter-region oscillation mode, which are 1 Hz–2 Hz and 0.1 Hz–0.7 Hz, respectively.

4 Example Analysis

Taking the United States West Power System(WSCC) 3-machine 9-bus system as an example [13]. G1, G2, and G3 adopt the classic fourth-order model. The rated capacity of the wind turbine is 2MVA, and the power factor is 1. All simulations are performed in PSAT.

4.1 Influence of Different Types of Wind Turbines on Small Signal Stability

4.1.1 Design of the Example

Designing of four kinds of examples are as follows: example A: The original 3-machine 9-bus system, no fan is connected. The examples B–D are connected to SCIG, DFIG, and DDPMSG on bus4. The three types of wind turbines all use the Weibull wind speed model, setting v = 15.00 m/s, $\rho = 1.225$ kg/m^3, taking C = 20, K = 2, the number of units is 100.

4.1.2 Characteristic Root Analysis

Through the eigenvalue analysis, it can be seen that compared with the example A and B, C, D after the wind turbines are added become unstable, but the eigen roots move to the left of the complex plane, which improves the stability of the system. Through the eigenvalue analysis, it can be seen that compared with the example A, and B, C, D after the wind turbines are added become unstable, but the eigen roots move to the left of the complex plane, which improves the stability of the system.

SCIG itself cannot suppress the power fluctuation caused by wind speed change and pitch angle change. DFIG and DDPMSG can eliminate the influence of wind speed change on the system through the decoupling control of active and reactive power (Table 1, Figs. 1 and 2) .

Observing the damping ratio, it can be known that the damping ratio of the inter-area oscillation mode is positive, and it is increased to different degrees compared with example A. The increase of the damping ratio improves the stability of the system. Compared with the other two fans, example C adds an oscillation mode of $-2.857 \pm j\, 0.70059$, which the DFIG is connected and makes the oscillations decay faster, the lesser the number of oscillations, the faster the new equilibrium point reached.

Table 1 Inter-regional oscillation damping characteristics of examples A–D

	Mode	Eigenvalue	Frequency	Damping ratio
A	1	$-0.44338 \pm j\, 1.2111$	0.20527	0.343783
	2	$-0.43894 \pm j\, 0.73953$	0.13687	0.510405
	3	$-0.4255 \pm j\, 0.49625$	0.10404	0.650917
B	1	$-0.45887 \pm j\, 1.173$	0.20046	0.364310
	2	$-0.43864 \pm j\, 0.73858$	0.13672	0.510632
	3	$-0.4263 \pm j\, 0.4969$	0.10420	0.651131
C	1	$-0.54209 \pm j\, 1.0346$	0.18589	0.464112
	2	$-0.44012 \pm j\, 0.73586$	0.13646	0.513298
	3	$-0.42631 \pm j\, 0.49662$	0.10417	0.651351
	4	$-2.857 \pm j\, 0.70059$	0.46817	0.971225
D	1	$-0.45156 \pm j\, 1.2146$	0.20623	0.348473
	2	$-0.43816 \pm j\, 0.73824$	0.13663	0.510392
	3	$-0.42632 \pm j\, 0.49701$	0.10421	0.651066

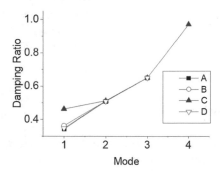

Fig. 1 Inter-area oscillation mode damping ratio when wind power penetration rate is different

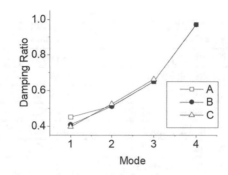

Fig. 2 Inter-area oscillation mode damping ratio when wind farm access distance is different

4.2 Influence of Different Access Distances of Wind Farms on Small Signal Stability of System

4.2.1 Design of the Example

The wind farm (fan is DFIG) is connected to the power system in bus4, where the wind farm capacity is 200 MVA, the power factor is 1, and the rest of the settings are consistent with the above example. The design examples are as follows: Example A–C: The distances to access the power system are 25 km, 15 km, and 5 km, respectively.

4.2.2 Characteristic Root Analysis

According to the eigenvalue analysis, when the access distance of the wind farm is 5 km, the system is stable. As the access distance is increasing, the system becomes unstable, and the visible distance has certain adverse effects on the stability of the system. Observing the damping ratio line graph, it can be seen that the damping ratios of modes 1, 2, and 3 have not changed too much. Compared with example A and B, a mode 4 is added to the example C, which plays a stabilizing role to stay system because of the DFIG. The calculation of example A is larger than the mode 4 damping ratio of example B by 0.001938, which makes it easier to achieve a small signal stability state. The smaller the visible distance, the better the stability of the small signal stability of the system. DFIG will lose its stability if it exceeds a certain range (Table 2).

Table 2 Inter-regional oscillation damping characteristics of examples A–C

	Mode	Eigenvalue	Frequency	Damping ratio
A	1	$-0.52817 \pm j\,1.0426$	0.18600	0.451910
	2	$-0.44035 \pm j\,0.73619$	0.13653	0.513326
	3	$-0.42635 \pm j\,0.4967$	0.10418	0.651326
	4	$-2.95630 \pm j\,0.70466$	0.48368	0.972748
B	1	$-0.48930 \pm j\,1.0913$	0.19035	0.409123
	2	$-0.43955 \pm j\,0.73773$	0.13667	0.511849
	3	$-0.42679 \pm j\,0.49732$	0.10430	0.651245
	4	$-3.7119 \pm j\,0.91706$	0.60853	0.970810
C	1	$-0.49470 \pm j\,1.1415$	0.19800	0.397641
	2	$-0.46423 \pm j\,0.75317$	0.14081	0.524705
	3	$-0.4911 \pm j\,0.55327$	0.11774	0.663838

4.3 Influence of Different Permeabilities of Wind Power on Small Signal Stability of System

4.3.1 Design of the Example

With the system in Sect. 4.2.1, the distance of the DFIG access system is 5 km, and the system is stable at this time. The design examples are as follows: Example A–E: The capacity of the wind farm is increased by 60, 80, 100, 120, and 140 MVA. The above examples all reduce the output of the corresponding thermal power units of G2 and G3.

4.3.2 Characteristic Root Analysis

According to the eigenvalue analysis, as the wind power permeability increasing, the characteristic root moves to the right side of the complex plane and the system becomes less disturbed and unstable. Observing the damping ratio of the oscillation mode, as the wind power permeability increases, the oscillation mode decreases and the dominant oscillation mode changes; But it is not obvious when the damping ratio of the same oscillation mode gradually decreases with the increase of the permeability (Table 3, Fig. 3).

Table 3 Inter-regional oscillation damping characteristics of examples A–E

	Mode	Eigenvalue	Frequency	Damping ratio
A	1	$-3.1639 \pm j\,0.88665$	0.52294	0.962904
	2	$-0.548 \pm j\,1.0551$	0.18922	0.460921
	3	$-0.44247 \pm j\,0.694$	0.13099	0.537596
B	1	$-3.2295 \pm j\,0.9931$	0.53774	0.955828
	2	$-0.56385 \pm j\,1.0464$	0.18918	0.474363
	3	$-0.45496 \pm j\,0.65194$	0.12653	0.572282
C	1	$-3.286 \pm j\,1.1465$	0.5539	0.944180
	2	$-0.56689 \pm j\,1.0094$	0.18425	0.489672
	3	$-0.46585 \pm j\,0.54539$	0.11416	0.649483
D	1	$-3.3875 \pm j\,1.3358$	0.57954	0.930284
	2	$-0.50531 \pm j\,0.95564$	0.17205	0.467442
E	1	$-3.4046 \pm j\,1.6982$	0.60553	0.894858
	2	$0.47452 \pm j\,0.94286$	0.16799	0.449554

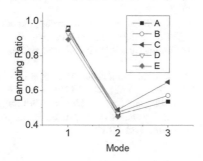

Fig. 3 Inter-area oscillation mode damping ratio when wind power penetration rate is different

5 Conclusions

In order to study the impact of wind farm access on the stability of small disturbances in the system, this paper starts with three different aspects: system access fan, wind farm access system distance, and wind farm penetration rate. The PSAT toolbox is used to compare and analyze the modified experiment on the WSCC 3-machine, 9-bus system. Through analysis, we can know that the above three aspects have a certain influence on the small signal stability of the system. Among them, the influence of DFIG is better than the other two units. In the case determined by the wind turbine, the closer the distance and the lower the wind power permeability, the more significant the stability of the system. Therefore, we can give priority to the doubly fed wind turbines in practice. When the long-distance transmission and wind power penetration rate are high, corresponding measures should be taken to improve the small signal stability of the power system, which is the next research work.

References

1. Domínguez-García, J.L., Gomis-Bellmunt, O., Bianchi, F.D., et al.: Power oscillation damping supported by wind power: A review. Renew. Sustain. Energy Rev. **16**(7), 4994–5006 (2012)
2. Xu, T.: Global wind power installed statistics in 2017//. Wind Energy Ind. **2018**, 7 (2018)
3. Liu, B., He, Z., Jin, H.: Wind power status and development trends. J. Northeast Dianli Univ. **36**(2), 7–13 (2016)
4. Wang, F.: Influence of wind power access to power system on frequency dynamic characteristics of power grid. Electron. Technol. Softw. Eng. **21**, 236–238 (2017)
5. He, P., Wen, F., Xue, Y., Ledwich, G., Li, S.: Impacts of different wind power generators on power system small signal and transient stability. Dianli Xitong Zidonghua Autom Electr. Power Syst. **37**(17), 23–29 (2013)
6. Wang, L.: Small signal stability analysis of wind power systems including DFIG. Peak Data Sci. **6**(05), 67–69 (2017)
7. Li, P.-Q., Wang, J.-F., Li, X.-R., et al.: Analysis on the impact of DFIG and DDSG wind plant on the small signal stability of the power system. J. Hunan Univ. (Nat. Sci.) **41**(01), 92–97 (2014)
8. Lin, C.-H., Su, Y.-L.: Study of influence on double-fed wind generator power system unit transient stability. Colliery Mech. Electr. Technol. **02**, 32–37 (2012)
9. Ren, Z.-Y., Zhang, S.: Effect of DDSG wind turbines to power system small-signal stability. Electr. Switchg. **55**(02), 57–60 (2017)
10. Li, Y., Qu, Y., Ma, S.: Impact of grid-connected wind turbine generators on small signal stability of power grid. Power Syst. Technol. **36**(08), 50–55 (2012)
11. Yang, X.-S.: Wind Power Technology and Wind Farm Project. Chemical Industry Press, Beijing (2012)
12. Hao, Y.-Z., Li, P.-Q., Li, X.-R., et al.: Analysing the impact of wind plant on power system transient stability. Proc. CSU-EPSA **24**(02), 41–46 (2012)
13. Kundur, P.: Power System Stability and Control. McGraw-Hill Inc., New York (1994)
14. Bazargan, M.: Small signal stability analysis with penetration of grid-connected wind farm of PMSG type. Autom. Electr. Power Syst. (2012)

Visual QR Code with Beautication and Self-recovery Based on Data Hiding

Huili Cai, Bin Yan and Jeng-Shyang Pan

Abstract For the problem of beautification and self-recovery of visual QR (Quick Response) code, this paper proposes two schemes. The first scheme uses data hiding and beautified QR code for self-recovery. By decoding this visual QR code, we can not only access the relevant links embedded in it but also restore the original image. However, the QR code may be maliciously tampered with by the attacker during transmission (e.g., redirected to a phishing website). For tampering protection and recovery of the QR code, the QR code is bound with the background image. This increases the security level when user scans a QR code. Experimental results demonstrate that the proposed algorithm is superior to the reference method in terms of visual quality.

Keywords Visual QR code · Beautification · Data hiding · Self-recovery

1 Introduction

As a common two-dimensional code, QR code can be well combined with mobile terminals such as smartphones. With the increasing demand for visual effects of QR codes in the field of media advertising, scholars have proposed to combine QR codes with data hiding. Data hiding is widely used in various fields of modern life [1, 2]. As an emerging technology, data hiding plays an increasingly important role in medical diagnosis, forensic identification, and copyright protection [3]. Using the part of the background image that is covered by the QR code as a watermark, one can design a self-recovery QR code. By decoding this QR code and extracting the

H. Cai · B. Yan (✉)
College of Electronics Communication and Physics, Shandong University of Science and Technology, Qingdao 266590, P. R. China
e-mail: yanbinhit@hotmail.com

J.-S. Pan
College of Computer Science and Engineering, Shandong University of Science and Technology, Qingdao 266590, P. R. China

© Springer Nature Singapore Pte Ltd. 2019
J.-S. Pan et al. (eds.), *Genetic and Evolutionary Computing*,
Advances in Intelligent Systems and Computing 834,
https://doi.org/10.1007/978-981-13-5841-8_13

hidden watermark, we can not only access the relevant links but also recover the original image [4].

As QR codes continue to be applied to various fields, people are beginning to consider improving the visual aesthetics of QR codes. Toshihiko et al. combined the existing QR code image directly with the background image [5, 6]. Considering that QR code decoding is only related to the middle pixel of the module, Visualead and LogoQ proposed a method to modify the middle pixel of the module [7]. Liu et al. proposed a beautified QR code where the supplementary message can be successfully recovered with very low computational complexity [8]. In the study of the combination of QR code and data hiding, Zhang and Yoshino [9] and Chung et al. [10] proposed to embed the QR code as an invisible watermark into the image. Liu et al. proposed a secure and visual improved QR code involving digital signature and watermarking techniques [11]. In 2011, Huang et al. proposed a scheme in which a QR code was used as a visible watermark and combined with reversible data hiding [4]. The scheme mainly considered increasing the data embedding capacity. However, the image after hiding the message had a poor visual effect, since part of the background image is covered by the QR image, as shown in Fig. 1.

This paper is devoted to improving the visual QR code to have better beautification and tampering protection ability. For these two problems, two schemes are proposed. In the first scheme, for better beautification effect, we propose to improve the quality of background image with a visual QR code instead of a standard QR code. In the second scheme, considering the existence of malicious attacks, the QR code and the background image are bound together to ensure that the QR code scanned by the user is safe.

The rest of this paper is organized as follows. In Sect. 2, we propose the generation of visual QR code. In Sect. 3, considering the security of the QR code, we propose tampering protection and recovery of the visual QR code. Through the analysis of experimental results, we show the effectiveness of the proposed algorithm in Sect. 4. Finally, the conclusions are given in Sect. 5.

Fig. 1 Image with QR code in Huang et al. [4]

2   The Generation of Visual QR Code

In this section, we generate a visual QR code to improve the quality of background image.

In Huang's algorithm, the QR code $Q(i, j)$ completely covers a part of the background image $I(i, j)$ according to Eq. 1, as shown in Figs. 1 and 2. Ω is a part of the background image which is covered by the QR code. Ω^C is the area of the background image excluding Ω.

$$\hat{I}(i, j) = Q(i, j), \quad \forall (i, j) \in \Omega, \tag{1}$$

where $\hat{I}$ is the watermarked image.

Traditionally, the watermarked image is generated according to Eq. 2.

$$\hat{H} = \mathrm{DH}(H; W) \tag{2}$$

where H is the host image, W is the watermark, DH is the data hiding algorithm, and $\hat{H}$ is the watermarked image. In Huang's algorithm, the watermarked image is generated according to Eq. 3.

$$\hat{I}(i, j) = \mathrm{DH}\left(\underbrace{I(i, j), (i, j) \in \Omega^C}_{H}; \underbrace{I(i, j), (i, j) \in \Omega}_{W}\right), \quad \forall (i, j) \in \Omega^C. \tag{3}$$

In our algorithm, the QR code with a part of the background image is generated by the following steps:

1. Generate a standard QR code $Q(i, j)$ according to the encoded content.
2. Embed the finder patterns, the alignment patterns, the timing patterns, and the format and version information into the background image.
3. Shorten the pixels of data modules of the QR code and fill them into the background image. Other pixels of the data modules are replaced with background image according to Eqs. 4 and 5.

Fig. 2   QR code and its corresponding region in background image

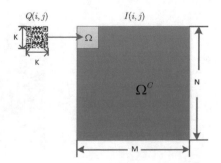

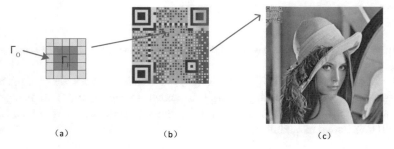

(a)   (b)   (c)

Fig. 3 A module in the background image. **a** A module. **b** The QR code with a part of background image Q_B. **c** Visual QR code

$$Q_B(i, j) = \begin{cases} I(i, j), & \forall (i, j) \in \Gamma_O; \\ Q(i, j), & \forall (i, j) \in \Gamma_I, \end{cases} \tag{4}$$

$$\Gamma = \Gamma_I \cup \Gamma_O, \tag{5}$$

where Γ is a module in the QR code, Γ_I is the inner region of the module, and Γ_O is the outer region of the module, as shown in Fig. 3.

Then, we use Eq. 6 to hide the message of QR modules. Here, the data hiding algorithm we used is least significant bits(LSB) replacement. Some watermarked visual images are shown in Fig. 4.

Fig. 4 Some watermarked visual QR codes. The visual QR code is embedded at the top-left corner of the background image

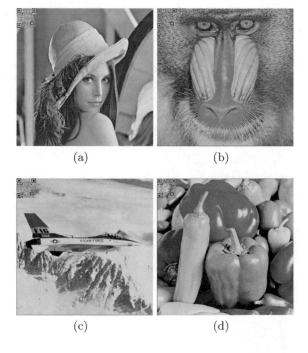

(a)   (b)

(c)   (d)

$$\hat{I}(i, j) = \mathrm{DH} \left(\underbrace{I(i, j), \forall(i, j) \in \Omega^C}_{H}; \underbrace{I(i, j), \forall(i, j) \in \Omega_{\mathrm{I}}}_{W} \right), \quad \forall(i, j) \in \Omega^C,$$

(6)

where Ω_{I} is the union of all Γ_{I} for all modules in the QR code.

3 Tampering Protection and Self-recovery

The QR code appears to be a meaningless image, and we can only know the message it encoded after it has been decoded by the relevant device. This will give criminals a chance to change the QR code content and change the destination URL to the unprotected website. Thus criminals will achieve the purpose of deducting user's funds or stealing user's personal information, which seriously affects user's property and information security. In Huang's algorithm, there is no binding between the QR code and the background image. A malicious attacker can change the content of the QR code at will. Considering malicious attacks, we have proposed the second scheme. In this paper, the QR code image is embedded into the background image. When the QR code content is tampered with, we can detect the tampered regions and restore the QR code image.

The watermark embedding, tampering detection, and image recovery processes are shown in Fig. 5 and Fig. 6, respectively. First, we take advantage of hashing algorithm MD5 to produce hash bits **h** and compress the QR code image into bits **s** using a source encoding algorithm SPIHT (Set Partitioning In Hierarchical Trees). Then, we combine these two results to generate watermark. Finally, we employ the embedding algorithm in [12] to generate the watermarked visual QR code. When we receive a visual QR code, the watermarking message is first extracted. After detecting tampered regions, we reconstruct the image. Finally, a recovered visual QR code is obtained.

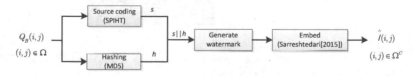

Fig. 5 The block diagram of watermark embedding

Fig. 6 The block diagram of tampering detection and image recovery

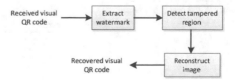

4 Experimental Results

In this section, we present the experimental results to verify the effectiveness of the proposed algorithm.

As can be seen from Fig. 1, Huang et al. combined the standard QR code with the background image. The QR code covers a region of the background image, which affects the visual effect of the background image. It can be seen from Fig. 7 that under the premise of ensuring that we can scan the QR code to jump to the corresponding link and restore the original image, the proposed algorithm has a significant improvement in visual quality compared to [4].

It can be seen from Table 1 that the proposed algorithm in this paper has a higher PSNR than [4]. The effectiveness of this paper's algorithm is validated.

In the experiment of tampering detection and recovery, the content of the original visual QR code we set is http://www.lenna.org. We change the contents of this visual QR code to http://www.lenna1.org. According to the block diagrams of Figs. 5 and

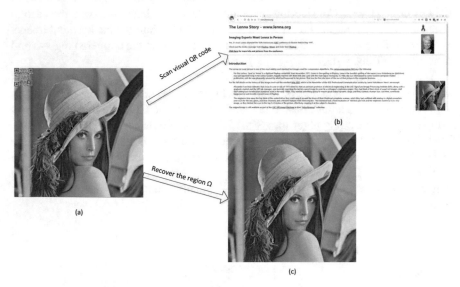

Fig. 7 Resulting images in the first scheme. **a** Watermarked visual QR code (the QR code is in the upper left corner) **b** Decoded message. **c** Recovered image

Table 1 Comparison of the PSNR between Huang et al. [4] and this paper

Algorithm	Huang et al. [4]	This paper
Image	With QR code	With QR code
Lena	22.81 dB	26.24 dB
Baboon	21.71 dB	23.96 dB
Airplane	22.08 dB	25.37 dB
Peppers	21.59 dB	26.58 dB

Fig. 8 Resulting images in the second scheme. **a** Original visual QR code. **b** Watermarked visual QR code. **c** Tampered visual QR code. **d** Recovered visual QR code

(a)   (b)

(c)   (d)

6, the generated visual QR code images are as shown in Fig. 8. The QR code in the original image as shown in Fig. 8b is modified to another QR code, as shown in Fig. 8c. Using the tempering detection algorithm, this modification is detected and the original QR image is recovered, as shown in Figs. 8d. So the user scanning Fig. 8b is prompted with the alert that he is scanning a tempered QR code, and the correct URL is recovered.

It can be seen from Table 2 that the PSNR of the watermarked visual QR code and the recovered visual QR code generated by this paper are both around 44 dB. This indicates that the quality of the image is high.

Table 2 The PSNR of watermarked visual QR code and recovered visual QR code

Image	Watermarked visual QR code	Recovered visual QR code
Lena	44.15 dB	44.08 dB
Baboon	44.14 dB	43.21 dB
Airplane	44.12 dB	43.84 dB
Peppers	44.14 dB	43.61 dB

5   Conclusions

In this paper, we introduced two schemes. The first scheme can improve the visual quality of the image with QR code compared to [4]. And the QR code is bound to the background image in the second scheme. When an attacker tampers with the contents of the QR code, we can detect tampered regions and restore the original QR code image. The second scheme enhances the security of the visual QR code.

Acknowledgements This work is supported by the National Natural Science Foundation of China (NSFC) (No. 61272432) and Shandong Provincial Natural Science Foundation (No. ZR2014JL044).

References

1. QR code standardization: http://www.denso-wave.com/qrcode/qrstandard-e.html
2. Kan, T.W., Teng, C.H., Chou, W.S.: Applying QR code in augmented reality applications. In: Proceedings of the 8th International Conference on Virtual Reality Continuum and its Applications in Industry, pp. 253–257 (2009)
3. Qu, D.C., Pan, J.S., Weng, S.W., Xu, S.Q.: A novel reversible data hiding method for color images based on dynamic payload partition and cross-channel correlation. J. Inf. Hiding Multimed. Signal Process. **7**(6) (2016)
4. Huang, H.C., Chang, F.C., Fang, W.C.: Reversible data hiding with histogram-based difference expansion for QR code applications. IEEE Trans. Consum. Electron. **57**(2), 779–787 (2011)
5. Wakahara, T., Yamamoto, N., Ochi, H.: Image processing of dotted picture in the QR code of cellular phone. In: 2010 International Conference on P2P, Parallel, Grid, Cloud and Internet Computing, pp. 454–458 (2010)
6. Wakahara, T., Yamamoto, N.: Image processing of 2-dimensional barcodes. In: 14th International Conference on Network-Based Information Systems, pp. 484–490 (2011)
7. Logoq. http://www.qrcode.com/en/codes/logoq.html/
8. Liu, S.J., Zhang, J., Pan, J.S., Weng, C.J.: A novel information embedding and recovering method for qr code based on module subdivision. J. Inf. Hiding Multimed. Signal Process. **9**(2), 515–522 (2018)
9. Zhang, S., Yoshino, K.: DWT-based watermarking using QR code. Sci. J. Kanagawa Univ. **19**, 3–6 (2008)
10. Chung, C.H., Chen, W.Y., Tu, C.M.: Image hidden technique using QR-barcode. In: International Conference on Intelligent Information Hiding and Multimedia Signal Processing, pp. 12–14 (2009)
11. Liu, S.J., Zhang, J., Pan, J.S., Weng, C.J.: SVQR: A novel secure visual quick response code and its anti-counterfeiting solution. J. Inf. Hiding Multimed. Signal Process. **8**(5), 1132–1140 (2017)
12. Sarreshtedari, S., Akhaee, M.A.: A source-channel coding approach to digital image protection and self-recovery. IEEE Trans. Image Process **24**(7), 2266–2277 (2015)

Cloud Service and APP for Drug–Herb Interaction

Lingyun Feng, Jui-Le Chen, Li-Chai Chen and Shih-Pang Tseng

Abstract Herb–drug interaction is one special kind of drug interactions between herb medicines and conventional drugs. In the East Asia, the herb medicines have been used several thousands of years. However, many people may mix using the herbs and the drugs. Sometimes, the mix used is dangerous. In this study, we have established a cloud database of herb–drug interactions including common danger information and helpful suggestions. On the front page of our cloud database, users can simply enter the key words such as the herb name or the drug name/international classification code to access the relative information. Doctors and pharmacists can also use this database enquiry system on a daily basis to find out whether their patients have any herb–drug interactions problems. This can ultimately improve the overall drug safety for patients. Furthermore, our system comes with a mobile application (APP) to provide more accessibility.

Keywords Drug–herb interaction · Cloud service · APP

1 Introduction

Orange technology introduced by Jhing-Fa Wang [5] brings up a novel paradigm that the development of technology should improve the happiness of human beings. Especially, the medical care and healthcare are both important research issues in the orange technology. The humans' life has been apparently extended in the recent decades. At the same time, the aged people are usually with some chronic diseases.

L. Feng
Software school, NingXia Polytechnic, Yinchuan, China

J.-L. Chen · S.-P. Tseng (✉)
Department of Computer Science and Entertainment Technology, Tajen University,
No.20, Weixin Rd., Yanpu Township, Pingtung County 90741, Taiwan
e-mail: tsp@tajen.edu.tw

L.-C. Chen
Department of Pharmacy, Tajen University, Pingtung, Taiwan

© Springer Nature Singapore Pte Ltd. 2019
J.-S. Pan et al. (eds.), *Genetic and Evolutionary Computing*,
Advances in Intelligent Systems and Computing 834,
https://doi.org/10.1007/978-981-13-5841-8_14

The probability of multiple diseases is increased in the aged people. Therefore, the aged people usually have multiple drugs to different diseases.

In the East Asia, the herb medicines have been used several thousands of years. The traditional herb medicines have already been researched and approved the effectiveness and limitations of modern technologies. However, many people may mix using the herbs and the drugs. Sometimes, it is fine and may be more effective, but sometimes the mix used is dangerous.

In this study, we have established a cloud database of herb–drug interactions including common danger information and helpful suggestions. On the front page of our cloud database, users can simply enter the key words such as the herb name or the drug name/international classification code to access the relative information. Doctors and pharmacists can also use this database enquiry system on a daily basis to find out whether if their patients have any herb–drug interaction problems. This can ultimately improve the overall drug safety for patients. Furthermore, our system comes with a mobile application (APP) to provide more accessibility.

The remainder of the paper is organized as follows. The related works are in Sect. 2. Section 3 provides a brief introduction about the prototype system. Section 4 demonstrates implementation of the prototype system. Conclusion is given in Sect. 5.

2 Related Works

Drug–drug interaction (DDI) is generally used to describe the pharmacokinetics and pharmacodynamics interaction effect of two or more drugs. Figure 1 shows the possible effect of drug–drug interactions. The DDI may decrease or increase the action of drugs, even cause the adverse effect. These can influence the treatment plan and bring the unexpected result [4]. In [2], the medical cost increased by DDI was analyzed in USA. It is an important issue to provide a rapid and correct DDI query service for reducing the risk of using multiple drugs.

Herb–drug interaction is one special kind of drug interactions between herb medicines and conventional drugs, such as shown in Fig. 2. In Fugh-Berman and Ernst [1], in general, the conventional drugs typically contain only one pharmacologically active ingredient. Because there may be multiple pharmacologically active

Fig. 1 Drug–Drug interactions [4]

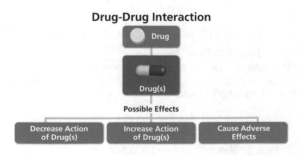

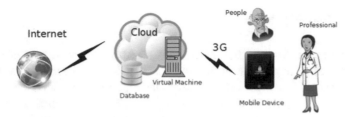

Fig. 2 System design

ingredients in one herb, the herb–drug interactions may be with more probability. Fortunately, the active ingredient density in the herb is usually lower than the conventional drug. However, some herb–drug interactions are clinically significant [3].

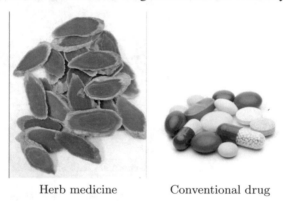

Herb medicine   Conventional drug

3 Design of System

Figure 2 shows the design of proposed system. In software engineering, rapid prototyping means that the prototypes of software applications, i.e., incomplete versions of the software program would be developed in the short time. The prototype system is provided for users' evaluation. The programmers can enhance the prototype according to the users' feedbacks. The improved functions and new functions would be supported in the next version. One version after one version, the prototype system can progressively satisfy the requirements.

In the rapid prototyping process, the users' opinions can be respected entirely. The rapid prototyping method is suitable for developing the inter-discipline application. In this study, the pharmacy and technology experts co-work and intercommunicate to design and implement the system.

4   Implementation

Because our application must be used in PC-based browser and smartphones, the HTML5 is used to implement in this study. HTML5, first released in 2008, is the newest version of HTML. HTML5 improves the support for the newest multimedia; and can be easier to read by humans and Web parsers, such as search engines. HTML5 also provide more interoperability to support complex interactive web application.

The cloud service implemented in this study is shown in http://163.24.224.87: 8081/druginfo.html This service is supported by the cloud system of Tajen University in Taiwan. The main interface is shown in Fig. 3. The user can submit one or two drug names or ATC codes, and the possible DDI result would be as shown in Fig. 4. If some DDI item is clicked, the system can provide more details about this DDI, such as shown in Fig. 5. Especially, the pharmacists can generally want to find the

Fig. 3 The main interface

Fig. 4 The DDI query result

Fig. 5 The details of DDI

forbidden/serious DDI information to adjust the drug used. Figure 4 shows that all the services can be used in the smartphones, including Android and iPhone.

Query result on APP Query detail on APP

5 Conclusion

Herb–drug interaction is one special kind of drug interactions between herb medicines and conventional drugs. In the East Asia, the herb medicines have been used several thousands of years. However, many people may mix using the herbs and the drugs. Sometimes, the mix used is dangerous. In this study, we have established a cloud database of herb–drug interactions including common danger information and helpful suggestions. On the front page of our cloud database, users can simply enter the key words such as the herb name or the drug name/international classification code to access the relative information. Doctors and pharmacists can also use this database enquiry system on a daily basis to find out whether if their patients have any herb–drug interaction problems. This can ultimately improve the overall drug safety for patients. Furthermore, our system comes with a mobile application (APP) to provide more accessibility. In the future, we will use the QR code reader and OCR technology to input the drug/herb information and examine the drug list in the prescription.

Acknowledgements This work was supported by Ministry of Science and Technology, Taiwan, under contact number MOST 106-2622-E-127-002 -CC2.

References

1. Fugh-Berman, A., Ernst, E.: Herb-drug interactions: review and assessment of report reliability. Br. J. Clin. Pharmacol. **52**(5), 587–595. https://bpspubs.onlinelibrary.wiley.com/doi/abs/10.1046/j.0306-5251.2001.01469.x
2. Hamilton, R.A., Briceland, L.L., Andritz, M.H.: Frequency of hospitalization after exposure to known drug-drug interactions in a medicaid population. Pharmacother. J. Hum. Pharmacol. Drug Ther. **18**(5), 1112–1120. https://onlinelibrary.wiley.com/doi/abs/10.1002/j.1875-9114.1998.tb03942.x
3. Hu, Z., Yang, X., Ho, P.C.L., Chan, S.Y., Heng, P.W.S., Chan, E., Duan, W., Koh, H.L., Zhou, S.: Herb-drug interactions. Drugs **65**(9), 1239–1282 (2005). https://doi.org/10.2165/00003495-200565090-00005
4. U.S. Department of Health and Human Services: Drug-drug-interaction. https://aidsinfo.nih.gov/understanding-hiv-aids/glossary/213/drug-drug-interaction, 01 Sept 2018
5. Wang, H.Y., Chen, B.W., Bharanitharan, K., Wu, J.S., Tseng, S.P., Wang, J.F.: Human-centric technology based on orange computing. In: 2013 International Conference on Orange Technologies (ICOT), pp. 250–251 (2013)

Part IV
Monitoring Applications and Mobile Apps

Design and Implementation of the Health Monitor for Aged People

Zhong-Jie Liu and Shih-Pang Tseng

Abstract The humans' life has been apparently extended in the recent decades. However, the cost to maintain the aged-people health is still growing up continually. It has been the most important issue how to apply the modern technology to construct a sustainable and affordable health maintenance organization (HMO). In this paper, we propose a work which applies the Internet of things (IOT) and connects to the community monitor center to improve the healthcare quality to the aged people. The system consists of mainly the coordinator and community monitor center. This system can monitor the health status of the elderly community. Combined with the healthcare professionals, it can provide the reasonable food and exercise plan for the aged people.

Keywords Healthcare · IOT · Smart house

1 Introduction

The concept of orange technology [6] is that the development of technology should be for the happiness of human beings. Especially, the medical care and healthcare are both important research issues in the orange technology. The humans' life has been apparently extended in the recent decades. The cost to maintain the aged-people health is still growing up steadily and continually. [1] It has been the most important issue how to apply the modern technology to reduce the healthcare cost and to construct a sustainable health maintenance organization.

Z.-J. Liu
School of Software, Changzhou College of Information Technology,
Changzhou, China

S.-P. Tseng (✉)
Department of Computer Science and Entertainment Technology, Tajen University,
No.20, Weixin Rd., Yanpu Township, Pingtung County 90741, Taiwan
e-mail: tsp@tajen.edu.tw

© Springer Nature Singapore Pte Ltd. 2019   139
J.-S. Pan et al. (eds.), *Genetic and Evolutionary Computing*,
Advances in Intelligent Systems and Computing 834,
https://doi.org/10.1007/978-981-13-5841-8_15

The healthcare resources, such as nursing professionals and medicine supply, are not always enough to satisfy the increasing requirement. To improve the efficiency of using healthcare resources has become an important research topic in the healthcare domain.

House is the major space of many human activities, such as living and working. Especially, the aged people have more time in the house. In the recent decades, *smart house* almost means the technologies which are used to provide more and better living facilities in the house. The aged people have more requirements about more convenient, comfortable, and healthy living. In this paper, we propose a work which applies the Internet of things (IOT) and connects to the community monitor center to improve the healthcare quality to the aged people.

The remainder of the paper is organized as follows. The related works are in Sect. 2. Section 3 provides a brief introduction about the prototype system. Section 4 demonstrates implementation of the prototype system. Conclusion is given in Sect. 5.

2 Related Works

In the first decade of the twenty-first century, the development of sensor and communication technologies enables the interconnecting among numerous and different sensors and actuators. This integration of sensor and communication technologies is called as Internet of Things (IOT) [2]. Several heterogeneous wired and wireless networks, such as ZigBee, Wi-Fi, and Bluetooth, may be used in the IOT construction. The concept of IOT has been quickly applied to the smart house, factory, campus, and so on.

Smart house, sometimes known as *House automation*, is applying new technologies, such as IOT, to make the housework and household activities more convenient, comfortable, healthy, green, economical, and so on. In order to integrate the house devices and services effectively and efficiently, a middleware framework is proposed by Helal et al. [3]. This middleware framework defines each sensor and actuator as various services in smart house. Tseng et al. [4] have been shown one smart house prototype based on Zigbee network and motion detection. This prototype provides the automatic control of the lighting, air-conditioning, and door locks. The health care is an important issue in the smart house domain. An agent-based architecture, proposed in [5], supports real-time monitoring of various health parameters and protects the privacy in the sharing information.

3 Design of System

Figure 1 shows the architecture of proposed system. First, the wireless Zigbee sensors correct the health data, such as blood pressure and blood oxygen. The coordinator is designed to control these sensors and the Zigbee network in the house.

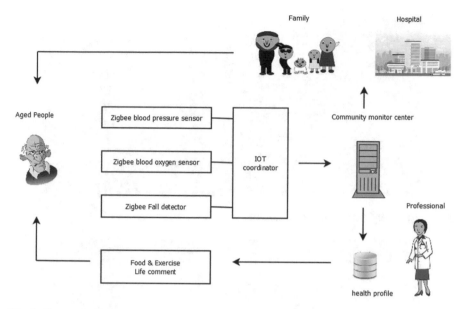

Fig. 1 System architecture

The coordinator transfers all the health data to the community monitor center. The community monitor center generates the health profile of the aged people to the healthcare professional. The healthcare professional can give some food/exercise life comments according to the health profile in order to the health status of the aged people in the long term. On the other hand, the emergency information extracted from the health data, like the fall situation, will be sent to the hospital and the family.

4   Implementation

The two main modules, the coordinator and the community monitor, are implemented in this study. The coordinator is implemented by the TI CC2530 chip. The CC2530 is a true system-on-chip (SoC) solution for Zigbee applications with low cost. Due to the limitations of the SoC processing capacity, there still is one PC in the house to provide the user interface and the Internet connectivity, as shown in Fig. 2. In the next stage, we will use more powerful SoC, such as MTK 7688, to directly provide the user interface and the Internet connectivity in one box.

The community monitor is implemented by the C# and SQL Server 2008 database. In addition, we use the VR-like to show the real-time health status of the aged people monitored in this community. Figure 3 shows the operational status of the community monitor.

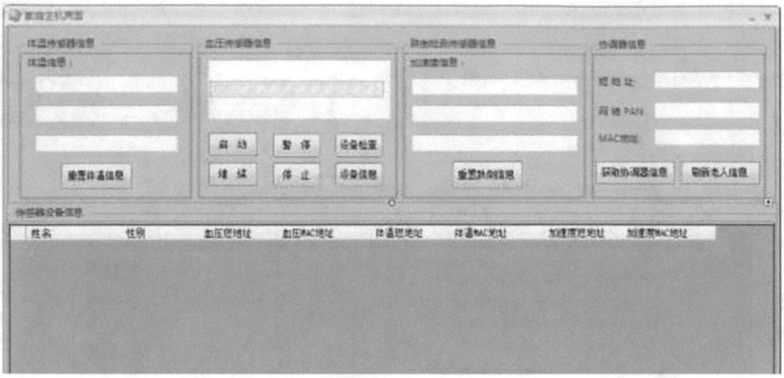

Fig. 2 The main interface of the coordinator

Fig. 3 The main interface of the community monitor center

5 Conclusion

The humans' life has been apparently extended in the recent decades. The cost to maintain the aged-people health is still growing up continually. It has been the most important issue of how to apply the modern technology to reduce the healthcare cost and to construct a sustainable health maintenance organization. In this paper, we propose a work which applies the Internet of things (IOT) and connects to the community monitor center to improve the healthcare quality to the aged people. The system consists of mainly the coordinator and community monitor center. This system can monitor the health status of the elderly community. Combined with the healthcare professionals, it provides reasonable food and exercise plan for the aged people.

In the future, we will use the more powerful SoC to improve the coordinator in order to reduce the complexity and cost. In addition, we will apply the technology acceptance model to verify the effectiveness of the proposed system.

Acknowledgements This work is supported by Top-notch Academic Programs Project of Jiangsu Higher Education Institutions (TAPP) under Grant No. PPZY2015A090.

References

1. Gregersen, F.A.: The impact of ageing on health care expenditures: a study of steepening. Eur. J. Health Econ. **15**(9), 979–989 (2014). https://doi.org/10.1007/s10198-013-0541-9
2. Gubbi, J., Buyya, R., Marusic, S., Palaniswami, M.: Internet of things (IoT): a vision, architectural elements, and future directions. Future Gener. Comput. Syst. **29**(7), 1645–1660 (2013). http://www.sciencedirect.com/science/article/pii/S0167739X13000241
3. Helal, S., Mann, W., El-Zabadani, H., King, J., Kaddoura, Y., Jansen, E.: The gator tech smart house: a programmable pervasive space. Computer **38**(3), 50–60 (2005)
4. Tseng, S., Li, B., Pan, J., Lin, C.: An application of internet of things with motion sensing on smart house. In: 2014 International Conference on Orange Technologies, pp. 65–68 (2014)
5. Wahaishi, A., Samani, A., Ghenniwa, H.: Smarthealth and internet of things. In: Geissbühler, A., Demongeot, J., Mokhtari, M., Abdulrazak, B., Aloulou, H. (eds.) Inclusive Smart Cities and e-Health, pp. 373–378. Springer International Publishing, Cham (2015)
6. Wang, H.Y., Chen, B.W., Bharanitharan, K., Wu, J.S., Tseng, S.P., Wang, J.F.: Human-centric technology based on orange computing. In: 2013 International Conference on Orange Technologies (ICOT), pp. 250–251 (2013)

A Uniform Methodology for Mobile and Desktop Web Applications

Chi-Tsai Yeh and Ming-Chih Chen

Abstract The popularity of mobile applications is high and continues to grow. These applications provide users with numerous services for entertainment, leisure, finance, study, life, and productivity, for example. However, the service providers of mobile applications also provide these services on other platforms, such as desktop-based web browsers. Thus, they suffer from the huge effort required to develop, maintain, and deploy their services in a large number of versions and formats over different platforms, from personal computers to mobile devices. To overcome such issues, this paper proposes a uniform development methodology for mobile/desktop applications for institutional repository applications to efficiently reduce effort. The paper demonstrates a single-page application using web technology to develop the desktop and mobile web applications simultaneously.

Keywords Cross-platform development · Web technology ·
Single-page applications

1 Introduction

As mobile applications become increasingly popular and the related technologies move into mainstream development, desktop applications still exist for the original services. Currently, there are many existing proposed frameworks to standardize cross-platform implementation.

Web technology is one of the accessible solutions to address this challenge for various devices, including smart phones, tablets, and even desktop computers. However, the user experience for desktops and mobile devices is substantially different.

C.-T. Yeh (✉) · M.-C. Chen
National Kaohsiung University of Science and Technology, Kaohsiung, Taiwan
e-mail: yehchitsai@gmail.com

C.-T. Yeh
Shih-Chien University, Kaohsiung, Taiwan

© Springer Nature Singapore Pte Ltd. 2019
J.-S. Pan et al. (eds.), *Genetic and Evolutionary Computing*,
Advances in Intelligent Systems and Computing 834,
https://doi.org/10.1007/978-981-13-5841-8_16

Software designers cannot use one set of code with different devices because of different screen sizes, input interfaces (such as a mouse and a touch panel), and operating behaviors (such as click and sweep).

To respond to this need, this paper demonstrates a uniform development methodology for institutional repositories (IRs). The latest web technologies, such as responsive web design (RWD), front-end ModelViewController (MVC), and back-end MVC, are combined with the proposed uniform development methodology. To achieve the goal in this work, the RWD uses Bootstrap, the front-end MVC uses AngularJS, and the back-end MVC adopts CodeIgniter.

The remainder of this research is organized as below. Section 2 describes related research concerning the technologies of cross-platform development, and web development. Section 3 depicts the concept of our proposed methodology and the system architecture. Section 4 demonstrates the implementation of our proposed solution. Finally, we summarize the conclusion in Sect. 5.

2   Related Works

The first problem is cross-platform development and the other is little modification across different platforms. We review the mobile development trend and the cross-platform technologies to find out the possible solutions in Sect. 2.1. Afterward, we summarize a variety of frameworks and tools out in Sect. 2.2 and the web front-end techniques in Sect. 2.3.

2.1   *Uniform Development Approaches for Multiple Mobile Platforms*

Hartmann et al. [1] arranged three possible solutions to overtake the problem of multiple platforms. First, this approach applies on compilation time. A cross-compiler builds the different execution application for the target operating system (OS). It decouples source code from its OS effectively. The second technique is to apply on a virtual machine (VM). It emulates the target application environment including OS on the multiple platforms. The most popular approach is to adopt web technologies to develop the application. The general web technologies consist of hypertext markup language (HTML), Cascading Style Sheets (CSS), and JavaScript. The same application runs on the standard web browser over the different platforms and looks as a native application.

The first approach, cross-compilation, possesses the advantages of performance, user experience, and arbitrarily accessing the resources of the devices, like memory, storage and 2D/3D graphics accelerators, and sensors. It is an adequate solution to CPU-intensive, highly interactive, or visually rich applications such as real-time interactive and augmented reality (AR). However, the technique of web application provides simplified deployment and immediate availability to apply for common content-based applications such as booking, mobile banking, e-books, e-mail, and so on.

The web approach has two types of implementation, a standalone mobile web browser and a native application within an embedded web browser. For example, the rendering engine of native application like the WebKit [2], which powers the iPhone and Android mobile browsers, acts as a bridge between web applications and device OS in the hybrid model. The communication between the web application and the native application usually uses JavaScript via custom-built application program interfaces (APIs). The hybrid technique brings the better solution: flexibility of web applications with system performance and feature richness of native applications.

2.2   Types of Cross-Platform Framework

According to make a clear distinction of what the cross-platform frameworks offer and their capabilities, Hartmann et al. [1] divided the cross-platform frameworks into the four high-level groupings as follows.

- Library: provide a self-contained toolkit with some specific functionality to the developer.
- Framework: offer a comprehensive toolkit, which includes a set of libraries, components and architecture guidelines, to develop/build/test a mobile application.
- Platform: provide a total solution with a set of frameworks, dependent tools, and services that allows the developer to build a mobile application and to configure, package, and distribute it to product market.
- Product/Service: provide a mature and self-contained service for the developer to integrate into a mobile application.

Apache Cordova [3] with an open-source mobile development framework allows the web developers to build their applications with the same code for different platforms. It adopts the common standard web languages such as JavaScript, HTML, and CSS, instead of Objective-C/Swift of iOS and Java/Kotlin of Android for cross-platform development.

Figure 1 shows a high-abstract view of Cordova application architecture. There are several components consisting of web app, HTML rending engine, and plugins to a Cordova application. Cordova provides a set of native application wrappers working on all major mobile platforms. The Web app part calls JavaScript functions via Cordova APIs and HTML APIs to access mobile resources in Cordova Plugins part such as storage, geolocation, contacts, etc. Due to its flexibility, straightforward architecture, and open source, it is quite popular among mobile developers mainly.

However, PhoneGap is also a well-known distribution of Apache Cordova. The developers can think that Apache Cordova is the build engine that supports PhoneGap, similar to how WebKit is the web render engine that supports Chrome or Safari.

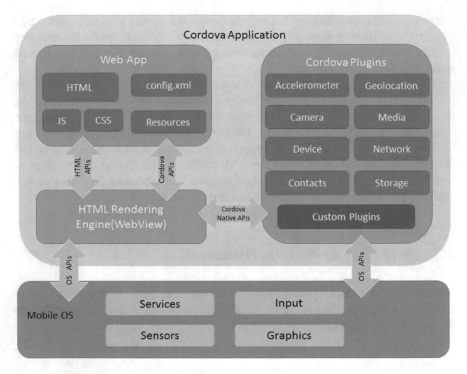

Fig. 1 The architecture of Cordova application

Adobe Systems adopts PhoneGap to provide Adobe PhoneGap Build service that uses one code to build the separate package from their own operating system in the cloud automatically. Therefore, the developers can write code once and run the application on any mobile devices [4–6].

2.3 Web Front-End Techniques

The disadvantage of Apache Cordova is lack of support for native user interface (UI) components. The famous UI frameworks like Sencha Touch [7], jQuery mobile [8], and Twitter Bootstrap [9] offer a supplement for the development of web applications. Specially, Bootstrap easily and efficiently scales your project with one code base, from phones to tablets to desktops.

On the other hand, AngularJS has introduced the technique of dynamic web development. This framework applied the concepts of two-way data-binding, routing, templating, front-end MVC, single-page application (SPA), etc. These concepts rescue the developers from writing boilerplate, so the developers could pay more effort on the core parts of final product [10].

3 The Proposed Uniform Development Methodology for Mobile/Desktop Applications

This section introduces the proposed method to develop web applications in mobile and desktop platform simultaneously. Figure 2 explains the deployment of front-end and back-end applications. For desktop user, the developers do not install any program on the user desktop/platform. The desktop users can access web server directly by using the web browsers (WB) such as Internet Explorer, Chrome, or Safari on Windows/Mac OS through personal computers, laptop, and even large screen LCD.

When the developers package this application as mobile app by using Apache Cordova and publish this app in app store such as Apple App Store, Google Play Store, or Windows Phone Store, the mobile app users access the web server via the mobile application downloaded from app store. Apache Cordova provides custom browser (CB), Webkit, as user interface to run on the various Operating Systems (OSs) such iOS and Android and devices.

The developers design the portal page (PP) as the first page for the users, write the developing JavaScript/CSS files (DevJS), and include the third party JavaScript/CSS files (ThirdJS) to enhance and improve the visual effect and functionality of the application. Apache Cordova provides the device-related JavaScript files (DrJS) to the developers to manage the mobile devices.

The developers store the portal page (PP), the developing JavaScript/CSS files (DevJS), the third party JavaScript/CSS files (ThirdJS) and HTMLs (views) in the web server shown in the left of Fig. 2 and stores the portal page (PP), the developing JavaScript/CSS files (DevJS), and the device-related JavaScript files (DrJS) shown in the right-bottom of Fig. 2. The PP and DevJS are the same between the web server and mobile device. All data and UI are sent from the web server whether the desktop user or the mobile app user.

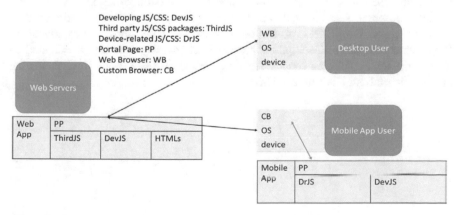

Fig. 2 The proposed methodology of the mobile/desktop development

...

```
switch(window.location.protocol) {
case 'http:':
case 'https:'://remote file over http or https – desktop user
        BASE_URL = location.href; // get_remote_baseurl()
        break;
case 'file:'://local file system executed – mobile app user
        BASE_URL = 'http://remote_web_server/IR/'; //set_remote_baseurl()
        // dynamice load device-related JS files
        includeJs("cordova.js");
        includeJs("js/index.js");
        break;
```

...

Fig. 3 An example of the pseudo code of the portal page

The last thing the developers need to consider is to solve that the PP should load the HTMLs from the remote web server. It means the developers should resolve cross-origin resource sharing (CORS) on the web server. The example code of the pseudo code of the portal page is shown in Fig. 3.

4 Experimental Result

This section demonstrates a uniform development methodology for institutional repositories (IRs) of university. A real searching result example of searching the related projects of a program actually executed is demonstrated in Fig. 4. Figure 4a, b are executed by a mobile device. Figure 4a is a web-based version and Fig. 4b is an App-based version. Figure 4c is a web-based version which is executed by a desktop computer. As a result, this undergraduate student can get the Android information of relevant topics and project titles. Moreover, the proposed work adopts the RWD which the different equipment have different displays and can be executed on different platform simultaneously. The proposed work designs and implements the application on both Android and iOS.

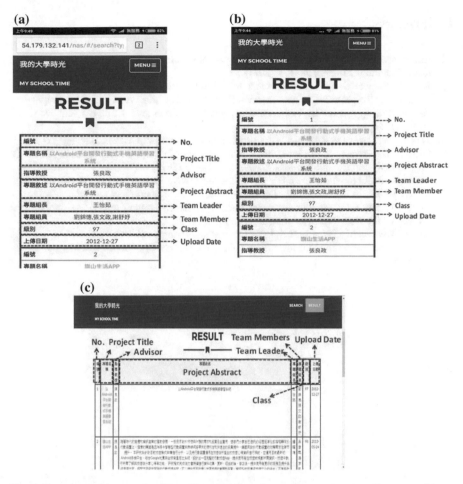

Fig. 4 A real searching result of searching related projects of a program actually executed by a mobile device and a desktop computer. **a** A web-based version executed by a mobile device; **b** an app-based version executed by a mobile; **c** a web-based version executed by a desktop computer

5 Conclusion

In this paper, the institution repository application demonstrated an efficient method to develop and maintain the application across the desktops and mobile devices. Moreover, we completed the package process to Google play/Windows phone store and the end users can download the mobile part of this application from the cloud. As a result, it can assist the end users easy to access the project works via any devices their owned.

2

References

1. Hartmann, G., Stead, G., DeGani, A.: Cross-platform mobile development. Mob. Learn. Environ. (Cambridge) **16**(9), 158–171 (2011)
2. WebKit: The webkit open source project. https://webkit.org/
3. Cordova, A.: Target multiple platforms with one code base. https://cordova.apache.org/
4. Easily create apps using the web technologies you know and love: Html, css, and javascript. https://phonegap.com
5. Take the pain out of developing mobile apps, adobe systems. https://build.phonegap.com/
6. LeRoux, B.: Phonegap, Cordova, and whats in a name? http://phonegap.com/blog/2012/03/19/phonegap-cordovaand-what-e2-80-99s-in-a-name/ (2012)
7. Touch, S.: Build mobile web apps with html5. https://www.sencha.com/products/touch/
8. jQuery Mobile: A touch-optimized web framework. https://jquerymobile.com/
9. Bootstrap: The most popular front-end framework for developing responsive, mobile first projects on the web. https://getbootstrap.com/
10. Fat, N., Vujovic, M., Papp, I., Novak, S.: Comparison of angularjs framework testing tools. In: Zooming Innovation in Consumer Electronics International Conference (ZINC), pp. 76–79. IEEE (2016)

Design of Greenhouse Wireless Monitoring System Based on Genetic Algorithm

Lijuan Shi, Shengqiang Qian, Lu Wang and Qing Li

Abstract Wireless node deployment is a key problem in wireless sensor network design. It has an important impact on network connectivity. In this paper, Arduino as a development platform, using ZigBee technology, sensors and LabVIEW to build a greenhouse environment monitoring system. This paper proposed a model to minimize the number of mobile nodes under wireless node connectivity constraints, and used genetic algorithm to optimize the distribution of mobile nodes. When the number of system nodes was large, the encoding region contraction mechanism based on dichotomy was proposed to improve the optimization speed of genetic algorithm. The upper computer interface of the system was friendly and easy to operate. The real-time data collected by the system was accurate and the system worked stably and was easy for long-term monitoring.

Keywords Genetic algorithm · Monitoring · Arduino · ZigBee · LabVIEW · Sensor

1 Introduction

Wireless sensor network (WSN) is a wireless network composed of a large number of perceptive sensors. The purpose of wireless sensor network is to collect, process, and transmit the monitoring information of perceived objects within the network coverage area. As the sensing range and transmission distance of each node are rather limited, multiple nodes must cooperate to form a network to ensure the network connectivity to complete the monitoring task. Node distribution directly determines the key aspects of normal operation of wireless sensor network. Effective deployment of wireless sensor nodes will solve the coverage problem and improve network connectivity [1].

Wireless sensor network is an application-based network. Different application backgrounds and application characteristics make the distribution of nodes present

L. Shi (✉) · S. Qian · L. Wang · Q. Li
Changzhou College of Information Technology, Changzhou, Jiangsu, China
e-mail: shilijuan@ccit.js.cn

© Springer Nature Singapore Pte Ltd. 2019
J.-S. Pan et al. (eds.), *Genetic and Evolutionary Computing*,
Advances in Intelligent Systems and Computing 834,
https://doi.org/10.1007/978-981-13-5841-8_17

153

different characteristics. There are two strategies for the distribution of nodes in WSN. One is random seeding, that is, nodes are randomly scattered in the monitoring area by means of airplane or projectile launching. Another method is planned placement, that is, the node is placed in the predetermined position by means of controlling the robot or manually installing. Random seeding is mainly used in complex and volatile application environments, such as battlefield monitoring, where the number of nodes must be dropped far more than is actually required. The distribution of these nodes is uncertain. Although the nodes are relatively dense, the network connectivity cannot be guaranteed. Many applications, such as forest monitoring and fire warning, allow pre-optimization of node locations, greatly reducing system costs [2].

The choice of sensor node deployment strategy depends on the specific application, so it is more meaningful to study in combination with the specific application. This paper used wireless sensor network to monitor greenhouse temperature, humidity, light intensity, and soil moisture. We mainly study the problem of wireless node distribution based on genetic algorithm, and improve the genetic algorithm to get better optimization results, so as to improve the stability and full coverage of the system.

2 System Design

2.1 System Design Plan

The application environment of this paper is greenhouse, and the monitoring environment of the measurement and control system is known. The system mainly consisted of wireless nodes and upper computer monitoring software. The wireless terminal nodes were placed according to the monitoring needs. The distribution distance of terminal nodes was generally far away. First, terminal nodes collected environmental parameters such as temperature and humidity, which were forwarded to the coordinator via the multi-hop routing node. Then, the coordinator sent the data to the PC through the serial port. The upper computer built the monitoring software by using LabVIEW graphics programming software. LabVIEW software communicated with the coordinator mainly through the VISA serial port. Finally, remote users could obtain relevant information through the Internet. The wireless sensor network scenario was shown in Fig. 1.

The task of wireless node distribution optimization was to make use of the mobility of routing nodes to dynamically adjust the location of mobile nodes, so that fixed terminal nodes could connect with each other through mobile routing nodes and send the collected data to base stations through multi-hop routing. Under the condition of guaranteeing the connectivity of the whole ZigBee network, this paper mainly used the improved genetic algorithm to optimize the deployment of routing nodes so as to minimize the number of nodes and achieve the purpose of reducing costs. We finally realized the greenhouse environmental monitoring.

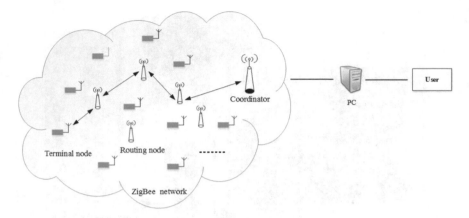

Fig. 1 Scene diagram of wireless sensor network in Greenhouse

2.2 Wireless Node Design

The wireless node was mainly designed by using Arduino and wireless communication module. The wireless node adopted Arduino open source hardware which has been developing rapidly in recent years as the control module. The wireless communication module selected the XBee wireless transmission module of DIGI in the United States which supports ZigBee network. According to different system monitoring requirements, wireless terminal node can configure temperature and humidity sensors, illumination sensors, and other different sensors to achieve a variety of parameters acquisition. Data was uploaded to the coordinator using routing node.

Different types of wireless nodes consisted of the same hardware except for different sensor modules. We can set the XBee module as the coordinator, routing node, and terminal node through the X-CTU software. The hardware circuit diagram of wireless node was shown in Fig. 2.

3 Wireless Node Distribution Optimization

3.1 Wireless Node Distribution Optimization Model

Wireless node set $T = \{w_1, w_2, \ldots, w_M, \ldots w_{M+N}\}$, where M was the number of fixed nodes, which was the terminal node in Fig. 1 and N was the number of mobile nodes estimated according to fixed nodes. The mobile node was the routing node in Fig. 1. We should find a subset C < T, so that the whole network is connected. C includes all fixed nodes and C is required to have the least number of nodes. We need to find a subset of C belonging to T to make the whole network connected. We

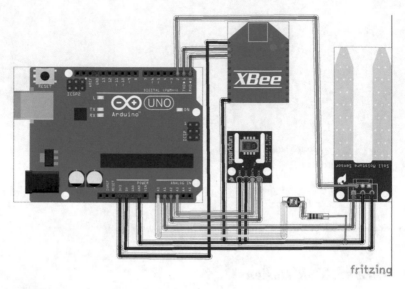

Fig. 2 The hardware circuit diagram of wireless node

required that set C include all fixed nodes, and C should have the least number of nodes [3, 4].

Problem description:

(1) The monitoring area was a two-dimensional plane: $\{S(x, y): 0 \leq x \leq \text{Width } 0 \leq y \leq \text{Length}\}$, parameter length was the length of the monitoring area, and parameter width is the width of the monitoring area.

(2) The number of fixed nodes was M, the parameters of each fixed node were the same, and the coordinates were known.

(3) According to the size of the monitoring area and the number of fixed nodes, N identical mobile nodes were estimated to ensure the connectivity of the whole network. Mobile nodes were represented by (x, y).

(4) The distance between two nodes was represented by dij, and RSSI parameter was used to measure the distance between nodes.

(5) Assume that the transmission range of each fixed node and mobile node was the same. The transmission range of wireless nodes was a circle with node coordinates as the center and r as the radius.

(6) All nodes were regarded as an undirected graph, and A was used to represent adjacency matrix of undirected graph. When $d_{ij} \leq r$, the wireless nodes i and j can communicate directly, $A_{ij} = 1$. When $d_{ij} > r$, $A_{ij} = 0$.

$$A = \begin{cases} A_{11}, A_{12}, A_{13}, \ldots, A_{1j}, \ldots A_{1M+N} \\ A_{21}, A_{22}, A_{23}, \ldots, A_{2j}, \ldots A_{2M+N} \\ A_{31}, A_{32}, A_{33}, \ldots, A_{3j}, \ldots A_{3M+N} \\ \cdots \\ A_{i1}, A_{i2}, A_{i3}, \ldots, A_{ij}, \ldots A_{iM+N} \\ \cdots \\ A_{T1}, A_{T2}, A_{T3}, \ldots, A_{Tj}, \ldots A_{TM+N} \end{cases} \tag{1}$$

In Formula 1, $0 \le i \le M + N, 0 \le j \le M$.

When using genetic algorithm to optimize nodes, the coding length should be fixed, that is, the number of mobile nodes should be determined beforehand. A large number of nodes were used for node optimization. When the nodes were redundant, the coordinates of the nodes coincided. At this point, multiple overlapping nodes can be regarded as one node, so the optimization problem with the minimum number of nodes can be transformed into the maximum problem of the number of overlapping nodes. In this way, the objective function was optimized by formula 2.

$$f_1 = \sum_{i=0}^{M+N-1} \sum_{\substack{j=0 \\ i \ne j}}^{M+N-1} (e^{-d_{ij}^2/b_D})/b_{M+N} \tag{2}$$

In formula 2, d_D was the base value of distance to determine the degree of overlap between two mobile nodes. The smaller the d_{ij}, the larger the value of f_1 and the greater the coincidence degree. The smaller the d_{ij}, the more accurate the overlapping judgment. b_{M+N} was the base value of the number of overlapping nodes, and its value depended on the entire monitoring area and the number of terminal nodes.

Penalty function method can be used for connected constraints. The penalty function is shown in formula 3.

$$f_2 = e^{-(P_{MN}-(M+N)^2)/b_{MN}} \tag{3}$$

In formula 3, P_{MN} was the number of elements in the adjacency matrix A that were not equal to 0. b_{MN} was a base value, depending on the size of $(M + N)^2$. when all wireless nodes in the network could connect to other nodes via mobile routing nodes, $P_{MN} = (M + N)^2$. At this point, the f_2 value equaled to the maximum value 1, which ensures the network connectivity. Thus, the actual number of routing nodes to be placed was equal to the pre-estimated number of mobile nodes N subtracted the number of overlapping nodes.

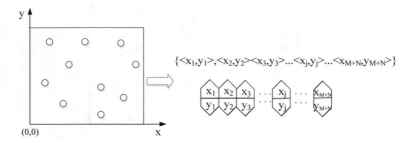

Fig. 3 Distribution coding mapping for wireless nodes

3.2 Selection of Fitness Function

The distribution optimization of wireless sensor network nodes was a multi-objective optimization problem, and a balance should be achieved between the number of wireless nodes and network connectivity. The total fitness function of this paper was the transformation weighted sum of the above sub-objective functions normalized. The total fitness function was shown in formula 4.

$$f = w_1 f_1 + w_2 f_2 \tag{4}$$

In formula 4, w_1 and w_2 were the weights corresponding to sub-objective functions, and w_1 and w_2 depended on the comprehensive requirements of network indicators. w_1 and w_2 should satisfy $w_1 + w_2 = 1$ conditions.

Aiming at the distribution problem of nodes in the wireless sensor network, a coding scheme based on node coordinates was adopted, and the position of the node in the target area was replaced by coordinates. We assumed that the chromosomes of each individual in the genetic algorithm space were made up of M + N genomes. Each genome represented the location of a node. The two genes in the genome represented the x-axis and y-axis coordinates of a wireless node, respectively. Coordinate (0,0) was the coordinates of the coordinator. The coding map of the wireless node distribution was shown in Fig. 3.

3.3 Selection Operation

Selection operation was to select some nodes from the wireless nodes according to a certain probability and used them as parents to reproduce offspring. The selection probability of general node was directly proportional to its adaptive value. The commonly used selection operators were fitness ratio method, random traversal sampling method, local selection method, and roulette selection method. In the roulette selection method, the selection probability of each individual was proportional to its fitness value, which was the simplest method. We used the roulette selection method. The roulette selection method calculated the individual's selection probability according to formula 5.

$$P_i = \frac{f_i}{\sum_{i=1}^{L} f_i} \tag{5}$$

In formula 5, f_i was the fitness of the i-th member of the population, and L was the population size.

3.4 Accelerating Genetic Manipulation

In order to accelerate the convergence speed, the method of reducing the number of coding bits should be adopted. However, this method lead to the imprecision of optimization results. In order to solve this problem, this paper adopted the variable interval acceleration genetic algorithm of dichotomy. First, we make an optimization in the initial search interval $[Length, Width]$. After the iteration, the node position coordinates were generated. Then, we used dichotomy to change the search interval to $[Length/2, Width/2]$ and recode. Finally, we ran the GA algorithm again. By accelerating circular iteration in this way, the variation interval of excellent nodes would be gradually reduced, which was closer to the best advantage. Until the optimal node satisfied the system requirements, the whole algorithm was finished.

The monitoring area was a two-dimensional plane: $\{S(x, y) : 0 \leq x \leq Width, 0 \leq y \leq Width\}$. The binary search interval was shown in Fig. 4.

After an evolutionary iteration, the search interval was divided into four compartments based on $Width/2$ and $Length/2$ axes. We determined which cell area each

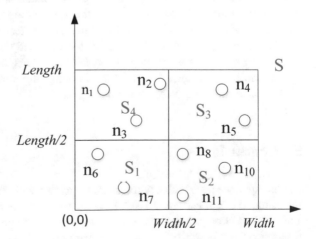

Fig. 4 Dichotomy search interval

Fig. 5 Deloyment of mobile
nodes

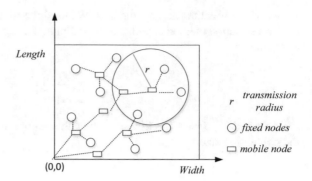

wireless node fell in. We judged which small area each wireless node was located in, and changed the search area of this node to its small area in the next optimization iteration. In this way, we divided the search interval into more cells as needed.

4 Simulation

The wireless node distribution simulation was set in the monitoring area of about 300 * 400 m. According to the need of greenhouse monitoring object, fixed nodes were equipped with temperature and humidity sensor, soil moisture sensor, and luminous intensity sensor. In the monitoring area, 10 fixed nodes were pre-placed, and the transmission radius of each node was 100 m. Through node distribution optimization, we finally selected mobile node deployment as shown in Fig. 5 for system temperature and humidity monitoring.

This system mainly completed the function test of the greenhouse wireless monitoring system in the training room. After the wireless node was powered on, the acquisition command was sent through the upper computer. The coordinator waked each terminal node through the ZigBee network. The terminal node uploaded the sensor data to the coordinator. The coordinator uploaded the data to the host computer monitoring system. After testing, the system could display the sensor data collected by the terminal node normally, and the data changed smoothly. The monitoring results of the upper computer were shown in Fig. 6.

5 Conclusion

In this paper, we have built a wireless monitoring system for greenhouse environment by using XBee, sensor, and LabVIEW of short-distance wireless communication. This system adopted wireless network to monitor the greenhouse environment. We focused on the optimization of the distribution of the routing nodes of the wire-

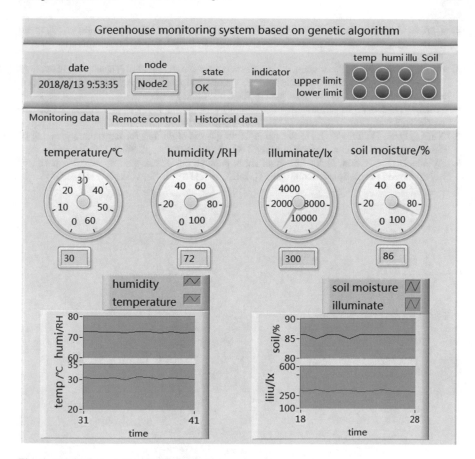

Fig. 6 Monitoring results of upper computer

less monitoring system by genetic algorithm, which enabled the ZigBee network to achieve connectivity, so as to better monitor the greenhouse environment.

Acknowledgements This work is supported by the Changzhou University higher vocational education research project under grant CDGZ2018047, Teaching reform of higher vocational education of CCIT under grant 2018CXJG10, University philosophy social science research fund project of Jiangsu Province under grant 2017SJB1822.

References

1. Gupta, S.K., Kuila, P., Jana, P.K.: Genetic algorithm approach for k -coverage and m -connected node placement in target based wireless sensor networks[J]. Comput. Electr. Eng. **56**, 544–556 (2015)

2. Tian, J., Gao, M., Ge, G.: Wireless sensor network node optimal coverage based on improved genetic algorithm and binary ant colony algorithm[J]. Eurasip J. Wirel. Commun. Netw. **2016**(1), 104 (2016)
3. Singh, A.K., Debnath, S., Hossain, A.: Efficient deployment strategies of sensor nodes in wireless sensor networks[J]. In: International Conference on Computational Techniques in Information and Communication Technologies, pp. 69–73. IEEE (2016)
4. Fouchal, H., Hunel, P., Ramassamy, C.: Towards efficient deployment of wireless sensor networks[J]. Secur. Commun. Netw. **9**(17), 3927–3943 (2016)
5. Xu, G., Plets, D., Tanghe, E., et al.: An efficient genetic algorithm for large-scale planning of dense and robust industrial wireless networks[J]. Expert Syst. Appl. **96**, 311–329 (2018)
6. Ayinde, B.O., Hashim, H.A.: Energy-efficient deployment of relay nodes in wireless sensor networks using evolutionary techniques[J]. Int. J. Wireless Inf. Networks **3**, 1–16 (2018)

Design and Implementation of Microbiological Fermentation Monitoring System

Jing Wang, Jun Wang and ShengXia Liu

Abstract A method of real-time monitoring of microbial fermentation system based on King-View software is presented in this paper. The paper introduces the hardware platform built with temperature sensor, pH sensor, DO dissolved oxygen sensor, and MSP430 chip. The field sensor signals are sent to the MSP430 processor through the collection circuit, and the analysis and processing are carried out. In the end, the host computer monitoring system developed by King-View software can be used to collect, display, and preserve the various parameters of the fermentation process in real time, and meet the precise control requirements in the fermentation process, so as to ensure the quality of the fermentation products.

Keywords Microbes · Sensors · MSP430 · King-View

1 Introduction

Microbial engineering, also known as microbial fermentation engineering, is a new technology that uses some biological functions of microbes to produce useful biological products for human beings, or microbes can be used directly to participate in and control certain industrial processes. Fermentation industry is a technology-intensive industry, which involves microbiology, biochemistry, chemical industry,

Fund projects: Construction Project of Brand Major in Jiangsu Universities (Project number: PPZY2015C237)

J. Wang (✉) · S. Liu
Changzhou College of Information Technology, Changzhou 213164, China
e-mail: 1920197865@qq.com; 13776859616@qq.com

S. Liu
e-mail: 9280818822@qq.com

J. Wang
Tiandi (Changzhou) Automation Co., Ltd, Changzhou 213015, China
e-mail: wangjun_info@yeah.net

automatic control technology, and computer technology. In the fermentation industry, the fermenter has developed from several liters to tens of tons, now hundreds of tons, or even thousands of tons. For such a large fermenting tank, if the operation is not properly controlled, it will cause a great economic loss of [1]. At present, the fermentation industry in China is in the key stage of upgrading and transformation. How to accurately control the process of fermentation reaction, improve the efficiency of reaction, and reduce the consumption of resources is the future trend of development [2].

Microbial fermentation is a complex biochemical reaction process. In order to improve the yield of the final product, we must ensure that there is a suitable microbial growth and metabolic environment in the whole process. On this basis, the control of the fermentation process must be optimized by [3]. Therefore, the control of fermentation environment is a very important link in the whole fermentation industry. In this paper, the microchip control board and King-View software are used to monitor the microbial fermentation system. This system is not only economical, simple and easy to use but also convenient for maintenance and reliable operation [4].

2 The Overall Realization Scheme

The fermentation process is more complex and involves the growth and metabolism of microorganism cells. It is a dynamic process with time-varying, randomness, and variability input and output. Therefore, in the process of fermentation, the main variables related to mass, such as temperature, stirring speed, pH value, oxygen amount, aeration amount, foam, and so on are used as the control [5]. This paper mainly focuses on temperature, PH, and dissolved oxygen.

- Temperature control: temperature control has the most suitable growth temperature for specific microorganisms. Therefore, temperature control is a very important growth environment parameter during microbial fermentation.
- PH control: pH value is another important environmental parameter for microbial growth. In the fermentation process, it must be strictly controlled, otherwise, it will seriously affect the microbial metabolism and the synthesis of metabolites.
- Control of dissolved oxygen concentration: oxygen is a necessary raw material for microbial growth. If oxygen supply is insufficient, the growth of microbes is seriously affected. In the process of fermentation, a certain amount of solution should be maintained. The main parameters affecting the DO value are air volume, stirring speed, and tank pressure.

After measuring data from all kinds of sensors, the A/D conversion module is used to receive and process data, and then through the RS485 serial communication mode to the monitoring system of the host computer to realize the real-time collection and query of the parameters of the fermentation system. The overall structure diagram of the system is shown in Fig. 1.

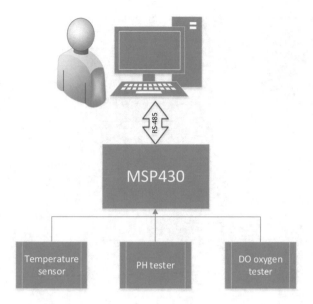

Fig. 1 Overall structure diagram of the system

3 Hardware Design

3.1 Design of Main Control Chip

The main control chip adopts MSP430 series single-chip microcomputer, which was launched by TI in 1996. MSP430 series single-chip microcomputer is a 16-bit single chip microcomputer, which adopts the reduced instruction set (RISC) structure and has rich addressing methods. This series of single chip not only has the advantages of strong processing ability, fast operation speed, and low power consumption, but also has the advantages of rich in—chip resources, convenient, and efficient development environment, and can meet the application of most data acquisition [6]. The minimum system of this single chip microcomputer is shown in Fig. 2.

3.2 Communication Module Design

RS485 bus communication mode is widely used in industrial control system, especially in small and medium data acquisition and control system, because of its simplicity and flexibility, simple hardware interface, easy realization of software, high cost performance, far transmission distance, low error rate, and strong anti-interference ability [7].

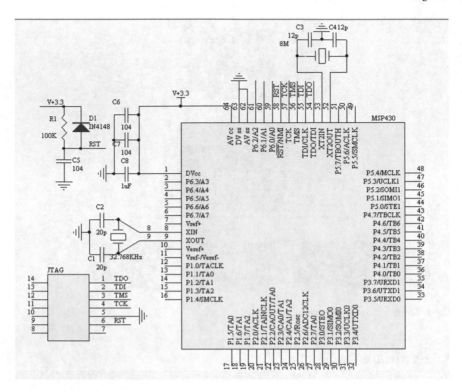

Fig. 2 Minimum system circuit diagram of single-chip microcomputer

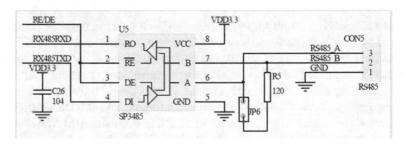

Fig. 3 RS485 module hardware schematic diagram

This monitoring system uses industrial computer (IPC), industrial computer access RS485 bus network through serial port RS232–RS485 bus, select series network structure or tree structure according to field application, and add RS485 router in proper case [8].

This design RS485 module uses SP3485 as the transceiver, which supports 3.3 V power, the maximum transmission speed up to 10 Mbit/s, support up to 32 nodes, and the output short circuit protection, the hardware circuit, as shown in Fig. 3 [9].

3.3   Selection and Application of Sensors

The temperature measurement module of this system adopts DS18B20 series digital temperature sensor. The sensor has the characteristics of small size, strong anti-interference ability, and high accuracy [9]. The packaged DS18B20 can be used in cable trench temperature measurement, blast furnace water cycle temperature measurement, boiler temperature measurement, temperature measurement in the machine room, temperature measurement in agricultural greenhouse, temperature measurement in clean room, temperature measurement in ammunition storehouse, and other non-limit temperature occasions. It is easy to use and has various packaging styles. It is suitable for all kinds of narrow space equipment digital temperature measurement and control field [10].

The Ph sensor uses the HQd series portable water quality analyzer of hash company in the United States. The hash HQd portable water quality analyzer has greater measuring flexibility, and the electrode replacement is easy to operate, and the electrode can be automatically identified. The multi-parameter water quality detector contains various Intelli CALTM electrodes, which can be used to measure pH, conductivity, dissolved oxygen (LDO), LBOD, ORP and parameters of sodium, ammonium, ammonia, fluoride, nitrate, chlorine, and so on.

The DO sensor is a DO6400 series of underwater point solution dissolved oxygen sensor, which contains large capacity electrolytic cell, easy to replace oxygen permeable film, and use reliable battery measurement technology to support long time work. Very suitable for aquaculture and biological fermentation monitoring applications, and these dissolved oxygen DO sensors only need minimal maintenance.

4   The Software Design of the Upper Computer

The upper computer part of the system is used by the King-View 6.53 software produced by Wellintech, which is an economical, open, and easy to expand product with short development cycle [11]. The software can be divided into three levels: control level, monitoring level, and management level. The monitoring layer to the lower connection control layer, to the upper connection management layer, it not only realizes the real-time monitoring and control of the site, but also completes the upload, configuration and other important functions in the automatic control system. It can make full use of the graphics editing function of Windows, make it convenient to form the monitoring picture, and display the state of the control equipment by animation. It has alarm window, real-time trend curve, and so on, which can facilitate the generation of various reports. It also has rich device drivers and flexible configuration mode, data link function [12].

The system host computer is mainly divided into four functional modules: user login module, communication equipment, data management equipment, and user

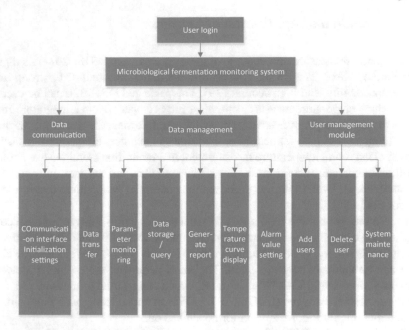

Fig. 4 System software function module

management equipment. Staff can register, log, realize data storage, query, add, and delete functions. The software function module is shown in Fig. 4.

4.1 Main Interface Design

After login, the administrator enters the main interface of the monitoring system, as shown in Fig. 5. Different administrators have different permissions, and senior administrators can set the port's COM port, baud rate, parity check, and other parameters, and can also set the standard value of temperature, dissolved oxygen, and pH value. Ordinary administrators can only perform query data, control switches, and other operations. When the actual temperature in the tank does not conform to the set temperature, the temperature can be adjusted by the cooling and heat making switch. At the same time, the amount of dissolved oxygen and the pH can be adjusted by the control of the stirring motor and the discharge valve, so that the whole fermenting tank is in the best working environment.

Through background database management, the system can also query historical data for various parameters, and facilitate management and analysis. At the same time, in order to analyze the change of parameters and trend more intuitively, the parameters can be presented in the form of curves. When the actual measured temperature is not consistent with the set value, the alarm indicator of the lower right

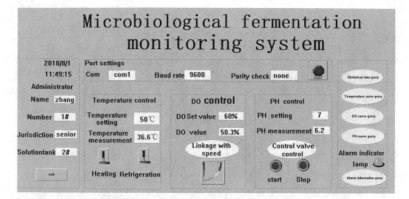

Fig. 5 Design of the main interface of the system

corner of the main interface will send out the alarm signal to remind the administrator to perform the related operation and to inquire the specific alarm information.

4.2 Curve Query Interface

The system can not only receive the value of the Fermor in real time, but the background database can generate the curve through the data sorting and other relevant tools. The administrator can query the temperature trend in a certain period of the tank through the date and the choice of the solution can, so as to make it easier to observe and analyze the numerical value more intuitively. Figure 6 is the temperature curve timetable of the 1# fermentor on August 1, 2018. At the same time, the administrator can carry out a curve query of the pH value and the amount of DO dissolved oxygen through other related operations.

4.3 Alarm Interface

Before the actual production, the standard parameter range will be set and stored in the database. When the measured value does not conform to the production requirements, the alarm indicator of the main interface will send out the red alarm signal. The administrator can query the detailed alarm information by clicking the alarm information inquiry button. As shown in Fig. 7, the alarm information can be observed according to the date and solution tank, and the corresponding parameters such as alarm time, date, limit value, and priority can be determined.

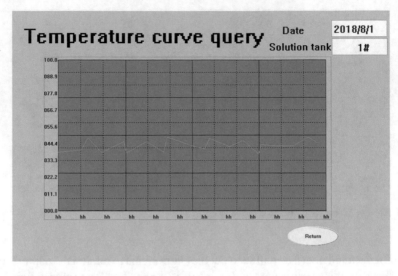

Fig. 6 Temperature curve query interface

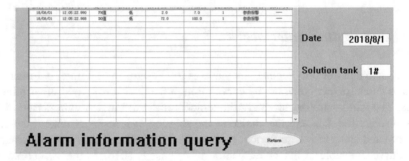

Fig. 7 Alarm information interface

5 Summary

With the development of microbial fermentation technology, more and more attention has been paid to the research and development of universal and miniaturized biological fermentation equipment [13]. The design and implementation of the microorganism fermentation monitoring system proposed in this paper, based on the embedded system as the core, combined with sensor technology, and used the King-View software platform to develop the upper computer monitoring interface. It can measure the parameters in each fermentor timely and accurately, understand the production status at this time, and control the chemical workshop remotely according to the production requirements. At the same time, using the automation control and information management, the workers on duty in the production workshop are free from the boring work, and the work efficiency is improved, and it has a strong practical [14].

References

1. Chen, J., Liu, L., Du, G., et al.: Optimization Principle and Technology of Fermentation Process, pp. 1–5. Beijing, Chemical Industry Agency (2009)
2. Ye, Q.: Principle of Fermentation Process, pp. 1–11. Beijing, Chemical Industry Press (2005)
3. Zhang, W., Guo, J.: Research progress of microbial fermentation process optimization. Guangdong Agric. Sci. (6) (2013)
4. Jing, W., Zhang, J., Shao, T.: Designed the on-line monitoring system for bus temperature rise based on WSN. Microprocessor (2) (2014)
5. Yan, X., Chen, Z., et al.: Remote monitoring system for the vibration signal of machine-room-less elevator. Process. Autom. Instrum. **8**, 47–49 (2011)
6. Zhang, J., Xi, W.: MSP series single chip microcomputer and its application. Instrum. Anal. Monit. (5) (2002)
7. Shen, H.: A network protocol based on RS485 bus and its implementation method. Appl. SCM Embed. Syst. **3**(6), 68–70 (2003)
8. Zhang, H., Kang, W.: Design of the data acquisition system based on STM32. In: First International Conference on Information Technology and Quantitative Management. Procedia Computer Science, vol. 17, pp. 222-228 (2013)
9. Zhang, J.: Intelligent temperature sensor DS18B20 and its application. Instrum. Technol. (4) (2010)
10. Ti, Y., Song, F., et al.: Design of NRF905 -based wireless temperature and humidity detection and transmission system. Control. Instrum. Chem. Ind. **4**, 404–407 (2011)
11. Ralstona, P., Grahamb, J., et al.: Cyber security risk assessment for SCADA and DCS networks. ISA Trans. **46**, 583–594 (2007)
12. Beijing Kunlun Tong Automation Software Technology Co., Ltd.: MCGS Embedded Version User Manual (2003)
13. Nan, Z., Yan, X., Dong, H., et al.: Study on real-time monitoring and control system for fermentation process. Meas. Control. Technol. **22**(7), 18–22 (2003)
14. Li, M.: Fermentation Process of Bacillus Subtilis B2 and Its Field Colonization and Growth Promotion Study on the Effect of Disease. Lanzhou, Gansu Agricultural University (2008)

Part V
Image and Video Processing

A Bi-level Image Segmentation Framework Using Gradient Ascent

Cheng Li, Baolong Guo, Xinxing Guo and Yulin Yang

Abstract In order to solve the problem of under-segmentation in traditional super-pixel methods, a new image segmentation framework is proposed, which is based on gradient ascent including Simple Linear Iterative Clustering (SLIC) superpixels and watershed algorithm. First, SLIC method is adopted to generate uniform superpixels, which are then determined whether under-segmentation occurs by a homogeneity criterion. In heterogeneous regions, an adaptive watershed algorithm processes a more precise division based on luminance histogram. Experimental results show that the bi-level framework has good performance on detail-rich regions, without significantly increasing the time complexity compared with conventional SLIC.

Keywords Segmentation · Superpixel · Watershed · Subdivision

1 Introduction

The procedure of image segmentation is to divide an image into several fragments without intersecting. As an integral part of digital image processing, it has been widely applied in machine vision [2], medical applications [5], and video compression [3]. Superpixels [6] can significantly improve the efficiency of segmentation, and it is a homogeneity description of texture, color, and other features in accordance with visual sense, which make it receive much attentions.

Segmentation based on superpixel can be generally divided into two categories: graph-based method and gradient ascent method. In a graph-based algorithm, super-pixels are produced by minimizing a cost function defined over the graph in which

C. Li · B. Guo (✉) · X. Guo · Y. Yang
Institute of Intelligent Control and Image Engineering, Xidian University,
No. 2 South Taibai Road, Shaanxi, Xian, China
e-mail: blguo@xidian.edu.cn

C. Li · B. Guo (✉) · X. Guo · Y. Yang
School of Aerospace Science and Technology, Xidian University,
No. 2 South Taibai Road, Shaanxi, Xian, China
e-mail: licheng0812@gmail.com

© Springer Nature Singapore Pte Ltd. 2019
J.-S. Pan et al. (eds.), *Genetic and Evolutionary Computing*,
Advances in Intelligent Systems and Computing 834,
https://doi.org/10.1007/978-981-13-5841-8_19

each pixel is regarded as a node. Felzenszwalb and Huttenlocher [4] propose an efficient graph-based approach through the minimum spanning tree, which shows relative precise adherence to image boundaries, but the procedure is unconscious and the patches are very irregular. The normalized cuts (Ncut) [9] algorithm minimizes a cost function globally by contour and texture information, and results in uniform superpixels. However, the accuracy and computing efficiency is very low, especially in dealing with large-scale images. Gradient ascent methods propose the idea of clustering and iteratively refining the process until it meets a predefined criterion. SLIC [1] adopts local K-means clustering to all pixels based on color and spatial distance, despite simplicity, it relies heavily on the initial number of superpixels by manual. Watershed [10] is a relative fast segmentation approach based on the topological theory with mathematical morphology, but the amount of superpixels and their compactness is out of control.

In this paper, a bi-level image segmentation framework is proposed, which combines SLIC and watershed. First, the image is partitioned to regular superpixels by SLIC, and a homogeneity criterion is put forward to define under-segmentation. An adaptive watershed algorithm is then proposed to subdivide the misclassified pixels. Finally, a more precise segmentation result is obtained by the two procedures in hierarchical. Experimental results demonstrate that the combination effectively improves the performance of conventional SLIC, and does not significantly increase the processing time.

2   Conventional Gradient Ascent Method

2.1   SLIC Superpixel Method

The principle of SLIC superpixel is very concise to understand, and the overall process contains four major steps, initialization, assignment, update, and postprocessing. The algorithm can be described as follows:

- The only explicit parameter k is assigned manually to determine the grid interval $S = \sqrt{N/k}$, where N is the total pixels in the Lab image to be partition;
- k initial cluster centers are represented as $C_k = [l_k, a_k, b_k, x_k, y_k]$, where C_k is composed of $C_k^c = [l_k, a_k, b_k]$ in color space and $C_k^s = [x_k, y_k]$ in 2D position;
- Each pixel i is assigned a label in accordance with the nearest cluster center C_k based on a distance measure $D(i, k)$ as

$$D(i, k) = \sqrt{\left(\frac{\| C_i^c - C_k^c \|}{N_c}\right)^2 + \left(\frac{\| C_i^s - C_k^s \|}{N_s}\right)^2}$$

where N_c and N_s are two constant to normalize color and spatial proximity respectively, and $\| \cdot \|$ represents the Euclidean distance.

- For each pixel i, a local k-means method is adopted to adjust the center and the labels of pixels in a $2S \times 2S$ region. This procedure goes until all pixels get new label and all centers update to C_k' as

$$C_k' = \frac{1}{n_k} \sum [C_i^c, C_i^s]$$

where n_k is the number of pixels in the cluster centered at C_k. This step is iterated until a predefined time is achieved.
- The isolate clusters are merged to its largest neighbor so that the connectivity between pixels is enforced.

2.2 Watershed Segmentation

Conventional watershed segmentation treats the gradient image as a topographic surface and the algorithm is based on region-growing. The implementation can be described as a flooding process. It detects the minima of gradients image and pixels at the minima will be flooded. Eventually, the image can be partitioned into catchment basins and watershed lines, which correspond to the homogeneous regions in theory.

A marker extraction strategy is introduced to moderate the extreme over segmentation in conventional watershed method [8]. The extracted markers are regarded as minima of gradient image and suppress all other gradient minima, then watershed algorithm is used on the modified gradient image for partition optimization.

In this paper, the strategy is adopted and a new adaptive marker extraction approach is proposed based on luminance histogram.

3 Bi-level Image Segmentation Framework

3.1 Homogeneity Criterion

As mentioned in Sect. 2.1, SLIC superpixel segmentation is fundamentally flawed due to its simplicity. As shown in Fig. 1b, once the conventional SLIC starts with insufficient initial cluster centers, it is more likely to be under-segmentation in some superpixels after iterations. Moreover, if some small clusters merge falsely, there will be many heterogeneous regions. For this problem, a feasible solution is to find the superpixels without uniformity and then subdivide them in a more precise level.

In order to keep the result of SLIC segmentation, and only sift the superpixels mentioned above, a homogeneity criterion is put forward to define under-segmentation by luminance information:

$$H(C_k) = \begin{cases} 1, \ \frac{1}{n_k} \sum_{i \in S_k} \| l_i - l_k \| < \sigma \cdot l_k \\ 0, \ else \end{cases}$$

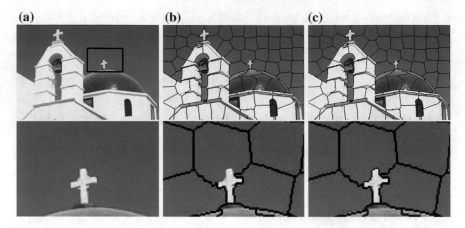

Fig. 1 The comparison of two results. **a** Test image. **b** The conventional SLIC. **c** The proposed method

where S_k represents the cluster centered at C_k, σ is a threshold factor for under-segmentation judgment, and in this paper, $\sigma = 0.15$. Thus, if $H(C_k) = 0$, superpixel labeled k is considered heterogeneous in luminance dimension and needs more precise segmentation.

3.2 Marker-Controlled Watershed Subdivision

The proposed framework adopts an adaptive marker-controlled watershed approach to operate the under-segmentation superpixels once they are selected. The improvement of the adaptive watershed consists in flooding the topographic surface from a previously defined set of markers, which are determined by luminance histogram.

In Fig. 2a, the sky and the cross are labeled in the same in SLIC, which causes this superpixel does not adhere to the boundary, so a histogram based on luminance

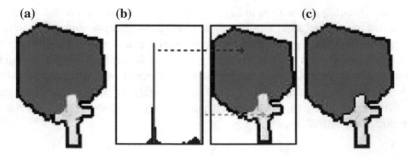

Fig. 2 Marker-controlled watershed subdivision. **a** Under-segmentation region. **b** Adaptive marker extraction. **c** Result of the proposed watershed method

channel is established in this region. As shown in Fig. 2b, if the statistical value of a bin is larger than half of the peak, the pixel which equals the corresponding luminance value is selected as a local minima for flooding. The conventional watershed is implemented and finally the possible subregions are divided precisely in Fig. 2c.

4 Experiment and Analysis

The experiments are performed on the Berkeley Segmentation Data Set and Benchmarks 500 (BSDS500) [7], where the images are all 481321 in size. The conventional SLIC method is used for comparison. Since the proposed framework can improve the under-segmentation by insufficient initial cluster centers, the only parameter is adjusted to control the result.

Figure 3 illustrates the visual comparison of both two algorithms with different k. Each image consists of two results with different k value of 100 and 200 respectively.

Fig. 3 Visual comparison of superpixels produced by conventional SLIC and the proposed method. In each image, the value of k in the upper left and lower right is 100 and 200 respectively. Alternating rows show each segmented image followed by a detail of the center of each image

Table 1 The comparison on processing time (ms)

Test image	$k = 100$		$k = 200$	
	SLIC	Proposed	SLIC	Proposed
Building	93.2	106.6	93.2	98.4
Men	93.1	108.4	93.2	97.3
Animal	93.1	102.9	93.1	96.0

It is obvious that the proposed method has a better performance, because the improved watershed segmentation method can be a supplement to conventional SLIC and subdivide the heterogeneous regions even they are irregular in contour.

Table 1 shows the processing time comparison of both two algorithms. For the three test images, the proposed method does not significantly increase the processing time compared with the conventional method when k is relatively small. On the other hand, if k is in a larger setting which results in over segmentation, the total time is approximately the same since the subdivision step almost does not work.

5 Conclusion

This paper proposes a bi-level framework for image segmentation, which combines SLIC and watershed to improve the performance of superpixel. A homogeneity criterion is put forward to define under-segmentation regions in conventional SLIC, and an adaptive watershed algorithm is introduced to subdivide the regions. After the two procedures in hierarchical, a more precise segmentation result is obtained. Experimental results demonstrate that the combination makes full use of luminance information in the framework and improves the performance on detail-rich regions with little additional time consumption.

Acknowledgements This work is supported by the National Natural Science Foundation of China (61571346). The research is also supported by the Fundamental Research Funds for the Central Universities and the Innovation Fund of Xidian University.

References

1. Achanta, R., Shaji, A., Smith, K., Lucchi, A., Fua, P., Ssstrunk, S.: SLIC superpixels compared to state-of-the-art superpixel methods. IEEE Trans. Pattern Anal. Mach. Intell. **34**(11), 2274–2282 (2012)
2. Arbelaez, P., Maire, M., Fowlkes, C., Malik, J.: Contour detection and hierarchical image segmentation. IEEE Trans. Pattern Anal. Mach. Intell. **33**(5), 898–916 (2011)
3. Boemer, F., Ratner, E., Lendasse, A.: Parameter-free image segmentation with SLIC. Neurocomputing **277**(2018), 228–236 (2017)

4. Felzenszwalb, P.F., Huttenlocher, D.P.: Efficient graph-based image segmentation. Int. J. Comput. Vis. **59**(2), 167–181 (2004)
5. Gao, M., Huang, J., Huang, X., Zhang, S., Metaxas, D.N.: Simplified Labeling Process for Medical Image Segmentation. Springer, Berlin, Heidelberg (2012)
6. Malik, J.: Learning a classification model for segmentation. In; ICCV, vol. 1, pp. 10–17 (2003)
7. Martin, D.R., Fowlkes, C., Tal, D., Malik, J.: A database of human segmented natural images and its application to evaluating segmentation algorithms and measuring ecological statistics. In: Proceedings of International Conference on Computer Vision, vol. 2, no. 11, pp. 416–423 (2001)
8. Salembier, P., Pardas, M.: Hierarchical morphological segmentation for image sequence coding. IEEE Trans. Image Process. **3**(5), 639 (1994)
9. Shi, J., Malik, J.: Normalized cuts and image segmentation. IEEE Trans. Pattern Anal. Mach. Intell. **22**(8), 888–905 (2000)
10. Vincent, L., Soille, P.: Watersheds in digital spaces: an efficient algorithm based on immersion simulations. IEEE Trans. Pattern Anal. Mach. Intell. **13**(6), 583–598 (1991)

Finger Vein Image Registration Based on Genetic Algorithm

Zilong Chen, Wen Wang, Lijuan Sun, Jian Guo, Chong Han and Hengyi Ren

Abstract Finger vein recognition technology is an emerging biometric identification technology that utilizes the distribution structure of venous blood vessels to achieve identification. The vein recognition process is divided into two parts: registration and recognition. In the registration process, the generation of registration template is particularly important. In order to obtain the registration template more accurately, this paper proposes a finger vein image registration algorithm based on genetic algorithm. The principle of the algorithm is to use the mutual information of two finger vein images as the fitness function of the genetic algorithm, and use the genetic algorithm to search for the optimal parameters of the rigid body transformation model. The experimental results show that the algorithm is effective and can achieve the registration of finger vein images within a short iteration.

Keywords Finger vein recognition · Image registration · Genetic algorithm

1 Introduction

The finger vein recognition technology performs the identity authentication of the individual by acquiring the finger vein structure. Due to its many advantages such as noncontact and living body detection, it has gradually become a research hotspot in recent years.

Z. Chen (✉) · W. Wang · L. Sun · J. Guo · C. Han · H. Ren
College of Computer, Nanjing University of Posts and Telecommunications,
210023 Nanjing, China
e-mail: 1029922679@qq.com

W. Wang
e-mail: wangw.www@qq.com

L. Sun · J. Guo · C. Han
Jiangsu High Technology Research Key Laboratory for Wireless Sensor Networks,
Nanjing University of Posts and Telecommunications, 210023 Nanjing, China

© Springer Nature Singapore Pte Ltd. 2019
J.-S. Pan et al. (eds.), *Genetic and Evolutionary Computing*,
Advances in Intelligent Systems and Computing 834,
https://doi.org/10.1007/978-981-13-5841-8_20

The finger vein recognition process mainly consists of two phases: the registration phase and the identification phase. In the registration process, first, we collect the user's finger veins through the shooting equipment. In order to contain more information of the finger vein to improving the accuracy of the recognition, we take three vein images for a registered finger continuously, and then these three images are fused to generate a registration template [1]. The original picture obtained from the shooting equipment contains a lot of noise (such as the device background), which is removed from the original picture in the preprocessing stage [2], and the area containing only the finger is obtained. This effective area is often referred to as the region of interest (ROI) [3]. Since the fingers of the collected person tend to be offset, rotated, etc., even if the ROI of the same finger has a large difference. Therefore, the image must be registered first in order to accurately achieve the fusion, and then get the correct finger vein template.

In terms of image registration, the existing research results are as follows: Sarit Chicotay et al. presented a novel two-phase GA-based image registration algorithm, whose main advantage over existing evolutionary IR techniques is that it provides a robust and automatic solution for a (quasi) fully affine transformation which is one of the most commonly used models for image registration [4]; Jun Zhang et al. proposed a novel coevolution-based coarse-to-fine registration method [5]. Compared with several other algorithms, the proposed algorithm is demonstrated to achieve high accuracy and robustness for various real remote sensing images even with significant gray-scale differences [6].

In the past few decades, many methods have emerged to make image registration more accurate. However, in the field of biometric identification, especially the direction of vein recognition, there is not much literature related to vein image registration based on genetic algorithm. This paper proposes a finger vein image registration based on genetic algorithm. The algorithm uses genetic algorithm to optimize the parameters of the transformation model. Finally, the effectiveness of the algorithm is verified by experiments.

2   Finger Vein Image Registration Related Principle

2.1   *Problem Model*

The essence of finger vein image registration is to use a transformation model (such as rigid body transformation, affine transformation, etc.) to make one image (reference image A) and another image (floating image B) reach a consistent position in space. In the world of computer, images are represented by a two-dimensional matrix, assuming that the gray values of image A and image B at a certain point $(\mathbf{x}, \mathbf{y})$ are $I_1(x, y)$, $I_2(x, y)$, then the mathematical relationship between the two images can be expressed as

$$I_1(x, y) = I_2(f(x, y)).$$ (1)

where f is a two-dimensional geometric transformation function.

2.2 Rigid Body Transformation

Generally, the spatial model used for image registration includes rigid body transformation, affine transformation, and nonlinear transformation. By analyzing the characteristics of the finger vein image, it is found that there are horizontal and vertical displacements and slight rotation between the images. Therefore, this paper chooses the rigid body transformation model as the space model. The rigid body transformation is a combination of two sub-transforms, translation and rotation. In two-dimensional image registration, the rigid body transformation contains three free parameters: the amount of translation in the horizontal and vertical directions and a rotation amount. Suppose the point before the transformation is (x_1, y_1), the transformed point is (x_2, y_2), and the rigid body transformation can be expressed as

$$\begin{pmatrix} x_2 \\ y_2 \end{pmatrix} = \begin{pmatrix} cos\theta & sin\theta \\ -sin\theta & cos\theta \end{pmatrix} \begin{pmatrix} x_1 \\ y_1 \end{pmatrix} + \begin{pmatrix} \Delta x \\ \Delta y \end{pmatrix}.$$ (2)

where θ is the rotation angle, Δx is the displacement in the horizontal direction, and Δy is the displacement in the vertical direction.

2.3 Mutual Information Measure

The similarity measure is a measure of the similarity between two images. It is used to evaluate the pros and cons of each transform model T, usually as the objective function of the search strategy.

Similarity measures commonly used in image registration include: inter-point distance method, histogram method, cross-correlation method, and mutual information method [7]. Since the mutual information method does not require too much preprocessing operation on the image, nor does it need to extract feature point information, considering the characteristics of the finger vein image, we use mutual information as the basis for registration.

The Mutual Information (MI) [8] concept comes from information theory and reflects the information correlation between two random variables. In the field of image processing, the mutual information of the two images reflects the degree of inclusion of information between them. It can be expressed as a mathematical formula:

$$MI(A, B) = H(A) + H(B) - H(A, B). \tag{3}$$

where H(A), H(B) represent the information entropy of the finger vein images A and B, respectively, and H(A, B) represents the joint entropy of the vein images A and B.

3 Image Registration Based on Genetic Algorithm

3.1 Genetic Algorithm Design

Genetic Algorithm (GA) [9] is a global optimization algorithm. It designs the selection, crossover, and mutation operator by imitating the "survival of the fittest" of the creatures in nature, achieving the goal of retaining excellent individuals and eliminating bad individuals. Genetic algorithm has wide adaptability, robustness, and many other advantages. Therefore, it is used to search for the optimal three parameters in the rigid body transformation model. The design of a specific genetic algorithm mainly includes three key parts: the coding method of the chromosome, the design of the fitness function, and the choice of termination conditions.

Coding method. The chromosomes are binary coded, each chromosome is 24 bits long, and the three 8-bit fields from low to high represent the rotation angle θ, the horizontal displacement Δx, and the vertical displacement Δy, as shown in Fig. 1.

Fitness function. In the genetic algorithm, the fitness function is used to evaluate the degree of excellence of the individual. In this algorithm, we use the mutual information between the two finger vein images to be registered as the fitness evaluation function. The fitness function is defined as follows:

$$F(t) = MI(A, B). \tag{4}$$

In the above formula, MI(A, B) represents the mutual information of the reference image A and the floating image B, which can be obtained by Eq. (3), and the larger the mutual information value, the higher the degree of registration. So we turn the registration problem into a multi-parameter optimization problem that finds the maximum value of F(t).

Termination condition. The genetic algorithm can be terminated by limiting the running time or specifying the number of iterations. Through many experiments, we

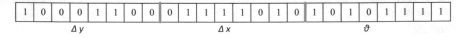

Fig. 1 Chromosome coding scheme

find that the iterative to 80 generations can basically find the optimal solution, so in this paper, we limit the maximum number of iterations to terminate the algorithm.

3.2   The Main Steps of This Algorithm

The algorithm framework based on genetic algorithm to achieve finger vein image registration is shown in Fig. 2. The detailed steps of this algorithm are as follows:

1. Read the two finger vein images A and B to be registered.
2. Set the number of populations N. The number of initialized populations should be moderately selected. If the population is too large, the efficiency of the algorithm

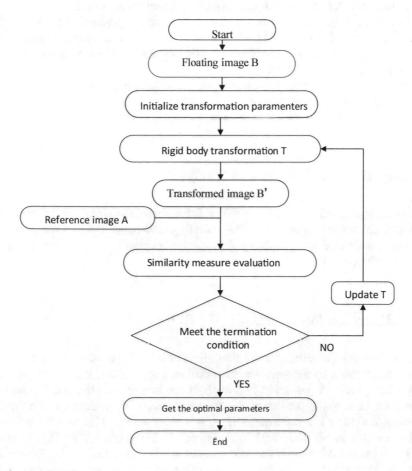

Fig. 2 Algorithm flowchart

will be reduced. If it is too small, the optimal solution may not be found. The population size of this algorithm is set to 50.

3. Binary coding of the population, the coding method is shown in Fig. 1.
4. To calculate the individual fitness value, this paper uses the mutual information between the two finger vein images as the fitness evaluation function. The calculation method is as shown in Eq. (4).
5. Selection. The purpose of the selection is to select good individuals from the current group, so that these excellent individuals have the opportunity to produce offspring. The algorithm adopts the "roulette selection method".
6. Crossover. Through step 5, select superior individuals, cross-operate these individuals, and generate offspring individuals. If the fitness value of the offspring individuals is greater than that of the parents, replace the parent individuals with the offspring individuals. Otherwise, do nothing. The algorithm uses a single-point crossover operator with a crossover probability p_m of 0.8.
7. Mutation. After the crossover, the individuals of the population are easily singularized and fall into local optimum, so mutations are needed to generate new individuals and maintain the diversity of the population. The mutation probability p_c of this algorithm is 0.01.
8. If the number of iterations reaches the maximum number of iterations, then the iteration is terminated and the optimal solution is obtained. Otherwise, the process jumps to step 4 and continues.

4 Experimental Results and Analysis

In this section, in order to evaluate the performance of the algorithm, we designed two different sets of experiments. The operating environment of all experiments in this paper is CPU: Corei5, Memory: 8 GB, software platform: matlab8.5, operating system: Windows 10.

4.1 Algorithm Robustness

In order to verify the validity of the algorithm realistically and objectively, we select two pictures taken by the same finger at different time nodes, the size of image is 513×256 pixels. As shown in Fig. 3a, b, it can be seen that the two pictures are inconsistent due to the position of the fingers placed by the collector at the time of shooting, and there is a significant offset. It is observed that Fig. 3a is more in line with the requirements of the registration template. Therefore, we use Fig. 3a as the reference image. After the registration, the image is shown in Fig. 3c. We can see that Fig. 3a, c is very close. Table 1. shows the variation of the parameters of the 5th, 50th, and 80th rigid body transformations respectively. Similarly, the parameters tend to be stable around 50 times, which is close to the optimal solution.

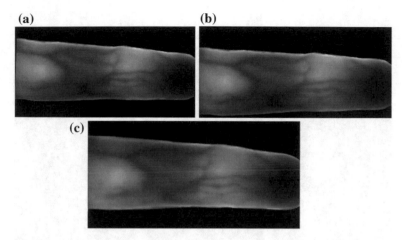

Fig. 3 Transformed result of input image. **a** and **b** Images from the same finger having pose variations. **c** Transformed image of (**b**)

Table 1 Parameter values for different iterations

Number of iterations	Vertical displacement Δy (pix)	Horizontal displacement Δx (pix)	Rotation angle θ (°)
5	−35.047	−0.127	1.702
50	−34.659	2.495	1.486
80	−34.664	2.809	1.715

4.2 Algorithm Comparison

In Experiment 2, the reference image is a finger vein image of a person, and the size of the image is 96 × 190 pixels as shown in Fig. 4a. We use Photoshop to move the reference image down by 3 pixels, right by 3 pixels, and rotate 2° clockwise to get a floating image, as shown in Fig. 4b. Because the transformation parameters are known, it is easy to verify the accuracy of the algorithm. In the above experimental environment, the genetic algorithm(GA), ant colony algorithm(ACO), and direct search method(DS) were used to optimize the registration parameters of Fig. 4a, b. The experiment was repeated 50 times to obtain the average value and average error of the registration parameters of the above three algorithms. The registration result obtained by the algorithm is shown in Fig. 4c. Table 2 shows the comparison between the different algorithms.

Through the above comparison, it can be found that the registration accuracy of the genetic algorithm is better than the ant colony algorithm, and the registration efficiency is much better than the direct search method, and has good performance.

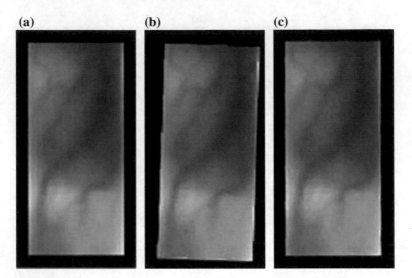

Fig. 4 Transformed result of input image. **a** and **b** Images from the same picture having pose variations. **c** Transformed image of (**b**)

Table 2 Comparison of registration result by different optimization algorithms

	Mean (pix)	Actual value (pix)	Error (pix)	Time (s)
GA	$(-2.88, -3.09, -2)$	$(-3, -3, -2)$	$(0.12, 0.09.0)$	51.62
ACO	$(-3.44, -2.68, -2.4)$	$(-3, -3, -2)$	$(0.44, 0.32, 0.4)$	65.33
DS	$(-3, -3, -2)$	$(-3, -3, -2)$	$(0, 0, 0)$	100.44

5   Conclusion

This paper proposes a finger vein image registration method based on genetic algorithm and rigid body transformation. The experimental results show that the algorithm can effectively correct the displacement and rotation generated during finger shooting, and achieve better results, which is beneficial to improve the recognition accuracy of the finger vein system. Therefore, the proposed algorithm has good practical value.

Acknowledgements This work is partly supported by the National Natural Science Foundation of China under Grant No. 61873131, 61702284 and 61572261.

References

1. Vakil, M.I., Malas, J.A., Megherbi, D.B.: Information theoretic approach for template matching in registration of partially overlapped aerial imagery. In: Aerospace and Electronics Conference, pp. 146–150. IEEE (2016)
2. Salhi, K., Jaara, E.M., Alaoui, M.T.: Pretreatment approaches for texture image segmentation. In: International Conference on Computer Graphics, Imaging and Visualization, pp. 221–225. IEEE (2016)
3. Tsai, C.M., Guan, S.S.: Identifying regions of interest in reading an image. Displays **39**, 33–41 (2015)
4. Chicotay, S., David, E., Netanyahu, N.S.: A two-phase genetic algorithm for image registration, pp. 189–190 (2017)
5. Zhang, J., Hu, J.: A novel registration method based on coevolutionary strategy. In: Evolutionary Computation, pp. 2375–2380. IEEE (2016)
6. Gou, Z., Ma, H.: An automatic registration based on genetic algorithm for multi-source remote sensing. In: International Conference on Control, Automation and Robotics, pp. 318–323. IEEE (2016)
7. Minvielle, P.: Fast Mutual information-based map model matching. In: IEEE International Geoscience and Remote Sensing Symposium. IEEE (2017)
8. Luo, H.Y.: Study on mutual information medical image registration based on ant algorithm. Int. J. Hybrid Inf. Technol. **8** (2015)
9. Huang, N.: Application research of genetic algorithm image enhancement. Comput. Simul. (2012)

A Review of Satellite Video on Demand with Evolutionary Computing

De Yao Lin and Xuehong Huang

Abstract As technology has developed, satellite TV has played an increasingly important role in the national economic construction. It has repeatedly provided outstanding services in situations such as earthquake relief, border posts, and foreign hotels by maintaining communication with the outside. Although technological progress has led to the development of both digital and analogue satellite TV, the former is much more expensive than the latter. Hence, there is likely to be some waste on some occasions if digital satellite TV is always chosen. This paper focuses on the technical characteristics of simulation and digital satellite TV, guiding people to adopt different design schemes for different situations, so as to maximize the benefits of satellite TV while having limited funds. This study also briefly analyzed the entire process of satellite TV design, installation, and debugging from a technical point of view to ensure that people can safely enjoy satellite TV. This paper also investigates the research in satellite video on demand and provides a review of applied evolutionary computing that improve video-on-demand service.

Keywords Satellite TV · Safe grounding · Video on demand · Evolutionary computing

1 Overview

Considering the value for money is essential when installing the TV. The scheme chosen should depend on the location. Analogue satellite TV can be used in places that cannot be covered by existing digital cable TV, such as border posts. On the other hand, digital satellite TV can be used in foreign hotels [1]. Receivers for satellite transmissions are usually installed outside. Lightning strikes often occur due to the inadequate design and installation of lightning protection facilities. In less severe

D. Y. Lin · X. Huang (✉)
National Demonstration Center for Experimental Electronic Information and Electrical
Technology Education, Fujian University of Technology, Fuzhou 350118, Fujian, China
e-mail: 423830961@qq.com

© Springer Nature Singapore Pte Ltd. 2019
J.-S. Pan et al. (eds.), *Genetic and Evolutionary Computing*,
Advances in Intelligent Systems and Computing 834,
https://doi.org/10.1007/978-981-13-5841-8_21

cases, the signal will be interrupted, but in highly severe cases, the equipment will be damaged or the safety of users and their property may be endangered. Therefore, careful and stringent checks should be carried out during the design and construction process for the safety of users. Not only should the design and construction be carried out in strict accordance with national standards and specifications, but a thorough examination should also be conducted to ensure that the grounding and lightning protection devices of the completed systems are correctly installed. To ensure safety, corrective measures should be adopted immediately if problems are found.

2 Digital Satellite TV

The down-converted satellite signal received by the satellite antenna is demodulated by the satellite receiver, after which the video and audio signals are output [2]. These signals are then inputted into the active channel duplexer after analogue-to-digital conversion, source coding, and channel coding modulation (in channel coding, the channel used should avoid the channel of the existing municipal channel) together with the municipal digital cable TV signal; these signals are then mixed into one channel for output. The end user will then be able to watch the broadcast using their existing set-top box [3].

2.1 Construction of a Small Analogue Satellite Cable TV System

The analogue cable TV system introduced in this paper can be used in mountainous or marginal mountainous areas where the digital cable TV system is not quite popular. It is very popular among users because of its low cost, stable and reliable system, and convenient maintenance.

2.2 Lightning Protection of Satellite TV

The antennas of satellite TV are usually installed on the roof or outside, which make it vulnerable to lightning strikes. Therefore, it is necessary to install lightning protection devices. The lightning protection device for satellite TV is illustrated in Fig. 1.

As shown in the figure, the grounding resistance of lightning protection devices should be less than 4 Ω. A lightning rod is adopted as the bus bar, and a tangent line is formed by the top of the lightning rod and the upper end of the satellite antenna. For the sake of safety, the included angle should be less than or equal to 37°. The

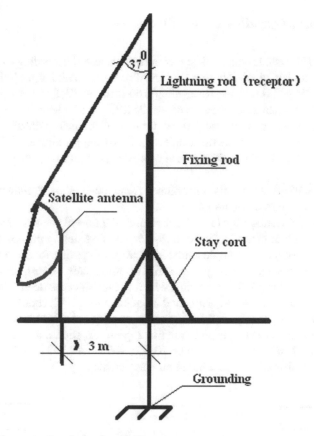

Fig. 1 Lightning protection device for satellite antenna

protection range of the lightning protection device is the cone formed by the circle around the lightning rod and its fixing rod. Lightning protection devices for other types of buildings are similar and will not be described in detail.

Direct lightning protection devices are generally composed of three parts: lightning rod, down lead, and grounding grid. Lightning rods, commonly known as lightning receptors, are mainly classified as direct and indirect lightning protection devices; they include lightning rods, wires, strips, and conductors. When a lightning rod is used as the lightning receptor for lightning protection, the lightning rod is connected to the ground by a wire to form an equipotential body with the grounding grid. The lightning receptor is connected to the ground by the conductor in advance, achieving lightning protection as a result of its own height. Two types of lightning are prevented from being induced in the protected object.

2.3 Design of Satellite Cable TV System

Small cable TV systems can be designed and constructed according to the requirements given in Cabled Distribution Systems Primarily Intended for TV and Sound Signals GB/T6510-96. The user level should conform to VHF (57–83 dB) and UHF (60–83 dB), and is generally between 65 and 75 dB. The user level in noninterference areas in remote mountainous regions should be approximately 65 dB, and in areas with strong signal interference, the user level should be approximately 75 dB. Typical front end process: A schematic diagram of a typical front end process is shown in Fig. 2.

Channel Settings. Channels with higher frequencies are recommended for the convenience of future expansion.

Design of Distribution System. A schematic diagram of the distribution system in a residential community is shown in Fig. 3. The first step in the design shown in Fig. 3a is the system diagram designed according to the distribution of users. The second step is to select the amplifier and its components according to the users' required level (if it is 70 dB). Materials with a network access license issued by the State Administration of Radio, Film, and Television (SARFT) should be selected.

As shown in Fig. 3a (Ui1 = 70 dB), for the amplifier at B0, an amplifier with KP = 20 dB and a three-way distributor attenuating −6 dB should be selected, i.e., the level of ports B1, B2, and B3 should be 70 + 20 − 6 = 84 dB. The following splitter models are not difficult to select based on these results.

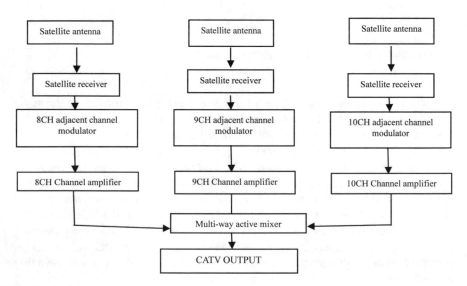

Fig. 2 Schematic diagram of a typical front end process

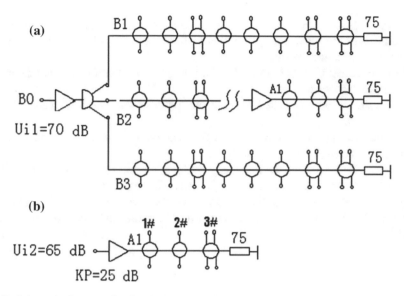

Fig. 3 Schematic diagram of a distribution system in a residential community

As for the selection of splitters, as shown in Fig. 3b, the signal level of A1 port $(U0) = Ui2 + KP = 65 + 25 = 90$ dB, and the user level is 70 dB, which is obtained by taking the median value. The line loss can be excluded if the circuit is short, then

1. Selection of 1# two-way splitter: The user attenuation of the 1# two-way splitter should be $90 - 70 = 20$ dB; hence, T01-220 should be selected.
2. Selection of 2# two-way splitter: For the user attenuation of the 2# two-way splitter, the insertion loss of the line caused by the 1# two-way splitter (-2 dB) should be considered. The attenuation at the user end of the 2# two-way splitter should be $90 - 70 - 2 = 18$ dB; hence, T01-218 should be selected.
3. Selection of the 3# four-way splitter: Similarly, the user attenuation of the 3# four-way splitter should be $90 - 70 - 4 = 16$ dB after considering the cumulative insertion loss of the line caused by the 1# and 2# two-way splitters (-4 dB); hence, T01-416 should be selected.

Maintenance of the Distribution System. The location of faults can be quickly identified as long as the signal level values of B0, B1, B2, B3, and A1 are monitored. If the only abnormal value is the level of the B3 port (e.g., it is only 23 dB, when the normal value should be $70 + 20 - 6 = 84$ dB), then the fault is between the B2 and B3 ports. This fault is generally caused by an open circuit of the signal cable. The other cases are similar.

Retrieval of Parameter Data and Adjustment of Elevation and Azimuth Angles. The spatial position of the satellite signals being received and the geographic position of the receiving place can be found online. Entering them into the satellite receiver allows the elevation angle and azimuth angle required by the satellite antenna to be calculated. These terms are defined as follows:

- Elevation angle of the antenna: angle between the axis of symmetry of the satellite antenna and the horizontal plane, that is, the included angle between the plumb line and the satellite plane.
- Azimuth angle: 0° at due north (i.e., 180° at due south) and increasing clockwise (from south to west).
- Polarization mode: H indicates horizontal polarization, and V indicates vertical polarization.
- Local frequency: frequency used for frequency conversion, 11,300 MHz, determined by the high-frequency head.
- Downstream frequency: the frequency at which satellite TV signals are transmitted from the satellite transponder to the ground receiving station.
- Symbol rate: it represents the quality of the transmitted image. The higher the symbol rate, the higher the image transmission rate and the better the image quality and vice versa.

3   Satellite Video on Demand

Video communication can be facilitated through various approaches, including decentralized video-on-demand [4] or hybrid video delivery service in conjunction with custom-designed video coding technique [5]. For digital video distributed over satellite, due to the infrastructure deployment, centralized broadcasting is typically designed, with focuses on modulation and coding scheme to improve the throughput [6]. With the popularity of video on demand, various video coding and delivery methods, including constant bit rate video download, variable bit rate video download, and variable bit rate video streaming, have been investigated [7]. Such mechanism can also be applied to moving targets, such as satellite video service on vehicles [8].

To overcome the challenge of network utility in satellite network, with balanced rate and delay profile, scalable video coding applied in a cross-layer setup has shown to yield superior video quality with reduced delay [9]. Other approaches, such as the investigation in QoS provisioning, also help improving satellite video services [10].

Satellite video on demand faces challenge subject to environmental conditions, including:

- **Rain attenuation**: Rain attenuation refers to attenuation in the satellite signal caused by dark clouds, rain, and fog blocking the route of the signal from the satellite to the antenna. Rain attenuation has a great influence on satellite signals in the Ku band due to their high frequency and short wavelength; these signals are greatly hindered and attenuated in rain and fog. To address these challenge, various attempts have been proposed in [11, 12].
- **Sun outage**: Sun outage occurs when the satellite is midway between the sun and the ground receiving station, when the sun, earth, and satellite move in a straight line during the satellite's orbit around the earth. At this moment, the parabolic antenna at the receiving end is aimed at the satellite, which results in interference from the sun. This kind of interference is called sun outage.

- **Satellite eclipse**: Satellite eclipse is similar to sun outage; the satellite enters the shadow area of the earth as it moves in a straight line around the earth such that the satellite and the sun are on the opposite sides of the earth. This phenomenon is called satellite eclipse.

4 Evolutionary Computing for Satellite Video on Demand

Evolutionary computation and evolution of things have found applications in solving real-world problems [13]. Various approaches, including genetic algorithm, ant colony optimization, and particle swarm optimization, are applied in applications such as association rule hiding for privacy preservation [14], computer-aided diagnosis system for diabetic retinopathy [15], resource-constrained project scheduling [16], and energy system management for micro-grids [17].

Evolutionary computing has also been applied in various researches on video on demand. The applied genetic algorithm for multicast routing protocol has been investigated to improve the HAP-Satellite multicast service, which can be applied in assuring multicast QoS [18, 19]. The ant-inspired algorithm has been applied for mini-community-based video-on-demand services [20]. For video-on-demand system management, Huang et al. have investigated electronic fraud detection using hybrid immunology-inspired algorithms [21]. Tang et al. have proposed a video placement strategy using the genetic algorithm and demonstrated an improved video-on-demand service [22].

Overall, studies have shown that evolutionary computing offers a promising solution for various aspects in video communication, which could be applied for satellite video on demand to facilitate improved video services.

5 Conclusion

Satellite TV has brought good news to people on specific occasions, established an international bridge of friendship and communication, and provided an effective and convenient window to the outside world. Only by designing and constructing systems in strict accordance with the specifications, people can safely enjoy the pleasure of modern technology. This paper reviewed the studies in satellite video on demand and also reviewed strategies in applied evolutionary computing that improve video-on-demand service.

References

1. Yao, Y.Y., et al.: Development Trend of Broadcast TV and Business Bundling Technology. Radio & TV Information, 03 (2010)
2. Kang, Y.: Development and Challenges of Gansu Broadcast TV Digital Service. Cable TV Technology, 11 (2006)
3. Zhong, X.D.: Analysis on the Development of Digital TV in China. China Media Technology, 02 (2013)
4. He, Y., Lee, I., Guan, L.: Distributed throughput maximization in P2P VoD applications. IEEE Trans. Multimed. (TMM) **11**(3), 509–522 (2009)
5. Lee, I., He, Y., Guan, L.: Centralized P2P streaming with MDC. In: Proceedings of Multimedia Signal Processing Workshop (MMSP), Shanghai, China, October (2005)
6. Cominetti, M., Morello, A.: Digital video broadcasting over satellite (DVB-S): a system for broadcasting and contribution applications. Int. J. Satell. Commun. **18**(6), 393–410 (2000)
7. Le-Ngoc, T., Tsingotjidis, P.: Provision of video-on-demand services via broadband GEO satellite systems. In: Mobile and Personal Satellite Communications 3, pp. 222–234 (1999)
8. Yang, T., et al.: Small moving vehicle detection in a satellite video of an urban area. Sensors **16**(9), 1528 (2016)
9. Pradas, D., Vazquez-Castro, M.A.: NUM-based fair rate-delay balancing for layered video multicasting over adaptive satellite networks. IEEE J. Sel. Areas Commun. **29**(5), 969–978 (2011)
10. Ma, T., Lee, Y.H., Winkler, S., Ma, M.: QoS provisioning by power control for video communication via satellite links. Int. J. Satell. Commun. Network. **33**(3), 259–275 (2015)
11. Imole, O.E., Walingo, T.: Call admission control for rain-impacted multimedia satellite networks. In: 2017 IEEE AFRICON, pp. 371–376 (2017)
12. Lee, Y.H., Winkler, S.: Effects of rain attenuation on satellite video transmission. In: 2011 IEEE 73rd Vehicular Technology Conference (VTC Spring), pp. 1–5 (2011)
13. Eiben, A.E., Smith, J.: From evolutionary computation to the evolution of things. Nature **521**(7553), 476–482 (2015)
14. Cheng, P., Lee, I., Lin, C.W., Pan, J.S.: Association rule hiding based on evolutionary multi-objective optimization. Intell. Data Anal. **20**(3), 495–514 (2016)
15. Mansour, R.F.: Evolutionary computing enriched computer-aided diagnosis system for diabetic retinopathy: a survey. IEEE Rev. Biomed. Eng. **10**, 334–349 (2017)
16. Merkle, D., Middendorf, M., Schmeck, H.: Ant colony optimization for resource-constrained project scheduling. IEEE Trans. Evol. Comput. **6**(4), 333–346 (2002)
17. Marzband, M., Yousefnejad, E., Sumper, A., Domínguez-García, J.L.: Real time experimental implementation of optimum energy management system in standalone microgrid by using multi-layer ant colony optimization. Int. J. Electr. Power Energy Syst. **75**, 265–274 (2016)
18. De Rango, F., Tropea, M., Santamaria, A.F., Marano, S.: An enhanced QoS CBT multicast routing protocol based on Genetic Algorithm in a hybrid HAP–Satellite system. Comput. Commun. **30**(16), 3126–3143 (2007)
19. Rango, F.D., Tropea, M., Santamaria, A.F., Marano, S.: Multicast QoS core-based tree routing protocol and genetic algorithm over an HAP-satellite architecture. IEEE Trans. Veh. Technol. **58**(8), 4447–4461 (2009)
20. Xu, C., Jia, S., Zhong, L., Zhang, H., Muntean, G.: Ant-inspired mini-community-based solution for video-on-demand services in wireless mobile networks. IEEE Trans. Broadcast. **60**(2), 322–335 (2014)
21. Huang, R., Tawfik, H., Nagar, A.: Electronic fraud detection for video-on-demand system using hybrid immunology-inspired algorithms. In: Artificial Immune Systems, pp. 290–303 (2010)
22. Tang, K.S., Ko, K.T., Chan, S., Wong, E.: Video placement in video-on-demand system using genetic algorithm. In: Proceedings of IEEE International Conference on Industrial Technology, pp. 672–676 (2000)

Graph-Regularized NMF with Prior Knowledge for Image Inpainting

Li Liu, Fei Shang, Siqi Chen, Yumei Wang and Xue Wang

Abstract The image inpainting problem can be converted to the matrix completion. A classical matrix completion method is based on matrix factorization. The product of two low-rank matrices fills in the missing regions. In this paper, we propose a novel matrix factorization framework to recover the images. Before decomposing the original matrix, approximation matrix as the prior knowledge is constructed to estimate the values of missing pixels. The estimation of the missing pixels can be obtained through resampling from the surface fitting the 3D projection points of the available pixels. To keep the latent geometrical structure between adjacent pixels, we modify the graph-regularized which allows the edge weights negative to decompose the approximation matrix. Experimental results of image inpainting demonstrate the effectiveness of the proposed method compared with the representative methods in quantities.

Keywords Graph-regularized NMF · Image inpainting · Hole filling · Approximation matrix

L. Liu (✉) · F. Shang · S. Chen · Y. Wang · X. Wang
School of Information Science and Engineering, Shandong Normal Univeristy,
Jinan 250014, China
e-mail: liuli_790209@163.com

F. Shang
e-mail: 862757715@qq.com

S. Chen
e-mail: 1832633513@qq.com

Y. Wang
e-mail: 15253167580@163.com

X. Wang
e-mail: wx13791080732@163.com

© Springer Nature Singapore Pte Ltd. 2019
J.-S. Pan et al. (eds.), *Genetic and Evolutionary Computing*,
Advances in Intelligent Systems and Computing 834,
https://doi.org/10.1007/978-981-13-5841-8_22

1 Introduction

Image inpainting is the problem of repairing parts of an image that have been damaged or partially occluded. When the image is regarded as a matrix, the image inpainting problem can be converted to the matrix completion. Matrix completion considers how to complete the matrix from an incomplete observation with both missing and corrupted entries. Many methods have been proposed to address the problem. One category of classical matrix completion methods are based on matrix factorization, in which the observed matrix is approximated by the multiplication of a thin matrix and a short matrix. Non-negative Matrix Factorization (NMF) is an effective technique to find two nonnegative and low-rank matrices whose multiplication provides a good approximation to the observed matrix.

In recent years, the framework of NMF has been proposed. Structure constraint NMF makes use of the intra-sample structures to facilitate the decomposition process [1]. Quadratic NMF derives multiplicative algorithms that monotonically decrease the approximation error under a variety of measures [2]. Penalized matrix factorization adds an additional penalty on the factorized components to yield interpretable factors and efficient convergence [3]. The Local NMF [4] imposes localization constraint to learn spatially localized, parts-based representation of visual patterns. Robust NMF decomposes the original matrix into sparse and low-rank components. The sparse component captures the outliers, and meanwhile, the low-rank component models the intrinsic structure of the data [5]. In addition, some variants of NMF have been proposed in order to enhance the sparsity of NMF. The projective NMF is developed to learn sparse representation by implicitly enforcing orthogonal constraint over the basis [6]. Nonnegative local coordinate factorization [7] is proposed to induce sparse coefficients via the local coordinate constraint. However, these methods neglect the geometric structure proven beneficial for various vision tasks, such as Graph-regularized NMF (GNMF) [8], Dual-regularized Multi-view NMF [9], Multiple graph-regularized NMF [10], Manifold-regularized discriminative NMF [11], and Mixed hypergraph-regularized NMF [12]. Since the sparse hypergraph inherits the merits of both the sparse representation and the hypergraph model, Sparse Hypergraph-regularized NMF enjoys more robustness and can better exploit the high-order discriminant manifold information for data representation [13].

Though various NMF methods are developed, image inpainting results based on these methods are unsatisfactory. Usually, the initial values of incomplete regions are set to zeros, which is quite different from the reality, so the multiplication of the two low-rank matrices cannot accurately express the real image. In this paper, we propose a novel matrix factorization framework to recover incomplete images. An approximation matrix is constructed to estimate the possible initial values in the missing regions as the prior knowledge before matrix factorization. The correlation between pixels around the missing regions is available, so we construct the fitting surface to approximate the 3D projection points of the available pixels and obtain the estimation values through resampling from the fitting surface. To keep the local intrinsic geometrical structure, modified graph-regularized NMF that allows the edge weights negative are utilized to decompose the approximation matrix. The multiplication of two low-rank matrices is used to depict the inpainted image.

The remaining content of this paper is organized as follows. In Sect. 2, we give the motivation of our proposed method. Section 3 briefly introduces the GNMF with prior knowledge for image inpainting. Experimental results and comparisons are given in Sect. 4. Finally, we draw a conclusion in Sect. 5.

2 Motivation

For the incomplete images, the image inpainting can be implemented through the matrix completion. The image and matrix are both two-dimensional, so we can establish the relationship between the image and its corresponding matrix. The incomplete image and its equivalent matrix are marked X and M, respectively. The matrix M can be defined as follows:

$$M = \begin{cases} g_{ij}, if(i, j) \in \Omega \\ 0, otherwise \end{cases} \tag{1}$$

where Ω denotes the pixels set in the complete regions, g_{ij} is the gray value of the ith row and jth column pixel.

The matrix completion is to recover the missing entries of the matrix M. Nonnegative matrix factorization is to find two nonnegative matrices whose product provides an approximation to the matrix M. The matrix M is decomposed to two nonnegative matrices $U = [u_{ik}] \in R_+^{M \times K}$ and $V = [v_{jk}] \in R_+^{N \times K} (K < \min(M, N))$. The loss function to quantify the approximation quality is defined as

$$\min_{U \in R_+^{M \times K}, V \in R_+^{K \times N}} \|M - UV^T\|_F^2 \tag{2}$$

where $\|\cdot\|_F$ denotes the Frobenius norm. The product of the matrix U and V is an approximation of the matrix M. The element of the matrix M is thus decomposed as a linear combination of K elementary components of the matrix U. The coefficients of the linear combination are given by the column of the matrix V.

The loss function is not convex when the matrix U and V are taken as variables. However, the loss function is convex in U when V is fixed and vice versa. The optimization problem in (2) can be optimized by iterative updating rules

$$U_{ij} = U_{ij} \frac{(MV)_{ij}}{(UV^TV)_{ij}}, V_{ij} = V_{ij} \frac{(M^TU)_{ij}}{(VU^TU)_{ij}}. \tag{3}$$

First of the iterative process, we initialize the matrix U and V randomly, and then update them according to the iterative update rules in formula (3), until the final condition is reached.

We would like to know the matrix M as precisely as possible. However, the only information available about M is the set of complete entries, which is a subset of the whole entries. Based on the non-negative matrix factorization, the missing pixels are obtained through the product of the matrix U and V. The completion matrix $\widehat{M}$ can be expressed as follows:

$$\widehat{M} = \begin{cases} g_{ij}, if(i,j) \in \Omega \\ \sum_{p=1}^{k} u_{i,p} v_{p,j}, otherwise \end{cases} \tag{4}$$

When the missing entries of the original matrix M in the formula (1) are filled with zeros, the repainting result is undesirable. Therefore, we explicitly exploit the full use of the known complete entries to acquire more ideal recovery and efficiency computation.

3 Proposed Method

The proposed framework for image inpainting consists of two steps: (i) constructing the approximation matrix as the prior knowledge of matrix factorization and (ii) decomposing the approximation matrix to recover the incomplete image. In this section, we describe the details of the process of the image inpainting.

3.1 Construction of the Approximation Matrix

The main objective of inpainting algorithm is to restore the unknown regions to create a pleasing and realistic feeling about the inpainted image. In the image inpainting, the missing pixels are estimated using the available pixels information with the assumption that the pixels in the known and unknown regions have similar geometrical structures. Texture information is the gray expression form of pixels. Different from several inpainting methods introduced based on texture information, we project the pixels to 3D subspace and take the gray value of every pixel as the height of the 3D point. The missing pixels of the image form the holes of 3D point clouds as shown in Fig. 1, and then the establishment of texture relationship between known and unknown regions are transformed to filling-in holes with feature preserving.

3.1.1 Construction of Approximation Surface

The pixels of the incomplete image are ordered in row and column, and are naturally arranged to form a horizontal line and a longitudinal line. The horizontal line acts

(a)   (b)

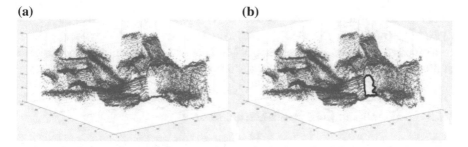

Fig. 1 The projection from the pixels to the points cloud. **a** Projection points cloud; **b** the hole in red contour

as the U axis, and the vertical line serves as the V axis. The UV plane is regarded as the parameterization plane and all the pixels are uniformly parametered in the range [0, 1]. The parameter of the pixel (i, j) is marked as (u_{ij}, v_{ij}), where i and j denote the ith row and jth column separately.

When the pixels are projected to 3D subspace, the estimation values of missing pixels can be obtained through resampling from the surface fitting the holes of projection points. In order to fill in holes, new points need to be added to the missing regions. Moving Least Square (MLS) is an approximation method developed for surface generation problems [14, 15]. In this method, continuous functions are generated from projection points by computing a weighted least squares approximation. The method is used for constructing meshless function.

For the sake of simplicity, we use x to represent the pixel (u, v) in the following. The height function $g(x)$ defined over a two-dimensional subspace is

$$g(x) = \sum_{j=1}^{m} p_j(x)a_j(x) = P(x)a(x) \tag{5}$$

Taking account into weight factors, the vector of unknown coefficients $a(x)$ is determined by minimizing the quadratic form through a weighted least square estimation as follows:

$$J = \sum_{i=1}^{N} w_i(x)(g(x_i) - g_i)^2 = \sum_{i=1}^{N} \frac{e^{-\alpha d_i^2(x)}}{d_i^2(x)}(g(x_i) - g_i)^2 \tag{6}$$

Here, $d_i(x)$ is the distance from the new position of the ith original point x_i, the parameter α controls the influence of vicinity features on the region to be resampled. The coefficients $a(x)$ can be obtained by solving

$$a(x) = (PW(x)P^T)^{-1}PW(x)g \tag{7}$$

where the matrices P and $W(x)$ are defined as $P = \left[P(x_1)^T, P(x_2)^T, \ldots, P(x_N)^T \right]^T_{N \times m}$, $W(x) = (diag(w_i(x)))$, $i = 1, 2, \ldots, N$.

Compared with least squares method, MLS compute one quadric surface for each evaluation point and fits higher order surfaces.

3.1.2 Estimation of the Approximation Matrix

Once a surface is reconstructed to fit the height field using MLS, new points for filling the hole can be obtained by resampling. For the missing pixel $\left(\widetilde{i}, \widetilde{j} \right)$ whose parameter is $(\tilde{u}_{ij}, \tilde{v}_{ij})$ in the parameterization plane, the gray value $\tilde{g}_{ij}$ can be calculated according to the formula mentioned in Sect. 3.1.1. When the gray values of all missing pixels are achieved, the approximation matrix can be defined as

$$\widetilde{M} = \begin{cases} g_{ij}, if(i, j) \in \Omega_+ \\ \tilde{g}_{ij}, if(i, j) \notin \Omega_+ \end{cases} \tag{8}$$

Compared with the matrix M with zeroes in the missing regions, the approximation matrix $\widetilde{M}$ is defined based on the correlation between the adjacent pixels and depicts the incomplete image more exactly. The approximation matrix $\widetilde{M}$ serves as the prior knowledge of the matrix factorization.

3.2 Local Graph-Regularized NMF

The approximation matrix is regarded as the prior knowledge to further complete the matrix factorization. By using the nonnegative constraints, NMF can learn a parts-based representation. However, NMF fails to discover the intrinsic geometrical structure of adjacent pixels. A natural assumption here could be that if two pixels are close in the original image, the filling pixels with respect to the new basis are also close to each other. The objective function is formulated as follows:

$$\min_{U \in R_+^{M \times K}, V \in R_+^{K \times N}} \left\| \widetilde{M} - U V^T \right\|_F^2 + \lambda Tr(V^T L V) \tag{9}$$

where $\widetilde{M}$ is the approximation matrix of the incomplete image, $Tr(\cdot)$ denotes the trace of a matrix, λ is the regularization parameter, and L is graph Laplacian.

To minimize the objective function, Eq. (11) can be rewritten as

$$Tr((\tilde{M} - U V^T)(\tilde{M} - U V^T)^T) + \lambda Tr(V^T L V)$$
$$= Tr(\tilde{M} \tilde{M}^T) - 2Tr(\tilde{M} V U^T) + Tr(U V^T V U^T) + \lambda Tr(V^T L V) \tag{10}$$

Let ψ_{ik} and ϕ_{kj} be the Lagrange multiplier for constraint $u_{ik} \geq 0$ and $v_{kj} \geq 0$ respectively, and $\Psi = [\psi_{ik}]$, $\Phi = [\phi_{kj}]$, the Lagrange L is

$$L = Tr(\widetilde{M}\widetilde{M}^T) - 2Tr(\widetilde{M}VU^T) + Tr(UV^TVU^T)$$
$$+ \lambda Tr(V^TLV) + Tr(\Psi U^T) + Tr(\Phi V^T) \tag{11}$$

Using the KKT conditions for the partial derivative of L with respect to U and V, we get the following updating rules:

$$U_{ij} = U_{ij}\frac{(\widetilde{M}V)_{ij}}{(UV^TV)_{ij}}, V_{ij} = V_{ij}\frac{(\widetilde{M}^TU + \lambda WV)_{ij}}{(VU^TU + \lambda DV)_{ij}}. \tag{12}$$

We can iteratively learn U and V via giving the initial guesses. The updating rules in Eq. (12) reduce to the updating rule of the standard NMF when λ is 0.

To preserve the local geometric structure of adjacent pixels, we represent any pixel as the linear combination of adjacent pixels and allow the edge weight negative. For each pixel p_i, we find its "up", "down", "left", and "right" four pixels as it neighbors $N(p_i)$ and put edges between pixel p_i and its neighbors, and then a signed data graph $\widetilde{G} = (V, \widetilde{E})$ is constructed. The weights of pixel p_i and its neighbors can be obtained by

$$\sum_{i=1}^{N}\left\| x_i - \sum_{x_j \in N(x_i)} w_{ij}x_j \right\|^2, s.t.w_{ij} = w_{ji} \tag{13}$$

The weight matrix from the Eq. (13) is symmetrical and the weight can be negative. Equation (13) pay more attention on the link relationship of four adjacent pixels, and the unsigned graph becomes signed graph.

4   Experiments and Discussions

In this section, we implement the described method and use it to recover the missing images. In all cases, the input to the algorithm consists of color and gray images. The influence factors are discussed and different methods are compared to verify the effectiveness of our approach. To evaluate the objective quality of the inpainted images, we use the Peak Signal-to-Noise Ratio (PSNR) via Mean Square Error (MSE).

According to the range of the missing regions, the incomplete images are divided into block-incomplete and dot-incomplete images. The characteristic of the block missing is local large loss, and for the latter, the missing pixels are distributed irregularly in the whole image. For the dot-incomplete images, the fitting surface is constructed to approximate the available projection points. Figure 2a is the dot-incomplete "Lenna" image, and Fig. 2b is the inpainting result using GNMF with

(a)   **(b)**

Fig. 2 Inpainting results for the dot-incomplete image. **a** Image with 40% missing entries; **b** inpainting result using MLS +GNMF

prior knowledge. In the experiment, the degree of the approximation surface is 3, and the rank K of the matrix factorization is 100.

To demonstrate the influence of the rank K, we use the gray scale "Lenna" image with missing blocks. For every missing block, different rank K is adopted to test the influence on the inpainted quality. Figure 3 shows the PSNR of every missing block when rank is 10, 20, 50, 100, and 200, respectively. From the results, we find that due to different local characteristic, every missing region should adopt different rank K to get higher inpainting quality. Therefore, if there is more than one missing regions, the different rank K should be taken instead of one same rank K. For every missing region, we reconstruct its local fitting surface and decompose its approximation matrix to two low-rank matrices, separately.

Similar to the gray image, three separate approximation matrices (R, G, and B channels) for the color images are constructed as prior knowledge. As mentioned in Sect. 3.2, in every color channel, the approximation matrix are decomposed to generate two low-rank matrices. The color image is recovered through combining the estimation values of three color channels. Figure 4 gives the inpainting results of color "Lenna" image with twelve missing regions.

Among different types of inpainting artifacts, Partial Differential Equation (PDE)-based and Exemplar-based methods based on texture information are representative. Several variations of PDE-based methods are introduced based on the flow of texture information in linear, nonlinear, isotropic or anisotropic directions. The Exemplar-based techniques optimize cost functions to fill the missing regions with the similar known regions. Figure 5a is the missing image with three incomplete blocks. Figure 5b is the inpainting result using PDC-based method [16]. Figure 5c is the inpainting result using one of Exemplar-based method [17]. Figure 5d is the inpainting results using the proposed method.

The PDE-based methods are not suitable for restoring large unknown regions [18], and the inpainting result in Fig. 5b is a blur in the missing regions. For the

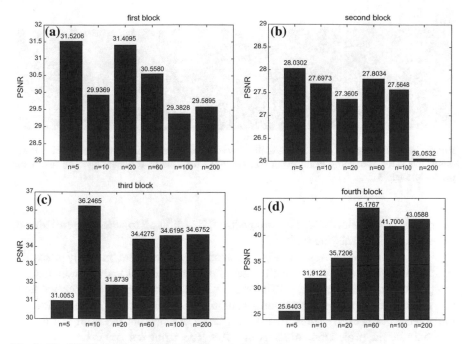

Fig. 3 The PSNR of different rank K for one missing block

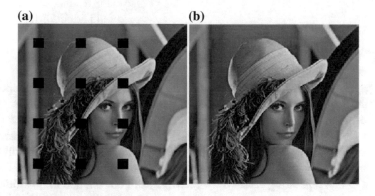

Fig. 4 Inpainting result of color image. **a** Block-incomplete color image; **b** inpainted image using MLS + GNMF

Exemplar-based method, it is crucial to find examples from the image and copy pixels from examples to missing regions. The image itself decides the inpainting quality. In Fig. 5c, no suitable similar patches are achieved to fill the missing regions in the trousers and scarf. The inpainting result using the proposed method in Fig. 5d is close to the original image, but there is still some loss of texture information. This is because the recovery of incomplete images is a process from scratch. Without any

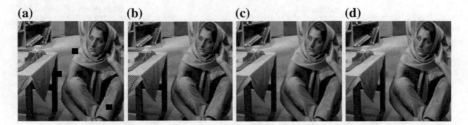

(a)   (b)   (c)   (d)

Fig. 5 Inpainting result of color image. **a** Block-incomplete image; **b** inpainting result using PDC-based method; **c** inpainting result using Exemplar-based methods; **d** inpainting result using the proposed method

instructive information except the image itself, it is difficult to obtain the full texture information of the missing regions.

From the inpainting results, we find another problem that in many cases the inpainted images have unconnected edges along the border of known and unknown regions as shown in Fig. 5b, c. The reason for this problem is that the methods are introduced based on texture information, while the texture is discontinuous. Our method project the pixels to 3D subspace and construct the fitting surface to approximate the projection points using MLS. The fitting surface is continuous, so the filling region along the border is smooth.

5 Conclusions

In this paper, a novel algorithm called GNMF with prior knowledge is proposed to recover the dot-incomplete and block-incomplete images. The value of the missing pixel is estimated and regarded as the prior knowledge to decompose the matrix. Modified GNMF which allows the edge weight negative is utilized to preserve the geometrical structure of adjacent pixels. The whole method is to make full use of the available information to recover images closer to the real images. Experimental results of image inpainting verify the effectiveness of our method in quantities.

Acknowledgements This work was supported by the National Natural Science Foundation of China (61702310 and 61772322).

References

1. Lu, N., Miao, H.: Structure constrained nonnegative matrix factorization for pattern clustering and classification. Neurocomputing **171**(C), 400–411 (2016)
2. Yang, Z., Oja, E.: Quadratic nonnegative matrix factorization. Pattern Recogn. **45**(4), 1500–1510 (2012)

3. Witten, D.M., Tibshirani, R., Hastie, T.: A penalized matrix decomposition, with applications to sparse principal components and canonical correlation analysis. Biostatistics **10**(3), 5–15 (2009)
4. Li, S.Z., Hou, X., Zhang, H., Cheng, Q.: Learning spatially localized parts-based representation. Comput. Vis. Pattern Recogn. **1**, 1–207 (2001)
5. Shen, B., Si, L., Ji, R., Liu, B.: Robust nonnegative matrix factorization via l1 norm regularization. In: IEEE International Conference on Image Processing, pp. 1204–2311 (2012)
6. Yuan, Z., Oja, E.: Projective nonnegative matrix factorization for image compression and feature extraction. In: Proceedings of 14th Scandinavian Conference on Image Analysis (SCIA), Springer, pp. 333–342 (2005)
7. Chen, Y., Zhang, J., Cai, D., Liu, W., He, X.: Nonnegative local coordinate factorization for image representation. IEEE Trans. Image Process. **22**(3), 969–979 (2013)
8. Cai, D., He, X., Han, J., Huang, T.S.: Graph regularized nonnegative matrix factorization for data representation. IEEE Trans. Pattern Anal. Mach. Intell. **33**(8), 1548–1560 (2011)
9. Luo, P., Peng, J.Y., Guan, Z.Y., Fan, J.P.: Dual regularized multi-view non-negative matrix factorization for clustering. Neurocomputing **294**(24), 1–11 (2017)
10. Wang, J.Y., Bensmail, H., Gao, X.: Multiple graph regularized nonnegative matrix factorization. Pattern Recogn. **46**(10), 2840–2847 (2013)
11. Guan, N., Tao, D., Luo, Z., Yuan, B.: Manifold regularized discriminative nonnegative matrix factorization with fast gradient descent. IEEE Trans. Image Process. **20**(7), 2030–2048 (2011)
12. Wu, W.H., Kwong, S., Zhou, Y., Jia, Y.H., Gao, W.: Nonnegative matrix factorization with mixed hypergraph regularization for community detection. Inf. Sci. **435**, 263–281 (2018)
13. Huang, S., Wang, H.X.: Improved hypergraph regularized Nonnegative Matrix Factorization with sparse representation. Pattern Recogn. Lett. **102**, 8–14 (2018)
14. Mashallah, M.F., Pourabd, M.: Moving least square for systems of integral equations. Appl. Math. Comput. **270**, 879–889 (2015)
15. Mehrabi, H., Voosoghi, B.: Recursive moving least squares. Eng. Anal. Bound. Elem. **58**, 119–128 (2015)
16. Telea, A.: An image inpainting technique based on the fast marching method. J. Gr. Tools. **9**(1), 23–34 (2004)
17. Criminisi, A., Perez, P., Toyama, K.: Region filling and object removal by exemplar-based image inpainting. Image Press. **13**(9), 1200–1212 (2004)
18. Guillemot, C., Meur, O.: Image inpainting: overview and recent advances. Signal Process. Mag. **31**(1), 127–144 (2013)

Collaborative Filtering Model Based on Time Context for IPTV Live Recommendation

Zhengying Hu, Yun Gao, Xin Wei and Fang Zhou

Abstract With the increase in the number of IPTV live channels, the users are faced with the problem of information overload and the degraded users' quality of experience (QoE). This paper combines the time context information with the collaborative filtering recommendation algorithm to design the IPTV live recommendation system (RS). Finally, the fusion recommendation model is proposed and its accuracy rate has a great improvement.

Keywords Live recommendation system · Time context · Collaborative filtering · Fusion model

1 Introduction

With the integration of tri-networks, IPTV interactive network television has become increasingly popular. The resources obtained by the users become more and more [1]. However, it may bring a serious problem. When the users face a huge amount

Z. Hu · Y. Gao · X. Wei (✉)
College of Telecommunications and Information Engineering,
Nanjing University of Posts and Telecommunications, Nanjing, China
e-mail: xwei@njupt.edu.cn

Z. Hu
e-mail: andy199588@foxmail.com

Y. Gao
e-mail: gaoyungreen@163.com

Y. Gao · X. Wei
National Engineering Research Center for Communication
and Information Technology (NUPT), Nanjing, China

F. Zhou
School of Electrical and Information Engineering, Anhui University of Technology,
Maanshan, China
e-mail: zf7782@126.com

© Springer Nature Singapore Pte Ltd. 2019   213
J.-S. Pan et al. (eds.), *Genetic and Evolutionary Computing*,
Advances in Intelligent Systems and Computing 834,
https://doi.org/10.1007/978-981-13-5841-8_23

of resources on the internet, they will find it difficult to find programs or videos that they are interested in [2]. This phenomenon not only reduces the utilization rate of the information, but also causes a decrease at the quality of user experience (QoE).

The recommendation system (RS) is an effective tool to improve QoE and solve the problem of information overload [3].

The recommendation algorithms most widely used are collaborative filtering algorithms [4–7]. Collaborative filtering algorithms include users-based collaborative filtering and items-based collaborative filtering. The author in [8] proposes an improved algorithm based on item-based collaborative filtering recommendation algorithm. Based on this paper and the real-time requirement of the live RS [9, 10], we make some improvements in this paper: (1) With the help of the users' tune-up habits in the context of time, we improve the items-based collaborative filtering algorithm. (2) We add the target users' records to increase the accuracy of prediction in the traditional users-based collaborative filtering algorithm. (3) Finally, we propose a fusion recommendation model, which can improve the accuracy rate of the RS.

The structure of this article is organized as follows. In Sect. 2, the analysis of user behavior is introduced. In Sect. 3, the adopted algorithms and its improvement points are introduced. In Sect. 4, the experimental results are given. Finally, the conclusions are contained in Sect. 5.

2   User Behavior Analysis

2.1   Data Preparation

The dataset used in the experiment comes from the IPTV users' set-top box. The data include the viewing records of 1,100 users in August 2016 and the size of data is 434 M. We use the data of the first 3 weeks for statistical analysis and use the experimental results of the last week as validation. The description of data is shown in Table 1.

Table 1 The description of data

Features	Meanings
Collect time	Collect time of set-top box
Start time	The start viewing time during collection interval
End time	The stop viewing time during collection interval

2.2 User Behavior Analysis

Continuous Channel Switching Process

First, the following two concepts need to be defined:

a. *switching process*: If the channels in the two adjacent records are same for one user and the difference between the end time of the first record and the start time of the next record does not exceed 2 s, we merge these two records into one record. Otherwise, they cannot be combined into one record (2 s mean is the delay time of the set-top box in this paper.).
b. *switching transition*: The process of channel switching will generate some transitional switching procedures, e.g., if someone presses the up or down button on the remote to switch channels, then the transitional switching process will occur.

In order to eliminate the influence of the transitional channel during switching process, we will regard a record whose watching duration is not less than 1 min as an effective viewing record in this paper.

Statistics of Time Segment

The user IDs of IPTV live are often not a single member, but more likely to be a family. Although their interest and lifestyles are different, the time periods of watching television remain unchanged everyday for one member of the same user ID. Therefore, the users' interest can be analyzed based on time context information.

According to the survey, 24 h can be divided into five time periods, i.e., [0, 5), [5, 12), [12, 16), [16, 20), and [20, 0). According to the statistical results, it can be seen that the behavior of one user ID during time periods is regular. For example, user A's records from Monday to Friday are usually in [20, 0) and [0, 5), while they usually watch TV in [12, 16) and [16, 20) at the weekend.

For a more detailed analysis of user behavior, we also analyze the viewing records hourly and find that most users watch the same channel within the same hour of everyday, but sometimes, there are some deviations. From our researches on channels, these deviations are due to some delays and advances in the schedule of daily programs. In order to reduce it, we decide to introduce a new time segment that is expanding 1 h before and after the recommendation time, which can take into account the changes of the time schedule on channels.

3 Channel Recommender System

3.1 Base RSs Based on Time Context

It is necessary to utilize the time context information selected by the sliding time window to reflect the features of real time. Some of the following variables should be defined:

- Δt: the time window before the time t (e.g., last 1 day before the time t);
- $S(t)$: the time granularity of the time t (e.g., the whole hour of the time t);

$$\text{rec}_{\text{channel}} = \text{top}_N(sorted(list(score_c))) \tag{1}$$

$$score_{u,c,t}^{love} = \sum T(u, c, [|t - \Delta t| \cap S(t), t)) \tag{2}$$

Equation (1) represents that we sort $list(x)$ in descending order and select the first N channels. Equation (2) represents the score of channel c for user u at time t, where $T(u, c, \tau)$ represents the total duration that user u watches channel c within period τ.

When the user switches channel, the RS starts working and it will provide the user with a recommendation list of N channels with the highest scores.

In the experiment, we will select different values of Δt and $S(t)$, then the accuracy rate of the different recommendation results will be compared.

Items-based Collaborative Filtering RS
Since the live content of each channel is constantly changing, it can be assumed that the one channel at different time everyday is different items.

The traditional items-based collaborative filtering RS algorithm needs to calculate the similarity between items. For the reason that the daily schedule of each channel does not change much, in this paper, we think that the channels watched by the user within the recommendation time period of the previous days are most similar to the channel watched at the recommendation time for the same user. For example, the channel A plays the third episode of a TV series today, and the fourth episode of the TV series will be played at the same time tomorrow. User B watches this TV series today, and then, the channel list that is most likely watched by him at this time tomorrow must include the channel A.

Users-based Collaborative Filtering RS
Since the sparseness of historical records and the changes of interest will lead to the decrease in the accuracy rate of RS, we introduce the target user's records to improve the accuracy of RS.

First, we use the cosine similarity of their channel-watching duration vector during a period $\tau(\tau = [|t - \Delta t| \cap S(t), t))$ to compute the similarity between user u and v at recommendation time t.

$$sim(u, v, t) = \frac{\sum_{i=1}^{n}(U_i \times V_i)}{\sqrt{\sum_{i=1}^{n}(U_i)^2} \times \sqrt{\sum_{i=1}^{n}(V_i)^2}} \tag{3}$$

where U represents the vector of user u at the time t, V represents the vector of user v at the time t, X_i represents the elements of vector X, which is the total time duration that the user x spends on watching the ith channel.

Then, we use the K-nearest neighbor algorithm to select the top-K users who are most similar to the target user (the K users include the target user, and K is set to 25).

Subsequently, we need to determine the input dataset of RS, and then use the input dataset to predict the scores of different channels, and return a list of channels with the highest scores as the recommendation result:

$$\text{Dataset}_{RS} = \sum_{u \in U_1} Data(u, [|t - \Delta t| \cap S(t), t]) + Data(\text{target}_u, [|t - \Delta t| \cap S(t), t)) \quad (4)$$

$$U_1 = U(K\text{-}nearest\text{-}neighbors) - U(target) \quad (5)$$

$$score^{love}_{target\_u,c,t} = count\left(\sum_{u \in U} u^{love}_{c,M}\right)) \quad (6)$$

where $Data(u, \tau)$ represents the records of user u within period τ, $U(x)$ represents the set of user x, $u^{love}_{c,M}$ represents a list of M channels with the highest love scores for one user, and $\sum_{u \in U} u^{love}_{c,M}$ represents a total channel set of all users.

3.2 Fusion RSs Based on Time Context

We design the fusion model based on the abovementioned time context RSs. Since the different combination orders have a great influence on the results of the fusion RS, we will use the algorithm shown in Fig. 1 to maximize the accuracy rate of the fusion RS.

First, we select the recommendation list with the highest accuracy rate on top-1 as the basic list, and then select complementary recommendation list based on accuracy and the diversity of the time and algorithms. Next, the used fusion method is that the content of a complementary recommendation list is selected to supplement the basic recommendation list, until the accuracy rate of the fusion recommendation list no longer has a larger increase trend when the length of the basic recommendation list is less than 10.

4 Result and Analysis

In the experiment, we choose different input data for RSs according to Table 2.

The evaluation indexes of a single recommendation and the entire RS for each recommendation model are given as follows:

$$score^{evaluate}_{one} = |N(u, c, t) \cap N(u, rec_{list}, t)| \cap |T(u, c, [t, t + \Delta t])\rangle \frac{\Delta t}{2}| \quad (7)$$

$$accuracy\_rate_{RS} = \frac{\sum score^{evaluate}_{one}}{N(rec)} \quad (8)$$

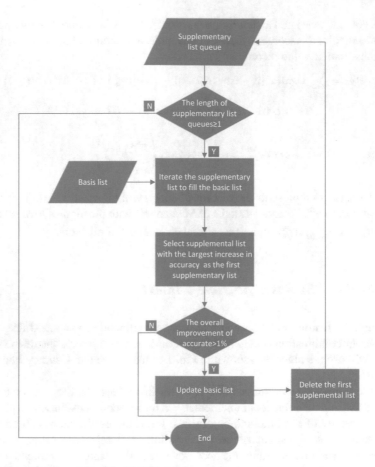

Fig. 1 Algorithm flowchart of fusion RS model

Table 2 Labels of the input data for RSs

	Entire of 1 day	One session	One hour	Expanding 2 h
Last week	a1	a2	a3	a4
Last 4 days	b1	b2	b3	b4
Last 3 days	c1	c2	c3	c4
Last 2 days	d1	d2	d3	d4
Last 1 day	e1	e2	e3	e4

where $N(u, c, t)$ represents the channel that the user watches at the recommendation time t, $N(u, rec_{list}, t)$ represents the recommendation list that the RS provides at the recommendation time t, $N(rec)$ represents the total number of times that the RS works. If $N(u, rec_{list}, t)$ contains $N(u, c, t)$ and $T(u, c, [t, t + \Delta t])$ is bigger than $\frac{\Delta t}{2}$, we will set $score_{one}^{evaluate}$ to 1. Otherwise, it will be set to 0 (Δt is set to 10 min).

4.1 Base RSs Performance

Items-based Collaborative Filtering RS Performance
In order to intuitively show the impact of Δt and S(t) on the accuracy rate of RS, we plot the four graphs in Fig. 2 to reflect the trend of accuracy rate.

From (a) (b) (c) in Fig. 2, we can observe the three findings: (1) The overall performance of the RS becomes better as $S(t)$ becomes smaller. It shows that the ability of the historical records selected to reflect the changes of the users' interest gradually becomes stronger. (2) When the $S(t)$ is set as the expanding 1 hour before and after the recommendation time, the top-N (when N is smaller than 5) channel recommendation list has the highest accuracy. It indicates that this time granularity can take into account the channel changes in the time schedule. (3) The accuracy rate of top-N (N is bigger, e.g., 10) channel recommendation list decreases as the Δt

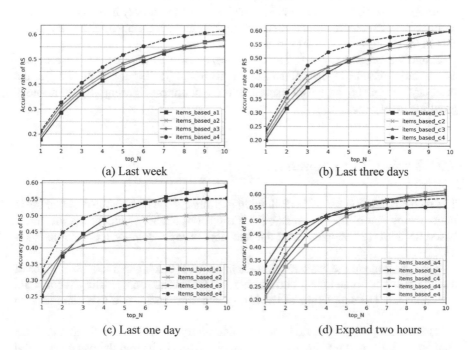

(a) Last week

(b) Last three days

(c) Last one day

(d) Expand two hours

Fig. 2 The accuracy rate of items-based collaborative filtering RS

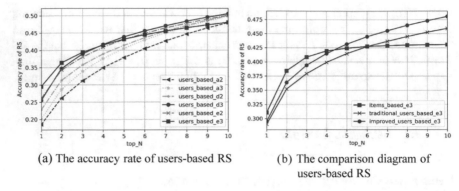

(a) The accuracy rate of users-based RS   (b) The comparison diagram of
                                                      users-based RS

Fig. 3 The accuracy rate of users-based collaborative filtering RS

decreases. This is due to the fact that the historical records are insufficient, which results in that the lengths of most top-10 channel lists are actually less than 10.

Figure 2d shows that the accuracy rate of top-N (when N is smaller than 3) channel recommendation list increases as Δt decreases, while the accuracy of top-10 channel list decreases. It reflects that users' recent records are more able to track users' interest changes, but the insufficient records lead to the fall of the accuracy rate of top-10.

Users-based Collaborative Filtering RS Performance

In this experiment, we do not consider the best time granularity in the above experiment because the records of the nearest neighbors at the recommendation time t have shown the changes of channel schedule.

From Fig. 3a, we can see that the influence of $S(t)$ and Δt on the accuracy rate of users-based RS is similar to the items-based RS. Figure 3b shows that the accuracy rate of the improved users-based collaborative filtering algorithm has a great improvement.

4.2 Fusion RS Performance

The recommendation list of the following two models can be combined to maximize the accuracy rate of the fusion RS: (1) Basic list: the label of input data for items-based RS is e4; (2) Supplemental list: the label of input data for users-based RS is d3.

The advantage of the basic recommendation list is that it can track the changes of users' interest to some extent and its shortage is that the users' history records are not sufficient. The supplemental list is opposite to it. So the two lists can make up each other.

The accuracy rate of fusion model achieves up to 58% when three candidate channels are given. And, the accuracy rate of the top-1 channel list has been raised by 3.41% and the accuracy rate of the top-10 channel list has been raised by 20.32%.

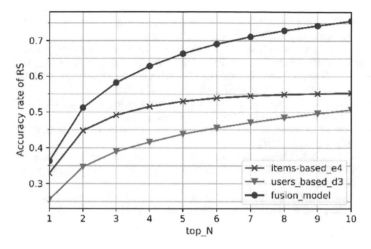

Fig. 4 The improvement of fusion RS

We conclude that the fusion model can greatly improve the accuracy rate of RS from Fig. 4.

5 Conclusion

In this paper, to optimize the accuracy rate of RS, we design the fusion recommendation model for live RS based on several base collaborative filtering algorithms. The results of the experimental show that our fusion mode has better performance than other competing models, which is also beneficial to improve users' QoE.

Acknowledgements This work is partly supported by the National Natural Science Foundation of China (Grants No. 61571240, 61671474), the Jiangsu Science Fund for Excellent Young Scholars (No. BK20170089), the ZTE program "The Prediction of Wireline Network Malfunction and Traffic Based on Big Data" and the Priority Academic Program Development of Jiangsu Higher Education Institutions.

References

1. Yu, C., Ding, H., Cao, H., Liu, Y., Yang, C.: Follow me: personalized IPTV channel switching guide. In: Proceedings of the 8th ACM on Multimedia Systems Conference, pp. 147–157. ACM, Taipei, Taiwan (2017)
2. Kwon, O.B.: "I know what you need to buy": context-aware multimedia-based recommendation system. Expert Syst. Appl. **25**(3), 387–400 (2003)
3. Prota, T., Bispo, A., Ferraz, C.: A literature review of recommender systems in the television domain. Expert Syst. Appl. **42**(22), 9046–9076 (2015)

4. Zhou, L., Wu, D., Dong, Z., Li, X.: When collaboration hugs intelligence: content delivery over ultra-dense network. IEEE Commun. Mag. **55**(12), 91–95 (2017)
5. Zhou, L., Wu, D., Chen, J., Dong, Z.: Greening the smart cities: energy-efficient massive content delivery via D2D communications. IEEE Trans. Ind. Inf. **14**(4), 1626–1634 (2017)
6. Xijun, Y.E., Gong, Y.: Study on diversity of collaborative filtering recommendation algorithm based on item category. Comput. Eng. **41**(10), 42–46 (2015)
7. Wu D., Zhou L., Cai Y.M., Qian Yi.: Collaborative caching and matching for D2D content sharing. IEEE Wirel. Commun. Mag. **25**(3), 43–49 (2018)
8. Deshpande, M., Karypis, G.: Item-based top-N recommendation algorithm. ACM Trans. Inf. Syst. **22**(1), 143–177 (2004)
9. Lee, S.I., Lee, S.Y.: Integration of user profiles and real-time context information reflecting time-based changes for the recommendation system. Int. J. Fuzzy Logic Intell. Syst. **8**(4), 270–275 (2008)
10. Hu, L., Song, G., Xie, Z., Zhao, K.: Personalized recommendation algorithm based on preference features. Tsinghua Sci. Technol. **19**(3), 293–299 (2014)

Part VI
Pattern Recognition

Virtual Reality Projection Alignment for Automatic Measuring Truck's Box Volume in Single Image

Wei Sun, Lei Bian, Peng-hui Li and Yue-cheng Li

Abstract A virtual reality projection alignment-based volume measurement using single image is presented to reduce the cost of truck's box volume measuring system. This paper uses the ASM algorithm to establish a virtual reality environment with the parameters of the camera. Then, the tank volume of the truck is determined in the virtual reality environment with by using the EPnP algorithm and the geometric correction calculation. Experiment results show that with the proposed method single image is involved to measure the volume of the truck's box automatically, complexity of the measuring system is low which leads to good real-time performance, and the measurement error is within 5%.

Keywords ASM · EPnP · Geometry correction · VR · Volume measurement

1 Introduction

Sufficing the requirements of truck box volume measurement, an automatic measurement method based on virtual reality technology [1] and single image is presented in this paper.

The remainder of the paper is organized as follows. Section 2 presents a detailed theoretical derivation and calculation of the proposed method. In Sect. 3, we summarize the main implementation steps of the method and give the overall block diagram. Experimental verification and error analysis are shown in Sect. 4. Finally, Sect. 5 concludes the whole work.

W. Sun (✉) · L. Bian · P. Li
School of Aerospace Science and Technology, Xidian University, Xi'an 710118, China
e-mail: wsun@xidian.edu.cn

Y. Li
Department of Neurological Surgery, University of Pittsburgh, Pittsburgh, PA 15213, USA

© Springer Nature Singapore Pte Ltd. 2019
J.-S. Pan et al. (eds.), *Genetic and Evolutionary Computing*,
Advances in Intelligent Systems and Computing 834,
https://doi.org/10.1007/978-981-13-5841-8_24

2   Volume Measurement Algorithm in VR Environment

Our proposed method takes advantage of the equation of ground surface where the truck stands and a virtual camera to establish the virtual reality environment. A further discussion of this method is below by starting with constructing the virtual reality environment.

2.1   Virtual Reality Environment

In our proposed method, we use OpenGL virtual cameras to determine the ratio between the virtual reality environment and the real environment. And, a virtual ground is set up in the virtual camera coordinate system.

The established virtual reality environment is shown in Fig. 1 with the relative positions of camera coordinate system, world coordinate system, screen coordinate system, and image coordinate system.

OpenGL fluoroscopic imaging process can be described as follows: a point in three-dimensional object is transformed into the image space coordinates with transformations of rotation, translation, and scaling devoted by the 3D model matrix M; then, the reality imaging results can be obtained by perspective projection matrix P. Finally, screen image coordinate of a three-dimensional scene with affine transformation F can be figured out, and a two-dimensional display of three-dimensional model is got. The process can be expressed as

$$
Z_w \begin{pmatrix} u \\ v \\ 1 \end{pmatrix} = \begin{pmatrix} \alpha_x & 0 & u_o & 0 \\ 0 & \alpha_y & v_o & 0 \\ 0 & 0 & 1 & 0 \end{pmatrix} \begin{pmatrix} R & T \\ 0^T & 1 \end{pmatrix} \begin{pmatrix} X_w \\ Y_w \\ Z_w \\ 1 \end{pmatrix} = KG \begin{pmatrix} X_w \\ Y_w \\ Z_w \\ 1 \end{pmatrix} \tag{1}
$$

Fig. 1 Virtual reality environment

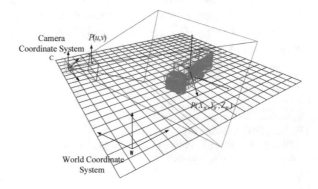

K is the matrix of intrinsic parameter of a camera, R and T are the external parameters of a camera, R is rotation matrix, T is translation matrix, and G is the matrix of camera external parameter. K and G are calculated by a flat panel calibration method [2].

2.2 Recognition of Truck Feature Points

ASM algorithm [3] has been successfully applied to face recognition which has a good result [4]. In this paper, ASM algorithm is used.

First, selecting 50 truck images as a training set is shown in Fig. 2, and manually labeling 34 feature points of each training image is given in Fig. 3a.

In order to eliminate the influence of trucks' relative position in different images, we use Procrustes method [5] to register shape vector as $H'_c = (z'_{c1}, o'_{c1}, z'_{c2}, o'_{c2}, \ldots, z'_{ck}, o'_{ck}, \ldots, z'_{cn}, o'_{cn})^T$; then, the principal component analysis is used with shape vector H'_c to obtain partial gray level information of each truck feature points, so an active shape model is established. Finally, using established

Fig. 2 Truck training samples

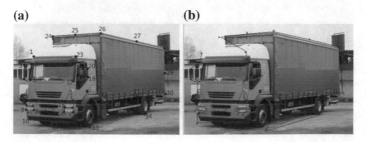

Fig. 3 Truck feature points

active shape model to identify the feature point x_k of truck image to be measured, the recognition result is shown in Fig. 3b.

2.3 Matching 3D Model with Truck Image

Some existing 3D model correction methods require manual operation [6–8], and in this paper, a method with initial pose estimation and local fine-tuning is presented.

Estimation initial pose of truck 3D model. We define a set λ representing 3D-2D correspondences. Selecting five 3D-2D correspondences from the set, where X_j belongs to the truck 3D model, x_j is in the image. $T \in \mathbb{R}^3$ is the truck 3D model translation. R and T are estimated by using EPnP algorithm [9]. Substituting into (2) these 10 points' coordinates and the virtual camera intrinsic parameters matrix K which is also the known parameters.

$$\varphi_j \left[x_j \; 1 \right]^T = K * \sum_{m=1}^{4} \alpha_{jm} X_j^T \tag{2}$$

The initial pose of truck 3D model is determined as shown in Fig. 4a.

Geometry correction of truck 3D model. As shown in Fig. 4a, the initial projection of the truck's 3D model and the truck in the image is not entirely consistent, and it is caused by the difference of geometry between the truck 3D model and the truck being measured. We use a set γ feature points to correct the geometry of the truck's 3D model. The feature points are matched with the projections of a set of control points. So, we can obtain the target points which are corresponding to the control points.

As shown in Fig. 5, an energy function is constructed to constrain projection points X'_i of X_i matched with unobstructed feature points x_i, we calculate the ray $R_i = K^{-1} * [x_i z_i]^T$, $z_i \in [0, 1]$, back-projected through each feature points, and the ray starts from virtual camera optical center and points to the virtual space of infinity ($z_i = 1$). $E_1(X)$ is the sum distance of points between X_i and the ray R_i:

(a)   **(b)**

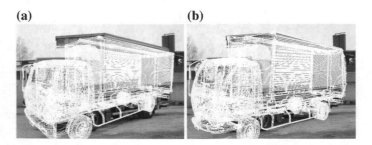

Fig. 4 3D truck model correction

Fig. 5 Control points matching

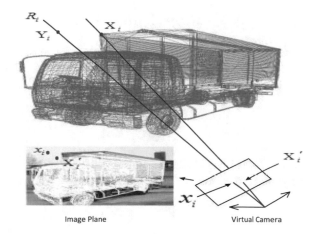

Image Plane   Virtual Camera

$$E_1(X) = \sum_{i=1}^{9} \left\| X_i - K^{-1} * [\, x_i \ z_i \,]^T \right\| \tag{3}$$

By minimizing the energy function $E_1(X)$, we calculate $Z_i = a_i$, matched target points are obtained. It was found that only using the minimized energy function to correct the truck 3D model's geometry will lead to structural distortion of the 3D model. Therefore, in this paper, we constrain this correction of geometry according to the truck 3D model facade geometric relationship. As shown in Fig. 6a, line segments which are determined by Y_i have the same geometric relationships, such as $Y_5 Y_6$ and $Y_6 Y_7$ are perpendicular, then $\vec{Y_5 Y_6} \cdot \vec{Y_6 Y_7} = 0$ can be given as

$$\left[K^{-1} * [(x_6 - x_5)(z_6 - z_5)]^T \right] \cdot \left[K^{-1} * [(x_7 - x_6)(z_7 - z_6)]^T \right] = 0 \tag{4}$$

Using the method discussed above to construct equality constraints, we get

(a)   (b)

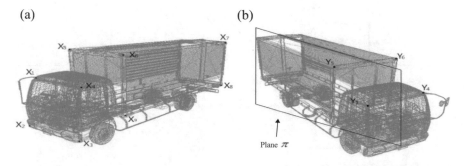

Plane π

Fig. 6 Control points and target points

$$g(z_i) = \sum (D_i z_i^{\varsigma_i} + F_i) = 0, \quad i = 1, 2, \dots, 9 \tag{5}$$

where D_i and F_i are constant factor, ς_i is the power of z_i. Using $g(z_i) = 0$ to constrain min $E_1(X)$, which can ensure that truck 3D model's distortion does not occur, then the Eq. (6) can be solved to get target points Y_i.

$$\begin{cases} \min E_1(X_i) \\ g(z_i) = 0 \end{cases}, \quad i = 1, 2, \dots, 9 \tag{6}$$

For some invisible feature points in truck picture, we solve them according to the obtained target points Y_i and the symmetry of truck's box, as shown in Fig. 6b.

3 Steps of Truck Volume Measurement

In summary, our proposed method comprises of the following steps:

1. Obtain the camera intrinsic matrix K through camera calibration, and calculate the equation for the ground surface where truck stands;
2. Then, set a virtual ground according to the equation of ground surface;
3. Obtain truck's feature points by ASM algorithm;
4. Select five 3D-2D correspondences from the truck 3D mode, use EPnP algorithm to estimate the truck 3D model's initial pose;
5. Use the projections of a set control points to match with a set feature point, then obtain the target points, solve invisible points by the symmetry of truck's box;
6. Control points and target points are used to correct geometry of the truck 3D model so that the truck 3D model can precisely match with the measured truck;
7. Measure the volume of the measured truck by calculating the volume of the truck 3D model.

Procedures of this method are depicted in Fig. 7.

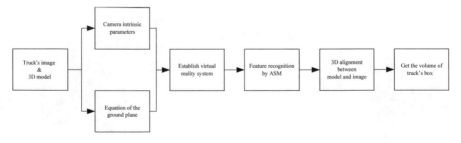

Fig. 7 Framework of volume measurement algorithm

(a)   (b)   (c)   (d)

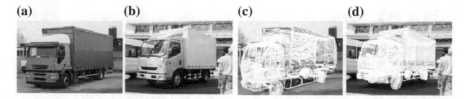

Fig. 8 Different images and matching results in our experiment

Table 1 Comparison measurement and the true volume value

Truck 1 (truck 2)	Actual value	Measurement value	Relative error (%)
Length (m)	6.20 (4.85)	5.90 (5.05)	4.84 (4.12)
Width (m)	2.30 (1.94)	2.41 (1.85)	4.78 (4.64)
Height (m)	2.40 (1.80)	2.29 (1.88)	4.58 (4.44)
Volume (m^3)	34.22 (16.94)	32.56 (17.56)	4.85 (3.66)

4 Experimental Results

To verify the effectiveness and feasibility of this method, we obtain multiple truck images with different certain angle, and 3D matching with truck image are implemented with volume measurement method proposed in this paper, the results are shown in Fig. 8, Fig. 8a, b are specification truck different from each other, Fig. 8c, d are the matched results between truck 3D models' projection and trucks under measuring.

Figure 8a, b of the truck box volume measurement results are shown in Table 1.

As Table 1 shows the measurement error of truck box' volume is less than 5%, which can meet the real application requirements of measurement accuracy. According to the results, the main factors affect the truck volume measurements are:

- Truck's information cannot be fully used by ASM algorithm, which produces inaccurate feature detection and subsequently affects measurement accuracy of truck volume;
- The alignment between the 3D model and image of truck under measurement cannot be perfect which leads the truck 3D models not completely consistent with measured truck, so measurement accuracy of truck volume is affected.

5 Conclusions

This method breaks the limitations of the existing measurement methods, which cannot automatically measure the volume of the truck.

Experiments showed that the error between truck's volume measured automatically by this method and the true value of truck's volume is less than 5%, which indicates that this method is suitable for truck volume measurement in the real application. ASM-based feature recognition and the geometry correction of the model are key steps in the proposed method, and in future studies, we will further study the ASM algorithm and the geometry correction algorithm of truck 3D model to improve measurement accuracy. At the same time, we will also extend our method to volume measurements for other objects.

Acknowledgements This work was supported by National Nature Science Foundation of China (NSFC) under Grants 61671356, 61201290.

References

1. Burdea, G.: Virtual reality technology an introduction. In: IEEE Computer Society, p. 307 (2005)
2. Sun, W., Sun, N., Guo, B., et al.: An auxiliary gaze point estimation method based on facial normal. Formal Pattern Anal. Appl. (2014)
3. Cootes, T.F., Taylor, C.J., Cooper, D.H., et al.: Active shape models-their training and application. Comput. Vis. Image Underst. **61**(1), 38–59 (1995)
4. Shan, S., Gao, W., Cao, B.: An improved active shape model for face alignment. In: Fourth IEEE International Conference on, 523–528. IEEE (2002)
5. Igual, L., Perez-Sala, X., Escalera, S., et al.: Continuous generalized procrustes analysis. Pattern Recogn. **47**(2), 659–671 (2014)
6. Nealen, A., Sorkine, O.: A sketch-based interface for detail-preserving mesh editing. ACM Trans. Gr. **24**(3), 1142–1147 (2005)
7. Alexa, O.S.M.: As-Rigid-As-possible surface modeling. In: Eurographics Symposium on Geometry Processing (2007)
8. Kholgade, N., Simon, T., Efros, A., et al.: 3d object manipulation in a single photograph using stock 3d models. ACM Trans. Gr. (TOG) **33**(4), 127 (2014)
9. Lepetit, V., Moreno-Noguer, F., Fua, P.: EPnP: an accurate $O(n)$ solution to the PnP problem. Int. J. Comput. Vis. **81**(2), 155–166 (2009)
10. Xu, C., He, Y., Khanna, N., et al.: Model-based food volume estimation using 3D pose. In: 2013 20th IEEE International Conference on, pp. 2534–2538. IEEE (2013)
11. Zhang, Z., Yang, Y., Yue, Y., et al.: Food volume estimation from a single image using virtual reality technology. In: IEEE Annual Northeast Bioengineering Conference, pp. 1–2 (2011)
12. Schaefer, S., Mcphail, T., Warren, J.: Image deformation using moving least squares. ACM Trans. Gr. **25**(3), 533 540 (2006)
13. Castillo-Castaneda, E., Turchiuli, C.: Volume estimation of small particles using three-dimensional reconstruction from multiple views. In: Lecture Notes in Computer Science, pp. 218–225 (2008)

License Plate Occlusion Detection Based on Character Jump

Wenzhen Nie, Pengyu Liu and Kebin Jia

Abstract The license plate location is the basis of the license plate occlusion detection. On the basis of the positioning, the key is to accurately determine the license plate occlusion. This paper proposes a license plate occlusion detection algorithm based on character jumping. For the positioned occluded license plate area, it is determined whether the license plate is occluded according to the number of license plate character jumps. The number of jumps of a normal license plate is greater than or equal to 14 times. If the license plate is blocked, the number of jumps is less than 14 times. The experimental results show that the detection effect of the license plate occlusion based on this algorithm has a good judgment result.

Keywords License plate positioning · Character jump · License plate occlusion determination

1 Introduction

In the Beijing area, due to the large population, and low public transportation sharing rate, the public demand for taxis is increasing, which directly leads to the increasingly complicated problems in the operation of taxis. At present, the types of violations that have been applied to the identification of off-site law enforcement means include: blocking license plates, smoking, and private pick-up. The existing research on license plates is mainly for the positioning of license plates and the identification of license plates, which are applied to highways, parking lots, and road toll stations. However, there is no specific application study for the driver to block the license

W. Nie · P. Liu (✉) · K. Jia (✉)
Faculty of Information Technology, Beijing University of Technology, Beijing 100124, China
e-mail: liupengyu@bjut.edu.cn

Beijing Laboratory of Advanced Information Networks, Beijing 100124, China
kebinj@bjut.edu.cn

Beijing Key Laboratory of Computational Intelligence and Intelligent System,
Beijing University of Technology, Beijing 100124, China

© Springer Nature Singapore Pte Ltd. 2019
J.-S. Pan et al. (eds.), *Genetic and Evolutionary Computing*,
Advances in Intelligent Systems and Computing 834,
https://doi.org/10.1007/978-981-13-5841-8_25

plate. Therefore, the license plate occlusion detection studied in this paper has great necessity and practical significance.

Literature [1, 2] proposed a method for introducing license plate occlusion and verified it. In [3], a recognition method for occlusion and defacement of character parts is proposed. These methods only recognize characters that are missing from the stroke. It does not apply to the occlusion or full occlusion of the license plate part of this article.

Based on the previous research, this paper proposes a method based on taxi license plate occlusion detection for railway stations, airports, etc., based on vehicle detection and license plate location. First, the surveillance video image is processed, and the taxi is detected from the video, thereby realizing the positioning of the license plate, thereby counting the number of jumps of the license plate character, and finally determining whether the license plate is occluded. The overall process is shown in Fig. 1.

Fig. 1 License plate occlusion detection process

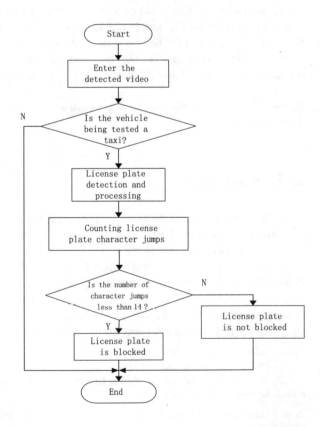

2 Taxi Detection

To identify a taxi from a social vehicle, we needed to understand the characteristics of the taxi. For Beijing taxis, its characteristics mainly include the following two aspects:

(1) The color in the middle of the body is yellow. According to the survey, the color of taxis in Beijing the middle of the body is yellow, which can be used to determine whether the vehicle is a taxi.
(2) The first two license plates are "Jing B". It is also possible to determine whether the vehicle is a taxi by identifying the first two characters.

However, the above two methods are not applicable to the detection of license plate occlusion. Through the analysis of illegal video, the method of using the body color is to detect the taxi from the side, and the license plate is in front of the car and behind the car, so it is not feasible. This article is to detect the occlusion of the license plate. If the license plate is blocked, it is not feasible to use the method of identifying "Jing B".

Therefore, in order to determine whether a taxi vehicle is included in the image, it is necessary to model multiple features of the taxi. Here, the Haar-like features [4, 5] proposed by Viola et al. are used as feature vectors for vehicle detection to characterize taxis. The eigenvalues are calculated as the sum of the gradation integrals of the constituent rectangular regions, as shown in formula (1).

$$\text{Feature} = \sum_{i \in I\{1,2,....N\}} \omega_i * RecSum(r_i) \tag{1}$$

where $\omega_i \in R$ is the weight of the rectangle; $RecSum(r_i)$ is the area gradation integral of the rectangle r_i; N is the number of rectangles formed.

This article uses 8000 positive samples and 24,000 negative samples for training. The positive samples are shown in Fig. 2. The test results for the taxi are shown in Fig. 3.

3 License Plate Positioning

The license plate location is to determine the position of the license plate from the detected taxi, and segment the position from the area. The effect of the license plate location directly affects the statistics of the number of subsequent license plate character jumps, thus affecting the judgment effect of the license plate occlusion.

A. Characteristics of the license plate area

The main features of the Beijing taxi license plate are mainly as follows:

Fig. 2 Car positive samples

Fig. 3 Taxi test result

(1) Character characteristics. The first two characters of the Beijing taxi license plate are "Jing B". The next five characters are composed of letters or numbers, and each character has a width of 45, a height of 90, a spacer width of 10, and a character interval of 12.

(2) Geometric features. The aspect ratio of the entire license plate is approximately 3:1.

(3) Grayscale variation characteristics. The gray levels in the image are different in the license plate because the internal gray level of the character itself and the license plate background color are uniform. Therefore, the horizontal line passing through the license plate presents a continuous wave, valley, and peak distribution.

3.1 License Plate Positioning Method

In order to meet the real-time requirements of license plate occlusion detection, the current mainstream method of license plate location is still based on grayscale maps. Some things will also add color information for positioning correction or fine positioning. The license plate location methods mainly have the following categories:

(1) A method based on texture features [6–8]. Commonly used textures have jumps, scans, projections, and so on. A common disadvantage of this type of method is that the stability of the features depends on the image quality, so there is usually a high demand for image quality.

(2) Based on mathematical morphology. This type of method focuses on the overall characteristics of the license plate and processes it directly on the binary map. This type of method combined with other methods can improve the anti-interference and positioning accuracy of license plate positioning.

(3) Vehicle license plate location method based on edge detection [9, 10]. The method uses the feature that the gray-level changes sharply at the edge of the license plate to detect the area where the license plate is located. It has high positioning accuracy, short reaction time, and can effectively remove noise, and is suitable for images containing multiple license plates.

(4) The method of deep learning [11, 12]. The machine learning method is used to train the license plate and finally find the best position in the license plate area. This method has a high accuracy of license plate location but requires a large number of positive and negative samples for training.

However, for occluded license plates, the texture features, color features, and edge features are reduced, so the above method is not applicable. Therefore, the AdaBoost algorithm is used in this paper, which belongs to machine learning. This method is used to train the license plate. For the detection of the occluded license plate, the occluded license plate is added to the training sample, with 7200 positive samples and 21,000 negative samples. And the test results are shown in Fig. 4.

Fig. 4 License plate positioning

4   Determination of License Plate Occlusion

4.1   License Plate Occlusion Determination Method

On the basis of vehicle detection and license plate location, the determination of license plate occlusion is realized. Therefore, there are three main ideas for determining license plate occlusion

(1) License plate character recognition. If the character can be accurately recognized, it can be determined that the license plate is not blocked, but if the license plate character cannot be recognized, it is determined that the license plate is blocked.

(2) The threshold value of the license plate characters is judged. For the positioned license plate, after the extraction and binarization, the proportion of the license plate character pixel points to the entire license plate is counted, thereby judging whether the license plate is occluded.

(3) The number of license plate character jumps [13]. In the current research, it is often determined whether the area is a license plate according to the number of jumps of the license plate characters.

Through the above analysis, if the method of character recognition is adopted, the training of letters, numbers and Chinese characters is required. For the method of the threshold value of the license plate characters, after the experiment, due to the different character structure, the threshold value is more sensitive, and more false detections will occur. Therefore, the number of character jumps is used to determine whether the license plate is occluded.

4.2   Determination of the Number of Character Jumps

a.   License plate image binarization

Binarization is the key to the number of statistical jumps. Therefore, this paper uses the maximum inter-class variance method to binarize the image.

In the ideal case of uniform illumination, no noise and no interference, the grayscale change in the license plate image is gentle, and it can be assumed that the character grayscale is g_1, the background grayscale is g_2, and $0 \le g_1, g_2 \le 255$. The proportion of character pixels is r_1, and the proportion of background pixels is r_2, then $0 \le r_1, r_2 \le 1$, and $r_1 + r_2 = 1$.

The average value of the license plate grayscale image is

$$M = r_1 * g_1 + r_2 * g_2, \quad g_1 < M < g_2 \tag{2}$$

The variance is calculated as follows:

$$C^2 = r_1 * (g_1 - M)^2 + r_2 * (g_2 - M)^2 \tag{3}$$

From formula (2)

$$(g_2 - M) = -\frac{r_1}{r_2}(g_1 - M) \tag{4}$$

So

$$C^2 = \frac{r_1}{r_2} * (g_1 - M)^2 \tag{5}$$

From this, the character grayscale is

$$g_1 = M \pm \frac{\overline{r_2}}{r_1} C \tag{6}$$

And the background grayscale is

$$g_2 = M \pm \frac{\overline{r_1}}{r_2} C \tag{7}$$

The ratio of the pixel points of the characters in the formula (6) and the formula (7) r_1 is unknown, but according to the statistical rule, the ratio of the pixel points of the characters in the license plate is from 0.3 to 0.4. Since the license plate area automatically positioned by image segmentation is substantially accurate near the four borders of the license plate, the standard license plate height and length ratio RatioLP is $140: 440 = 0.3182$, r_1 is close to the RatioLP. Since the required threshold

Fig. 5 License plate image
binarization

is obtained here, r_1 is approximated by RatioLP, and the error of the coarse threshold
T1 (T2) is about 10 pixels.

The binarization result of the license plate image is shown in Fig. 5.

b.   Character jump statistics

Whether the license plate is occluded can be determined by the method based on the
grayscale jump.

The algorithm steps are as follows:

(1)   In order to accurately determine whether the license plate is blocked, the
      extracted license plate area is reduced by a certain ratio to ensure that the license
      plate frame is removed to prevent false detection. Use to indicate the line number
      numbered from the bottom up from the binarized license plate image.
(2)   Count the total number of grayscale jumps in the i-th row, that is, the number
      of pixels whose pixel gray value changes from 0 to 255 or from 255 to 0, which
      is denoted as. If is greater than or equal to 14 times, the license plate is normal.
      Otherwise, it is determined that the license plate is blocked.

The reason why the character jump method can effectively determine whether
the license plate is occluded is that the inherent feature of the license plate causes a
grayscale jump between the character and the background.

c.   License plate occlusion judgment result

The results of the experiment are shown in Table 1. The experimental results are
shown in Fig. 6.

Table 1   Experimental results

Video source	Total number of license plates	Correct rate (%)	False detection rate (%)	Missed detection rate (%)
T2 second floor police station exit	118	78	17	10
T1 social lane	56	80	15	8

Fig. 6 License plate judgment result

5 Conclusions

This paper designs and implements the detection of license plate occlusion. A taxi license plate occlusion detection method for airports, railway stations, and other places is proposed. The experimental results show that the method can effectively detect common license plate occlusion problems, such as full occlusion and partial occlusion. It has not yet entered the practical application stage, but laid the foundation for further research.

Acknowledgements This paper is supported by the Project for the National Natural Science Foundation of China under Grants No. 61672064, the Beijing Natural Science Foundation under Grant No. 4172001, the China Postdoctoral Science Foundation under Grants No. 2016T90022, 2015M580029, the Science and Technology Project of Beijing Municipal Education Commission under Grants No. KZ201610005007, Beijing Municipal Education Committee Science Foundation under Grants No. KM201810005030, and Beijing Laboratory of Advanced Information Networks under Grants No. 040000546617002, Beijing Municipal Communications Commission Science and Technology Project under Grants No. 2017058.

References

1. Xiujuan, L.I., Hebiao, Y.A.N.G., Ying, W.E.I.: Partially occluded license plate location method based on multi-layer edge constraint and region merging. Softw. Guide **14**(7), 196–199 (2015)
2. Nianfeng, S.: Intelligent detection of illegal vehicles involved in card-based images. Nanjing Normal University (2013)

3. Yu, W.: Research on license plate recognition algorithm based on PCA and grid features. Xidian University (2013)
4. Yongquan, C., Ying, C., Xuesan, C.: Vehicle detection and tracking algorithm based on adaboost classifier. Comput. Technol. Dev. **27**(9), 165–168 (2017)
5. Yulin, D.: Research and implementation of license plate detection based on adaboost algorithm in OpenCV. J. Guangxi Teachers Univ. (Nat. Sci.) **28**(1), 109–112 (2011)
6. Wenfeng, L., Hongying, Z.: License plate location method based on texture features. Microcomput. Appl. **3**, 41–43 (2014)
7. Qian, X.: License plate location algorithm based on image preprocessing and texture features. Electron. Des. Eng. **22**(16), 13–17 (2014)
8. Yang, J., Feihu, Q.I.: A license plate locating approach based on shape and texture characteristics. Comput. Eng. **32**(2), 170–172 (2006)
9. Ning, L., Yanlei, X., Ning, L., et al.: License plate location method based on mathematical morphology and color features. J. Gr. **35**(5), 774–779 (2014)
10. Lili, L., Xingwu, W.: Vehicle license plate location algorithm based on Sobel operator edge detection and mathematical morphology. MODERN Electron. Technol. v.38 **445**(14), 98–100 (2015)
11. Jing, F.: Vehicle detection and license plate location based on deep learning. Jiangxi University of Science and Technology (2017)
12. Haiyan, L., Furong, C.: A complex environment license plate location method based on deep learning text detection. Mod. Comput. Prof. Edit. **33**, 10–14 (2017)
13. Fei, G., Kaicheng, M., Zhenggao, H., et al.: Research on license plate location method based on grayscale jump and character interval mode. J. Comput. Measure. Control **24**(4), 219–221 (2016)

Reversible Watermarking Based on Adaptive Prediction Error Expansion

Qi Li, Bin Yan, Hui Li and Jeng-Shyang Pan

Abstract In traditional prediction error expansion (PEE) based reversible watermarking (RW), the watermark bits are embedded into the two peaks of the global prediction error histogram. This scheme ignores some bins in local histogram. To improve the utilization of prediction error, a method based on locally adaptive PEE is proposed. The original image is divided into two regions. In the first region, the image is divided into several blocks and the local prediction error histogram of each block is obtained after checkerboard prediction. Then, the two peaks in the local histogram of each block are used for watermark embedding. In the second region, the least significant bit (LSB) replacement is used to embed the auxiliary information and compressed positioning map. It is verified that, under the same image distortion, this method provides higher embedding capacity.

Keywords Reversible watermarking · PEE · Checkerboard prediction ·
Global histogram · Local histogram

1 Introduction

With the increasing number of data transmitted on the network, the security and integrity of data have received widespread attention. This requires that the data needs to be embedded with additional information, such as authentication information, copyright information, etc. What's more, the embedded information needs to be extracted completely and the original data needs to be restored losslessly at the receiving end. Using reversible watermarking(RW) technology, we can not only

Q. Li · B. Yan (✉)
College of Electronics Communication and Physics, Shandong University of Science
and Technology, Qingdao 266590, People's Republic of China
e-mail: yanbinhit@hotmail.com

H. Li
College of Computer Science and Engineering, Shandong University of Science
and Technology, Qingdao 266590, People's Republic of China

© Springer Nature Singapore Pte Ltd. 2019   243
J.-S. Pan et al. (eds.), *Genetic and Evolutionary Computing*,
Advances in Intelligent Systems and Computing 834,
https://doi.org/10.1007/978-981-13-5841-8_26

embed additional data into the carrier data, but also extract this additional data and fully recover the carrier to achieve the purpose for authenticating the content [1, 2]. At present, according to different embedding methods, RW algorithms are mainly divided into three categories: lossless compression [3], histogram shifting [4–8], and difference expansion [9, 10]. Although the research and application of RW has made great progress, improving the embedding capacity while ensuring the quality of watermarked image is always a difficult problem.

To hide secret information, the earliest RW technique utilized available redundant space through lossless compression. Celik et al. proposed an algorithm to find room for RW by compressing the least significant bit (LSB) [3]. However, this compression method is limited by the complexity of the image content. The pixel value histogram shifting [4] algorithm has less embedding distortion, but the embedding capacity is limited due to the fixed number of selected peak points. To overcome this shortcoming, Tian proposed an algorithm based on difference expansion [9], which provides an embedding rate of 0.5 bpp (bits per pixel) for single-layer embedding. Thodi et al. proposed an algorithm based on prediction error expansion (PEE) [11], which uses the difference between the predicted value and the original pixel value to embed the secret information. Both embedding capacity and quality of watermarked image have been improved. Since the histogram shifting algorithm does not have a high utilization rate of the local area of the original image, it is difficult to obtain a high embedding capacity. Li et al. adopted a local optimization method to select pixel modules with less error to embed, which further improved the quality of watermarked image [12]. Wang et al. proposed an algorithm based on dynamic PEE [13]. According to different embedding capacity, the algorithm chooses different prediction errors to embed secret information under the premise of small distortion. For images with simple textures, the algorithm effectively improves the quality of watermarked image, but for images with more complex textures, the quality is not significantly improved.

In order to improve the embedding capacity, this paper proposes a RW algorithm based on adaptive PEE. The basic idea is to make full use of the error histogram. To this end, the watermark is embedded using the two peaks of the local histogram which corresponds to multiple histogram bins in the global histogram. Since more than two peaks can be used for embedding, it has a higher embedding capacity.

This paper is organized as follows: Sect. 2 introduces the checkerboard prediction method. Section 3 presents the RW algorithm using adaptive PEE proposed in this paper. The experimental results are given in Sect. 4. Finally, we conclude this paper in Sect. 5.

2 Checkerboard Prediction

The checkerboard prediction [14] method uses 25% of the pixels in the original image to predict the remaining 75% of the pixels. As shown in Fig. 1a, each small square represents a pixel. The prediction process is divided into two steps. In the first step,

(a)

O	2	O	2	O	2	O	2
2	1	2	1	2	1	2	1
O	2	O	2	O	2	O	2
2	1	2	1	2	1	2	1
O	2	O	2	O	2	O	2
2	1	2	1	2	1	2	1
O	2	O	2	O	2	O	2
2	1	2	1	2	1	2	1

(b)

$x_{i-1,j-1}$	$x_{i-1,j}$	$x_{i-1,j+1}$
$x_{i,j-1}$	$x_{i,j}$	$x_{i,j+1}$
$x_{i+1,j-1}$	$x_{i+1,j}$	$x_{i+1,j+1}$

Fig. 1 The process of checkerboard prediction: **a** Pixel position of checkerboard prediction. **b** The neighborhood pixels of $x_{i,j}$

the position marked "1" is predicted by the position marked "O". In the second step, the position marked "2" is predicted with the positions marked "O" and "1".

The prediction process is performed in a local 3×3 neighborhood, as shown in Fig. 1b. The pixel $x_{i,j}$ is predicted from the neighborhood pixels, where i, j represents the position of the pixel. The predicted pixel $\hat{x}_{i,j}$ in the first step is calculated using the pixel $x_{i-1,j-1}$ in the upper left corner, the pixel $x_{i-1,j+1}$ in the upper right corner, the pixel $x_{i+1,j-1}$ in the lower left corner, and the pixel $x_{i+1,j+1}$ in the lower right corner, as shown in Eq. 1. The predicted pixel $\hat{x}_{i,j}$ in the second step is calculated using the upper pixel $x_{i-1,j}$, the lower pixel $x_{i+1,j}$, the left pixel $x_{i,j-1}$, and the right pixel $x_{i,j+1}$, as shown in Eq. 2.

$$\hat{x}_{i,j} = \begin{cases} \left[\frac{x_{i-1,j-1}+x_{i+1,j+1}}{2}\right], & \text{if } |d_{45}| > |d_{135}|, \\ \left[\frac{x_{i-1,j+1}+x_{i+1,j-1}}{2}\right], & \text{if } |d_{45}| < |d_{135}|, \\ \left[\frac{x_{i-1,j-1}+x_{i+1,j+1}+x_{i-1,j+1}+x_{i+1,j-1}}{4}\right], & \text{otherwise,} \end{cases} \quad (1)$$

$$\hat{x}_{i,j} = \begin{cases} \left[\frac{x_{i,j-1}+x_{i,j+1}}{2}\right], & \text{if } |d_{90}| > |d_{180}|, \\ \left[\frac{x_{i-1,j}+x_{i+1,j}}{2}\right], & \text{if } |d_{90}| < |d_{180}|, \\ \left[\frac{x_{i,j-1}+x_{i,j+1}+x_{i-1,j}+x_{i+1,j}}{4}\right], & \text{otherwise,} \end{cases} \quad (2)$$

where $[x]$ returns the nearest integer to x. d_{45}, d_{135}, d_{90} and d_{180} measure the change of pixels in the 45° direction, the 135° direction, the 90° direction and the 180° direction, respectively

$$|d_{45}| = |x_{i-1,j+1} - x_{i+1,j-1}|,$$

$$|d_{135}| = |x_{i-1,j-1} - x_{i+1,j+1}|,$$

$$|d_{90}| = |x_{i-1,j} - x_{i+1,j}|,$$

$$|d_{180}| = |x_{i,j-1} - x_{i,j+1}|.$$

3   Proposed Algorithm

In the RW algorithm based on PEE, the image quality will be degraded while modifying a large number of prediction errors. To improve the quality of watermarked image, this paper proposes a RW method based on adaptive PEE. We divide the image into several small image blocks, and predict the pixels in each block to obtain prediction errors. The prediction error histogram can be more fully utilized using local prediction error peaks than selecting the prediction error peaks for the entire image. Higher embedding capacity can be achieved with the same quality of watermarked image.

The embedding and extraction algorithms are described in detail below.

3.1   Embedding

1. *Divide area*: Divide the original image into two areas R_1 and R_2. The area R_1 contains the first 99% rows of the original image, and the area R_2 contains the remaining 1% rows. If it is not an integer number of rows, round it off.
2. *Create positioning map*: To prevent overflow after pixel modification, positions (75% of original pixel positions in the checkerboard prediction) with pixel values of 0 and 255 in R_1 are marked as "0" and the remaining positions are marked as "1". Run Length Coding is used to losslessly compress the positioning map. A compressed positioning map of length N_{CPM} is obtained.
3. *Pixel prediction*: The checkerboard prediction method is used to predict pixels $x_{i,j}$ marked "1" on the positioning map. The predicted value is denoted by $\hat{x}_{i,j}$. The pixels marked as "0" on the positioning map remain unchanged. Using the predicted value, we obtain the prediction error $e_{i,j} = x_{i,j} - \hat{x}_{i,j}$.
4. *Generate prediction error histogram*: Divide R_1 into nonoverlapping blocks with the same size, $\{S_1, S_2, \ldots, S_k, \ldots, S_I\}$ (I represents the total number of blocks, $k = 1, 2, \ldots, I$). The prediction error is calculated and the histogram of the prediction error in each block is constructed.

5. *Modify prediction errors and embed secret information*: In each block, the prediction error pair (m_k, n_k) (two prediction error peaks) is selected. The secret information is embedded in the m_k bin or the n_k bin: the prediction error larger than n_k is increased by 1, and the prediction error smaller than m_k is reduced by 1, and the prediction error in the range $(m_k < e_{i,j} < n_k)$ is not modified. The modified prediction error $\tilde{e}_{i,j}$ is

$$\tilde{e}_{i,j} = \begin{cases} e_{i,j}, \text{ if } m_k < e_{i,j} < n_k, \\ e_{i,j} + b, \text{ if } e_{i,j} = n_k, \\ e_{i,j} - b, \text{ if } e_{i,j} = m_k, \\ e_{i,j} + 1, \text{ if } e_{i,j} > n_k, \\ e_{i,j} - 1, \text{ if } e_{i,j} < m_k, \end{cases} \tag{3}$$

where $b \in \{0, 1\}$ is the embedded secret information. The pixels containing secret information are obtained as $x'_{i,j} = \hat{x}_{i,j} + \tilde{e}_{i,j}$.

6. *Embed auxiliary information and compressed positioning map*: The length of auxiliary information (I and two prediction peaks of each block) is denoted by N_{AI}. The total length of the auxiliary information and the compressed positioning map is $B = N_{AI} + N_{CPM}$. Suppose the LSB of the first B pixels in R_2 is S_{LSB}. Embed S_{LSB} into R_1 with the secret information using the above method. The auxiliary information and the compressed positioning map are embedded into the first B pixels of R_2 using the LSB replacement method.

Figure 2 is an illustration of the embedding allocation of the method. The data embedded in R_1 is the secret information and the original LSB stream S_{LSB} of R_2, and the data embedded in R_2 is the auxiliary information and the compressed positioning map.

Since different cover image may have different size and content, the number of pixels to be marked in R_1 is different. So the size of the compressed map is different. The different numbers I of blocks also cause a change in the number of auxiliary information. If the auxiliary information increases, the percentage of the original image occupied by R_2 may increase, such as 2, 3%.

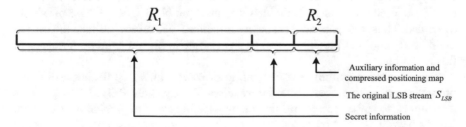

Fig. 2 Allocation of regions for embedding

3.2 Extraction and Image Recovery

1. *Divide area*: Same as the first step of Sect. 3.1.
2. *Extract auxiliary information and compressed positioning map*: The LSBs in the first B pixels of R_2 are extracted to obtain auxiliary information and the compressed positioning map. Decompress the positioning map.
3. *Pixel prediction*: Based on the auxiliary information and the positioning map, the checkerboard prediction method is used to predict the pixel value again. The predicted pixels $\hat{x}_{i,j}$ are obtained.
4. *Extract secret information*: Calculate the modified prediction error $\tilde{e}_{i,j} = x'_{i,j} - \hat{x}_{i,j}$. Determine the secret information embedded in each block based on the modified prediction error

$$b = \begin{cases} 0, & \text{if } \hat{e}_{i,j} \in \{m_k, n_k\}, \\ 1, & \text{if } \hat{e}_{i,j} \in \{m_k - 1, n_k + 1\}. \end{cases} \tag{4}$$

Let the bit stream extracted from the k-th block be $\mathbf{b}_k$, $k = 1, 2, \ldots, I$. All extracted bit stream is spliced into $\mathbf{b}$, where $\mathbf{b} = \mathbf{b}_1 || \mathbf{b}_2 || \cdots || \mathbf{b}_I$, $\mathbf{b}_1 || \mathbf{b}_2$ represents sequence splicing. The prediction errors can be restored

$$e_{i,j} = \begin{cases} \hat{e}_{i,j} - 1, \text{if } \hat{e}_{i,j} \geq n_k + 1, \\ \hat{e}_{i,j} + 1, \text{if } \hat{e}_{i,j} \leq m_k - 1, \\ \hat{e}_{i,j}, \text{if } m_k < \hat{e}_{i,j} < n_k. \end{cases} \tag{5}$$

5. *Pixel recovery*: The original pixels in R_1 are restored as $x_{i,j} = e_{i,j} + \hat{x}_{i,j}$. The LSBs in the first B pixels of R_2 are separated from $\mathbf{b}$, and then, the original B pixels of R_2 can also be restored.

4 Experimental Results and Analysis

Four grayscale images with size 512×512 are selected: Lena, Barbara, Airplane, and Boat. When the image is divided into multiple blocks, some edge pixels are discarded. Thus, the embedding capacity is reduced. Therefore, we take the case of dividing R_1 into four blocks ($I = 4$) as an example to illustrate the performance of this algorithm.

The proposed algorithm and the algorithm of Wang et al.'s [13] all choose different prediction errors to embed secret information. The dynamic PEE of Wang et al.'s algorithm is to embed secret information by selecting a pair of prediction errors with less embedding distortion based on the given embedding capacity. The proposed method selects a pair (m_k, n_k) with the highest frequency appearing in the prediction error histogram in each block for embedding. The local prediction error peak is different from the global peak, and the prediction error value selected for embedding in each block is also different. The local histogram peaks correspond to a plurality of different prediction errors on the global histogram. Therefore, the utilization of the prediction error is effectively improved. The embedding capacity and the quality of the entire image are also improved.

Table 1 shows PSNR between the image containing the secret information and the original image under the same embedding capacity. The improved algorithm has a value of PSNR that is more than 50 dB, using $I = 4$. When embedding the same secret information, the PSNR of the proposed algorithm is higher than that of Wang et al.'s. For the images with simple textures, Lena and Airplane, the increased PSNR

Table 1 Comparison of PSNR (dB) with [13] under the same embedding capacity

Test images	Ref. [13]	Proposed	Increased
Lena	49.8852	50.0689	0.1837
Barbara	49.7073	50.7409	1.0336
Airplane	49.6268	50.0474	0.4206
Boat	49.0837	50.0760	0.9923

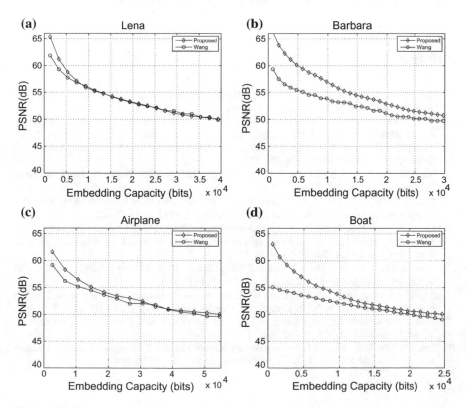

Fig. 3 PSNR comparison for: **a** lena, **b** barbara, **c** airplane, **d** boat

is less. For images with more complex textures, Barbara and Boat, the increased PSNR is about 1 dB. The local prediction error histogram peak in the more complex texture image is different from the global prediction error histogram peak, so the selection of the local prediction error peak has better hiding effect.

A line graph of the embedding capacity and the PSNR is shown in Fig. 3, for each of the testing images. It can be seen from the figure that the PSNR of the proposed algorithm is higher than that of Wang et al.'s under the same embedding capacity.

5   Conclusion

This paper proposes a RW method based on adaptive prediction error expansion. The peaks of local prediction error histogram are selected to embed the secret information. Not only the prediction error is fully utilized to increase the embedding capacity, but also the quality of the image containing secret information is improved.

Acknowledgements This work is supported by the National Natural Science Foundation of China (NSFC) (No. 61272432) and Shandong Provincial Natural Science Foundation (No. ZR2014JL044).

References

1. Yang, C.-Y., Lin, C.-H., Hu, W.-C.: Reversible data hiding by adaptive IWT-coefficient adjustment. J. Inf. Hiding Multimed. Signal Process. **2**(1), 24–32 (2011)
2. Yang, C.-Y., Lin, C.-H., Hu, W.-C.: Reversible data hiding for high-quality images based on integer wavelet transform. J. Inf. Hiding Multimed. Signal Process. **3**(2), 142–150 (2012)
3. Celik, M.U., Sharma, G., Tekalp, A.M., et al.: Lossless generalized-LSB data embedding. IEEE Trans. Image Process. **14**(2), 253–266 (2005)
4. Ni, Z.C., Shi, Y.Q., Ansari, N., et al.: Reversible data hiding. IEEE Trans. Circuits Syst. Video Technol. **16**(3), 354–362 (2006)
5. Lin, C.C., Tai, W.L., Chang, C.C.: Multilevel reversible data hiding based on histogram modification of difference images. Pattern Recognit. **41**(12), 3582–3591 (2008)
6. Qu, D.-C., Pan, J.-S., Weng, S.-W., et al.: A novel reversible data hiding method for color images based on dynamic payload partition and cross channel correlation. J. Inf. Hiding Multimed. Signal Process. **7**(6), 1194–1205 (2016)
7. Tsai, P., Hu, Y.C., Yeh, H.L.: Reversible image hiding scheme using predictive coding and histogram shifting. Signal Process. **89**(6), 1129–1143 (2009)
8. Li, X.L., Zhang, W.M., Gui, X.L., et al.: Efficient reversible data hiding based on multiple histograms modification. IEEE Trans. Inf. Forensics Secur. **10**(9), 2016–2027 (2015)
9. Tian, J.: Reversible data embedding using a difference expansion. IEEE Trans. Circuits Syst. Video Technol. **13**(8), 890–896 (2003)
10. Hu, Y.J., Lee, H.K., Li, J.W.: DE-based reversible data hiding with improved overflow location map. IEEE Trans. Circuits Syst. Video Technol. **19**(2), 250–260 (2009)

11. Thodi, D.M., Rodriguez, J.J.: Expansion embedding techniques for reversible watermarking. IEEE Trans. Image Process. **16**(3), 721–730 (2007)
12. Li, X.L., Yang, B., Zeng, T.Y.: Efficient reversible watermarking based on adaptive prediction-error expansion and pixel selection. IEEE Trans. Image Process. **20**(12), 3524–3533 (2011)
13. Wang, C., Li, X. L., Yang, B.: Efficient reversible image watermarking by using dynamical prediction-error. In: 17th International Conference on Image Processing, vol. 119, no. 5, pp. 3673–3676 (2010)
14. Rad, R.M., Wong, K.S., Guo, J.M.: A unified data embedding and scrambling method. IEEE Trans. Image Process. **23**(4), 1463–1475 (2014)

A New Comprehensive Database for Hand Gesture Recognition

Lanyue Pi, Kai Liu, Yun Tie and Lin Qi

Abstract At present, limited hand gesture databases are made available for reference and research, and these databases are small in scale with limited variants of gestures. In response to this problem, we have established a large-scale dynamic gesture database called LR-DHG database. The sensors for collecting data include Leap Motion and RealSense. The data formats include video frame, depth images, color images, and hand-joint point coordinates. In this paper, we describe in detail the recording work of this database, and use subjective evaluation methods to conduct preliminary tests on it, resulting in a higher recognition rate. This database provides more valuable reference and research data for human–computer interaction.

Keywords Gesture · Database · Subjective evaluation

1 Introduction

Gestures as a means of information carrier are one of the common body language used for communication and interaction. Because of the fast interaction and accurate expression, gestures are widely used in sign language and human-computer interaction systems [1]. Gesture databases are necessary for reliable testing of gesture recognition algorithms.

However, the current number of gesture databases is small, and the content and data are relatively simple. It is not good for experiments and research for training models and algorithm verification [2]. In [3], a total of 18 participants used Leap Motion to acquire 378 dynamic gestures related to television control research, including 21 TV control actions for free control of the TV. But each hand movement is done only once by the participants. In [4, 5], the researchers realized the research and

L. Pi (✉) · K. Liu · Y. Tie · L. Qi
Zhengzhou University, 450000 Zhengzhou, People's Republic of China
e-mail: 270570494@qq.com

Y. Tie
e-mail: ieytie@zzu.edu.cn

© Springer Nature Singapore Pte Ltd. 2019
J.-S. Pan et al. (eds.), *Genetic and Evolutionary Computing*,
Advances in Intelligent Systems and Computing 834,
https://doi.org/10.1007/978-981-13-5841-8_27

development of the sign language by using Leap Motion. The two gesture databases are static, but the same problem is that there are fewer data samples, which is not beneficial to the inaccuracy of the training model. In [6] the author used Leap Motion and Kinect devices to identify the American handmade alphabet based on the position and orientation of the fingertips, and then sent these features to a multi-class support vector machine (SVM) classifier to identify their gestures. But they only focus on static gestures rather than dynamic gestures. Therefore, a large and representative database is needed for researchers to evaluate different methods.

We worked hard to develop a new dynamic gesture database. The database has a wide variety of data types, including Video frame, RGB-D image, and 3D Skeleton coordinates. Two sensors are used: Leap Motion and RealSense. The former outputs a gesture model in real time according to Time Flight Technology (TOF) in three-dimensional coordinates, and the recognition rate can reach 0.01 mm [7]; The latter is a depth camera. This database will provide a representative experimental platform for the comparative study of different gesture recognition methods.

There are many commonly used gesture recognition algorithms, but there are few attempts to use genetic and evolutionary algorithms [8] for gesture recognition. In the literature [9], the author proposes an recognition method based on adaptive genetic algorithm. The method improves the recognition rate and ensures real-time performance. In the literature [10], the author discusses the feasibility of combining genetic algorithm with neural network to improve recognition rate. In the literature [11], the researcher proposes an evolutionary optimization algorithm to solve the problem that the classifier performs poorly in the gesture dataset with multiple attributes, and provides a basis for the evolutionary algorithm in the classification of gestures.

2 Describe Our Database

Access a large amount of relevant literature [1, 3, 12, 13], we summarized a set of theoretical basis for selecting gestures:

- Gesture is a continuous set of actions (dynamic gestures);
- The frequency of selected gestures is generally high in daily use;
- For similar actions, the difference between the use of a different number of fingers of the definition of the gesture (such as grab and pinch);
- Inverse gestures also have practical significance (such as grip and grip-release);
- To consider the use of both hands;
- To consider the direction and angle issues;
- To involve the definition of gestures based on virtual interaction.

According to the above criteria, we finally defined 27 gestures. Each gesture is shown in Table 1.

We use the Leap Motion Software Development Kit (SDK) together with the HTC vive and Unity3D game engines and C# scripts to capture image information and

Table 1 LR-DHG database

No.	Gesture	Name	No.	Gesture	Name
1		Grab	15		Twist anticlock-wise (finger)
2		Grab-release	16		Twist clockwise (hand)
3		Grip	17		Twist anticlock-wise (hand)
4		Grip-release	18		Rise up
5		Pinch	19		Downward
6		Pinch-release	20		Left waved
7		Side pinching	21		Right waved
8		Side pinch-release	22		Waved
9		Throw	23		Grab with both hands
10		Single finger press	24		Panning with both hands

(continued)

Table 1 (continued)

No.	Gesture	Name	No.	Gesture	Name
11		Click	25		Rotating with both hands
12		Finger direction	26		Zoom with both hands
13		Finger select	27		Continuous Grab-Twist-Throw
14		Twist clockwise (finger)			

Fig. 1 Recording process

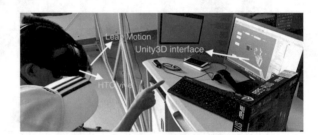

bone joint point coordinates from the gestures in the 3D space returned by the Leap Motion in real time and store it (Fig. 1).

RealSense is a depth camera, slightly larger than the Leap Motion. It has an infrared camera on the front for measuring distance. We use the RealSense official SDK to record each participant's gestures and store them in video format. With each gesture's video we can capture the frame and turn it into a depth image. Figures 2 and 3 illustrate two types of data for "Grab" gesture.

At the end of the recording work, 88 people participated. For each gesture, each participant performed 10 times. This database we called LR-DHG Database (Leap-Motion and RealSense Dynamic Hand Gesture Database).[1,2]

---

[1] Available upon request. Please contact ieytie@zzu.edu.cn.

[2] This study has been conducted with ethic approval obtained from Prof. Lin Qi from Zhengzhou University. Participants are recruited from the CVPR lab and all participants have given their consent to use the dataset and disclose information relevant for research in this study.

Fig. 2 "Grab" with
RealSense

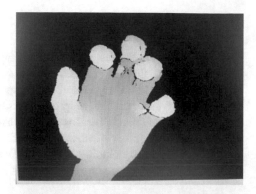

Fig. 3 "Grab" with Leap
Motion

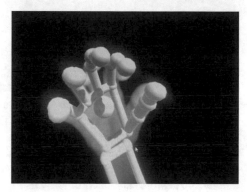

3 Experiment

Considering that each gesture is performed by different participants, in order to verify whether the gesture can accurately express the given meaning, we use a subjective evaluation method to verify its accuracy [14].

On the subjective evaluation of the experimental link we can be divided into four steps: 1. evaluation object; 2. evaluation conditions; 3. evaluation methods; 4. data processing.

3.1 Evaluation Object

The evaluation object refers to a gesture database that participates in the subjective evaluation of dynamic gestures.

3.2   Evaluation Conditions

The evaluation conditions refer to the observers, equipment, and environment involved in the evaluation process.

Observers used in a large number of experiments have not been trained. Initial training before evaluation, to help observers understand the purpose of the test and determine the evaluation score. In order to make the subjective evaluation statistically meaningful, the number of observers participating in the score is not too small. We chose 50 people, including 8 women. Data have been collected using identical equipment and under the same environment in the laboratory.

3.3   Evaluation Method

Sorted method. The sorted method is to require the observers to observe all dynamic gesture samples and rank them according to the degree of recognition. In addition to the last gesture (Continuous Grab-Twist-Throw), there are a total of 26 gestures in the database. Due to the large number and the temporal nature of human memory, we do not limit the number of times an observer watches each gesture. The statistical results obtained after the test are shown in the Table 2.

We listed the top five and the bottom five gestures. From the table, we can see that the top gestures are all commonly used gestures in daily life, and the movements are relatively single, so the recognition is higher. The lower ranking is some strange gestures. It should be pointed out that since some gestures in our database are based on relevant criteria and are not conventional gestures, the observer does not know what it means to be watching for the first time, so we will give the observers some simple training before starting. In addition, we give relevant hints during the observation process.

Table 2 Sorted evaluation results (Top 5 and Bottom 5)

Rank	Gesture	Rate (%)
1	Grab	96
2	Grip	92
3	Throw	90
4	Rise up and downward	88
5	Waved	86
22	Single finger press	66
23	Panning with both hands	56
24	Zoom with both hands	52
25	Finger point	44
26	Finger select	38

Table 3 Score setting

Score	Description
5 Accurately recognition	Correctly express the meaning of gestures
4 Probably recognition	Correctly stated but not completely accurate
3 Tough recognition	Only the approximate meaning
2 Deviation recognition	The expression is biased but not completely deviated
1 Not recognition	The expression completely deviated

Table 4 Score results (Top 5 and Bottom 5)

Rank	Gesture	Mean	Variance
1	Grab	4.7200	0.2873
2	Throw	4.6600	0.4963
3	Grip	4.5000	0.5532
4	Waved	4.4200	0.5714
5	Twist clockwise (hand)	4.3400	0.6731
22	Single finger press	3.5200	1.0302
23	Zoom with both hands	2.9600	1.2673
24	Panning with both hands	2.7200	1.2764
25	Finger direction	2.4800	1.3567
26	Finger select	2.1800	1.0894

Graded score method. First set a level interval, such as 1–5, and then the observer selects the corresponding score after watching each gesture sample as shown in the Table 3.

According to this rating scale, we count the score of each gesture in Table 4.

From the above table, it can be seen that the mean of the top five gestures is relatively high, and the variance is gradually increasing, while the mean of the bottom five is only in the median line, and the variance is also relatively large. This is because the larger the mean is, the easier it is to recognize the gesture, so the difference in their scores is smaller and the variance is smaller; the smaller the mean is, the larger their score difference is, so the variance is larger.

3.4 Data Processing

After data statistics are completed, we process the data. The scores (means and variances) for each gesture are calculated based on the following two formulas

Fig. 4 Linear fitting of
mean and variance

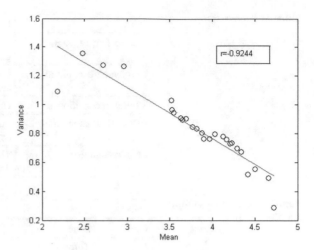

$$E_n = \frac{1}{n} \cdot \sum_{i=1}^{n} x_i \tag{1}$$

$$S_n = \frac{1}{n} \cdot \sum_{i=1}^{n} (x_i - E_n)^2 \tag{2}$$

where $n = 50$, $i = 1, 2, 3, \ldots, 50$, and x_i indicates the score of the i-th observer on this gesture (1–5). E_n represents the mean and S_n represents the variance.

Figure 4 shows a linear fit of the mean and variance of the 26 gestures. The correlation coefficient r $= -0.9244$, the correlation is very high. Remove the first four points and the last point, and other points are linear. The higher the recognition degree is, the smaller the score difference is; the lower the recognition gesture is, the larger the score difference is given by each observer.

4 Conclusion

In this paper, we describe our database in detail. We collected 23,760 dynamic gesture samples with a total data volume of nearly 400 GB. Then we experimented on whether the gesture can subjectively express the meaning it represents, and adopt the method of subjective evaluation. The final experimental results also demonstrate the rationality of the meaning of the gestures we define.

Our database can test a variety of gesture recognition methods. In the next work, we fuse the features extracted from the two data formats, and send the data frames into the classifier for training and modeling. Finally, we conduct a comparative study of each method.

Acknowledgements This work was supported in part by "the National Natural Science Foundation of China under Grant No. 61331021".

References

1. Kelly, S.D., Manning, S.M., Rodak, S.: Gesture gives a hand to language and learning: perspectives from cognitive neuroscience, developmental psychology and education. Lang. Linguist. Compass **2**(4), 569–588 (2008)
2. Kanade, T., Cohn, J.F., Tian, Y.: Comprehensive database for facial expression analysis. In: Proceedings Fourth IEEE International Conference on Automatic Face and Gesture Recognition. IEEE, Grenoble (2000)
3. Zaiți, I.-A., Pentiuc, Ş.-G., Vatavu, R.-D.: On free-hand TV control: experimental results on user-elicited gestures with Leap Motion. Pers. Ubiquitous Comput. **19**, 821–838. Springer, London (2015)
4. Mapari, R.B., Kharat, G.: Real time human pose recognition using leap motion sensor. In: 2015 IEEE International Conference on Research in Computational Intelligence and Communication Networks, IEEE, Kolkata (2015)
5. Khelil, B., Amiri, H.: Hand gesture recognition using leap motion controller for recognition of Arabic sign language. In: 3rd International Conference ACECS'16, pp. 233–238, Proceedings of Engineering & Technology (2016)
6. Opromolla, A., Volpi, V., Ingrosso, A., Fabri, S., Rapuano, C., Passalacqua, D., Medaglia,. C.M.: A Usability Study of a Gesture Recognition System Applied During the Surgical Procedures. DUXU, vol. 9188, pp. 682–692. Springer, Cham (2015)
7. Leap Motion Controller. (Accessed on 10 November 2014). https://www.leapmotion.com
8. Chakraborty, U.K.: Genetic and evolutionary computing. Inf. Sci. **178**, 4419–4425, Association for Computing Machinery (2008)
9. Wang, X., Bao, H.: Gesture recognition based on adaptive genetic algorithm. J. Comput. Aided Des. Comput. Gr. **19**, 1056–1062; China Acad. J. Electron. Publ. House (2007)
10. Chaudhary, A., Raheja, J.L.: Intelligent approaches to interact with machines using hand gesture recognition in natural way: a survey. Comput. Sci. Eng. Surve. **2**, 122–132, Cornell University Library (2011)
11. Palacios-Alonso, M.A., Brizuela, C.A.: Evolutionary learning of dynamic naive Bayesian classifiers. Autom. Reason. **45**, 21–37, Springer, Netherlands (2010)
12. Marin, G., Dominio, F., Zanuttigh, P.: Hand gesture recognition with leap motion and kinect devices. In: 2014 IEEE International Conference on Image Processing, pp. 1565–1569. IEEE, Paris (2014)
13. Argyros, A.A., Lourakis, M.I.A.: Vision-based interpretation of hand gestures for remote control of a computer mouse. In: Computer Vision in Human-Computer Interaction, pp. 40–51. Springer, Heidelberg (2006)
14. Cui, L.C.: Do experts and naive observers judge printing quality differently. Image Qual. Syst. Perform. **5294**, 132–145. SPIE, California (2004)

Fast Algorithms for Poisson Image Denoising Using Fractional-Order Total Variation

Jun Zhang, Mingxi Ma, Chengzhi Deng and Zhaoming Wu

Abstract In this paper, a new Poisson image denoising model based on fractional-order total variation regularization is proposed. To obtain its global optimal solution, the augmented Lagrangian method, the Chambolle's dual algorithm and the primal-dual algorithm are introduced. Experimental results are supplied to demonstrate the effectiveness and efficiency of the proposed algorithms for solving our proposed model, with comparison to the total variation Poisson image denoising model.

Keywords Poisson denoising · Fractional-order total variation ·
Augmented Lagrangian method · Dual algorithm · Primal-dual algorithm

1 Introduction

In photon-counting devices such as computed tomography, magnetic resonance imaging, and electronic microscopy, photon noise is very common. Different from the Gauss white noise, photon noise is signal-dependent. This is because photon noise is directly dependent on the number of photons which depends on the scene brightness. Since the number of photons follows a Poisson distribution, photon noise is also called Poisson noise. The mean and variance of this kind of noise are equal.

J. Zhang · C. Deng (✉) · Z. Wu
Jiangxi Province Key Laboratory of Water Information Cooperative Sensing
and Intelligent Processing, Nanchang Institute of Technology, Nanchang 330099,
Jiangxi, China
e-mail: dengcz@nit.edu.cn

J. Zhang
e-mail: junzhang0805@126.com

Z. Wu
e-mail: zmwunit@foxmail.com

J. Zhang · M. Ma
College of Science, Nanchang Institute of Technology, Nanchang 330099,
Jiangxi, China
e-mail: xixi06291106@163.com

© Springer Nature Singapore Pte Ltd. 2019   263
J.-S. Pan et al. (eds.), *Genetic and Evolutionary Computing*,
Advances in Intelligent Systems and Computing 834,
https://doi.org/10.1007/978-981-13-5841-8_28

The smaller the peak value of the image is, the higher the intensity of Poisson noise is. Therefore, the denoising and restoration models proposed for Gaussian white noise are not suitable to remove Poisson noise. So it is very important to study Poisson denoising models and their algorithms.

Recently, Poisson denoising variational regularization models have been widely investigated. These regularization terms include Tikhonov regularization, total variation (TV) regularization, higher order regularization, and so on. For TV Poisson denoising model [3], researchers have proposed a number of fast and efficient numerical methods, for instance, the dual algorithm [6], the augmented Lagrangian method [2] and the split Bregman method [4]. In [4], the TV Poisson deblurring model has also been studied. The TV Poisson denoising model performs very well for preserving edges while removing noise. However, it often causes staircase effects in flat regions. Furthermore, small details such as textures are often filtered out with noise.

In recent years, more and more attention has been paid to variational regularization using fractional-order derivatives. This fractional-order total variation (FOTV) has been successfully applied to Gaussian noise image restoration [7], multiplicative noise removal [9], image decomposition [5] and other fields. Inspired by the successful applications of FOTV, in this paper, we propose a new FOTV Poisson denoising model to overcome the drawbacks of TV Poisson denoising model. Furthermore, we propose three fast and efficient algorithms to solve this model. Numerical results show that our proposed model has superior performance over TV Poisson denoising model in recovering texture images, and our proposed algorithms also converge faster.

This paper is organized as follows. Section 2 introduces the definition of the fractional-order gradient and divergence. Meanwhile, we propose the new Poisson image denoising model based on the FOTV in this section. The augmented Lagrangian method, the Chambolle's dual algorithm and the primal-dual algorithm for solving this model are presented in Sects. 3, 4 and 5, respectively. Numerical experiments are presented in Sect. 6 to justify the effectiveness and efficiency of our proposed methods. Finally, Sect. 7 concludes the paper.

2 The Proposed Model

2.1 Fractional-Order Gradient and Divergence

We first give some notations and definitions. Suppose that our images will be two-dimensional matrices of size $N \times N$. Then the Euclidean space $\mathbb{R}^{N \times N}$ is denoted by X. For $\forall\, u \in X$, the discrete fractional-order gradient is defined by Zhang et al. [8]

$$\nabla^a u := [(\nabla^a u)_{i,j}]_{i,j=1}^N = ([(D_1^a u)_{i,j}]_{i,j=1}^N, [(D_2^a u)_{i,j}]_{i,j=1}^N), \tag{1}$$

with

$$
(D_1^a u)_{i,j} = \sum_{k=0}^{K-1} (-1)^k C_k^a u_{i-k,j},
$$
$$
(D_2^a u)_{i,j} = \sum_{k=0}^{K-1} (-1)^k C_k^a u_{i,j-k},
$$

(2)

where K stands for the number of pixels used to compute the fractional-order gradient, $C_k^a = \frac{\Gamma(a+1)}{\Gamma(k+1)\Gamma(a+1-k)}$ and $\Gamma(x)$ is the Gamma function.

For $\forall\ p = (p^1, p^2) \in Y = X \times X$, the discrete fractional-order divergence is defined by Zhang et al. [8]

$$
\text{div}^a p := [(\text{div}^a p)_{i,j}]_{i,j=1}^N,
$$

(3)

with

$$
(\text{div}^a p)_{i,j} = (-1)^a \sum_{k=0}^{K-1} (-1)^k C_k^a (p_{i+k,j}^1 + p_{i,j+k}^2).
$$

(4)

There is a relation that $(\nabla^a)^* = \overline{(-1)^a \text{div}^a}$.

2.2 FOTV Poisson Denoising Model

The TV regularization model is one of the most successful models in image processing. It can preserve edges very well while denoising. However, while using it for Poisson noise reduction, there are still some defects. Therefore, in order to overcome the staircase effects and better maintain texture details, a Poisson denoising model based on FOTV is presented below

$$
\min_{u>0} \| \nabla^a u \|_1 + \alpha \int_\Omega (u - f \log u) dx.
$$

(5)

3 The Augmented Lagrangian Method

In this section, we propose the augmented Lagrangian method (ALM) to solve the FOTV Poisson denoising model (5).

It is easy to see that the model (5) can be reformulated as follows:

$$
\min_{u>0,\mathbf{p}} \| \mathbf{p} \|_1 + \alpha \int_\Omega (u - f \log u) dx,
$$
$$
\text{s.t.}\quad \mathbf{p} = \nabla^a u.
$$

(6)

The augmented Lagrangian functional of (6) is

$$\mathcal{L}(u, \mathbf{p}; \boldsymbol{\lambda}) = \parallel \mathbf{p} \parallel_1 + \langle \boldsymbol{\lambda}, \mathbf{p} - \nabla^a u \rangle + \frac{\mu}{2} \parallel \mathbf{p} - \nabla^a u \parallel_2^2 + \alpha \int_{\Omega} (u - f \log u) dx,$$

where $\mathbf{p}$ is auxiliary vector, and $\boldsymbol{\lambda}$ is Lagrangian multiplier. Then we solve model (5) by ALM as follows:

$$\begin{cases} u^{k+1} = \arg\min_u \mathcal{L}(u, \mathbf{p}^k; \boldsymbol{\lambda}^k), \\ \mathbf{p}^{k+1} = \arg\min_{\mathbf{p}} \mathcal{L}(u^{k+1}, \mathbf{p}; \boldsymbol{\lambda}^k), \\ \boldsymbol{\lambda}^{k+1} = \boldsymbol{\lambda}^k + \mu(\mathbf{p}^{k+1} - \nabla^a u^{k+1}). \end{cases} \tag{7}$$

Thus we only need to consider the above two subproblems. The Euler–Lagrange equation for the u-subproblem is

$$-(\nabla^a)^* \boldsymbol{\lambda}^k - \mu(\nabla^a)^* (\mathbf{p}^k - \nabla^a u) + \alpha(u - f)/u^k = 0.$$

Here, the u in the denominator is fixed as the previous iteration u^k. Under circulant or reflective boundary condition assumption, we can use FFT to solve this subproblem

$$u^{k+1} = \mathcal{F}^{-1} \left(\frac{\mathcal{F}[(\alpha f + u^k (\nabla^a)^* \boldsymbol{\lambda}^k)/\mu + u^k (\nabla^a)^* \mathbf{p}^k]}{\alpha/\mu + u^k \mathcal{F}((\nabla^a)^*) \mathcal{F}(\nabla^a)} \right). \tag{8}$$

Note that the closed form solution of the $\mathbf{p}$-subproblem is known. It is

$$\mathbf{p}^{k+1} = \text{Shrinkage}(\nabla^a u^{k+1} - \frac{\boldsymbol{\lambda}^k}{\mu}, \frac{1}{\mu}), \tag{9}$$

where Shrinkage is the shrinkage operator.

4 The Dual Algorithm

This section is devoted to present a new dual algorithm to solve model (5).

Just like the EM-TV and the gradient descent algorithm for TV-based Poisson denoising model [3, 6], we solve the following models: for $k = 0, 1, \ldots, M$,

$$\min_u \parallel \nabla^a u \parallel_1 + \frac{\alpha}{2} \int_{\Omega} \frac{(u - f)^2}{u^k} dx. \tag{10}$$

We can easily see that (10) is equivalent to the following constrained optimization problem:

$$\min_{u,\mathbf{z}} \parallel \mathbf{z} \parallel_1 + \frac{\alpha}{2} \int_{\Omega} \frac{(u-f)^2}{u^k} dx,$$

$$\text{s.t.} \quad \mathbf{z} = \nabla^a u.$$

The corresponding Lagrangian functional is

$$\mathcal{L}(u, \mathbf{z}; \boldsymbol{\lambda}) = \parallel \mathbf{z} \parallel_1 + \frac{\alpha}{2} \int_{\Omega} \frac{(u-f)^2}{u^k} dx + \langle \boldsymbol{\lambda}, \nabla^a u - \mathbf{z} \rangle.$$

Then we get the dual functional

$$
\begin{aligned}
q(\boldsymbol{\lambda}) &= \inf_{u,\mathbf{z}} \mathcal{L}(u, \mathbf{z}; \boldsymbol{\lambda}) \\
&= \inf_{u}[\frac{\alpha}{2} \int_{\Omega} \frac{(u-f)^2}{u^k} dx + \langle (\nabla^a)^* \boldsymbol{\lambda}, u \rangle] + \inf_{\mathbf{z}}[\parallel \mathbf{z} \parallel_1 - \langle \boldsymbol{\lambda}, \mathbf{z} \rangle] \\
&= -\sup_{u}[\langle -(\nabla^a)^* \boldsymbol{\lambda}, u \rangle - \frac{\alpha}{2} \int_{\Omega} \frac{(u-f)^2}{u^k} dx] - \sup_{\mathbf{z}}[\langle \boldsymbol{\lambda}, \mathbf{z} \rangle - \parallel \mathbf{z} \parallel_1] \\
&= -\frac{u^k}{2\alpha}[(\nabla^a)^* \boldsymbol{\lambda}]^2 + f[(\nabla^a)^* \boldsymbol{\lambda}] - \chi_K(\boldsymbol{\lambda}),
\end{aligned}
\tag{11}
$$

where $\chi_K(\boldsymbol{\lambda})$ is a characteristic function on $K = \{\boldsymbol{\lambda} \mid \parallel \boldsymbol{\lambda} \parallel_{L^\infty} \le 1\}$, namely,

$$\chi_K(\boldsymbol{\lambda}) = \begin{cases} 0, \boldsymbol{\lambda} \in K, \\ +\infty, \text{ otherwise.} \end{cases}$$

Furthermore, from (11), the solution u of problem (10) is simply given by

$$u = f - \frac{u^k[(\nabla^a)^* \boldsymbol{\lambda}]}{\alpha}. \tag{12}$$

Thus the dual problem of model (10) is

$$
\begin{aligned}
\max_{\boldsymbol{\lambda}} q(\boldsymbol{\lambda}) &= -\min_{\boldsymbol{\lambda}} \frac{u^k}{2\alpha}[(\nabla^a)^* \boldsymbol{\lambda}]^2 - f[(\nabla^a)^* \boldsymbol{\lambda}] + \chi_K(\boldsymbol{\lambda}), \\
&= -\min_{\parallel \boldsymbol{\lambda} \parallel_{L^\infty} \le 1} \frac{u^k}{2\alpha} \parallel (\nabla^a)^* \boldsymbol{\lambda} - \frac{\alpha f}{u^k} \parallel_2^2.
\end{aligned}
\tag{13}
$$

According to the Karush–Kuhn–Tucker conditions, there exists a Lagrange multiplier $\alpha_{i,j} \ge 0$ such that

$$[\nabla^a((\nabla^a)^* \boldsymbol{\lambda} - \frac{\alpha f}{u^k})]_{i,j} + \alpha_{i,j} \lambda_{i,j} = 0. \tag{14}$$

In either of the cases, $\alpha_{i,j} = \mid [\nabla^a((\nabla^a)^* \boldsymbol{\lambda} - \frac{\alpha f}{u^k})]_{i,j} \mid$. Then we can use the semi-implicit gradient descent algorithm to solve (14).

The whole dual algorithm is summarized as follows:

$$
\begin{cases}
\boldsymbol{\lambda}^{k,0} = \boldsymbol{\lambda}^k (\boldsymbol{\lambda}^0 = \mathbf{0}), \\
\text{for } l = 0, \ldots, N-1, \\
\dfrac{\boldsymbol{\lambda}^{k,l+1} - \boldsymbol{\lambda}^{k,l}}{\tau} = -[\nabla^a((\nabla^a)^* \boldsymbol{\lambda}^{k,l} - \dfrac{\alpha f}{u^k})] - \mid [\nabla^a((\nabla^a)^* \boldsymbol{\lambda}^{k,l} - \dfrac{\alpha f}{u^k})] \mid \boldsymbol{\lambda}^{k,l+1}, \\
\boldsymbol{\lambda}^{k+1} = \boldsymbol{\lambda}^{k,N}, \\
u^{k+1} = f - \dfrac{u^k}{\alpha}(\nabla^a)^* \boldsymbol{\lambda}^{k+1}.
\end{cases}
\tag{15}
$$

5   The Primal-Dual Algorithm

In this section, we transform the proposed model (5) into a minimax problem and then solve it by a primal-dual algorithm. Primal-dual algorithms have been widely applied to compute the saddle-points of minimax problems. Here, we will apply the primal-dual algorithm proposed by Chambolle and Pock in [1] to find the saddle-point of our minimax problem.

The primal problem is given by

$$
\min_u F(Ku) + G(u),
\tag{16}
$$

where K is a continuous linear operator with induced norm $\parallel K \parallel$. By using the property of convex conjugate $(F^*)^*(Ku) = F(Ku)$, it is easy to get the primal-dual formulation of (16)

$$
\min_u \max_{\mathbf{p}} \langle Ku, \mathbf{p} \rangle - F^*(\mathbf{p}) + G(u),
\tag{17}
$$

where $F^* : Y \rightarrow [0, +\infty]$ and $G : X \rightarrow [0, +\infty]$ are convex, proper and lower semicontinuous functions. Then the Arrow-Hurwicz primal-dual algorithm is summarized as follows:

$$
\begin{cases}
\mathbf{p}^{k+1} = \arg\min_{\mathbf{p}} -\langle Ku^k, \mathbf{p} \rangle + F^*(\mathbf{p}) + \dfrac{\parallel \mathbf{p} - \mathbf{p}^k \parallel_2^2}{2\sigma}, \\
u^{k+1} = \arg\min_u \langle Ku, \mathbf{p}^{k+1} \rangle + G(u) + \dfrac{\parallel u - u^k \parallel_2^2}{2\tau}.
\end{cases}
\tag{18}
$$

As G is uniformly convex, we propose the accelerated Arrow-Hurwicz primal-dual algorithm

$$\begin{cases} \mathbf{p}^{k+1} = \arg\min_{\mathbf{p}} -\langle Ku^k, \mathbf{p}\rangle + F^*(\mathbf{p}) + \dfrac{\parallel \mathbf{p} - \mathbf{p}^k \parallel_2^2}{2\sigma_k} \\[2mm] \qquad = \arg\min_{\mathbf{p}} F^*(\mathbf{p}) + \dfrac{\parallel \mathbf{p} - (\mathbf{p}^k + \sigma_k Ku^k) \parallel_2^2}{2\sigma_k}, \\[2mm] u^{k+1} = \arg\min_{u} \langle Ku, \mathbf{p}^{k+1}\rangle + G(u) + \dfrac{\parallel u - u^k \parallel_2^2}{2\tau_k} \\[2mm] \qquad = \arg\min_{u} G(u) + \dfrac{\parallel u - (u^k - \tau_k K^*\mathbf{p}^{k+1}) \parallel_2^2}{2\tau_k}, \\[2mm] \theta_k = 1/\sqrt{1 + 2\gamma\tau_k}, \ \sigma_{k+1} = \sigma_k/\theta_k, \ \tau_{k+1} = \tau_k\theta_k. \end{cases} \tag{19}$$

5.1 FOTV Poisson Denoising

For the FOTV Poisson denoising model, we see that $F(Ku) = F(\nabla^a u) = \parallel \nabla^a u \parallel_1$ and $G(u) = \alpha \int_{\Omega} (u - f \log u)dx$. Here, G is strictly convex. Hence, we can make use of the accelerated primal-dual algorithm (19). Since

$$F^*(\mathbf{p}) = \begin{cases} 0, & \text{if } \parallel \mathbf{p} \parallel_\infty \le 1, \\ +\infty, & \text{otherwise,} \end{cases}$$

the $\mathbf{p}$-subproblem has a closed form solution. Its solution can be given by

$$\mathbf{p}_{i,j}^{k+1} = \begin{cases} \tilde{\mathbf{p}}_{i,j}, & \mid \tilde{\mathbf{p}}_{i,j} \mid \le 1, \\ \dfrac{\tilde{\mathbf{p}}_{i,j}}{\mid \tilde{\mathbf{p}}_{i,j} \mid}, & \text{otherwise,} \end{cases} \quad \tilde{\mathbf{p}} = \mathbf{p}^k + \sigma_k \nabla^a u^k. \tag{20}$$

For the u-subproblem, its solution can be obtained by solving the corresponding Euler–Lagrange equation. Then we have

$$u^{k+1} = \frac{\tilde{u} - \alpha\tau_k}{2} + \sqrt{\left(\frac{\tilde{u} - \alpha\tau_k}{2}\right)^2 + \alpha\tau_k f}, \tag{21}$$

where $\tilde{u} = u^k - \tau_k(\nabla^a)^*\mathbf{p}^{k+1}$.

So we can conclude that all the subproblems in the proposed primal-dual algorithm have closed form solutions. Owing to these, our algorithm is efficient.

6 Numerical Experiments

In this section, we test our proposed algorithms for FOTV Poisson denoising model. For simplicity of presentation, we denote the newly developed algorithms as FOTVALM, FOTVDA, and FOTVPDA, respectively. To show the superiority of our

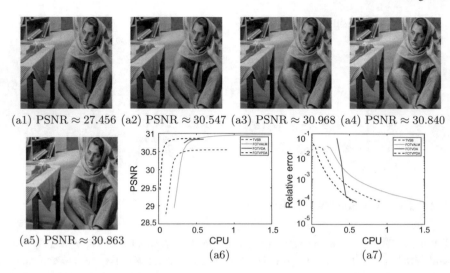

(a1) PSNR $\approx$ 27.456 (a2) PSNR $\approx$ 30.547 (a3) PSNR $\approx$ 30.968 (a4) PSNR $\approx$ 30.840

(a5) PSNR $\approx$ 30.863

(a6)

(a7)

Fig. 1 **a1** Noisy barbara image; **a2–a5** restored images by TVSB, FOTVALM, FOTVDA and FOTVPDA, respectively; **a6–a7** plots of PSNR and relative error

model, we compare our results with those of the split Bregman method for TV Poisson denoising model [4] denoted as TVSB. In experiments, we add Poisson noise by applying the matlab routine **poissrnd**$(I/\eta) \times \eta$. For quantitative comparison, we use the peak signal-to-noise ratio (PSNR) given by

$$\text{PSNR} = 10 \log_{10} \frac{255^2}{\frac{1}{MN} \sum_{i=1}^{M} \sum_{j=1}^{N} (u_{i,j} - I_{i,j})^2}, \tag{22}$$

where $u_{i,j}$ and $I_{i,j}$ denote the pixel values of the restored image and original clean image, respectively. The stopping criterion of algorithms is $\frac{\|u^{k+1}-u^k\|_F}{\|u^{k+1}\|_F} < 10^{-4}$.

In the experiment, we choose the "barbara" image with rich texture information of size 512×512 as the test image. While adding Poisson noise, $\eta = 1$. For the TVSB, the values of parameters are $\alpha = 24.8$, $\mu = 0.012$. For our three algorithms, we fix $K = 20$, $a = 2.25$, $\alpha = 46.5$. Especially, we set $\mu = 0.007$ for our FOTVALM, $\tau = 0.011$ for our FOTVDA and $\sigma_0 = 0.007$, $\tau_0 = 16$, $\gamma = 0.186$ for our FOTVPDA. The denoising results are listed in Table 1. Meanwhile, the restored images are shown in Fig. 1.

From Table 1, we can easily see that the PSNR and SSIM values of our three algorithms by FOTV model are higher than the TVSB. And our FOTVALM is the best of these three algorithms. The primal-dual algorithm is faster than the ALM and the dual algorithm. Compared with TV Poisson denoising model, our FOTV model can overcome the staircase effects, which can be seen through the flat ground area of the denoised images in Fig. 1. Furthermore, from Fig. 1, we find that our model can

Table 1 Numerical results

Image	Algorithm	Iter.	CPU(s)	PSNR	SSIM
Barbara 512 × 512	TVSB	35	4.953	30.547	0.922
	FOTVALM	37	7.265	30.968	0.925
	FOTVDA	9	3.406	30.840	0.923
	FOTVPDA	20	2.937	30.863	0.924

well preserve textures. In practice, this property is more obvious through observation of residual images. We can conclude from the plots of PSNR and relative error that our algorithms are convergent.

7 Conclusions

We proposed a new FOTV Poisson denoising model and presented three fast and efficient algorithms to solve it. The experimental results show that our model is better in terms of PSNR than the TV based Poisson denoising model. Moreover, our primal-dual algorithm and dual algorithm is faster.

Acknowledgements The research has been supported by the Science and Technology Project of Jiangxi Provincial Department of Education (GJJ161111, GJJ171015), the NNSF of China grants (61865012), the CSC (201708360066), and the NSF of Jiangxi Province (20161BAB202040, 20151BAB207010).

References

1. Chambolle, A., Pock, T.: A first-order primal-dual algorithm for convex problems with applications to imaging. J. Math. Imaging Vis. **40**(1), 120–145 (2011)
2. Figueiredo, M.A.T., Bioucas-Dias, J.M.: Deconvolution of Poissonian images using variable splitting and augmented Lagrangian optimization. In: IEEE/SP 15th Workshop on Statistical Signal Processing 2009, pp. 733–736 (2009)
3. Le, T., Chartrand, R., Asaki, T.J.: A variational approach to reconstructing images corrupted by Poisson noise. J. Math. Imaging Vis. **27**(3), 257–263 (2007)
4. Liu, X., Huang, L.: Total bounded variation Poissonian images recovery by split Bregman iteration. Math. Method. Appl. Sci. **35**(5), 520–529 (2012)
5. Ma, L., Zeng, T., Li, G.: Hybrid variational model for texture image restoration. E. Asian J. Appl. Math. **7**(3), 629–642 (2017)
6. Sawatzky, A., Brune, C., Muller, J., Burger, M.: Total variation processing of images with Poisson statistics. In: International Conference on Computer Analysis of Images and Patterns, pp. 533–540. Springer, Heidelberg (2009)
7. Zhang, J., Chen, K.: A total fractional-order variation model for image restoration with nonhomogeneous boundary conditions and its numerical solution. SIAM J. Imaging Sci. **8**(4), 2487–2518 (2015)

8. Zhang, J., Wei, Z., Xiao, L.: Adaptive fractional-order multi-scale method for image denoising. J. Math. Imaging Vis. **43**(1), 39–49 (2012)
9. Zhang, J., Wei, Z., Xiao, L.: A fast adaptive reweighted residual-feedback iterative algorithm for fractional-order total variation regularized multiplicative noise removal of partly-textured images. Signal Process. **98**, 381–395 (2014)

Part VII
Intelligent Multimedia Tools and Applications

An Adaptive Model Parameters Prediction Mechanism for LCU-Level Rate Control

Zeqi Feng, Pengyu Liu, Kebin Jia and Kun Duan

Abstract In this paper, an adaptive model parameters prediction mechanism is proposed to take the place of parameter updating method based on the experience value in HEVC. And, normalized mutual information is exploited to guide the model parameters of α and β prediction. Experimental results show that the proposed algorithm controls the rate error within 0.1%. Compared with HM16.9, it further improves average 0.03% bit rate accuracy. Meanwhile, the proposed algorithm yields average 1.10% BDBR reduction and 0.05 dB BDPSNR enhancement without introducing additional computation. And it demonstrates less bit rate fluctuation, which achieves better adaptability for HEVC in real-time transmission.

Keywords HEVC · Rate control · Model parameters

1 Introduction

With the development of network communication and multimedia technologies, video applications have become widespread. However, the network environment for video transmission is very complicated in practical, especially in the condition of limited channel bandwidth resources or switching frequent communication links. In order to transmit video data efficiently in heterogeneous networks, rate control

Z. Feng · P. Liu (✉) · K. Jia (✉)
Faculty of Information Technology, Beijing University of Technology, Beijing 100124, China
e-mail: liupengyu@bjut.edu.cn

K. Jia
e-mail: kebinj@bjut.edu.cn

Z. Feng · P. Liu · K. Jia · K. Duan
Beijing Laboratory of Advanced Information Networks, Beijing 100124, China

Z. Feng · P. Liu · K. Jia · K. Duan
Beijing Key Laboratory of Computational Intelligence and Intelligent System,
Beijing University of Technology, Beijing 100124, China

© Springer Nature Singapore Pte Ltd. 2019
J.-S. Pan et al. (eds.), *Genetic and Evolutionary Computing*,
Advances in Intelligent Systems and Computing 834,
https://doi.org/10.1007/978-981-13-5841-8_29

becomes an indispensable video coding technology [1], which makes it able to match the channel capacity for transmission and fault tolerance.

Since the H.264/AVC video coding standard [2], the researchers have focused on improving its network adaptability. They strive to seek the best balance between video quality and bandwidth utilization by reasonable rate control technique. As an upgrade for H.264/AVC, high-efficiency video coding (HEVC) [3] pays more attention to high-definition (720P, 1080P) and ultra-high-definition (4 K, 8 K) video transmission. Unfortunately, there is no rate control algorithm proposal in the early Joint Collaborative Team on Video Coding (JCT-VC) meeting for HEVC. Until the 8th JCT-VC meeting, the R-Q model in JCTVC-H0213 proposal [4] is developed, which makes up for this gap. Subsequently, B Li et al. [5] find that the quantization parameter (QP) is no longer the only factor that determines bit rate, and the correlation between Lagrange multiplier λ and bit rate has strong robustness for HEVC. Therefore, at the 11th JCT-VC meeting, rate control algorithm based on $R - \lambda$ model is adopted in the JCTVC-K0103 proposal. This algorithm effectively improves bit rate accuracy as well as the rate-distortion performance of reconstructed video, which is added to the test model HM10.0 and its upgraded version.

Even then, HEVC has not stagnated. The research on stronger rate-distortion performance and higher bit rate accuracy is still in full swing. First, various research studies [6, 7] have been carried out to improve rate-distortion performance. But they lost bit rate accuracy, which wastes channel resources. Second, some rate control algorithms focused on bit rate accuracy control. Li et al. [8] start from the model parameters updating process and optimize model parameters α and β updating by texture orientation analysis in large coding unit (LCU) level. Although this algorithm achieves better performance for rate control, the analysis of texture inevitably consumes additional time in encoding, which makes HEVC detrimental to real-time application.

Therefore, in this paper, an adaptive model parameters prediction mechanism is proposed, which takes the place of parameter updating method based on experience value in HEVC. And normalized mutual information is exploited to guide the model parameters α and β prediction. The proposed algorithm achieves synchronous lifting in bit rate accuracy and rate-distortion performance without introducing additional computation.

The remainder of the paper is organized as follows. Section 2 analyses the LCU-level rate control benchmark scheme in HEVC and its deficiency. Section 3 proposes an adaptive parameters prediction mechanism. Extensive experimental results are demonstrated and illustrated in Sect. 4, followed by the conclusion in Sect. 5.

2   LCU-Level Rate Control for HEVC

In the JCTVC-K0103 proposal, rate control algorithm has two phases for HEVC. First, target bits are allocated to different levels, and then λ and QP is determined by $R - \lambda$ model.

For LCU-level rate control, the target bits of each LCU is given by

$$T_{curLCU} = \frac{T_{curPic} - Bit_{head} - Code_{LCU}}{\sum\limits_{NotCodedLCUs} \omega_i} \cdot \omega_{curLCU} \qquad (1)$$

where T_{curPic} is the target bits of current encoding frame. Bit_{head} denotes the bits of header estimated from the encoding frame. $Code_{LCU}$ is the total bits for the encoded LCUs of the current frame. ω_i and ω_{curLCU} denote the bit allocation weights of the ith LCU and current encoding LCU, respectively.

Then, λ and QP are carried out based on $R - \lambda$ model as follows.

$$\lambda = \alpha \cdot bpp^\beta \qquad (2)$$

$$QP = 4.2005 \ln \lambda + 13.7122 \qquad (3)$$

where α and β are model parameters related to video content, which derive from, respectively, updating α and β of colocated LCU at the frame that the same frame level as the current frame. And, this frame is closest to the current frame as shown in Fig. 1. For example, α and β of each LCU at the eighth frame are calculated by, respectively, updating α and β of colocated LCU at the fourth frame. For B frames (the first, the second, and the fourth frames), the initial values of α and β are 3.2003 and -1.367, respectively. For I frame (the zeroth frame), the initial values of α and β are 6.7542 and 1.7860, respectively. It is noted that low delay (LD) is the encoding configuration, which is suitable for real-time application. The following experiment is carried out under the LD configuration in this paper.

The model parameters updating is specified by

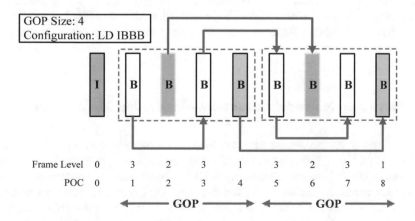

Fig. 1 Diagram of model parameters updating

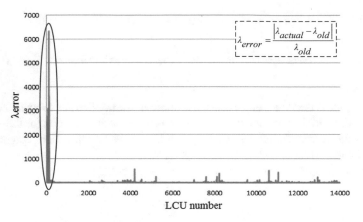

Fig. 2 λ error of "BasketballPass" (QP = 30)

$$\lambda_{\text{actual}} = \alpha_{old} \cdot bpp_{actual}^{\beta_{old}} \tag{4}$$

$$\alpha_{new} = \alpha_{old} + \delta_\alpha \cdot (\ln \lambda_{old} - \ln \lambda_{actual}) \cdot \alpha_{old} \tag{5}$$

$$\beta_{new} = \beta_{old} + \delta_\beta \cdot (\ln \lambda_{old} - \ln \lambda_{actual}) \cdot \ln bpp_{actual} \tag{6}$$

where δ_α and δ_β are experience value that set as 0.1 and 0.05, respectively. bpp_{actual} is actual bits.α_{old}, β_{old} and λ_{old} are the values that determining QP. α_{new} and β_{new} denote updated values.

α and β are constantly updated to adjust λ, which brings actual bits closer to target bits eventually [9]. It could be analyzed that a large error of λ will be generated, namely the bit rate accuracy is lower in the adaptation period of α and β. Preliminary experimental results also proved the above analysis as shown in Fig. 2. Therefore, in order to alleviate this problem, an adaptive model parameters prediction mechanism is proposed in this paper.

3 An Adaptive Model Parameters Prediction Mechanism

Since α and β are model parameters related to video content, an assumption is proposed that if two frames are identical, model parameters of colocated LCUs at the two frames would be same. However, for HEVC benchmark rate control algorithm, the reference frame selection is limited to (5) and (6) in model parameters updating and this frame may not be the best association frame that has little difference from the current frame. Therefore, how to determine the best association frame and predict model parameters based on known encoded information in the best association frame becomes the key to the proposed method.

3.1 Best Association Frame Selection

It is well known that video sequences have temporal correlation. However, the closest frame may not be the most similar to the current frame for different video sequences. Herein, to select appropriate best association frame, normalized mutual information [10] is adopted as the similarity measure among frames, which has achieved good performance in shot segmentation. The greater the normalized mutual information value, the higher the similarity. As shown in Fig. 3, four video sequences with different characteristics are tested to measure the normalized mutual information between the 27th frame and its previous N (N = 1, 2, 3, 4, ..., 9, 10, 11, 12, ... , 23, 24, 25, and 26) frames. The four video sequences are listed in Table 1.

Experimental results show that, for natural scene video sequences, the normalized mutual information of the current frame and its previous first frame is the largest, namely the similarity is the highest. The previous first frame would be selected as the best association frame. For the video sequences with shot mutation, the normalized mutual information will generate a jump at the time of shot mutation. Thus, I frame should be inserted and the model parameters should be initialized in time to prevent inaccurate prediction. And for reciprocating motion video sequences, the normalized mutual information exhibits regular changes. The periodicity of motion should be considered to determine the best association frame flexibly.

Fig. 3 Normalized mutual information for different video sequences

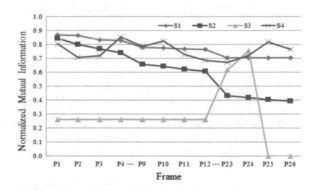

Table 1 The four video sequences with different characteristics

No.	Sequences	Characteristics	Source
S1	FourPeople	Natural scene, slight motion	HEVC test video sequences
S2	BasketballPass	Natural scene, Drastic motion	HEVC test video sequences
S3	SlideShow	Existence of shot mutation	HEVC test video sequences
S4	Mechanics	Reciprocating motion	Internet [11]

In this paper, the proposed algorithm is mainly for natural scene video. And the best association frame of the current frame is the previous first frame.

3.2 Model Parameters Prediction

In HEVC, λ is the slope of the R-D curve essentially. The relationship is as follows:

$$D(R) = CR^{-K} \tag{7}$$

$$\lambda = -\frac{\partial D}{\partial R} = CK \cdot R^{-K-1} \tag{8}$$

where C and K are related to video content. D and R denote distortion and bit rate, respectively.

Then α and β are defined as

$$\alpha = C \cdot K \tag{9}$$

$$\beta = -K - 1 \tag{10}$$

Based on (7) and (8), C and K can be derived as follows:

$$C = \frac{D}{R^{-\frac{\lambda \cdot R}{D}}} \tag{11}$$

$$K = \frac{\lambda \cdot R}{D} \tag{12}$$

After encoding each LCU at the best association frame, bit rate and distortion are achieved. Based on (9), (10), (11), and (12), this known information are utilized to predict the colocated LCU at the current frame as follows:

$$\alpha'_{new} = \frac{\lambda_{actual}}{\text{bpp}_{actual}^{\frac{\lambda_{actual} \cdot \text{bpp}_{actual}}{d_{actual}} - 1}} \tag{13}$$

$$\beta'_{new} = -\frac{\lambda_{actual} \cdot \text{bpp}_{actual}}{d_{actual}} - 1 \tag{14}$$

where λ_{actual} and bpp_{actual} are the same as the parameters in (4). d_{actual} is actual distortion of the encoded LCU at the best association frame. α'_{new} and β'_{new} are the predicted model parameters for the colocated LCU at the current frame.

Note that, all parameters involved in the proposed algorithm come from known encoded information without the parameter update process based on experience value and it does not introduce additional computation.

4 Experimental Results

The proposed algorithm has been tested on the HEVC test model version 16.9 (HM 16.9) under low delay (LD) IBBB configuration. To evaluate the performance of the proposed algorithm, extensive test video sequences are adopted from 416×240 to 1920×1080. For each sequence, the LCU size is 64×64, the GOP size is 4, and the frame number is 100. QPs set 25, 30, 35, 40. Other encoder parameters are configured as the HEVC common test conditions by JCT-VC.

First, the bit rate accuracy is calculated by

$$\Delta E = \frac{\left|bits_{RC} - bits_{Target}\right|}{bits_{Target}} \times 100\% \tag{15}$$

where ΔE denotes bit rate error, $bits_{RC}$ and $bits_{Target}$ are actual bits and target bits, respectively. Bit rate accuracy of the proposed algorithm and HM16.9 is compared as shown in Table 2. It can be seen that the proposed algorithm controls the rate error within 0.1%. And, it further improves 0.03% bit rate accuracy compared with HM16.9.

Then, in order to validate gain or loss of the proposed algorithm, BDBR and BDP-SNR [12] are measured to testify the rate-distortion performance. Table 2 also shows the comparison between the proposed algorithm and HM16.9 in rate distortion. Compared with HM 16.9, the proposed algorithm yields average 1.10% BDBR reduction and 0.05 dB BDPSNR enhancement. Experimental results prove that the proposed algorithm performance is affected by motion characteristics. For slight motion video sequences, the encoding performance is better, such as "FourPeople", "Johnny", and "KristenAndSara", where the two frames have less difference. Therefore, it also confirms the assumption that if two frames are identical, model parameters of colocated LCUs at the two frames would be the same in the Sect. 3.

Table 2 Rate-distortion performance and bit rate accuracy between the proposed algorithm and HM16.9 rate control algorithm

Class	Sequences	BDPSNR (dB)	BDBR (%)	$\Delta E_{HM16.9}$ (%)	$\Delta E_{proposed}$ (%)
B	BasketballDrive	0.01	−0.36	0.02	0.02
C	BasketballDrill	0.01	−0.26	0.02	0.01
	PartyScene	0.05	−1.21	0.05	0.03
D	BasketballPass	0.04	−0.77	0.03	0.02
	BlowingBubbles	0.01	−0.28	0.10	0.07
E	FourPeople	0.07	−1.58	0.48	0.41
	Johnny	0.10	−2.66	0.02	0.02
	KristenAndSara	0.08	−1.66	0.16	0.03
Average		0.05	−1.10	0.11	0.08

Based on the analysis above, the proposed algorithm achieves synchronous lifting in bit rate accuracy and rate-distortion performance without introducing additional computation compared with HM16.9. And compared with HM16.9 rate control algorithm, the bit rate fluctuation of the proposed algorithm is less. It enhances the adaptability of HEVC in real-time transmission.

5 Conclusions

In this paper, an adaptive model parameters prediction mechanism for LCU-level rate control based on HEVC is proposed. Experimental results show that the proposed algorithm achieves synchronous lifting in bit rate accuracy and rate-distortion performance without introducing additional computation compared with HM16.9. Meanwhile, it demonstrates less bit rate fluctuation, which achieves better adaptability for HEVC in real-time transmission. At present, the proposed algorithm mainly aims at the natural scene video sequences with regular change. Further, we will optimize the rate control algorithm for the shot mutation and reciprocating motion video sequences without only rule.

Acknowledgements This paper is supported by the Project for the National Natural Science Foundation of China under Grants No. 61672064, the Beijing Natural Science Foundation under Grant No. 4172001, the China Postdoctoral Science Foundation under Grants No. 2016T90022, 2015M580029, the Science and Technology Project of Beijing Municipal Education Commission under Grants No. KZ201610005007, Beijing Municipal Education Committee Science Foundation under Grants No. KM201810005030, and Beijing Laboratory of Advanced Information Networks under Grants No. 040000546617002, Beijing Municipal Communications Commission Science and Technology Project under Grants No. 2017058.

References

1. Ramanand, A.A., Ahmad, I., Swaminathan, V.: A survey of rate control in HEVC and SHVC video encoding. In: IEEE International Conference on Multimedia & Expo Workshops (ICMEW), pp. 145–150 (2017)
2. Wiegand, T., et al.: Overview of the H.264/AVC video coding standard. IEEE Trans. Circ. Syst. Video Technol. **13**(7), 560–576 (2003)
3. Sullivan, G.J., Ohm, J.R., Han, W.J., et al.: Overview of the high efficiency video coding (HEVC) standard. IEEE Trans. Circ. Syst. Video Technol. **22**(12), 1649–1668 (2012)
4. Choi, H., Jand, N., Yoo, J.: Rate control based on unified RQ model for HEVC. JCT-VC of ITU-T and ISO/IEC, JCTVC-H0213. San Jose (2012)
5. Li, B., Li, H., Li, L., et al.: Rate control by R-lambda model for HEVC. Jt. Collab. Team Video Coding (JCT-VC) of ITU-T SG, K0103 (2012)
6. Si, J., Ma, S., Gao, W.: Efficient bit allocation and CTU level rate control for high efficiency video coding. In: Picture Coding Symposium, pp. 89–92. IEEE (2014)
7. Gao, W., Kwong, S., Zhou, Y., et al.: SSIM-based game theory approach for rate-distortion optimized intra frame CTU-level bit allocation. IEEE Trans. Multimed. **18**(6), 988–999 (2016)

8. Li, B., Zhou, M., Zhang, Y., et al., Model parameters estimation for CTU level rate control in HEVC. IEEE Multimed. 1–1 (2018)
9. Li, B., Li, H., Li, L., et al.: λ domain rate control algorithm for high efficiency video coding. IEEE Trans. Image Process **23**(9), 3841–3854 (2014)
10. Stanford, Evaluation of clustering. https://nlp.stanford.edu/IR-book/html/htmledition/evaluation-of-clustering-1.html (2009)
11. Jin, Y.: The principle of visual machinery: a classic reciprocating motion mechanism. http://www.iqiyi.com/w_19rv29u8eh.html (2017)
12. Bjontegaard, G.: Calculation of average PSNR differences between RD-curves. Doc. VCEG-M33 ITU-T Q6/16, Austin, TX, USA, 2–4 April 2001 (2001)

An Image Encryption Scheme Based on the Discrete Auto-Switched Chaotic System

Chunlei Fan, Kai Feng, Xin Huang and Qun Ding

Abstract Aiming at finite computational precision would make digital chaotic binary sequences into a short period sequence, so that the security of multimedia encryption information based on chaotic cipher is not guaranteed. In this paper, we designed a discrete auto-switched chaotic system with a simple structure and good performance in order to increase the period length of chaotic sequences. The auto-correlation and complexity experiments of chaotic sequences were performed to demonstrate the good performance of the sequence. Finally, the simulation experiment results show that this image encryption scheme is both reliable and secure.

Keywords Chaotic system · Binary sequences · Image encryption · Computational precision

1 Introduction

Chaos theory has been widely studied in nonlinear dynamics field. Chaotic signals are widely used in secure communication and cryptography because of their good performance such as positive Lyapunov exponents, ergodicity, sensitivity to initial conditions, topological transitivity, and unpredictability [1–5]. Since Matthews put forward the idea of chaotic encryption, the research on chaotic cryptography and chaotic secure communication schemes has gradually become a hot topic in the field of international electronic communications. Chaotic cryptography has the advan-

C. Fan · K. Feng · X. Huang · Q. Ding (✉)
Electronic Engineering College of Heilongjiang University, Harbin, China
e-mail: Qunding@aliyun.com

C. Fan
e-mail: Chunlei_fan@aliyun.com

K. Feng
e-mail: vfengkai@126.com

X. Huang
e-mail: hxenjoying@163.com

© Springer Nature Singapore Pte Ltd. 2019
J.-S. Pan et al. (eds.), *Genetic and Evolutionary Computing*,
Advances in Intelligent Systems and Computing 834,
https://doi.org/10.1007/978-981-13-5841-8_30

tages of simple structure and large keyspace, so it has more advantages in real-time encryption. Especially, for multimedia information such as video and image, chaotic cipher can show its superiority. In recent years, in order to better solve the security transmission of digital image, some scholars put forward a series of image security encryption scheme. For instance, Yang et al. [6] proposed a color image encryption scheme based on Logistic map over the finite field Z(N). Jin et al. [7] put forward a robust digital image watermarking scheme using Logistic and RSA encryption. Li et al. [8] came up with an image encryption scheme based on chaotic Tent map. Fan et al. [9] proposed an embedded Ethernet interface based on chaotic stream cipher.

However, for some discrete one-dimensional chaotic systems, such as Logistic, Tent Sine mapping, etc. Performance of the chaotic sequence generator will be affected by the limited computational precision, which makes the discrete chaotic sequences emerge some short-period phenomena. Aiming at this problem, Du et al. [10] proposed a novel chaotic key sequence generator based on double K-L transform. It can effectively improve the complexity and period length of Logistic chaotic sequence. Xiang et al. [11] used the sequences generated by two chaotic systems to inter-perturb these sequence values and the control parameter in order to obtain the discrete chaotic sequences with good performance. Chen et al. [12] designed a new chaotic sequence generator based on the novel interacting neural networks and the multiple chaotic systems with the purpose of enhancing the performance of chaotic sequences. However, these solutions are too complex to be implemented in hardware. In view of this, we designed a discrete auto-switched chaotic system with a simple structure and good performance. The security and performance analysis of chaotic binary sequence and image encryption show good pseudo-randomness of binary sequences.

2 A New Discrete Auto-Switched Chaotic System

The section introduces the generation method of auto-switched chaotic mapping with the purpose of overcoming the negative effect of limited computational precision on discrete chaotic mapping. We used three discrete chaotic mapping (Logistic, Tent, and Sine mapping) to generate auto-switched chaotic system.

2.1 Brief Introduction of Chaotic System

Logistic chaotic mapping is a classical model of the nonlinear dynamic system. It is represented by the following differential equation:

$$x_{n+1} = \mu x_n(1 - x_n), \mu \in (0, 4], x_n \in (0, 1) \tag{1}$$

Fig. 1 The cobweb diagram of the new auto-switched chaotic system

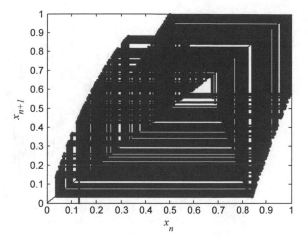

where μ is called as branch parameter, when μ belongs to [3.5699, 4], Logistic mapping work on chaotic state and presents complex dynamic characteristics. The discrete dynamic equation of Tent mapping can be expressed as the follows:

$$x_{n+1} = \begin{cases} x_n/a & 0 \le x_n \le a \\ (1 - x_n)/(1 - a) & a \le x_n \le 1 \end{cases} \tag{2}$$

Parameter a represents external control parameter and $0 < a < 1$. When $a = 0.5$, Eq. (2) is a standard Tent mapping. Further, the mathematical expression of Sine mapping is as follows:

$$x_{n+1} = \lambda \sin(\pi \cdot x_n) \tag{3}$$

where λ is control parameter and $\lambda \in [0, 6]$. When $\lambda \ge 0.87$, the chaotic system enters chaotic state and presents random distribution characteristics. In this paper, we set $\mu = 4$, $a = 0.5$ and $\lambda = 1$. The Eq. (3) is improved to $|\lambda \sin(\pi \cdot x_n)|$ with the purpose of unifying the value range of x_{n+1}. Moreover, when the state variables x_n of the auto-switched chaotic system respects the following conditions $x_n \in (0, 1/3)$, the new discrete auto-switched chaotic system runs on a Logistic mapping. Similarly, when the control conditions become $x_n \in [1/3, 2/3)$, it runs on Tent mapping. Finally, when the conditions become $x_n \in [2/3, 1)$, the auto-switched chaotic system work on Sine mapping. The cobweb diagram of the new auto-switched chaotic system is shown in Fig. 1. In addition, the chaotic time series of the auto-switched chaotic system is shown in Fig. 2.

Fig. 2 The chaotic time series of the new auto-switched chaotic system

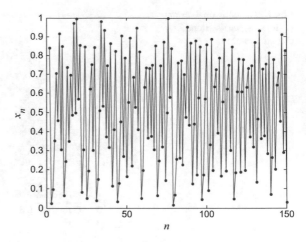

2.2 Quantization Method of Chaotic System

For the above auto-switched chaotic system, we adopt the binary quantization method to quantize discrete chaotic real value sequences. The mathematical expression of the quantization method is defined as follows:

$$Q_n = \begin{cases} 0 \ x_n < c \\ 1 \ x_n \geq c \end{cases} \tag{4}$$

where c represents quantization threshold with $c = 0.5$, and Q_n is the quantized binary sequence.

2.3 Performance Analysis of Chaotic Binary Sequence

Autocorrelation Analysis. Autocorrelation test [13] is a significant property of pseudo-random binary sequences. Suppose z_n is a binary sequence, $R_z(m)$ and N denote autocorrelation function and the length of binary sequence. Further, the autocorrelation function is defined as follows:

$$R_z(m) = \frac{1}{N} \sum_{n=0}^{N-1} z_n z_{n+m} \tag{5}$$

Based on the Eq. (5), autocorrelation test can be performed with auto-switched chaotic binary sequence. MATLAB simulation results are shown in Fig. 3. In this experiment, the sequences length and computational precision are 10^4 and 32. It can

Fig. 3 The simulation diagram of autocorrelation test

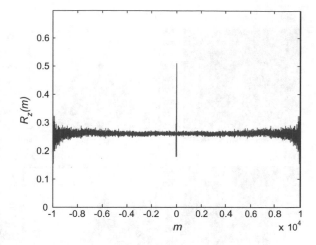

Table 1 The MPE value of chaotic binary sequences

Scale factor s	MPE value
3	0.86721
4	0.88514
5	0.90738
6	0.93019

be seen from the diagram that the autocorrelation function of the sequence is closer to the δ function, which shows good randomness.

Complexity Analysis. Multi-scale permutation entropy (MPE) [14] has the advantages of high robustness and fast computational speed. It is widely applied in the measurement of binary sequence complexity. In this section, we perform MPE analysis for the above-chaotic binary sequence. The parameters of MPE have embedding dimension m, delay factor τ, and scale factor s. In this experiment, we set $m = 3$, $\tau = 2$ and $s \in [3, 6]$, respectively. The experimental results are shown in Table 1. As can be seen from Table 1, MPE values of auto-switched chaotic binary sequences are closer to 0.9 and show good sequence complexity.

3 Image Encryption with an Auto-Switched Chaotic Sequence

In this section, the Fruits and Baboon gray images with a size of 256 * 256 are encrypted by the auto-switched chaotic binary sequence. The encryption method is synchronous stream cipher. The security analysis of image encryption is shown below.

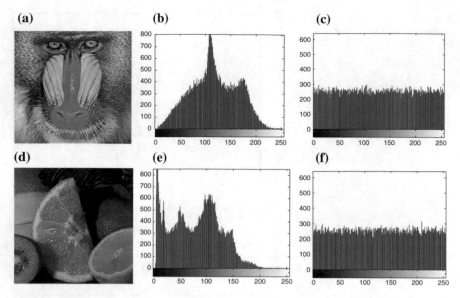

Fig. 4 Histogram test with: **a** plain image of Baboon; **b** histogram of plain image of Baboon; **c** histogram of encrypted image of Baboon; **d** plain image of Fruits; **e** histogram of plain image of Fruits; **f** histogram of encrypted image of Fruits

3.1 Histogram Analysis

The histogram of an image shows the distribution of pixel values. The histogram is an important statistical feature of the image. A uniformly distributed histogram is an indication of good quality for the encrypted image. The results of the gray histogram are shown in Fig. 4. As can be seen from the figure, the pixel values of the ciphertext images are evenly distributed. It can resist statistical attack well.

3.2 Correlation Analysis

Correlation analysis refers to the analysis of two variables with correlation in order to measure the correlation degree of two variables. In this section, 1000 pairs of adjacent pixels are randomly selected from the encrypted and decrypted the image Fruits and Baboon. The mathematical equation for the correlation of adjacent pixels is shown as follows:

$$\rho_{xy} = \frac{\text{cov}(x, y)}{\sqrt{D(x)D(y)}} \tag{6}$$

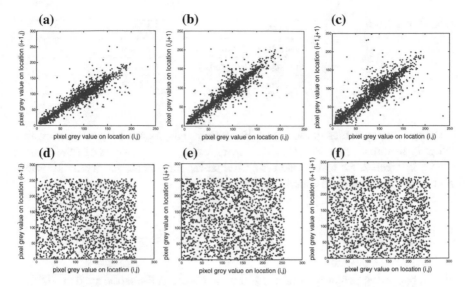

Fig. 5 The correlation plots of two adjacent pixels with: **a** horizontal correlation of plain image of Fruits; **b** vertical correlation of plain image of Fruits; **c** diagonal correlation of plain image of Fruits; **d** horizontal correlation of cipher image of Fruits; **e** vertical correlation of plain image of Fruits; **f** diagonal correlation of cipher image of Fruits

where $D(x) = \frac{1}{N} \sum_{i-1}^{N} (x_i - E(x))^2$, $E(x) = \frac{1}{N} \sum_{i=1}^{N} x_i$, x_i, and y_i represents the gray values of two adjacent pixels. Where N denotes the number of randomly selected adjacent pixels. For correlation analysis of Fruits image, the experimental results are shown in Fig. 5. As can be seen from the figure, the correlation of cipher images is much lower than that of plain images, which shows good performance.

3.3 PSNR Analysis

The peak signal-to-noise ratio (PSNR) is an objective criterion for evaluating images, and its mathematical equation is given as follows:

$$PSNR = 10 \log_{10}(L^2/MSE) \tag{7}$$

$$MSE = \frac{1}{M^2} \sum_{i=1}^{M} \sum_{j=1}^{M} (I'(i, j) - I(i, j))^2 \tag{8}$$

where $M * M$ is the size of image, $I'(i, j)$ and $I(i, j)$ represent the pixel value of encrypted and original images, respectively. MSE is mean squared error and L is

Table 2 PSNR test

Methods	PSNR value
Our scheme for Baboon	8.2360
Our scheme for Fruits	8.0143
Reference [15]	8.4300

the range of gray values in the image. For PSNR analysis, the better the encryption effect is, the smaller the PSNR of the image becomes. The results of PSNR test is shown in Table 2. As can be seen from the table, our scheme has a smaller PSNR value, which shows good encryption effect.

4 Conclusion

In this paper, we proposed a new discrete auto-switched chaotic system. Further, the performance of chaotic binary sequences is analyzed through MPE and auto-correlation. MATLAB simulation results show the good security and randomness of the chaotic binary sequence. On the basis of the above chaotic sequence generator, an image encryption scheme is designed to encrypt digital images. Finally, related simulation experiments show that the proposed scheme is both secure and feasible.

Acknowledgements This work was supported by the Natural Science Foundation of China (No. 61471158) and the "modern sensing technology" innovation team project of Heilongjiang province (No. 2012TD007).

References

1. Feigenbaum, M.J.: Quantitative universality for a chaos of nonlinear transformations. J. Stat. Phys. **19**(1), 25–52 (1978)
2. Li, T.Y., Yorke, J.A.: Period three implies chaos. J. Am. Math. Mon. **82**(10), 985–992 (1975)
3. Matthews, R.: On the derivation of a chaotic encryption algorithm. J. Cryptologia. **3**(1), 29–41 (1989)
4. Liu, C.Y., Ding, Q.: Complexity analysis and research based on the chaotic system of sample entropy. J. Netw. Intell. **3**(3), 162–169 (2018)
5. Fan, C.L., Ding, Q.: ARM-embedded Implementation of H.264 selective encryption based on chaotic stream cipher. J. Netw. Intell. **3**(1), 9–15 (2018)
6. Yang, B., Liao, X.F.: A new color image encryption scheme based on logistic map over the finite field Z(N). J. Multimed. Tools Appl. **77**(16), 21803–21821 (2018)
7. Liu, Y., Tang, S.Y., Liu, R., et al.: Secure and robust digital image watermarking scheme using logistic and RSA encryption. J. Expert Syst. Appl. **97**, 95–105 (2018)
8. Li, C.H., Luo, G.C., Qin, K., et al.: An image encryption scheme based on chaotic tent map. J. Nonlinear Dyn. **87**(1), 127–133 (2017)
9. Fan, C.L., Liu, S.Y., Ding, Q.: Design of embedded ethernet interface based on chaotic stream cipher. J. Inf. Hiding Multimed. Sig. Process. **6**(3), 2073–4212 (2015)

10. Du, B.X., Geng, X.L., Chen, F.Y., Pan, J., Ding, Q.: Generation and realization of digital chaotic key sequence based on double K-L transform. Chin. J Electron. **22**(1), 131–134 (2013)
11. Xiang, F., Qiu, S.S.: Stream cipher design based on inter-perturbations of chaotic systems. Acta Phys. Sin-Ch. Ed. **57**(10), 6132–6138 (2008)
12. Chen, T.M., Jiang, R.R.: New hybrid stream cipher based on chaos and neural networks. Acta Phys. Sin-Ch. Ed. **62**(4), 040301 (2013)
13. Fan, C.L., Zhang, Q., Sun, H.R., Song, B.B., Ding, Q.: Designing network encryption machine based on Henon chaotic key algorithm. J. Appl. Anal. Comput. **6**(4), 1126–1134 (2016)
14. Bandt, C., Pompe, B.: Permutation entropy: a natural complexity measure for time series. Phys. Rev. Lett. **88**, 174102 (2002)
15. Yin, Q., Wang, C.H.: A new chaotic image encryption scheme using breadth-first search and dynamic diffusion. Int. J. Bifurcat. Chaos. **28**, 1850047 (2018)

A Robust On-Road Vehicle Detection and Tracking Method Based on Monocular Vision

Ling Xiao, Yongjun Zhang, Jun Liu and Yong Zhao

Abstract In this paper, we propose a new framework for vehicle detection and tracking. Multi-features are used in the vehicle detection algorithm, which can be divided into two main steps: generation of candidates using features such as the shadow and vertical edge, and verification of the candidates using HOG and SVM. In the vehicle tracking algorithm, the RGB model and orientation histogram are used to represent the object feature, and the mean shift is employed to search the mode of the potential object rapidly in a neighborhood frame, which obtains the preliminary tracking results. Then, we use ORB feature matching and correction methods to adjust the preliminary tracking results. The improved Mean-Shift tracking results and the ORB correction results are then fused by linear weighted, which obtains the final results of the tracking. Experimental results demonstrate that the proposed approach is robust and validate in complicated real scenes.

Keywords Vehicle detection · Vehicle tracking · Vertical edge · Mean shift · ORB

L. Xiao · Y. Zhang (✉)
Key Laboratory of Intelligent Medical Image Analysis and Precise Diagnosis of Guizhou Province, College of Computer Science and Technology, Guizhou University, Guiyang, China
e-mail: zyj6667@126.com

L. Xiao
e-mail: xl201304@163.com

J. Liu · Y. Zhao
School of Electronic and Computer Engineering, Shenzhen Graduate School of Peking University, Shenzhen, China
e-mail: liujun@sz.pku.edu.cn

Y. Zhao
e-mail: zhaoyong@pkusz.edu.cn

© Springer Nature Singapore Pte Ltd. 2019
J.-S. Pan et al. (eds.), *Genetic and Evolutionary Computing*,
Advances in Intelligent Systems and Computing 834,
https://doi.org/10.1007/978-981-13-5841-8_31

295

1 Introduction

The on-road vehicle detection and tracking, as one of the important steps in intelligent driver assistance systems, have been a topic of great interest. At present, a large number of vehicle detection algorithms [1–3] have been proposed and almost every visual vehicle detection system follows two basic steps: Hypothesis Generation (HG) which hypothesizes the locations in images, where vehicles might be present and Hypothesis Verification (HV) which classifies the candidates into vehicles and non-vehicles. The HG step used the knowledge-based, stereo-based, and motion-based methods [4]. Usually, the prior knowledge of vehicle includes shadow [5, 6], edge [7], and symmetry [8]. Vehicle tracking is the process of predicting the probable location of vehicles in future frames according to the detected vehicle location in former frames. The detection performance in term of time and accuracy can be greatly improved through vehicle tracking. The commonly used tracking algorithms include Kalman Filters [9], Particle Filters [10], or Mean Shift [11].

Based on vehicle detection, vehicle tracking is used to predict the probable location of vehicles in future frames according to the detected vehicle location in former frames. Mean shift is a kernel-based tracking algorithm which is first used by Comaniciu, etc. [12]. Kalman filter [13] predicts the position of the vehicle in the next frame by building a state space model of the vehicle. Particle filter has good performance in tracking vehicle in complex background via solving Bayesian probability [14]. In terms of tracking of vehicles, Mean Shift is an important tracking algorithm, which is not sensitive to occlusion, object rotation, distortion, and moving background. The drawback of Mean Shift is that it tracks the object only by the color information. When the background has the similar color with the object, the target of tracking can be lost. Literature [15] combines the orientation histogram and HIS model to enhance the Mean-Shift tracking algorithm. Literature [16] presents a SIFT features based mean shift algorithm. SIFT features are used to correspond the region of interests across frames. Meanwhile, mean shift is applied to conduct similarity search via color histograms. The probability distributions from these two measurements are evaluated in an expectation–maximization scheme so as to achieve maximum likelihood estimation of similar regions.

In this paper, we propose an effectively and robustly method to detect and track vehicles. The detection algorithm first uses the shadow feature to get the ROI of vehicles. Then vertical edges of vehicles are used to get the candidates of vehicles. After that, the classifier-based HOG + SVM is used to verify the targets. After the detection, the proposed tracking method is used to track the targets. After tracking 100 frames, we verify the image within the bounding box until the verification fails.

2 Vehicle Detection

2.1 Shadow Detection

A distinctive feature of vehicles is the shadow underneath them. Though the intensity of shadow depends on the illumination, which in turn depends on the weather, it is always presented on the road. More importantly, it is darker than any other areas on an asphalt paved road. We can use the shadow to get the possible position of the vehicles. We use the method to extract the shadow is that the grayscale histogram because the grayscale histogram can reflect the whole image grayscale distribution well. The grayscale of shadow locates on the lower part of the grayscale distribution. We can use the threshold which is the sum of the lowest of the grayscale and the 10% of the width of the grayscale histogram to detect the shadow regions underneath vehicles. Then we can find the bottom line of the shadow. Because the top of a vehicle has no consistent cues, we can assume that the aspect ratio of any vehicle is 1.2 and get the ROI. Figure 1a is a vehicle image from Caltech Cars (Rear) [17]. Black regions in Fig. 1b are shadow regions segmented by the threshold.

2.2 Vertical Edge Detection of Vehicle

The other salient feature that appears at the vehicle is the vertical edge. The vertical edge often appears on the left and the right side of the vehicle. So, the vertical edge feature can be used as an important feature for the vehicle detection. In the first step, we obtain the ROI through the shadow underneath vehicles. Then in the ROI, we use a Sobel operator to find strong vertical edges. To localize left and right position of a vehicle, the vertical gradient value of each pixel is projected in the vertical direction. As shown in Fig. 2, we search the local maxima peaks of the vertical profile and define the column that has the maxima as the left or right border of a vehicle.

Fig. 1 Vehicle shadow detection   **(a)**   **(b)**

(a)   (b)   (c)

Fig. 2 Vehicle vertical edge detection

3 Vehicle Tracking

3.1 Improved Mean-Shift Tracking Algorithm

Most Mean-Shift tracking schemes use the color histogram to represent the target. In our experiment, we combine the color histogram with local gradient orientation to increase the stability of the object feature model. The three components of RGB histogram of the target are divided into eight bins, respectively.

We also divide the orientation histogram into eight bins. Combining the color with the texture distribution, the features of the target model q can be presented as a vector of $8 \times 8 \times 8 \times 8$ dimensions.

The target model q corresponding to the target region is as shown follows.

$$q = \{q_u\}_{u=1\ldots m}$$
$$q_u = C \sum_{i=1}^{n} k\left(\|x_i\|^2\right) \delta[b(x_i) - u] \tag{1}$$

where q_u is the probability of feature u in target model and m is the number of feature spaces. δ is the Kronecker delta function and $b(x_i)$ associates the pixel x_i to the histogram bin. $k(x)$ is an isotropic kernel profile and the value of C is shown below.

$$C = 1 / \sum_{i=1}^{n} k\left(\|x_i\|^2\right) \tag{2}$$

The candidate model $p(y)$ of the candidate region is shown below.

$$p(y) = \{p_u(y)\}_{u=1\ldots m}$$
$$p_u(y) = C_h \sum_{i=1}^{n_h} k\left(\left\|\frac{y - x_i}{h}\right\|^2\right) \delta[b(x_i) - u]$$

$$C_h = 1/\sum_{i=1}^{n_h} k\left(\left\|\frac{y - x_i}{h}\right\|^2\right) \tag{3}$$

where $p_u(y)$ is the probability of feature u in candidate model. h is the bandwidth and C_h is a normalization function.

A metric based on the Bhattacharyya coefficient is defined between the two normalized histograms $p(y)$ and q to calculate the similarity of the target model and the candidate model.

$$\rho(p(y), q) = \sum_{u=1}^{m} \sqrt{p_u(y)q_u} \tag{4}$$

The distance between $p(y)$ and q is defined as shown follows.

$$d[p(y), q] = \sqrt{1 - \rho[p(y), q]} \tag{5}$$

We use the target location y_0 in the previous frame to initialize the iterative optimization process. The linear approximation of (4) is obtained using Taylor expansion around $p_u(y_0)$, which greatly reduces the error of tracking and improve the accuracy. The mean-shift tracking algorithm can find the most similar region in the new frame to the object.

3.2 Integrate Improved Mean Shift and ORB

The tracking algorithm combines the improved Mean Shift with ORB feature matching and the final result is linear weighted them. The method of ORB feature extracting and matching can be found in Literature [18]. The algorithmic process is shown in Fig. 3.

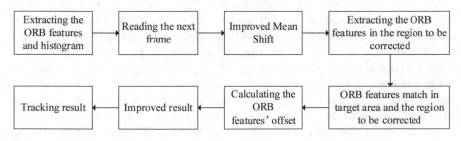

Fig. 3 Tracking algorithm

The feature points in search area and model area can be defined as $p' = \{(x_i', y_i'), V_i'\}$ and $p = \{(x_i, y_i), V_i\}$, $i = 1, 2, ..., N$. The method of arithmetic mean is used to calculate the offset of the feature points, which is shown in (6).

$$(\Delta x, \Delta y) = \frac{1}{N} \sum_{i-1}^{N} \left[(x_i', y_i') - (x_i, y_j) \right] \qquad (6)$$

Supposing that the result of the improved Mean Shift is $rect\_ms = \{x_0, y_0, width, height\}$, where (x_0, y_0) is the center of the rectangle, $width$ is the diameter in the vertical direction, and $height$ is the diameter in the horizontal direction. Combining the (6), the result is as follows:

$$rect\_orb = rect\_ms + (\Delta x, \Delta y) = [x_0 + \Delta x, y_0 + \Delta y, width, height] \qquad (7)$$

The final result can be defined as

$$rect = \alpha \times rect\_ms + (1 - \alpha) \times rect\_orb \qquad (8)$$

α is the weight coefficient, and we can let the α equal to the Bhattacharyya coefficient. In this way, it will large when the result of improved Mean Shift is better and small reversely.

4 Experimental Result

In this section, we conduct serials of experiments based on VS2010 and OpenCV to verify the performance. A part of vehicle images for testing is from the public test library Caltech Cars (Rear) [18]. The rest of vehicle images are self-collected in the real environment.

In the first frame of input video, we use the salient features mentioned above to detect the regions of vehicle candidates, and then extract the HOG features of the candidates to determine that they are vehicles or non-vehicles using the trained classifier. Parts of results of the detection algorithm are shown in Fig. 4. The left of Fig. 4 is the result of Caltech Rear public vehicle images. The right part of Fig. 4 in the result on self-collected vehicle images. In Fig. 4, the experiments show that our method can detect vehicles at different distances well. Meanwhile, even if there is more than one vehicle in the image, our method can recognize them accurately. More important, different types of vehicles such as bus, car, and truck can all be detected.

After detection, we use the improved mean shift and ORB matching to tracking the vehicles. The tracking period we design is 100 frames, and then, we verify the image within the bounding box until the verification fails. To test the effectiveness of our algorithm, we run the method on our video tests. The tracking results of one sequence are illustrated in Fig. 5. Clearly, the lighting in this video changes heavily. Besides,

Fig. 4 Detection results on Caltech Rear public vehicle images and our data set

(a)

(b)

Fig. 5 **a** The result of origin mean shift, **b** the result of our method

the camera vehicle will across an overhead bridge which increases the complexity of the tracking. Figure 5a is the result of the original mean shift. We can clearly see that the target is quickly lost because of the changing lighting. However, the proposed approach can handle bad illumination environment, according to the experimental results list in Fig. 5b. The target can still be tracking when it drives into, under and out of the overhead bridge. In other words, the proposed tracking algorithm has not been influenced by the sharp change of the illumination.

5   Conclusions

This paper describes a vehicle detection and tracking algorithm. There are two major contributions in this paper. We propose a new vehicle detection algorithm, and then an improved Mean-Shift tracking algorithm based on the RGB model and gradient orientation is combined with ORB features to track vehicles robustly in several distracted real environments. In the experiment part, public collections and our own test images are used to evaluate our algorithm, and the experiment results are encouraging. However, to generalize our algorithm, there are still several problems to solve, such as detecting vehicles using car lights at night. We will continue to research the problems mentioned above to improve the performance of vehicle detection and tracking.

Acknowledgements This work was supported by the Joint Fund of Department of Science and Technology of Guizhou Province and Guizhou University under grant: LH [2014]7635, Research Foundation for Advanced Talents of Guizhou University under grant: (2016) No. 49, Key Supported Disciplines of Guizhou Province—Computer Application Technology (No. QianXueWei-

HeZi ZDXX[2016]20), Specialized Fund for Science and Technology Platform and Talent Team Project of Guizhou Province(No. QianKeHePingTaiRenCai [2016]5609), and the work was also supported by National Natural Science Foundation of China (61462013, 61661010).

References

1. Chen, Z., Wang, C., Luo, H., et al.: Vehicle detection in high-resolution aerial images based on fast sparse representation classification and multiorder feature. IEEE Trans. Intell. Transp. Syst. **17**(8), 2296–2309 (2016)
2. Hsieh, C.H., Hung, M.H., Weng, S.K.: Visual object tracking based on color and implicit shape features. J. Inf. Hiding Multimed. Signal Process. **9**(1), 198–210 (2018)
3. Zhao, D.N., Guo, D.J., Lu, Z.M., Luo, H.: Tracking multiple moving objects in video based on multi-channel adaptive mixture background model. J. Inf. Hiding Multimed. Signal Process. **8**(5), 987–995 (2017)
4. Bebis, G., Sun, Z., Miller, R.: On-road vehicle detection review. IEEE Trans. Pattern Anal. Mach. Intell. 694–711 (2006)
5. Han, S., Han, Y., Hahn, H.: Vehicle detection method using Haar-like feature on real time system. In: Proceedings of World Academy of Science, Engineering and Technology, pp. 455–459 (2009)
6. Chan, Y.M., Huang, S.S., Fu, L.C., Hsiao, P.Y., Lo, M.F.: Vehicle detection and tracking under various lighting conditions using a particle filter. IET Intel. Transp. Syst. **6**(1), 1–8 (2012)
7. Lin, B., Lin, Y., Fu, L., et al.: Integrating appearance and edge features for sedan vehicle detection in the blind-spot area. IEEE Trans. Intell. Transp. Syst. **13**(2), 737–747 (2012)
8. Vojir, T., Noskova, J., Matas, J.: Robust scale-adaptive mean-shift for tracking. Pattern Recogn. Lett. **49**(3), 250–258 (2014)
9. Chen, L., Bian, M., Luo, Y., et al.: Real-time identification of the tyre–road friction coefficient using an unscented Kalman filter and mean-square-error-weighted fusion. Proc. Inst. Mech. Eng. Part D J. Automob. Eng. **230**(6) (2016)
10. Dong, F., Liu, Z., Kong, D., et al.: Adapting the sample size in particle filters through KLD-sampling. Int. J. Robot. Res. **22**(12), 985–1003 (2016)
11. Liu, T., Zheng, N., Zhao, L., Cheng, H.: Learning based symmetric features selection for vehicle detection. IEEE Intelligent Vehicles Symposium, Las Vegas, Nevada, USA, pp. 124–129 (2005)
12. Ramesh, V., Comaniciu, D., Meer, P.: Kernel-based object tracking. IEEE Trans. Pattern Anal. Mach. Intell. 564–577 (2003)
13. O'malley, R., Glavin, M., Jones, E.: Vision-based detection and tracking of vehicles to the rear with perspective correction in low-light conditions. IET Intell. Trans. Syst. https://doi.org/10.1049/iet-its.2010.0032
14. Takeuchi, A., Mita, S., McAllester, D.: On-road vehicle tracking using deformable object model and particle filter with integrated likelyhoods. In: Intelligent Vehicles Symposium, pp. 1014–1021. IEEE, Piscataway (2010)
15. Huan, S., Shun-ming, L., Jian-guo, M., et al.: A robust vehicle tracking approach using mean shift procedure. In: 5th International Conference on Information Assurance and Security, pp. 741–744. IEEE, Piscataway (2009)
16. Zhou, H., Yuan, Y., Shi, C.: Object tracking using SIFT features and mean shift. Comput. Vis. Image Underst. **113**(3), 345–352 (2009)
17. Caltech Cars (Rear). http://www.vision.caltech.edu/html-files/archive.html
18. Rublee, E., Rabaud, V., Konolige, K., et al.: ORB: an efficient alternative to SIFT or SURF. In: 2011 IEEE International Conference on Computer Vision (ICCV), pp. 2564–2571. IEEE (2011)

An AC Power Conditioning with Robust Intelligent Controller for Green Energy Applications

En-Chih Chang, Rong-Ching Wu and Chun-An Cheng

Abstract In this paper, a high-performance AC power conditioning is developed by using a robust intelligent controller. The moving sliding mode control (MSMC) ensures the sliding mode occurrence from an arbitrary initial state. Once a highly uncertain perturbation occurs, the MSMC has chatter problem, thus leading to high voltage distortion and performance deterioration of AC power conditioning. To weaken the chatter, the BPSO is employed to optimally tune the MSMC gains for achieving good steady state and transience. Using the proposed controller, the robustness of the AC power conditioning is effectively enhanced, and low distorted output voltage can be obtained against load disturbances. Experiments are given to demonstrate the efficacy of the proposed controller. Because the presented proposed controller provides better tracking exactness and convergence rate, this paper will be an applicable reference to the researchers of correlative robust control, evolutionary algorithm, and green energy applications.

Keywords AC power conditioning · Moving sliding mode control (MSMC) · Chatter · Binary particle swarm optimization (BPSO) · Green energy applications

1 Introduction

AC power conditioning is extensively applied in green energies such as photovoltaic energy, wind energy, and hydrogen fuel cells [1]. The AC power conditioning performance depends on the inverter-LC filter unit, whose task is to transform the DC-to-AC voltages. In order to gain low distorted output voltage and fast transient

E.-C. Chang (✉) · R.-C. Wu · C.-A. Cheng
Department of Electrical Engineering, I-Shou University, Kaohsiung City, Taiwan, ROC
e-mail: enchihchang@isu.edu.tw

R.-C. Wu
e-mail: rcwu@isu.edu.tw

C.-A. Cheng
e-mail: cacheng@isu.edu.tw

© Springer Nature Singapore Pte Ltd. 2019
J.-S. Pan et al. (eds.), *Genetic and Evolutionary Computing*,
Advances in Intelligent Systems and Computing 834,
https://doi.org/10.1007/978-981-13-5841-8_32

response under uncertain perturbations, feedback controllers can be employed. The proportional integral (PI) or proportional–integral–derivative (PID) controller has been often used, but it cannot obtain fast and stable output voltage under nonlinear interferences. Many advanced control methods are also used such as deadbeat control, repetitive control, and wavelet transform technique. But, these methods are difficult to realize and their algorithms are complex. Studied in [2] have displayed that sliding mode control (SMC) is a robust tracking control method because it offers insensitivity to parameter variations, rejection of external load disturbances, and fast dynamic response. The sliding mode-controlled AC power conditioning can be discovered in publications, but these adopt the fixed sliding hyperplanes and are sensitive to uncertain perturbations before the sliding mode occurs [3, 4]. In order to tackle such problem, the moving sliding mode control (MSMC) is used to pass arbitrary initial circumstances, and thereafter moved toward a pre-specified sliding hyperplane by rotating/shifting [5, 6]. The sensitivity to system uncertainties will be lessened through the shortening of the reaching phase, and then the exact tracking response can be yielded. However, the chatter still exists when a severely uncertain disturbance is applied. The chatter causes unmodeled high-frequency plant dynamics and sometimes, it may lead to the unstable system. Binary particle swarm optimization (BPSO) algorithm is one of the popular evolutionary algorithms, and has been broadly applied to many areas of science and engineering [7–11]. The MSMC control gains can be optimally determined by BPSO, and then the chatter can be lessened while the MSMC still provides the shortening of the reaching phase. By combining MSMC and BPSO, the proposed controller generates a closed-loop AC power conditioning with low harmonic distortion, fast transient response and chatter reduction. Experimental results have shown the effectiveness of the proposed controller.

2 System Modeling

A regular AC power conditioning for green energy applications is represented as Fig. 1. It is composed of DC power supply V_{dc}, resistance load R, full-bridge inverter, and LC filter. The LC filter and connected load can be regarded as the plant of the closed-loop system. The proposed controller is employed to design a 1 KW 60 Hz AC power conditioning.

The plant dynamics can be written as

$$\dot{x}_{p2} = -a_{p1}x_{p1} - a_{p2}x_{p2} + b_p u_p + \phi \tag{1}$$

where $x_{p1} = v_o, x_{p2} = \dot{v}_o, a_{p1} = \frac{1}{LC}, a_{P2} = \frac{1}{RC}, b_p = \frac{K_{pw}}{LC}, K_{pw}$ stands for inverter gain, u_p signifies the plant input, and ϕ is the uncertain perturbation.

The model system is provided through the sinusoidal function with f_o Hz as follows:

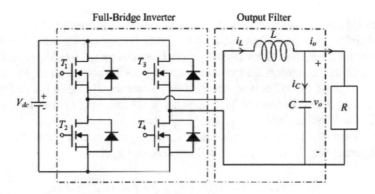

Fig. 1 Block diagram of AC power conditioning

$$\dot{x}_{m2} = -a_{m1}x_{m1} - a_{m2}x_{m2} + b_m u_m \tag{2}$$

where $x_{m1} = v_r$, $x_{m2} = \dot{v}_r$, $a_{m1} = (2\pi f_o)^2$, $a_{m2} = 0$, $b_m = 1$, u_m denotes the reference input.

Define $e_1 = x_{m1} - x_{p1}$ and then subtract (1) from (2), the error state equation yield

$$\dot{e}_1 = e_2 = \dot{x}_{m1} - \dot{x}_{p1} = x_{m2} - x_{p2} \tag{3}$$

$$\dot{e}_2 = -a_{m1}e_1 - a_{m2}e_2 + (a_{p1} - a_{m1})x_{p1} + (a_{p2} - a_{m2})x_{p2} + b_m u_m - b_p u_p - \phi \tag{4}$$

The moving sliding hyperplane can be selected as

$$\sigma = c_r(t)e_1 + e_2 - \beta_s(t) \tag{5}$$

where $c_r(t) = Ft + G$, $\beta_s(t) = Mt + N$, and F, G, M, and N are constants.

The control law u_p can be expressed as

$$u_p = u_e + u_s \tag{6}$$

where u_e symbols the equivalent control, which can handle the system dynamics on the sliding hyperplane, and u_s indicates the sliding control that is used to repress the interferences. The equivalent control u_e can be derived by $\dot{\sigma}|u_p = u_e = 0$. The sliding control u_s can be designed as

$$u_s = \eta_1 e_1 + \eta_2 e_2 + \eta_3 x_{p1} + \eta_4 x_{p2} + \eta_5 u_m + \eta_n \tag{7}$$

From (6) and (7), the $\eta_1, \eta_2, \eta_3, \eta_4, \eta_5$ and η_n will be obtained to satisfy a sliding-action existence $\sigma\dot{\sigma} < 0$. Nevertheless for green energy applications, the MSMC as shown in (7) has the chatter phenomenon. It is because the loading is probably a high-uncertainty condition, thus causing inexact tracking performance. Namely, the output voltage of the AC power conditioning cannot exactly equal the desired sine wave. To lessen the chatter, the MSMC gains of the shown in (7) can be optimally tuned by the BPSO algorithm illustrated in (8) and (9).

The (8) and (9) display the evolution models of a particle. The speed and position of each particle can be renewed while flying toward target.

$$V_i^{t+1} = \rho_0 V_i^t + \rho_1\kappa_1(X_i^{pb} - X_i) + \rho_2\kappa_2(X_i^{gb} - X_i) \tag{8}$$

$$X_i^{t+1} = \begin{cases} 0, & \text{if rand} \geq Sig(V_i^{t+1}) \\ 1, & \text{if rand} < Sig(V_i^{t+1}) \end{cases} \tag{9}$$

where ρ_0, ρ_1, and ρ_2 indicate variables, and κ_1, κ_2 are random numbers, V_i denotes present flying speed, X_i is present position, X_i^{pb} represents local best position, X_i^{gb} stands for global best position, and Sig signifies sigmoid function that transforms the particle velocity to the probability $1/1 + e^{-V_i^{t+1}}$.

3   Experimental Results

The experimental verification of the proposed controller is accomplished on an AC power conditioning with specifications in the following: $V_{dc} = 220$ V, output voltage $= 110$ V$_{rms}$, 60 Hz, switching frequency $= 10$ kHz, $L = 1$ mH, $C = 30$ μF, and rated load $= 12$ Ω. Figure 2 shows the waveform for the proposed controller under the step load from no load to full load at a 90° firing angle. The small voltage sag with good transient can be noticed and after the transience, the voltage waveform reverts to high steady-state exactness. But, the waveform obtained using the orthodox SMC plotted in Fig. 3 appears as a large voltage sag and presents a slow recovery time. Figures 4 and 5 depict the tracking errors for the proposed controller and orthodox SMC, respectively. The proposed controlled AC power conditioning illustrates that tracking error quickly converges to zero without the occurrence of the oscillation. The orthodox sliding mode-controlled AC power conditioning implies the arrival of the sliding mode in a long time, thus yielding convergent instability. As compared to orthodox SMC, the proposed controller produces higher tracking exactness, lower harmonic distortion, and quicker convergence speed.

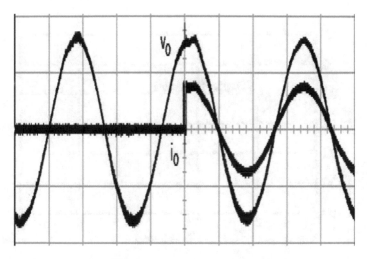

Fig. 2 Proposed controller under step change in load (100V/div; 20A/div; 5 ms/div)

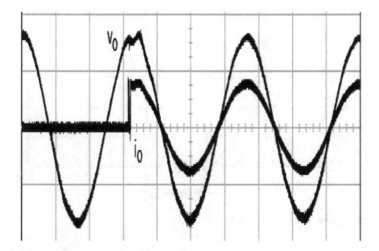

Fig. 3 Orthodox SMC under step change in load (100 V/div; 20A/div; 5 ms/div)

4 Conclusions

A BPSO-based MSMC to reject uncertain disturbances has been proposed and the system tracking performance is improved. The MSMC is capable of shortening the reaching phase, however if there is a severe nonlinear load, the chatter exists. For the sake of lessening the chatter, the BPSO is thus employed to optimally tune the

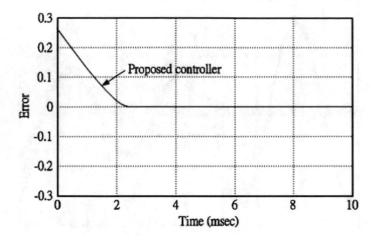

Fig. 4 Tracking error for the proposed controller

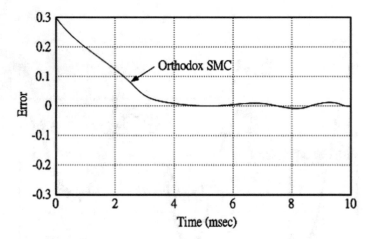

Fig. 5 Tracking error for the orthodox SMC

MSMC gains. The combination of MSMC and BPSO is advantageously used in AC power conditioning to carry out a high reliable performance in transience and steady state. Experiments have been reported to prove the applicability of the proposed controller.

Acknowledgements This work was supported by the Ministry of Science and Technology of Taiwan, R.O.C., under contract number MOST107-2221-E-214-006.

References

1. Ibrahim, D., Adnan, M., Haydar, K.: Progress in Sustainable Energy Technologies: Generating Renewable Energy. Springer International Publishing, Switzerland (2014)
2. Itkis, U.: Control Systems of Variable Structure. Wiley, New York (1976)
3. Malesani, L., Rossetto, L., Spiazzi, G., Zuccato, A.: An AC power supply with sliding-mode control. In: Proceedings of IEEE International Conference on Industry Applications Society Annual Meeting, pp. 623–629 (1993)
4. Sarinana, A.: A novel sliding mode observer applied to the three-phase voltage source inverter. In: Proceedings of European. Conference on Power Electronics, and Applications, pp. 1–12 (2005)
5. Geng, J., Sheng, Y.Z., Liu, X.D.: Second-order time-varying sliding mode control for reentry vehicle. Int. J. Intell. Comput. Cybern. **6**(3), 272–295 (2013)
6. Li, L., Zhang, Q.Z., Rasol, N.: Time-varying sliding mode adaptive control for rotary drilling system. J. Comput. **6**(3), 564–570 (2011)
7. Alireza, S., Mozhgan, A.: Binary PSO-based dynamic multi-objective model for distributed generation planning under uncertainty. IET Renew. Power Gener. **6**(2), 67–78 (2012)
8. Lin, C.J., Chern, M.S., Chih, M.C.: A binary particle swarm optimization based on the surrogate information with proportional acceleration coefficients for the 0-1 multidimensional knapsack problem. J. Ind. Prod. Eng. **33**(2), 77–102 (2016)
9. Chen, X.Y., Peng, X.Y., Li, J.B., Peng, Y.: Overview of deep kernel learning based techniques and applications. J. Netw. Intell. **1**(3), 83–98 (2016)
10. Fournier-Viger, P., Lin, C.W., Kiran, R.U., Koh, Y.S., Thomas, R.: A survey of sequential pattern mining. Data Sci. Pattern Recogn. **1**(1), 54–77 (2017)
11. Wu, C.M., Gong, H.Q., Yang, J.H., Song, Q.H., Wang, Y.J.: An improved FOA to optimize GRNN method for wind turbine fault diagnosis. J. Inf. Hiding Multimed. Signal Process. **9**(1), 1–10 (2018)

Part VIII
Single Processing

A Distributing Method for Insufficient Supply in P2P Video Broadcast System

Zhiming Cai, Xuehong Huang and Yiwen Ou

Abstract The video on demand applications is becoming more and more popular with the increasing in network bandwidth. However, there still remain some challenges including utilizing the resources efficiently, maximizing the throughput, and minimizing the delivery time. In this paper, we propose a genetic algorithm R-2DCGA for video distributing. It focuses on makespan and load balancing and can be used for insufficient supply. Experiments have been conducted to evaluate the effectiveness and performance. The experimental results show that R-2DCGA achieves greater ability to reduce makespan and keep load balancing compared to 2DCGA.

Keywords Video distribution · Insufficient supply · Genetic algorithm

1 Introduction

The Video on Demand (VoD) allows distributed consumers to have access to a set of servers, where videos are stored as files and transmitted to the consumers over high speed network [1]. Consumer satisfaction is the key factor in the success of such service [2]. On the contrary, the consumers' increasing demands also bring traffic congestion. Peer-to-peer networking technique is a vast potential to overcome many constraints in such content distribution networks [3]. Even in P2P video systems, there still remains some challenges including churn of peers, high transmission delay, and high bandwidth heterogeneity. To improve reliability of service scheduling, makespan must be minimized [4]. And aiming at utilizing the resources effi-

Z. Cai (✉) · Y. Ou
School of Information Science and Engineering, Fujian University of Technology,
Fuzhou 350118, China
e-mail: caizm@fjut.edu.cn

Z. Cai · X. Huang
National Demonstration Center for Experimental Electronic Information and Electrical
Technology Education, Fujian University of Technology, Fuzhou 350118, China

© Springer Nature Singapore Pte Ltd. 2019
J.-S. Pan et al. (eds.), *Genetic and Evolutionary Computing*,
Advances in Intelligent Systems and Computing 834,
https://doi.org/10.1007/978-981-13-5841-8_33

ciently, maximizing the throughput, and minimizing the request rejection rate, load balancing becomes another crucial factor. He et al. [5] propose an optimized video multicast scheme over wireless ad hoc networks to jointly optimize the source rate and the routing scheme. Gaber et al. [6] suggest a predictive and content-aware load balancing algorithm to meet such objectives. Bideh et al. [7] propose a receiver-side scheduler which attempts to request video frames from providers which can deliver them in a shorter time. Zhang et al. [8] present a server-aided adaptive video streaming scheme to cope with the bandwidth heterogeneity. Al-Habashna et al. [9] focus on improving the throughput of video transmission in cellular networks. Chen et al. [10] propose an algorithm called Zebra to speed up delivery with limited bandwidth. In this paper, we present a genetic algorithm for video distributing. This method has a good performance in reducing makespan and balancing load.

2 A Revised 2D Chromosomes Genetic Algorithm

Based on the 2D chromosome genetic algorithm (2DCGA) presented in [11], we introduce R-2DCGA for multi-objective optimization, which is implemented as follows:

2.1 Conditioned 2D Chromosome

Suppose there are m consumers and n video providers. The relationship between the consumers and providers can be expressed by a matrix

$$R = \begin{bmatrix} r_{11} & r_{12} & \cdots & r_{1n} \\ r_{21} & r_{22} & \cdots & r_{2n} \\ \cdots & \cdots & \cdots & \cdots \\ r_{m1} & r_{m2} & \cdots & r_{mn} \end{bmatrix} \tag{1}$$

where r_{ij} is 0 or 1, $i = 1, 2, 3, \ldots, m, j = 1, 2, 3, \ldots n$. When $r_{ij}=1$, it means provider j can supply the video the consumer i request. Figure 1 gives an example which every provider supplies insufficient videos for consumers.

If we select one 1 randomly in each row of R and compose a new matrix x, as shown in Fig. 1c, which we call **conditioned 2D chromosome**, then we can get NP conditioned 2D chromosomes, denoted as x_k

$$x_k = \begin{bmatrix} c_{11k} & c_{12k} & \cdots & c_{1nk} \\ c_{21k} & c_{22k} & \cdots & c_{2nk} \\ \cdots & \cdots & \cdots & \cdots \\ c_{m1k} & c_{m2k} & \cdots & c_{mnk} \end{bmatrix} \tag{2}$$

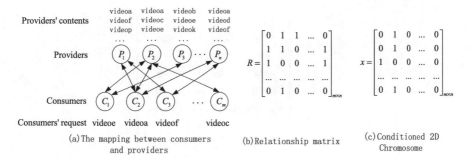

Fig. 1 Relationship between consumers and providers

where $c_{ijk} \in \{0, 1\}$, $\sum_{j=1}^{n} c_{ijk} = 1$, $k = 1, 2, 3, \ldots, NP$ (NP is the population of chromosomes. The maximum is $NP_{\max} = \prod_{i=1}^{m} \left(\sum_{j=1}^{n} r_{ij} \right)$).

2.2 Scheduling Time

As mentioned above, the bandwidth B between the consumer and the video provider can be detected in real time. Given the size of the video requested by the consumer S, the time the scheduling will last t_s can be estimated by S/B. The scheduling time of videos makes up a matrix

$$T_s = \begin{bmatrix} t_{11} & t_{12} & \cdots & t_{1n} \\ t_{21} & t_{22} & \cdots & t_{2n} \\ \cdots & \cdots & \cdots & \cdots \\ t_{m1} & t_{m2} & \cdots & t_{mn} \end{bmatrix} \tag{3}$$

where $t_{ij}(i = 1, 2, \ldots, m; j = 1, 2, \ldots, n)$ is the scheduling time when consumer i requests the video from provider j. We consider the situation of insufficient supply, that is, one cannot get the video what he wants from every provider. The element t_{ij} is set to zero when provider j cannot supply the video for the consumer i.

2.3 Evolution of Revised 2D Chromosomes

The evolution of the conditioned 2D chromosomes includes the operations of crossover and mutation.

Operation 1: Crossover. Let x_p, x_q be conditioned 2D chromosomes. The crossover between them is defined as

$$x_v = \begin{bmatrix} c_{11p} & c_{12p} & \cdots & c_{1np} \\ c_{21p} & c_{22p} & \cdots & c_{2np} \\ \cdots & \cdots & \cdots & \cdots \\ c_{r1p} & c_{r2p} & \cdots & c_{rnp} \\ c_{(r+1)1q} & c_{(r+1)2q} & \cdots & c_{(r+1)nq} \\ c_{(r+2)1q} & c_{(r+2)2q} & \cdots & c_{(r+2)nq} \\ \cdots & \cdots & \cdots & \cdots \\ c_{m1q} & c_{m2q} & \cdots & c_{mnq} \end{bmatrix} \tag{4}$$

That is, the crossover chromosome is composed of the upper r rows of x_p and the bottom $(m - r)$ rows of x_q.

Operation 2: Mutation. Let x_k be a conditioned chromosome. Select two rows of x_k randomly, and in each row we move the 1 to the new position where there is 1 in relationship matrix R at the same row. In fact, for conditioned 2D chromosomes, copying two rows from other chromosomes at the same positions will achieve the same effect if the population of chromosomes is big enough.

2.4 Fitness Functions

In video broadcasting applications, the consumers and the providers have different anticipations. For the former, they hope to get the video as quickly as possible. So the longest video transmitting time of any consumer, namely makespan should be as short as possible. For the latter, they may pay more attention to the load balancing among themselves. Such a multi-objective optimization can be processed by the genetic chromosomes which evolved by the adaptive values. We define two fitness functions f_1 and f_2 to generate adaptive values.

$$f_1(x_k) = 1 / \max_j \left\{ \sum_{i=1}^{m} (t_{ij} \cdot c_{ijk}) | j = 1, 2, 3, \ldots, n \right\} \tag{5}$$

$$f_2(x_k) = \frac{1}{\left(\sum_{j=1}^{n} \left| 1 - L_{rj} \right| \right)} \tag{6}$$

The first function f_1 is used to evaluate the makespan. In function f_2, L_{rj} is the relative load of transmitting of provider j. In theory, the providers who have the same contents should share the video distribution tasks in proportion to their transmitting capabilities [6]. That is, the load of transmitting of provider j should be

$$L_j = \sum_{i=1}^{m} \left(\frac{B_j \cdot r_{ij}}{\sum_{q=1}^{n} B_q \cdot r_{iq}} \cdot S_i \right) \quad j = 1, 2, \ldots n \tag{7}$$

where B_j is the bandwidth the provider j, S_i is the size of the video the consumer i request, and r_{ij} is the element of relationship matrix R. Assuming the total requested size of videos forwarded to provider j is W_j ($W_j = \sum_{i=1}^{m} S_i \cdot c_{ijk}$) and the relative load of transmitting of provider j is defined as $L_{rj} = W_j/L_j$. There are three situations for L_{rj}: $0 < L_{rj} < 1$, $L_{rj} = 1$ and $L_{rj} > 1$, which are called under load, full load, and overload, respectively. Therefore, function f_2 is used to evaluate the total deviations from full load for all the providers.

2.5   Procedure of R-2DCGA

The evolution is an iteration process aiming to search the optimal individual (chromosome). It involves the operations of selection, crossover, and mutation.

Step 1: Initialization. Given two minus values v_1 and v_2 as the initial adaptive values of f_1 and f_2 respectively, and x_o to record the optimal individual. Generate the initial conditioned chromosome population $X = \{x_k | k = 1, 2, \ldots, NP\}$, as described in Sect. 2.1. Given evolution generations NG, weights λ_1, λ_2, crossover and mutation probability p_c, p_m, and initialize loop counter $loop = 0$.
Step 2: Select fitness function $f_i(i = 1, 2)$ based on p_i

$$p_i = \frac{\lambda_i}{\lambda_1 + \lambda_2} \tag{8}$$

Then compute the adaptive values of all individuals with the selected fitness function.
Step 3: Select parents based on roulette wheel. Cross parents with **Operation 1** based on the probability of p_c, and mutate parents with **Operation 2** based on the probability of p_m.
Step 4: Search optimal individual

if f_i ($i = 1$ or 2) is selected then

if f_i ($i = 1$ or 2) is selected then

   for each $k \in \{1, 2, \ldots, NP\}$

      if $f_i(x_k) > v_i$, then

         $x_o = x_k$

      endif

   endfor

Step 5: $loop = loop + 1$. If $loop < NG$, go to **Step 2**; Else go to **Step 6**.
Step 6: Output x_o.

3 Experiments and Evaluations

We compare R-2DCGA and 2DCGA [11]. The evaluations are carried out in Matlab. For R-2DCGA, we consider the situation of insufficient supply. Suppose the number of video providers $n = 6$, the number of consumers $m = 80$, $NP = 100$, $NG = 80$, and $B_j(j = 1, 2, \ldots, n)$, $S_i(i = 1, 2, \ldots, m)$ are generated randomly ranging from (20,100) Mbps and (500,3000) MB. The relationship matrix R is built randomly and the conditioned chromosomes are generated as Sect. 2.1. For 2DCGA, we suppose any consumer can get any video from any provider. Its chromosomes are generated randomly. We investigate two cases: (1) case 1. $\lambda_1 = 0.7$, $\lambda_2 = 0.3$ (namely giving priority to reducing makespan). (2) case 2. $\lambda_1 = 0.3$, $\lambda_2 = 0.7$ (namely considering meeting full load of transmitting first). Tests are repeated 10 times.

In case 1, we focus on makespan. As shown in Fig. 2a, R-2DCGA has much smaller makespan than 2DCGA in most occasions, which indicates the better performance in reducing makespan. While in case 2, we define the load as W_j/B_j, which considers all the video size that provider j needs to transmit and the bandwidth it has. Figure 2b shows R-2DCGA has much lower standard deviation of load than 2DCGA, which suggests the former is good at assigning the video distributing jobs.

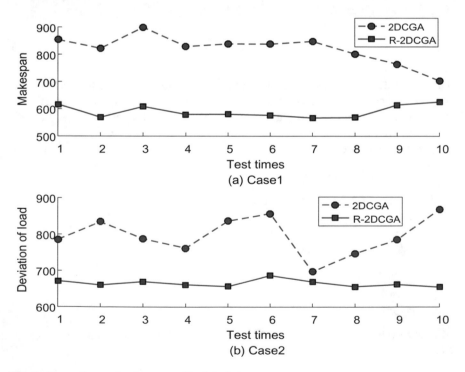

Fig. 2 Comparisons of makespan and load deviation

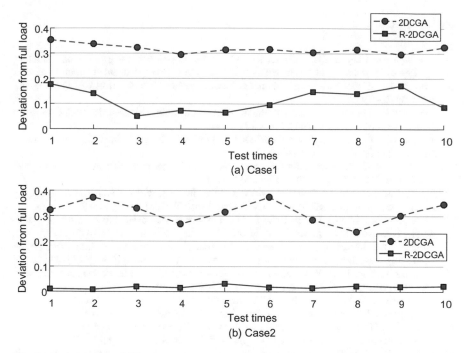

Fig. 3 Deviation from full load

Generally, we hope all the providers work in full load so that they can get more profits and the consumers will receive high quality experience. Figure 3 shows that R-2DCGA produces less deviation from full load in every test compared with 2DCGA. Moreover, when the cases changed from 1 to 2, the deviation of R-2DCGA has decreased nearly to zero while 2DCGA drops slightly. It shows R-2DCGA has a better ability for optimization.

4 Conclusions

In video on demand applications, it is both difficult and unnecessary for all the providers supply all the video for every consumer. This paper proposes a distributing method for such insufficient supply. We present a new algorithm R-2DCGA, which revised from 2DCGA. Experiment results show that R-2DCGA has a better performance at reducing makespan and keeping load balance. Compared to 2DCGA,R-2DCGA shows a better optimization ability and seems to keep the providers working in full load.

Acknowledgements This work is supported by Major Scientific and Technological Projects of Fujian, China (Grant No. 2013HZ0001-4).

References

1. Coelho Freire Batista, C.E., Salmito, T.L., Cunha Leite, L.E. et al.: Big videos on small networks—a hierarchical and distributed architecture for a video on demand distribution service. In: International Conference on Multimedia Services Access Networks (MSAN), Orlando, FL, USA, pp. 15–19 (2005)
2. Develder, C., Lambert, P., van Lancker, W., et al.: Delivering scalable video with QoS to the home. Telecommun. Syst. **49**(1), 129–148 (2012)
3. Lee, I., He, Y., Guan, L.: Centralized P2P Streaming with MDC. In: IEEE Workshop on Multimedia Signal Processing (MMSP) (2005)
4. Cai, Z., Chen, C.: Task scheduling based on degenerated monte carlo estimate in mobile cloud. Int. J. Grid Distrib. Comput. **7**(1), 179–196 (2014)
5. He, Y., Lee, I., Guan, L.: Optimized video multicasting over wireless ad hoc networks using distributed algorithm. IEEE Trans. Circuits Syst. Video Technol. **19**(6), 796–807 (2009)
6. Gaber, S.M.A., Sumari, P.: Predictive and content-aware load balancing algorithm for peer-service area based IPTV networks. Multimed. Tools Appl. **70**(3), 1987–2010 (2014)
7. Bideh, M.K., Akbari, B., Sheshjavani, A.G.: Adaptive content-and-deadline aware chunk scheduling in mesh-based P2P video streaming. Peer-to-Peer Netw. Appl. **9**(2), 436–448 (2016)
8. Zhang, J.-F., Niu, J.-W., Wang, R.-G., et al.: Server-aided adaptive live video streaming over P2P networks. J. Signal Process. Syst. **59**(3), 335–345 (2010)
9. Al-Habashna, A., Wainer, G.: Improving video transmission in cellular networks with cached and segmented video download algorithms. Mob. Netw. Appl. **23**(3), 543–559 (2018)
10. Chen, Y.-F., Huang, Y., Jana, R., et al.: Towards capacity and profit optimization of video-on-demand services in a peer-assisted IPTV platform. Multimed. Syst. **15**(1), 19–32 (2009)
11. Cai, Z., Chen, C.: Demand-driven task scheduling using 2D chromosome genetic algorithm in mobile cloud. In: International Conference on Progress in Informatics and Computing Shanghai China, pp. 539–545 (2014)

An Application of Support States from Speech Emotions in Consensus Building

Ning He and Yang Liu

Abstract People propose various ideas and opinions in public organization and conferences. In order to reach an agreement among the participants, discussion is essential. During a discussion, difference in the processing of forming an agreement will affect the conclusion. So major statements should be analyzed for building a consensus. However, in the process of forming consensus, a proper understanding of the statements acting as the basis among various discussants, is necessary. In this study, the relationship between the consciousness of the conversation and the state of support is discovered through the speech emotional perception of the conversation. Also, the state of support—supportive, negative and unknown, are inferred. In addition to listening experiments on the emotions and support states, the application of the analysis of the speech emotion recognition process is discussed on the basis of verifying the emotion of speech and the dependence of the support state. The accuracy of the average objective recognition rate can be increased to 75% in the formation of the consensus during speech conversation.

Keywords Emotion recognition · MFCC · Consensus building · Support state

1 Introduction

The discussion is essential for opinion-holders to seek agreement with each other when they are involved in a public organization or within an enterprise environment. The process of clarifying the explicitization of tacit knowledge through discussion, and making a mutual agreement between them is called consensus building. The difference in the consensus-building process will affect the conclusion. Therefore, it is necessary to analyze participants' utterances.

N. He (✉) · Y. Liu
Changzhou College of Information Technology, Ming xin Road 22 #,
Wujin District, Changzhou, Jiangsu Province, China
e-mail: 89039650@qq.com

© Springer Nature Singapore Pte Ltd. 2019
J.-S. Pan et al. (eds.), *Genetic and Evolutionary Computing*,
Advances in Intelligent Systems and Computing 834,
https://doi.org/10.1007/978-981-13-5841-8_34

In the process of reaching a consensus, it is necessary to correctly get the agreement and disagreement information in various utterances, which is necessary for the discovery of "clue". Stakeholders, basically, communicate with each other orally. However, their utterances are consciously controlled; hence, if only the textual record converted from the utterances is used, much information will be lost. Therefore, this study is based on the correct estimation of the speaker's intentions from the speech. In this research, emotion data contained in human voice in the dialogue of consensus building discussion are analyzed for the estimate of three intents, positive, negative and unknown.

Among the existing researches on consensus building, most of which emphasis on language analysis techniques such as text analysis, keyword extraction analysis, and language statistics-based estimation. However, in this research, in addition to text analysis of dialogue, features, such as fundamental frequency, duration, sound intensity, tempo, namely, emotional features of the speech are utilized.

Also, the hypothesis that "a strong dependence between emotion and support state" is made to verify its validity through "speech recognition in discussion". By combining text analysis methods and speech recognition of consensus building which automatically analyzing the state of support among debater, it is possible to improve the accuracy of finding remarkable utterance in a consensus building process.

2   Related Work on Consensus Building

Nowadays, there are many theories and methods for consensus building. Since 1999, more than thousand pages of "Consensus Building handbook" have been issued. As we known, consensus building will go through a more complicated procedures. In many open source projects, consensus building that was originally formed in the form of conferences can be gradually turned to computer-assisted [1]. The group can guarantee the equal rights of the members through the structured discussion method, and gather the knowledge and information of different group members to improve the acceptability of consensus building [2]. Other active of consensus research area is in computational statistics. They consider that the algorithm performs distributed computing on the partitioned data, and then makes comprehensive reasoning on the data set by Monte Carlo simulation [3, 4]. Pathmaker's data analysis tools provide assistance to the conference moderator, creative thinking and consensus building [5]. Schuman proposed a method of decision-making meeting from the angle of group decision. Polish System Research Institute built a computer support system called "Mediator" based on expert advice group and examined nine different convergence methods. In addition, many Japanese and American companies provided technology and consultation software such as AHP-based GDSS or MRV on consensus formation and group decision-making.

Microblog here is used to form the discussion environment for consensus building. In conferences and dialogues, when participants repeatedly enter wash-up questions, drill-down questions, suggestion questions and summary questions, the accuracy with which consensus can be formed within a certain time is significantly increased.

The research on the quantitative analysis of consensus building process in the tree structure is proposed [6]. In that study, a method of tracking "clue" is used by measuring changes in information entropy. Microblog comments in the tree structure are generally widespread, and comments about microblogs with speech intent are used in microblog reviews. In that commentary, fields can be marked as "speaker", "commentator", "content" and "intention".

When constructing an agreed microblog comment tree, the speaker's intention must be clear. So it is necessary to evaluate the commentator's response and the relationship between his speech and intention. The state of content, whether it is support for against the speaker should be analyzed. Estimation of utterance intention can be performed by using Negotiation Process Model. The evaluation which represents the position of the commentator to the speaker will be "positive", "negative" and "unknown".

3 Proposal

In human consensus-building dialogue, basically, speaking language advances interaction. Language is manifested information and is consciously controlled. Therefore, text information often contains falsehoods. In this section, a method of consensus-building process analysis using speech emotion recognition is outlined firstly. Then, the explanation of speech emotion recognition, that the relation between emotion and support status is carried out. Finally, an algorithm for estimating the support state will be described.

3.1 Outline of the Proposed Method

This study presents a method for analyzing the protocol formation process of speech emotion recognition as shown in Fig. 1. Six kinds of emotions "happy", "anger", "sadness", "dislike", "fear" and "neutral" are recognized for speech data in the discussion. Three support states "positive", "negative" and "unknown" are obtained. Then, the relation between emotion and supportive condition is examined.

3.2 Speech Emotion Recognition

In previous study on speech emotion recognition, the most important issue to be solved in human–computer interactions by Stanford University's Reeves and Nass research is called "emotional intelligence". In 1990, at the Massachusetts Institute of

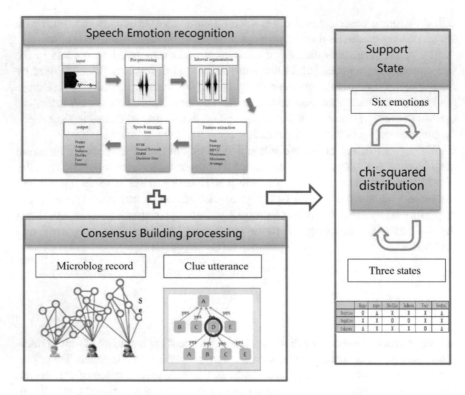

Fig. 1 Outline of the proposal

Technology's Multimedia Research Institute, an emotion editor was build. That is, the machine can recognize emotions extracted from various signals and train them to respond appropriately. In 1996, researchers at Nihon Seikei University advocated the concept of emotional space and built a special speech emotion model [7]. In recent years, modeling of speech emotion recognition using neural network, hidden Markov model is applied.

To obtain the overall emotion of a long conversation, long one should be divided into short parts at first. For recognizing emotion data, extraction of feature quantity called MFCC will be described. Continuous speech recognition in Hidden Markov Model using speech data including these emotions will be applied next.

Hidden Markov Model (HMM) is a statistical model widely used in speech recognition, automatic annotation, phonetic conversion, and other natural language processing applications. After a long-term development, especially in speech recognition, it has become a common statistical tool. HMM can achieve global optimization by analyzing transition probability, performance probability, and co-occurrence probability. Baum–Welch algorithm is utilized here which can solve the parameter estimation problem of HMM, that is, the parameter estimation problem of unknown state sequence of HMM without labeling in speech.

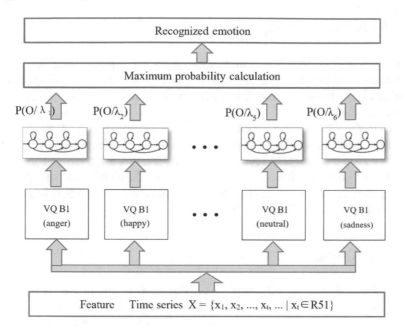

Fig. 2 Emotion recognition system by HMM

In this study, the emotion recognition unit is designed with patterns of time series $X = \{x_1, x_2, ..., x_t, ... \mid x_t \in R51\}$ of 12-dimensional feature quantity of MFCC explained before. The HMM is a nondeterministic model in which states transit according to preset transition probabilities for a given input.

Figure 2 shows the configuration of the emotion recognition system by HMM. The system consists of eight corresponding to six types of emotions, and a Viterbi algorithm recognition unit for collecting outputs from each HMM.

The input to the system is a time series data of feature quantities, but vector quantization (VQ) is performed before inputting to HMM, and the observation sequence is $O = o_1, o_2, ..., o_{51}$. Here, VQ by emotion category is used as VQ. Every VQ is prepared a codebook Bi for each HMM λ_i of each category C_i (i = 1, 2, 3, 4, 5, 6). A symbol sequence is generated for each category. We use Baum–Welch algorithm for HMM learning and Viterbi algorithm for recognition. In the calculation of the likelihood P (O | λ) at the time of recognition, a per-emotion category is used, so a penalty function that takes into consideration the distance between the codebook of VQ and the time series of feature quantities in probability calculation is introduced. In the maximum probability estimation part, the likelihood P (O | λ_i) of each HMM is set as input and the category of emotion corresponding to HMM giving the maximum value max P (O | λ_i) is taken as the final recognition result.

3.3 Association of Support and Affective State

Even if it is a word of praise on literal meaning, for example "it is wonderful", depending on the way of saying, there is a case to convey the intention of reproach. Or, even if it is a blame word on literal meaning like "foolish", it may be the case to convey the affection of love.

In consensus-building process not only the linguistic meaning of utterances but also the emotions involved are important elements to convey information. Therefore, from the six kinds of emotions recognized in Sect. 3.2, in order to estimate the state of support among debaters, it is necessary to examine the relation between emotions and the support state. Using the chi-squared test method, the relation between emotion and the support state will be described.

3.3.1 Support State

Support state has various definitions. The well-known definition is to agree with certain opinions or assertions, and to encourage it. Either from the aspect of cognition or emotion, it is regarded as an internal mental state. The state of support for the partner in discussion is assumed that "the attitude or emotion of the responding partner is equal in one's attitude and feeling". This is different from sympathy.

Support state in this research is defined for utterance Y which responds to the preceding utterance X during discussion, and whether the speaker of utterance Y has the same opinion or not. It holds the respect to the topic content of utterance X and is classified as follows, positive, negative, or unknown.

3.3.2 Relationship of Chi-Square

The hypothesis that the state of support can be estimated from emotion is tested by chi-square test. In the binomial test, a simple occurrence probability of whether an event occurs or not is targeted. However, when the observed event is classified into three or more categories, the sample distribution has a theoretical distribution that chi-square test is used to validate. Based on the crosstabulation table (division table) for the bivariate A and B, it is tested whether there is a relation between the bivariate. When actually used, it is necessary to examine from the viewpoint whether the difference of the factors defined for variable A affects the distribution of data corresponding to each factor of variable B, that is, whether there is a difference in data distribution between groups. Since the χ^2 distribution is used for the determination, it has the common name "χ^2 test".

3.4 Estimation of Support State in Emotion

Taken as a whole, the ultimate goal of the proposed method of this research is to estimate the supporting state of the utterance using the emotional sequence in the dialogue.

In this study, we estimate the supporting state by a simple algorithm. Here, based on the relation taken in the relation test method in Sect. 3.3, we estimate the support state from emotion. Emotions are consolidated into three categories, which are "positive", "negative", and "unknown". Collect all the sections of one conversation and get the statistical number of occurrences in each category. After that, each statistic number is normalized, and the result is used as output of the classifier.

As shown in code results, the support state of "positive" is related to the emotion of "joy", "fear" and "neutral"; "negative" is related to "anger", "dislike"; "Unknown" is related to "sadness" by calculation.

4 Experiment

Using the methods in the previous section, we analyze the emotions contained in the speech and further estimate the support state by experiments and evaluate the results.

In subjective judgement of speech corpus, data analysis was performed for each speaker using data in closed environment with total 228 sentences in 100 dialogues of speakers. As a result of the experiment, the subjective discrimination rate of the emotion was about 85–100% and the average rate 93.7% was obtained. Also, the discrimination ratio of the support state was about 85–100%, and the subjective discrimination ratio close to 93% was obtained on average. Looking at the results individually, 100% of the subjective discrimination ratio in anger emotional and supporting state "positive" is particularly high compared with those of the others.

Then, experiments are conducted using the method of speech emotion recognition described in the proposed method. Here, 600 corpuses of 1080 samples are targeted for training and 480 set as the corpus of emotional voice conversation are for recognition test. Emotions were recognized from 80 emotions using MATLAB, and the results are shown in Table 1.

We assumed that the state of support can be estimated by the emotion of speech, and we examined by chi-squared test whether these are independent or have any correlation.

In this experiment, we test the relevance of emotions and supportive status in 422 remarks of a controversial meeting. On the contrary, subjective discrimination of each utterance was made, and the results are shown in Fig. 3.

Based on the emotion recognition model and the relevance test, we carry out the estimation experiment of the support condition. Here, the data to be used is a dialogue speech corpus of 228 sentences total of 100 dialogues of the speakers. A corpus that can be used for general purpose has constructed a no-emotion emotional

Table 1 Difffernet kinds of recognized emotion

Recognized emotion	Happy	Anger	Dislike	Sadness	Fear	Neutral	Total
Happy	**56**	8	9	0	6	1	80
Anger	2	**56**	10	3	2	7	80
Dislike	0	10	**68**	0	2	0	80
Sadness	2	5	8	**62**	2	1	80
Fear	2	1	0	0	**77**	0	80
Neutral	2	7	1	5	1	**64**	80

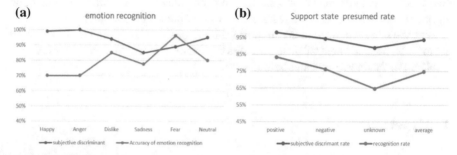

Fig. 3 a Comparison subjective discrimination rate and emotion recognition accuracy **b** Comparison of subject discrimination and recognized support state rate

speech corpus and a dialogue speech corpus. Moreover, subjective discrimination rate of each corpus is 93.7%. In addition, a speech emotion recognition experiment is performed using this corpus, and the average accuracy rate reached 75%.

5 Conclusions

Our experimental results show that emotion and support status is related. Based on this relationship, estimation experiments of emotional supportive state was conducted, and the average accuracy rate reaches 74.76%. In other words, the hypothesis that the support state can be estimated from emotion clues in speech is positive. For the case of "neutral", there is no sufficient support for the dependence relationship, so reestimate linguistic support status from the "neutral" feelings need to be devised, as a future improvement. It is expected that better results will be obtained using classifiers for estimating supportive states from emotions.

Acknowledgements Supported by a grant from Natural science fund for colleges and universities in Jiangsu Province (No. 17KBJ520002) and Top-notch Academic Programs Project of Jiangsu Higher Education Institutions (TAPP) under Grant No. PPZY2015A090.

Ethical Approval All procedures performed in studies involving human participants were in accordance with the ethical standards of the institutional and/or national research committee and with the 1964 Helsinki declaration and its later amendments or comparable ethical standards.

References

1. Farnham, S., Chesley, H.R., McGhee, D.E., Kawal, R., Landau, J.: Structured online interactions: improving the decision making of small discussion groups. In: Proceedings of CSCW (2000)
2. Scott, S.L. Comparing consensus Monte Carlo strategies for distributed Bayesian computation. Brazilian Journal of Probability and Statistics. (2017)
3. Scott, S.L., Blocker, A.W., Bonassi, F.V., Chipman, H.A., George, E.I., McCulloch, R.E.: Bayes and big data: the consensus monte carlo algorithm. In: Bayes 250 (2013). http://research.google.com/pubs/pub41849.html
4. Ko, A.J., Chilana, P.K.: Design, discussion, and dissent in open bug reports. In: Proceedings of the iConference (2011)
5. Moghaddam, R.Z., Bailey, B., Poon, C.: IdeaTracker: an interactive visualization supporting collaboration and consensus building in online interface design discussions. In: Proceedings of INTERACT (2011)
6. Albino, N., Asunción M., Antonio B., José B.: Speech Emotion Recognition Using Hidden Markov Models Eurospeech—Scandinavia (2001)
7. Chennoukh, S., Gerrits, A., Miet, G., Sluijter, R.: Speech enhancement via frequency extension using spectral frequency. Proceedings of the ICASSP, Salt Lake City, vol. 5 (2001)

An Adaptive Algorithm of Showing Intuitively the Structure of a Binary Tree

Pin-Ju Ye, Chun-Guo Huang and Yuan-Wang Hu

Abstract The adaptive software development(ASD) method is an agile development method based on complex adaptive system theory. It develops adaptive abilities through methods such as candidate program library, dynamic display selection, semantic self-description and self-test, and establishes expert system through the study of business rules, achieves intelligent adaptable software by decision theory, diagnosis, recovery and other methods. This paper presents an adaptive algorithm of showing intuitively the structure of a binary tree, the algorithm module is plug and play, and can be inserted directly into any program source code related to the binary tree without any modification. This technology can improve the efficiency of system design, reduce the workload of program maintenance and has some innovation and practicability.

Keywords Binary tree · Adaptive · Semantic · Self-description · Plug and play introduction

1 Introduction

The adaptive software development method is a software development method based on adaptive theory. It brings new ideas to the software development field and is motivated by adapting to the changes of user needs. It is a powerful method to develop software with certain adaptability.

P.-J. Ye (✉)
School of Software, Changzhou College of Information Technology, Changzhou
213164, China
e-mail: shirlyonline@163.com

C.-G. Huang
School of Computer Engineering, Jiangsu University of Technology, Changzhou
213001, China

Y.-W. Hu
School of Electronic and Electrical Engineering,
Changzhou College of Information Technology, Changzhou 213164, China

© Springer Nature Singapore Pte Ltd. 2019
J.-S. Pan et al. (eds.), *Genetic and Evolutionary Computing*,
Advances in Intelligent Systems and Computing 834,
https://doi.org/10.1007/978-981-13-5841-8_35

The adaptive problem first got attention in the field of artificial intelligence, and made great progress in the research direction of genetic algorithm and neural network. Through truly adaptive research from the perspective of software adaptability, as well as the US DARPA will support adaptive hardware and software series included in the study plan. In recent years, due to the popularity and development of embedded software and the Internet, research on adaptive software has become a hot topic in the software industry [1].

This paper presents an adaptive algorithm of showing intuitively the structure of a binary tree, the algorithm module is plug and play, and can be applied in any program associated with a binary tree without any changes.

2 Algorithm Conception

2.1 Establish Display Model

Determine the display range according to the maximum width of the screen. We define the minimum display length for each node, and the last row uses the minimum spacing. A full binary tree of 6 height appears in a 6 * 63 matrix. In order to make the display matrix in the middle, the upper left base coordinates of the matrix will be as (9, 6), as shown in Fig. 1.

2.2 Paging Implementation

The width of the display frame above is fixed. When we want to show the relatively large layer number of binary tree, it does not meet our needs, so we must consider paging processing. For excess layers, the sixth layer corresponding node is the root node for paging display, and a group of six layers and so on, until fully displayed.

There are two issues in the paging process, which are given as follows:

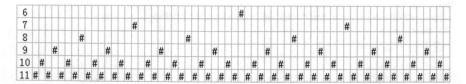

Fig. 1 The display model of binary tree

(1) When there are multiple pages to be displayed, then the sixth floor of the node which has child nodes are numbered for easy user selection display.
(2) Memory paging display case is required by constructing a stack to deal with [2].

2.3 General Discussion

To consider the characteristics of intelligent adaptation [3], so this algorithm module can be inserted into any program source code associated with a binary tree.

Storage structure of binary tree is generally binary chain structure, so we must first predefine node types. Requirements for the contents of the hinge point of the display, so node name needs to unified by the macro definition. Finally, it can cause different display functions because of the specific node is not the same type of content. Therefore, the display function names also requires to be unified by predefined macros [4]. To ensure the versatility of the algorithm modules, advance definition and macro definition should be prepared before designing algorithms.

3 Algorithm Implementation

3.1 Display Frame

The display model of binary tree in Fig. 1 shows the specific display coordinates of each node as follows:

(1) The Coordinate for the root node of the binary tree are (40, 6);
(2) Determine the node spacing x_h

From the row number for the node, the spacing between rows of nodes is obtained as

$$x_h = \mathrm{pow}(2, 12 - y) \tag{1}$$

Among them, x is the node spacing, y is the row number for the node.

(3) The first node coordinates of each line x_a

The distance from the first node of each line to the left baseline of rectangular display box (x = 8) is half of this line spacing, it follows that the coordinates of each node of the line are

$$x_a = 8 + \mathrm{pow}(2, 12 - (+ + y))/2 \tag{2}$$

Among them, x is the abscissa of the display node, y is the row of nodes, i.e. the relative ordinate of the node.

(4)   Other nodes' coordinates

Other nodes' coordinates are calculated according to the previous node coordinate. Ordinate unchanged, the abscissa is the node spacing of the current layer with the addition of the abscissa of the previous node.

(5)   Determination to change layer

Determine whether the abscissa of the successor node is beyond the rectangular display box.

If so, the change to the first node of next level, and gives the new coordinates.

3.2   Filling the Empty Branches of Non-full Binary Tree

Due to the use of the hierarchical traversal, traversal algorithm will stream the nodes of the binary tree by layer. But when the binary tree is not a full binary tree, you can not judge and display the case of empty branches. This will cause a lot of obstacles for accurate positioning of the binary tree display. So to solve this problem, we use a homemade special node to fill the binary tree, making it become a full binary tree. When the traversal algorithm access to a node which has no left or right child, then the node is put into the team as a child node. The two children of this special node are full, therefore, the part below this node can be made into an "empty tree" consisting of homemade nodes in order to fill the binary tree. When they read this particular node from the queue, it does not display any information but calculates the coordinates of the next node directly.

3.3   End Control

Because the display rectangular box is a limited space, and the binary tree can be infinitely extended filling by the above algorithm, it is necessary to carry out the end of the control display [5]. Seen by a hierarchy traversal: After completion of a layer removed from the queue, access to the next layer also completes, so we set up a counter, when the total number of dequeue nodes reach the $2^{(n-1)}$ (n is the total number of layers) of total number of each layers except the last level, the show is end.

3.4 Pagination

Based on the above displayed on a single page, we need to consider display algorithm processing of a binary tree more than the height of 6.

(1) Display method

After the end of a page is displayed, the algorithm checks flag Zyfy (Lord have pagination) variable. If the variable is non-zero, the user is prompted to have a page to continue displaying and interrupt the user's selection. The user can enter a node number of the last layer which should be further displayed, after entering the number, the algorithm puts the root of this page into the stack, and includes the selected node as the root of the new recall display function ReDisp_tree(), to display the appropriate sub-pages, and prompts the user to have the upper page [6]. When you input the back command (88), read out a node from the stack as a new root redisplay. When you enter the exit command (888), exit the entire display algorithm.

(2) Determine whether there is pagination

The paging level was successfully solved by introducing the stack. The main issue to be resolved is how the algorithm determines whether there is a binary tree node which has not been displayed. We will put this judgment in the node display function ShowNode_tree(). When the last layer of nodes (*y == 11) is displayed, if there are children at the node, look at the child's situation (baby's value: 1—has left children, 2—has the right child, 3—has two children. Corresponding ASCII characters (☺ ☻ ♥)are displayed in the sixth layer node, expressing their child's case), and calculate the serial number of the nodes in the layer, the above information is then displayed while the node number is returned to the calling function [7]. Parameter returns to the function ReDisp_tree(), If the parameter is not zero, the variable assignment Zyfy 1, which indicates that there is part that is not shown in this display. At the same time, according to the parameters returned, this node is put into its corresponding number baskets (fy [n]), ready to the user's call.

3.5 Versatility

Predefine PR_Tree as binary tree node type, for example, typedef BTNode PR_Tree;
    Macro define Ncontent as variable name for the node content, for example, #define Ncontent item;
    The FWLX is macro defined as selective output type (gpn\gpstr\gpc), for example, #define FWLX gpstr.

```
void gpn(int m, int n, int a)
/* Print digital */
void gpc(int m, int n, char a[])
/*print string*/
void gpstr(int m, int n, char a)
/*print characters*/
```

4 Examples of Application

4.1 Multi-page Display Binary

Construction of a binary tree in the experiment, using preorder and inorder traversal to display the structure of this binary tree [8]. The preorder traversal sequence is abdhiqwmrecfijopsuxyvt, the inorder traversal sequence is wqlhmrdbeaifojxuysvpct. Although we can see the structure of binary tree through the preorder and inorder traversal sequence of a tree, but it requires us very familiar with the two traversal methods of binary tree first of all [9].

Here we are using the following algorithm to construct a simple binary tree.

```
For (k=0;k<m_num; k++)
{ If ((k+1)*2-1<m_num) q[k].lchild=&q[(k+1)*2-1];
     else q[k].lchild=NULL;
if((k+1)*2<m_num) q[k].rchild=&q[(k+1)*2];
     else q[k].rchild=NULL;
}
```

After we use the display module of the binary tree in this paper, the binary tree will be displayed in front of us as the following graph's intuitive form. Figure 2 is a display of homepage, Fig. 3 is paging of the sections below the nodes. Such structure of binary tree is displayed intuitively, and this module can be inserted into any program code directly. Thus, the algorithms such as the number of points on the leaf node, exchange around subtree [10], height of a binary tree can be written in straight forward [11].

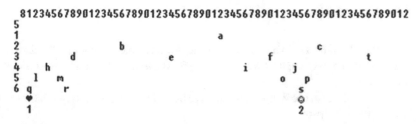

Fig. 2 Show the homepage

```
Please enter the serial number of the node to display:88
```

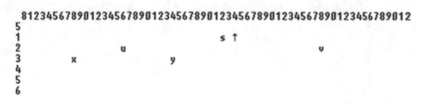

Fig. 3 Show the pagination under the nodes

4.2 The Application in the Binary Sort Tree

In the data structure learning, the binary sort tree is difficult. In the process of constructing binary sort tree, we must constantly observe the establishment of a binary tree, such as the verification of search, insert, delete, etc., algorithms [12]. If we use the original traversal methods to see binary tree, it would be very troublesome. But if we use display module program of binary tree in this paper, changes can be easily observed in the binary tree. Figure 4 is the binary sort tree before deleting its nodes, we will now delete node 26, as shown in Fig. 5, and observe the result of operation by displaying the binary tree again. We can see that the node 26 is deleted, and the node 44 has replaced its location, as shown in Fig. 6. This proves the success of this deletion.

5 Conclusions

Adaptive software is a software which can automatically adapt to changes in demand, in_depth understanding in the problem domain and technical progress, etc., environmental change factors. Adaptive software stresses that by detecting changes in

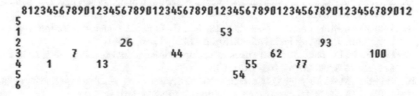

Fig. 4 The binary sort tree before deleting a node

Fig. 5 Delete the node 26

```
Please enter the contents to delete:26
OK!!
```

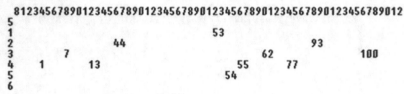

Fig. 6 The new binary sort tree after deleting a node

demand and the environment, adjusting and modifying the development plan, it makes the software evolving to adapt to changes throughout the software life circle. The adaptive algorithm in VC environment of showing intuitively the structure of a binary tree, it's plug and play and can be applied in any program associated with a binary tree without any changes, it will have certain value.

Acknowledgements This work is supported by Top-notch Academic Programs Project of Jiangsu Higher Education Institutions (TAPP) under Grant No. PPZY2015A090.

References

1. Cohen, J., Rodrigues, L.A., Duarte, E.P.: Parallel cut tree algorithms. J. Parallel Distrib. Comput. **109**(11), 1–14 (2017)
2. Wang, Q., Liu Z.: A cost-sensitive decision tree optimized algorithm based on adaptive mechanism. In: 2017 3rd International Conference on Artificial Intelligence and Industrial Engineering (AIIE2017), pp. 170–175. DEStech Publications (2017)
3. Li, J., Wang, Y.: A new fast reduction technique based on binary nearest neighbor tree. Orig. Res. Artic. Neurocomputing **149**(2), 1647–1657 (2015)
4. Maßberg, J.: Generalized Huffman coding for binary trees with choosable edge lengths. Inf. Process. Lett. **115**(4), 502–506 (2015)
5. Xiaofang, J., Mengxuan, L., Min, S., Libiao, J., Xianglin, H.: Research on the adaptive hybrid search tree anti-collision algorithm in RFID system. High Technol. Lett. **22**(1), 107–112 (2016)
6. Jinguang, S., Suhong, W.: Collision Detection optimization algorithm based on classified traversal. J. Comput. Appl. **35**(1), 194–197 (2015)
7. Shu-qin, H., Hai Z.: Research on deleting node algorithm of binary sort tree. J. Tonghua Norm. Univ. 12, 46–48 (2014)
8. Wen-yi, L., Ming-dao, Y., Xian-zhao, L., Jun, H.: Optimal evidence selection method using binary tree. J. Anyang Inst. Technol. Struct. **13**(3), 50–53 (2014)
9. Haraburda, D., Tarau, P.: Binary trees as a computational framework. Orig. Res. Artic. Comput. Lang. Syst. Struct. **39**(4), 163–181 (2013)
10. Jing-shan M.A., Yu-ping, Q.: Level traversal binary tree in sequence storage and its application. J. Bohai Univ. Sci. (Natural Science Edition) **2**, 172–176 (2013)
11. Zulhisam, M., Riaza, R.: Learning binary tree algorithm using 3-d visualization: an early results. Orig. Res. Artic. Procedia Soc. Behav. Sci. **90**(10), 388–395 (2013)
12. Xing-bo, W.: Study on non-recursive and stack-free algorithms for preorder traversal of complete binary trees. Comput. Eng. Des. **9**, 3078–3081 (2011)

Single Actor Pooled Steganalysis

Zichi Wang, Zhenxing Qian and Xinpeng Zhang

Abstract This paper considers a more practical situation for pooled steganalysis that only a single actor is observed, so that the steganalyst needed to analyze the actor independently without comparing with other actors. We propose a pooled steganalysis method for this situation. For a given actor that has emitted a number of images, feature sets are extracted from each image, respectively, and then feed to a binary classifier popularly used in single object steganalysis. Combining all the results output by the classifier, a final decision is made ensemble to label the given actor as "guilty" or "innocent" with the minimal detection error. Experimental results show that the proposed method is effective for single actor pooled steganalysis.

Keywords Image · Pooled steganalysis · Steganography

1 Introduction

Steganography aims to transmit secret data through public channels without drawing suspicion by hiding the secret data into digital media [1]. On the contrary, steganalysis aims to disclose the secret transmission by analyzing the media on public channels [2]. Both technologies have been developed rapidly in the past decades. Modern steganalytic methods use supervised machine learning to investigate the models of the covers and the stegos, and then try to identify the stegos [3, 4].

Different from the abovementioned single object steganalysis, pooled steganalysis is also a significant area of information security, which aims to identify the steganographers among other innocent actors [5–9]. Such a situation is completely different from classifying an individual object as cover or stego. The embedding methods used by the steganographer and the amount of payload are unknown. As shown in Fig. 1, existing pooled steganalysis methods suppose that the steganalyst

Z. Wang · Z. Qian (✉) · X. Zhang
School of Communication and Information Engineering,
Shanghai University, Shanghai 200444, China
e-mail: zxqian@shu.edu.cn

© Springer Nature Singapore Pte Ltd. 2019
J.-S. Pan et al. (eds.), *Genetic and Evolutionary Computing*,
Advances in Intelligent Systems and Computing 834,
https://doi.org/10.1007/978-981-13-5841-8_36

Fig. 1 Multiple actors
pooled steganalysis

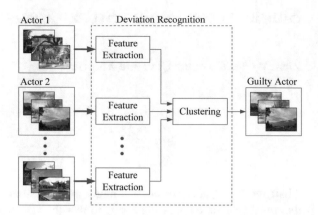

is monitoring a number of actors, with multiple innocent actors and some potential
steganographers. To determine who is guilty, it is assumed that the steganographers
are significantly deviated from the majority innocent actors. Based on this assumption, the steganographers can be recognized by unsupervised clustering algorithms
because the actors' deviation is evidence of their guilt. We called this kind of steganalysis as MAP (multiple actors pooled) steganalysis since it must deal with multiple
actors.

The pooled steganalysis problem is first proposed in [10]. After 5 years, a solution
for MAP steganalysis is proposed using clustering [5]. It calculates distances between
every pair of actors and uses a clustering algorithm to find the actor who deviates
the most from the rest, and then the actor is labeled as steganographer. Afterward,
an improved method is proposed in [6] by replacing hierarchical clustering with a
so-called local outlier factor, which achieves greater accuracy of detection. In [7],
a more complete MAP steganalysis method is designed. First, steganalytic features
are extracted from all objects. Then, the distances between every pair of actors are
calculated. Finally, the actors that deviate from the majority are identified using an
anomaly measure computed from the distances. In [8], an optimal function combining
outputs of many single object detectors is proposed. Recently, a new MAP steganalysis method is proposed using high-order joint features and clustering ensembles [9].
The features are extracted using co-matrices to indicate the dependencies of image
content. With these features, a number of hierarchical sub-clusterings are trained and
integrated as a clustering ensemble based on the majority voting strategy. In this way,
the optimal decision on a suspicious actor is made.

The steganalyst must deal with multiple actors in MAP steganalysis. In this paper,
we discuss a more practical situation of pooled steganalysis that only one actor is
observed, in which it is needed to judge the given actor is "guilty" or not without
the comparisons with other actors. We called this kind steganalysis as SAP (single
actor pooled) steganalysis. As shown in Fig. 2, the process of SAP steganalysis is
aiming at single actor. The steganalyst must analyze the actor independently. In this
situation, existing MAP steganalysis methods are ineffective.

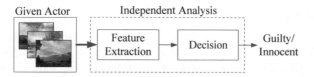

Fig. 2 Single actor pooled steganalysis

This paper proposes an SAP steganalysis method. Existing feature extraction algorithms are used to obtain the feature sets of all the images emitted by a given actor. With the help of a binary classifier popularly used in single object steganalysis, each image is assigned a binary result: "stego" or "clear". Combining all the results output by the classifier, a final decision is made ensemble for the actor: "guilty" or "innocent".

2 Proposed Method

The sketch of the proposed method is illustrated in Fig. 3. A binary classifier of single object steganalysis is trained first. Given an actor emitted a number of images, existing feature extraction algorithms are used to obtain the feature sets. Then, the extracted feature sets are feed to the trained binary classifier. After that, all the images are labeled as "stego" or "clear" by the classifier. Combining all the results output by the classifier, a final decision is made ensemble on the given actor to determine whether the actor is "guilty" or not with the minimal detection error.

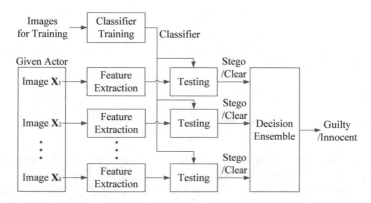

Fig. 3 Sketch of proposed method

2.1   Deductions of Binary Classifier

To make a decision on the given actor, a binary classifier is trained to label the captured images. For a given image, there are two possible results outputs of a binary classifier: "stego" or "clear". Since the detection accuracy of the binary classifier is not 100% usually, there are two errors existed: false alarm rate and missed detection rate. The relationships between the two errors and the probability of the two possible results outputted by the binary classifier are deducted in this subsection.

For the training of a binary classifier, denote the number of total images used for training as n, in which the number of cover images as n_c, and the number of stego images as n_s. In the process of judgement, denote the number of misjudged cover images as n_{cs}, the number of misjudged stego images as n_{sc}, the number of correctly judged cover images as n_{cc}, and the number of correctly judged stego images as n_{ss}. For a given image, the probability of being judged as cover and stego by the classifier can be calculated as P_c and P_s in Eqs. (1) and (2), respectively.

$$P_c = \frac{n_{cc} + n_{sc}}{n} = \frac{n_{cc}}{n} + \frac{n_{cs}}{n} - \frac{n_{cs}}{n} + \frac{n_{sc}}{n} = P_{MD} - P_{FA} + \frac{n_c}{n} \tag{1}$$

$$P_s = \frac{n_{cs} + n_{ss}}{n} = \frac{n_{cs}}{n} + \frac{n_{ss}}{n} + \frac{n_{sc}}{n} - \frac{n_{sc}}{n} = P_{FA} - P_{MD} + \frac{n_s}{n} \tag{2}$$

where P_{FA} and P_{MD} are the false alarm rate and missed detection rate, respectively.

The number of cover images used for training is equal to the number of stego image usually. That means $n_c = n_s = n/2$. Therefore, the relationships of P_c, P_s, P_{FA}, and P_{MD} can be written as

$$P_c = P_{MD} - P_{FA} + \frac{1}{2} \tag{3}$$

$$P_s = P_{FA} - P_{MD} + \frac{1}{2} \tag{4}$$

Based on this binary classifier, we designed an SAP steganalysis method which will be introduced in the next subsection.

2.2   Decision on a Given Actor

Assume that there are k images transmitted by the given actor and captured by the steganalyst. Some of them may contain additional data. It is sure that the given actor is steganographer when at least one of the k images is stego. However, because of the false alarm rate of classifier, the images been judged as stego may be clear. In this case, the actor that has been judged as steganographer may be innocent. Therefore, it is necessary to define a suitable strategy for the decision on the given actor.

The decision strategy for a given actor in the proposed method is: regard the actor as a steganographer when at least r images are judged as stego $(1 \leq r \leq k)$; otherwise, regard the actor as innocent actor. The value of r is determined as follows:

For the given actor, denote the probability of being judged as steganographer by the steganalyst as P'_s. According to the decision rule, it can be calculated as

$$P'_s = \sum_{i=r}^{k} C_k^i P_s^i P_c^{k-i} \tag{5}$$

where C_k^i represents mathematical combination.

Denote the probability of containing additional data for each image as p ($p = 0$ for a normal actor, and $0 < p \leq 1$ for a steganographer). The false alarm rate P'_{FA} and missed detection rate P'_{MD} can be calculated as

$$P'_{FA} = (1 - p)^k P'_s = (1 - p)^k \sum_{i=r}^{k} C_k^i P_s^i P_c^{k-i} \tag{6}$$

$$P'_{MD} = [1 - (1 - p)^k](1 - P'_s) = [1 - (1 - p)^k](1 - \sum_{i=r}^{k} C_k^i P_s^i P_c^{k-i}) \tag{7}$$

Then, the detection error P'_E of the pooled steganalyst is

$$P'_E = \frac{P'_{FA} + P'_{MD}}{2} = [(1 - p)^k - \frac{1}{2}] \sum_{i=r}^{k} C_k^i P_s^i P_c^{k-i} - \frac{1}{2}(1 - p)^k + \frac{1}{2} \tag{8}$$

There are k possible values of P'_E corresponding to $r = 1, 2, ..., k$, respectively. To achieve good detectability, the value of P'_E should be minimized. In other words, the minimal value among the k possible values of P'_E should be found. We calculate all the k values of P'_E to find the minimal one since the value of k is very limited. After that, the corresponding r is used in the proposed decision strategy. Thus, the value of r can be determined as

$$r = \arg \min_{1 \leq r \leq k} \{P'_E\} \tag{9}$$

To calculate P'_E, the values of P_c and P_s can be calculated using Eqs. (3) and (4). However, for the steganalyst, it is difficult to know the value of p which is determined by the actors. Here, we heuristically approximate the value of p using P'_s. That means

$$p = \sum_{i=r}^{k} C_k^i P_s^i P_c^{k-i} \tag{10}$$

Probability P'_s represents the possibility of an actor been judged as steganographer. The logic behind Eq. (10) is that an actor would make some embedding operations if the actor is a steganographer, and so the images transmitted by the actor are possibly contain additional data.

3 Experimental Results

3.1 Experiment Setup

The image datasets employed in our experiments are UCID [11] which contains 1338 uncompressed color images sized 512×384, and BOSSbass ver. 1.01 [12] which contains 10000 uncompressed grayscale images sized 512×512.

We use dataset UCID for the training of the binary classifier introduced in subsection III.A. All the images in UCID are transformed into grayscale images and compressed into JPEG with quality factor QF = 75 and QF = 95 to be adopted as covers. The popular spatial steganography algorithms SUNIWARD [13], WOW [14], and HILL [15] are used to embed secret data on the 1338 uncompressed images, respectively. The JPEG steganography algorithms JUNIWARD [13], UED [16], and UERD [17] are used for the embedding on the obtained JPEG images. The payloads are set as 0.1, 0.2, 0.3, 0.4, and 0.5 (bpp: bit per pixel for an uncompressed image or bpnzac: bit per nonzero AC coefficient for a JPEG image). Then, the features are extracted using SRM [18] and SRMQ1 [18] for spatial images and DCTR [19], GFR [20] for JPEG images. Finally, the ensemble classifier [21] is employed to measure the property of feature sets. One half of the cover and stego feature sets are used for training, while the remaining sets are used for testing. Trained binary classifiers are used to make decisions on captured images.

The dataset BOSSbass ver. 1.01 is used for the simulation of actors. As mentioned in subsection III.B, there are k images transmitted by the given actor and captured by the steganalyst. So, there are $2 \times \lfloor 10000 / k \rfloor$ actors in total, where $\lfloor \cdot \rfloor$ is a downward rounding function. Half of them are steganographers and the other half are normal actors. Denote the number of images containing additional data for an actor as s ($s = 0$ for a normal actor, and $1 \leq s \leq k$ for a steganographer). The employed steganography and feature extraction algorithms are also the same with those used in the training of the binary classifier. We use the detection error P'_E to evaluate the performance of SAP steganalysis method. The performance is evaluated using an average of P'_E over ten random tests. The lower value of P'_E means higher detection accuracy.

3.2   Detection Accuracy on Actors

The detection performance of the proposed method with SRM and SRMQ1 against HILL, WOW, and SUNIWARD for $k = 10$ and $s = 8$ is shown in Fig. 4.

The results indicate that the proposed method is effective for single actor pooled steganalysis. With the increasing of payload, the detection accuracy of the proposed method is increased correspondingly. Actually, the performance of binary classifier of single object steganalysis is increased with the increasing of payload. Therefore, the detection accuracy of the proposed method can be satisfactory as long as the binary classifier of single object steganalysis performs well enough.

Considering the performance with different values of s and k, Fig. 5a shows the comparisons of same ratio of k and s, and Fig. 5b shows the comparisons of different values of s with the same k. It can be seen from Fig. 5a that the detection accuracy is in inverse proportion to the value of k with the same ratio of s and k. Figure 5b shows that a large s avail steganalysis when the value of k is fixed, which is reasonable.

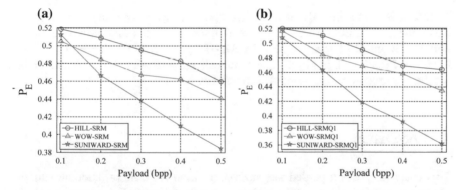

Fig. 4 Detection errors of pooled steganalysis using **a** SRM and **b** SRMQ1 against HILL, WOW, and SUNIWARD for $k = 10$ and $s = 8$

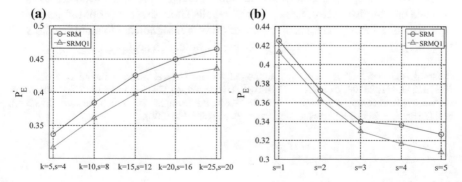

Fig. 5 Detection errors of pooled steganalysis using SRM and SRMQ1 against SUNIWARD with payload 0.5 bpp and different values of k and s: **a** $s/k = 0.8$ and **b** $k = 5$

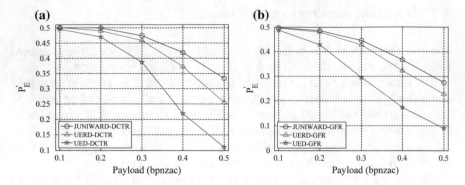

Fig. 6 Detection errors of pooled steganalysis using: **a** DCTR and **b** GFR against JUNIWARD, UERD, and UED for QF = 75

3.3 Performance on JPEG Images

To further verify the effectiveness of the proposed method, we also designed several experiments on JPEG images. The detection performance of the proposed method with DCTR and GFR against JUNIWARD, UED, and UERD with $k = 10$ and $s = 8$ for QF = 75 is shown in Fig. 6. The results also indicate that the proposed method is effective for single actor pooled steganalysis.

4 Conclusion

This paper proposes a pooled steganalysis method for a more practical situation of pooled steganalysis that only a single actor is observed. For a given actor that has emitted a number of images, a binary classifier popularly used in single object steganalysis is employed to label all the emitted images. Then, an ensemble decision is made on the given actor. Experimental results show that the proposed method is effective for SAP steganalysis. For further study, it is significant to combine the SAP steganalysis with MAP steganalysis for higher detection accuracy.

Acknowledgements This work was supported by the Natural Science Foundation of China (U1736213, U1536108, 61572308, 61103181, U1636206, 61373151, and 61525203), the Natural Science Foundation of Shanghai (18ZR1427500), the Shanghai Dawn Scholar Plan (14SG36), and the Shanghai Excellent Academic Leader Plan (16XD1401200).

References

1. Wang, Z., Zhang, X., Yin, Z.: Joint cover-selection and payload-allocation by steganographic distortion optimization. IEEE Signal Process. Lett. **25**(10), 1530–1534 (2018)
2. Böhme, R.: Advanced Statistical Steganalysis. Springer Science & Business Media (2010)
3. Ye, J., Ni, J., Yi, Y.: Deep learning hierarchical representations for image steganalysis. IEEE Trans. Inf. Forensics Secur. **12**(11), 2545–2557 (2017)
4. Li, B., Li, Z., Zhou, S., Tan, S., Zhang, X.: New steganalytic features for spatial image steganography based on derivative filters and threshold LBP operator. IEEE Trans. Inf. Forensics Secur. **13**(5), 1242–1257 (2018)
5. Ker, A., Pevný, T.: A new paradigm for steganalysis via clustering, media watermarking, security, and forensics III. Int. Soc. Opt. Photonics **7880** (2011)
6. Ker, A., Pevný, T.: Identifying a steganographer in realistic and heterogeneous data sets, media watermarking, security, and forensics. Int. Soc. Opt. Photonics **8303** (2012)
7. Ker, A., Pevny, T.: The steganographer is the outlier: realistic large-scale steganalysis. IEEE Trans. Inf. Forensics Secur. **9**(9), 1424–1435 (2014)
8. Pevny, T., Nikolaev, I.: Optimizing pooling function for pooled steganalysis. In: Proceedings of the IEEE International Workshop on Information Forensics and Security, Binghamton, Rome, Italy, Nov 2015
9. Li, F., Wu, K., Lei, J., Wen, M., Bi, Z., Gu, C.: Steganalysis over large-scale social networks with high-order joint features and clustering ensembles. IEEE Trans. Inf. Forensics Secur. **11**(2), 344–357 (2016)
10. Ker, A.: Batch steganography and pooled steganalysis. In: International Workshop on Information Hiding. Springer, Berlin, Heidelberg (2006)
11. Schaefer, G., Stich, M.: UCID—An uncompressed colour image database. In: Proceedings Conference on Storage and Retrieval Methods and Applications for Multimedia, San Jose, CA, USA, Jan, pp. 472–480 (2004)
12. Bas, P., Filler, T., Pevný, T.: Break our steganographic system: the ins and outs of organizing boss. In: Proceedings of the 13th International Conference on Information Hiding, Prague, Czech Republic, May, pp. 59–70 (2011)
13. Holub, V., Fridrich, J.: Digital image steganography using universal distortion. In: Proceedings of the First ACM Workshop on Information Hiding and Multimedia Security, New York, NY, USA, June, pp. 59–68 (2013)
14. Holub, V., Fridrich, J.: Designing Steganographic Distortion Using Directional Filters. In: Proceedings of the IEEE International Workshop on Information Forensics and Security, Binghamton, NY, USA, pp. 234–239, Dec 2012
15. Li, B., Wang, M., Huang, J., Li, X.: A new cost function for spatial image steganography. In: Proceedings IEEE International Conference on Image Processing, Paris, France, Oct, pp. 4206–4210 (2014)
16. Guo, L.J., Ni, J.Q., Shi, Y.Q.: Uniform embedding for efficient jpeg steganography. IEEE Trans. Inf. Forensics Secur. **9**(5), 814–825 (2014)
17. Guo, L.J., Ni, J.Q., Su, W.K., Tang, C.P., Shi, Y.Q.: Using statistical image model for JPEG steganography: uniform embedding revisited. IEEE Trans. Inf. Forensics Secur. **10**(12), 2669–2680 (2015)
18. Fridrich, J., Kodovsky, J.: Rich models for steganalysis of digital images. IEEE Trans. Inf. Forensics Secur. **7**(3), 868–882 (2012)
19. Holub, V., Fridrich, J.: Low complexity features for jpeg steganalysis using undecimated DCT. IEEE Trans. Inf. Forensics Secur. **10**(2), 219–228 (2014)
20. Song, X.F., Liu, F.L., Yang, C.F., Luo, X.Y., Zhang, Y.: Steganalysis of adaptive jpeg steganography using 2D gabor filters. In: Proceedings of the 3rd ACM Workshop on Information Hiding and Multimedia Security, New York, USA, June, pp. 15–23 (2015)
21. Kodovsky, J., Fridrich, J., Holub, V.: Ensemble classifiers for steganalysis of digital media. IEEE Trans. Inf. Forensics Secur. **7**(2), 432–444 (2012)

Part IX
Smart City and Grids

Field Analysis and Flux-Weakening Control of PMSM Used in Vehicle

Zhong-Shu Liu

Abstract Permanent magnet synchronous motor (PMSM) is widely used in the electric vehicle industry due to its advantages of low noise, small volume, large power density, small moment of inertia, and high efficiency. A three-phase permanent magnet synchronous motor (PMSM) used in vehicle was designed in this paper. Using Ansoft Maxwell software the electromagnetic field of PMSM was analyzed by finite element method, and the conclusion of optimum design of vehicle motor was made. The paper also discussed how to meet the driving requirements of electric cars within the limit of vehicular power supply voltage and speed range by using the method of flux-weakening control.

Keywords Vehicle, PMSM, field · Control · Flux weakening

1 Introduction

At present, the problems such as air pollution and energy shortage are attracting more and more attention worldwide. As the emission of automobile exhaust makes these problems more and more serious, the development of electric vehicles has become the focus of attention in automobile industry. Electric vehicles make the use of energy diversified and efficient, and achieve the purpose of reliable, balanced, and pollution-free utilization of energy. From the perspective of environmental protection, electric vehicles are a zero-emission vehicle, which will greatly reduce air pollution. In addition, electric vehicles are easier to achieve precise control than conventional fuel vehicles, and intelligent transportation systems may initiatively come to true through electric vehicles [1]. Among all kinds of vehicles, motor, permanent

Z.-S. Liu (✉)
School of Information Science and Engineering, Fujian University of Technology,
Fuzhou 350118, China
e-mail: lzs@fjut.edu.cn; 1241891962@qq.com

The Key Laboratory for Automotive Electronics and Electric Drive of Fujian Province,
Fujian University of Technology, Fuzhou 350118, China

© Springer Nature Singapore Pte Ltd. 2019
J.-S. Pan et al. (eds.), *Genetic and Evolutionary Computing*,
Advances in Intelligent Systems and Computing 834,
https://doi.org/10.1007/978-981-13-5841-8_37

magnet synchronous motor (PMSM) stands out because of its advantages such as high efficiency, high power density, small moment of inertia, and fast response, and becomes an ideal driving motor. Next, the optimum design of permanent magnet synchronous motor used in vehicles, electromagnetic field analysis, and flux-weakening speed control will be discussed in detail from several aspects.

2 Optimum Design of Permanent Magnet Synchronous Motor for Vehicle

2.1 Basic Structure and Principle

The structure of permanent magnet synchronous motor (PMSM) for three-phase eight poles 30 kW vehicle is shown in Fig. 1. The stator core and stator winding are the same as the three-phase asynchronous motor. The rotor structure adopts the built-in magnetic steel type, and the permanent magnet is distributed along the circumference direction, N and S poles are arranged alternately [2]. After connecting the stator winding to the three-phase alternating current supply, the rotating magnetic field is generated and the permanent magnetic field generated by the rotor magnetic steel works together to pull the rotor into synchronization, and the rotor and stator magnetic fields are maintained in the synchronous speed [3]. As the name suggests, it is called permanent magnet synchronous motor (PMSM). In particular, in order to enhance the cooling effect of large-capacity motor, improve the efficiency of the motor. The motor designed in this paper adopts the cooling method of water internal cooling, that is, a circulating water channel is set between the motor frame and the stator core, which effectively takes away heat and achieves better cooling effect than the traditional air cooling method.

2.2 Main Parameters

The main performance indicators of PMSM for vehicle designed in this paper are as follows: rated voltage 452 V, rated power 30 kW, peak power 60 kW, rated speed 4000 r/min, maximum speed 9000 r/min, NdFeB permanent magnet.

The design was conducted by using Ansoft Maxwell software, and the results are shown in Table 1.

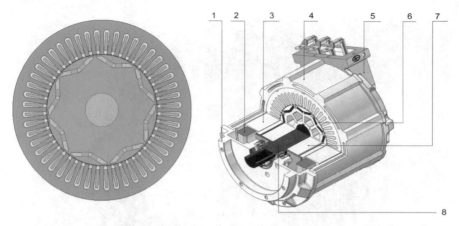

1. Front end cover 2.Winding 3. Stator core 4. Water-cooled channel
5. Back end cover 6. Magnetic steel 7. Rotor core 8. Shaft

Fig. 1 Structure of PMSM for vehicle

Table 1 Structure parameters of the prototype

Parameter	Value	Parameter	Value
The stator outer diameter	230/mm	The rotor outer diameter	149.2/mm
The stator inner diameter	150/mm	The rotor inner diameter	42/mm
Stator core length	120/mm	The air gap length	0.4/mm
The thickness of the magnet	6/mm	Slot number	48
The width of the magnet	18/mm	Pole	8

2.3 Stator Winding Design

The design of PMSM stator windings can be either single-layer winding or double-layer winding. The single-layer windings are characterized by simple process, convenient inlay, and high slot utilization [4]. After comprehensive consideration, single-layer winding was adopted in the prototype, and the winding distribution is shown in Fig. 2.

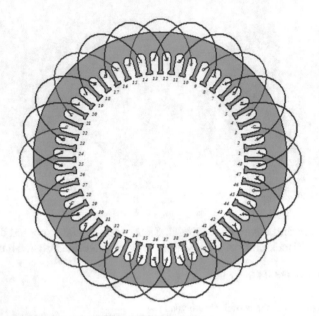

Fig. 2 Distribution diagram of winding of PMSM for vehicle

3 Analysis of Internal Electromagnetic Field of PMSM

3.1 Mesh Generation

Now, we will analyze the prototype by using two-dimensional finite element method, which is a numerical calculation method based on discretization [5]. It is particularly suitable for the calculation of the electromagnetic field distribution in the interior of the PMSM. We can use Maxwell 2D software package for grid subdivision. Figure 3 is the grid subdivision diagram of the prototype.

3.2 Analysis of Internal Electromagnetic Field in No-Load

The flux density cloud diagram of the PMSM when it is in no-load is shown in Fig. 4, and it can be analyzed by using Ansoft Maxwell software. It can be seen from Fig. 4 that the magnetic density of the stator yoke reaches 2.437T when the motor is in no-load. At this time, the motor's iron core is already saturated. The purpose of this design is to improve the utilization rate of ferromagnetic materials, so as to increase the power density of the motor [6]. It can be seen from Fig. 4 that the radial magnetic field of motor air gap is distributed symmetrically along the circumference direction of the motor, which indicates that the ferromagnetic material of the motor designed

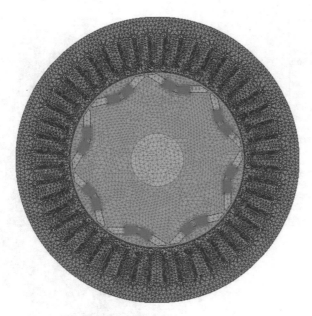

Fig. 3 The grid subdivision of PMSM for vehicle

this time has a higher utilization rate and a higher power density. The results show that the prototype fully meets the design requirements.

The motor is used for electromechanical energy conversion through air gap. The parameters and performance of the motor are based on the calculation of air gap magnetic field [7]. The distribution of air gap magnetic field can be intuitively reflected by the air gap magnetic density curve. Regarding the center of the shaft as the center, through calculation, the air gap flux density curve can be shown in Fig. 5. The vertical coordinate of Fig. 5 is the flux density corresponding to different positions along the circumference of the air gap, while the horizontal coordinate is the circumferential arc length of air gap [8].

3.3 Analysis of Internal Electromagnetic Field in Load

Similarly, by using Ansoft Maxwell software, the flux density cloud diagram of the PMSM when it is in load is shown in Fig. 6, and the air gap flux density curve can be shown in Fig. 7.

It can be seen from Fig. 6 that the saturation degree of the motor's iron core increases while it is in load, the magnetic density value of the stator yoke increases to 2.607 T, and the electrical center line of the air gap magnetic field is offset compared with no-load, which is caused by the distortion of air gap magnetic field arising from the armature reaction [9]. It can be seen from Fig. 7 that, although the saturation

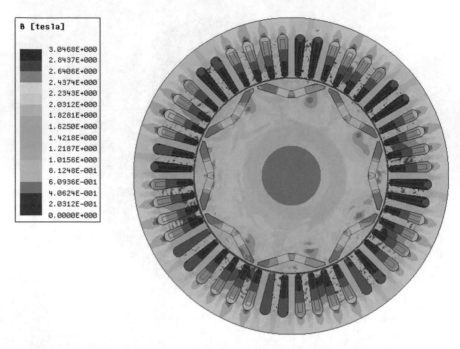

Fig. 4 The flux density cloud diagram of the PMSM in no-load

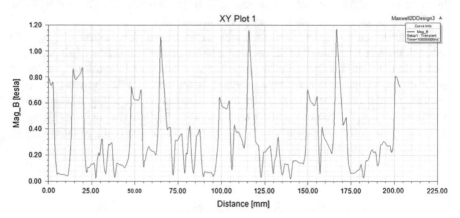

Fig. 5 The air gap magnetic density curve of the PMSM in no-load

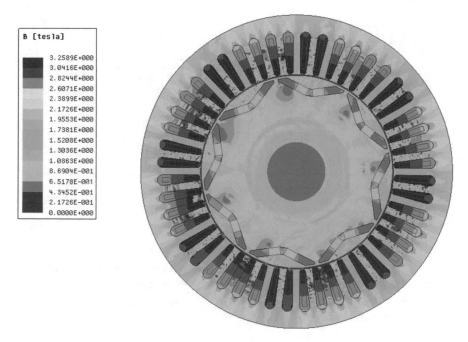

Fig. 6 The flux density cloud diagram of the PMSM in load

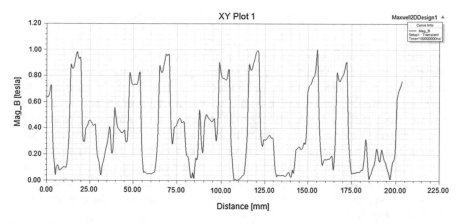

Fig. 7 The air gap magnetic density curve of the PMSM in load

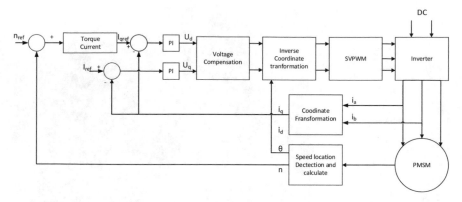

Fig. 8 PMSM flux-weakening control system

degree of the loaded iron core is intensified, the peak value of the gap magnetic
density of the air gap in load is smaller than that of no-load [10], which indicates
that the straight-axis armature reaction plays a role in demagnetization of the main
magnetic flux when loaded.

4 Flux-Weakening Speed Control of PMSM for Vehicle

The speed range of PMSM used in vehicle should be wider than motor used in
industry. When the motor is running in the rated speed and below, it outputs constant
torque. That is, the motor runs in the constant torque region [11]. When the motor
is running above the rated speed, it runs in the constant power region. At this point,
as the speed increases, the electromagnetic torque of the motor would be restricted
because of the stator voltage (it cannot be greater than the rated voltage), the speed
of motor cannot reach higher speed further, so it cannot meet the requirement of the
driving motor of vehicle [12]. In this case, we can use the method of flux-weakening
control. Specifically, in the actual control process, we use vector control to realize
the flux-weakening control of the PMSM, the block diagram of the control system
is shown in Fig. 8.

The basic concept of vector control theory is to take the rotating rotor flux space
vector as the parameter coordinate, the stator current is decomposed into two mutually
orthogonal components, one is in the same direction as flux, which represents stator
current excitation component, and it is called straight-axis component (id). Another
orthogonal to flux linkage, which represents the stator torque component of the
electric circulation [13], it is called quadrature-axis component (iq). To control them,
respectively, we could obtain a good dynamic characteristic like DC motor.

As we know, the back electromotive force of PMSM is directly proportional to
the rotating speed. When the motor terminal voltage is equal to the maximum output

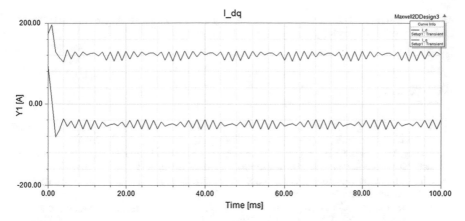

Fig. 9 id and iq current curves of PMSM for vehicle

voltage of the inverter as the speed increases, the motor stator current cannot be increased any more. At this time, the speed is called turning speed [14]. If the motor speed needs to be further increased, it can only be achieved by adjusting the current id and current iq. Specifically, when the high speed of motor measured is greater than the turning speed, meanwhile the stator voltage is greater than the maximum output voltage, the motor terminal voltage is greater than the maximum output voltage, the motor enters a state of flux weakening, which means that the current (id) should be increased and current (iq) should be reduced [15]. In this way, we can ensure that the stator current would not exceed the limit value, and this control method is called flux-weakening speed control. When the motor is running at high speed and the speed n equals 5074 r/min, the current curve of id and iq is shown in Fig. 9. As demagnetization current, id made the motor to run in the state of flux-weakening speed regulation, so as to achieve the purpose of speed expansion.

5   Conclusion

In this paper, the optimum design of permanent magnet synchronous motor (PMSM) for vehicle was carried out first according to the characteristics of it by using Ansoft Maxwell software, and then the finite element analysis of electromagnetic fields of the PMSM was conducted under no-load condition. Finally, the flux-weakening control strategy of the PMSM at high speed was discussed. The design and control of the prototype indicate that its performance is superior, and it can maintain high operation efficiency no matter in the low-speed high torque region or the high-speed constant power region. The research and development of permanent magnet synchronous motor (PMSM) used in vehicle are of great significance to the electric vehicle industry.

Acknowledgements This research was supported by the Science and Technology Project Fund of Fuzhou Science and Technology Bureau (Grant No. 2016-G-57).

References

1. Chen, N., Zhang, Y., Gui, W., et al.: Flux-weakening control for interior permanent magnet synchronous motor drive system. Control. Theory Appl. **30**(6), 717–723 (2013)
2. Huang, Y.-F., Hsu, C.-H.: Energy efficiency of dynamically distributed clustering routing for naturally scattering wireless sensor networks. J. Netw. Intell. **3**(1), 50–57 (2018)
3. Zhou, H., Chen, L., Liu, G., et al.: Flux-weakening strategy for improving pmsm dynamic performance. Electr. Mach. Control. **18**(9), 23–29 (2014)
4. Liu, C., Ding, Q.: Complexity analysis and research based on the chaotic system of sample entropy. J. Netw. Intell. **3**(3), 162–169 (2018)
5. Wu, Y.-P., Peng, X.-Q., Jiang, Q., Ruan, K.: An improved method of flotation froth image segmentation based on watershed transformation. J. Inf. Hiding Multimed. Signal Process. **8**(1), 238–247 (2017)
6. Zhilin, R., Junfeng, X., Qijun, C.: Control strategy for permanent magnet synchronous motor in fundamental voltage linearization. Electr. Mach. Control. **20**(2), 52–60 (2016)
7. Yang, J., Li, H.-Z., Li, W., Liu, X.-B.: Research on task allocation of geographic location related mobile sensing system. J. Inf. Hiding Multimed. Signal Process. **8**(1), 19–30 (2017)
8. Wen, J., Cao, B.: Research on flux-weakening control of drive system for electric vehicle. Small Spec. Electr. Mach. **43**(3), 49–51 (2015)
9. Yin, X., Tang, L., Wang, N.: Extremely randomized clustering forest based scene recognition algorithm in mobile devices. J. Inf. Hiding Multimed. Signal Process. **7**(2), 243–253 (2016)
10. Jiang, C., Hao, H., Shen, J.: Analysis of base speed determination of permanent magnet synchronous motor with flux-weakening control. Micromotors **49**(2), 1–5 (2016)
11. Zhao, Y., Zhang, J.: Study on electronic differential control system of independent in-wheel motor drive electric vehicle. J. Syst. Simul. **20**(18), 4767–4771 (2008)
12. Shen, X., Zhao, Q., Xu, P.: Control research on maximum power of the direct-drive permanent magnet synchronous rotor. J. Electr. Power **29**(1), 28–31 (2014)
13. Ye, Q.-Z., Zhang, M.-L., Lin, M.-Y.: Design of a high sensitivity and non-contact planar capacitance sensor. J. Inf. Hiding Multimed. Signal Process. **7**(5), 949–959 (2016)
14. Lin, B., Sun, D., He, Y.: High speed operation range extension for direct torque controlled permanent magnet synchronous motors. Electr. Mach. Control. **18**(9), 9–16 (2014)
15. Qiu, X., Huang, W., Bu, F., Yang, J.: Efficiency optimization of IPMSM direct torque control system used in electric vehicles. Trans. China Electrotech. Soc. **30**(22), 42–48 (2015)

Effects of Different Factors on the Visibility in Kaohsiung Area Using Hierarchical Regression

Chang-Gai Lee and Wen-Liang Lai

Abstract The main purpose of this research is to investigate the impact of temperature, humidity, wind velocity, wind direction, etc. on aerosols and visibility. It then uses inferential statistics techniques, such as principal component analysis, stepwise regression analysis, and hierarchical regression, to examine the effects of various components as part of the secondary pollutants on the visibility of the Kaohsiung area. The research encompasses data from 1999 to 2004, where air-pollution-related data were obtained from QianZhen Air Monitoring station, while meteorology-related data were obtained from the Kaohsiung Station of Central Weather Bureau. The results show, during the particular period, that the average visibility of the Kaohsiung area was 6.4 ± 3.6 km and the visibility, on average, increased by 0.33 km each year. Annual average of $PM_{2.5}$ concentration dropped 0.79 $\mu g/m^3$ each year. During this particular period, 68.74% of the time was under haze conditions while 69.83% of the time the visibility was lower than 10 km. The results show that low visibility of the Kaohsiung area is primarily caused by haze. Some average concentrations during the period are listed herein, $PM_{2.5} = 35.93$ $\mu g/m^3$, nitrate $= 4.10$ $\mu g/m^3$ (i.e., 11.40% of $PM_{2.5}$), sulfate $= 9.59$ $\mu g/m^3$ (constitute 26.68% of $PM_{2.5}$), organic carbon $= 8.49$ $\mu g/m^3$ (i.e., 23.64% of $PM_{2.5}$), and elemental carbon $= 2.64$ $\mu g/m^3$ (i.e., 6.58% of $PM_{2.5}$). The nitrate amount and nitrate-to-sulfate ratio of $PM_{2.5}$ are 11.40% and 0.43, respectively, which are lower than those found in previous researches in Kaohsiung area. During the same period, the monthly average OC/EC values were all more than 2.2, indicating there was a great potential of the formation of secondary aerosols in the Kaohsiung area, which was also confirmed by the calculated secondary organic carbon (SOC) and the monthly average of SOC/OC exceeding 85%. In addition, a model was built via multivariate linear regression and hierarchical regression to investigate the factors that can potentially affect the visibility of the Kaohsiung area. It has been found that the regional visibility is more or less inversely correlated

C.-G. Lee (✉)
Department of Tourism, Tajen University, Pingtung County, Taiwan, ROC
e-mail: maxlee@tajen.edu.tw

W.-L. Lai
Graduate Institute of Environmental Management, Tajen University,
Pingtung County, Taiwan, ROC

© Springer Nature Singapore Pte Ltd. 2019
J.-S. Pan et al. (eds.), *Genetic and Evolutionary Computing*,
Advances in Intelligent Systems and Computing 834,
https://doi.org/10.1007/978-981-13-5841-8_38

to pollutants such as sulfate, nitrate, elemental carbon, organic carbon, NOx, NH3, $PM_{2.5}$, and PM_{10}. All the potential factors, categorized into air pollutants and climate properties, were further analyzed using principal component analysis, which reveals the two categories belonging to two different groups. Based on multivariate linear regression analysis results, as of climate properties, only humidity has a large effect on the way how air pollutants affect/interact visibility. Accordingly, a model built via hierarchical regression found that the visibility in the Kaohsiung area is mainly impacted by nitrate (-0.62), followed by elemental carbon (-0.41) and then sulfate (-0.30). This implies that the visibility attenuation in Kaohsiung area could be mainly caused by photochemical reaction as well as combustion. Through path analysis, it is also discovered that the regional visibility in summer time is higher than that in winter time, which is likely due to different photochemical reaction rates caused by temperature variation.

Keywords Visibility · Aerosol · Big data · Hierarchical regression · Path analysis

1 Introduction

Over the past two decades, atmospheric visibility in the urban area of Kaohsiung has gradually worsened. Studying air quality and meteorological records allows for a better understanding of the nature of the problem and for the identification of trends. Climatological observations conducted in urban Kaohsiung between 1979 and 2004 were analyzed along with critical air pollutants monitored from 1999 to 2004 in order to establish the relationship between atmospheric visibility and major air pollutants and meteorological parameters in this urban area.

The air quality data in this study obtained from over 20,000 sets recorded between 1999 and 2004 at the Chengtserng Air Quality Monitoring Station, which is one of the stations of Taiwan Air Quality Monitoring Network (TAQMN), with 10-meter height at the rooftop of a three-story building located at Central Kaohsiung. The TAQMN stations monitor air pollutants according to reference or equivalent United States Environmental Protection Agency methods [1]. Five criteria air pollutants are continuously monitored at the stations using instruments manufactured by Thermo Environmental, USA. The air pollutants monitored are sulfur dioxide (SO_2), carbon monoxide (CO), ozone (O_3), nitrogen oxides (NO_x), and PM_{10}. According to the record reported by ROCEPA, the instrument detection limits were 0.03 ppbv for SO_2, 0.01 ppmv for CO, 0.08 ppbv for O_3, 0.05 ppbv for NO_x, and 0.17 $\mu g\ m^{-3}$ for PM_{10}, respectively.

2 Methodology

The meteorological data in this study is obtained from the Kaohsiung Meteorology Station, which is the site for visibility observations. Visibility was observed at a height of 14 meters, on the rooftop of a meteorology observation station. Visibility observations require that an observer scans the horizon for predetermined sets of objects. Obviously and easily identifiable structures and objects, such as tall building, towers, stacks, and mountain ridges, make up these sets. The furthest distance at which it is possible to see more than half of the set of objects is recorded as the prevailing visibility. Visibility observations are made in the 16 compass-point directions.

At different times of the day and at different times in the period of 1979–2004, different observers conducted the observations. It is unavoidable that different observers might have different capacities to see distant objects. According to rules set by the World Meteorological Organization (WMO), observations are always made by two observers, both specially trained, who arrived at a consensus for each observation, so guarding against such individual differences in eyesight. Since the procedures followed constantly throughout the whole period, the risk of variation from non-observer factors could be reduced to a minimum. The standard deviation between different observers in estimating visibility, around 0.18 km, could therefore be ignored. Since visibility is influenced by both haze aerosol and natural obstructions to vision, such as rain, fog, and snow, the visibility data collected during rainfall were discarded from this study. After screening, the available visibility data still amounted to more than 110,000 observations over the period of 1979–2004.

3 Results and Discussions

The climate of Kaohsiung is generally warm and humid, with a mean temperature of 24.6 ± 4.5 °C and an average relative humidity of $79 \pm 10\%$. Relative humidity varies between 60 and 90%. However, the higher relative humidity is usually observed during the summer and early autumn seasons. The average wind speed in this 26-year period is 2.5 ± 1.6 m s^{-1}. Figure 1 shows the typical monthly variations in visibility for this 26-year period. All the mean monthly visual ranges exceed 10 km between April and September, while the mean monthly visual ranges are always below 10 km between October and March of next year. In addition, the monthly visibility increases by a factor of two from 8 km in December and January to 16 km in July. Further, if we subdivide a whole year into four identifiable seasons: from December to February (winter), from March to May (spring), from June to August (summer), and from September to November (autumn), seasonal changes from the highest to the lowest visibility are found to be in the sequence of summer, spring, autumn, and winter.

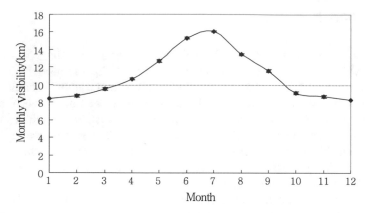

Fig. 1 Monthly variations of visibility in urban Kaohsiung during the period of 1979–2004

Figure 2 shows the typical diurnal variation in visibility and meteorological factors (temperature, relative humidity, and wind speed) in the period of 1979–2004. The general trends were easily identified from this mean data. The plot of mean diurnal variation in visibility shows that the highest visibility occurred in the afternoon (1–5 p.m.), while the lowest was observed in the early morning (2–8 a.m.). Meanwhile, the lowest relative humidity was found to be at 2 p.m., and then increased in the afternoon and on through the night (from 2 p.m. to 5 a.m.), declined in the morning and early afternoon. Due to the early morning radiation fogs and high RH disperse, visibility increased as expected from 5 a.m. till 2 p.m., and then declined gradually till 5 a.m. of the next day. It is obvious that the diurnal patterns of visibility and relative humidity show apparently opposite trends. Furthermore, mean wind speed increased from 9 a.m., reached the highest at 2 p.m. daily, and then declined onwards till midnight. Because the wind speed could enhance the air movement and improve the atmospheric dispersion, it is reasonable that wind speed shows almost the same variation pattern as that of visibility. Finally, diurnal variation in temperature is noticeably less than that of relative humidity and wind speed. Even though, we still can find that the highest visibility occurred with the highest temperature and the lowest RH at 2 p.m. daily. This means that under stronger wind speed and higher temperature condition, the atmospheric dispersion will be enhanced, and therefore resulting in higher visual ranges.

Since the four seasons in southern Taiwan are not able to be clearly identified, based on the critical 10 km of monthly mean visibility in Fig. 1, we split one whole year into two identifiable half years: the clear half year represents the months of April to September (with mean visibility >10 km), while the hazy half year represents October to March (with mean visibility <10 km). Further, Figs. 3, 4, and 5 show the plots of those data summarized. Winds blew predominantly from the north in hazy half year, but mostly from the south and west in clear half year. Poor visibility was always associated with the influence of northerly winds blowing particulate matter from Nan-zi, Jen-wu, and Ta-shei industrial-park areas to downtown Kaohsiung in

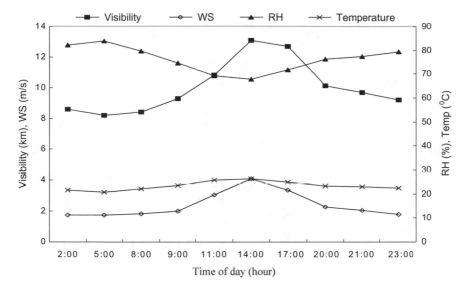

Fig. 2 Diurnal variations of visibility with meteorological factors in urban Kaohsiung during the period of 1979–2004

hazy half year. Figure 3 shows, as expected, the lowest mean visibility of only 8–9 km occurred with three wind directions of NNW, N, and NE for the entire period. On annual average, the highest mean visibility was found to be 16.7 km when the wind was blown from the south–south–west (SSW) direction, as shown in Fig. 4. This means the visibility is more influenced by relatively cleaner maritime air masses blown from the southwest, particularly during the warm periods of summer time (with an average visibility of 17.3 km, shown in Fig. 5).

As shown in Table 1, except for PM_{10}, the concentrations of criteria air pollutants (i.e., CO, NO_2, SO_2, and O_3) in urban Kaohsiung do not exceed WHO guidelines and other air quality standards (e.g., the National Air Quality Standard (NAAQS) in the United States of America). The average PM_{10} concentration (70.8 µg m^{-3}) violates the proposed annual PM_{10} regulation value (65 µg m^{-3}) of NAAQS in Taiwan. Accordingly, the pollution level of particulate matter was markedly high in urban Kaohsiung during the period of 1999–2004. Further, Table 2 summarizes the occurrence frequency and percentage of criteria air pollutants violating the daily regulation values of NAAQS in each year of 1999–2004. Of the five criteria air pollutants, PM_{10} and O_3 are the two air pollutants most responsible for the episodes of Kaohsiung air quality. Levels in excess of the 24 h PM_{10} standard of 150 µg m^{-3} are occasionally reported and tend to occur from October through February of each year. Such events are also generally associated with visibility deterioration linked to photochemical smog.

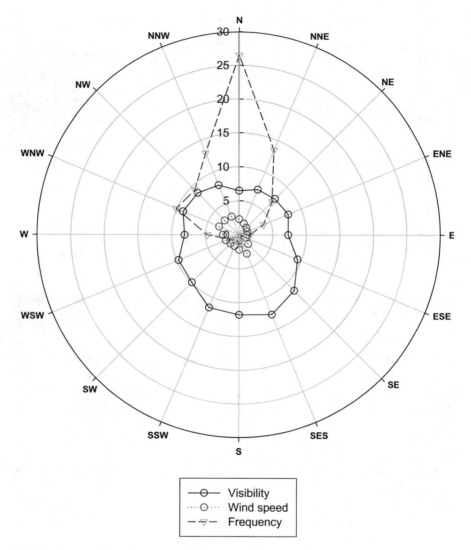

Fig. 3 Mean visibility and wind speed based on wind direction for hazy half year during the period of 1979–2004 in urban Kaohsiung

Diurnal variations of five criteria air pollutants in urban Kaohsiung over the period of 1999–2004 are illustrated in Fig. 6. It was clear that all the pollutant levels exhibited typical diurnal variation patterns for urban air pollution. A single daily peak for SO_2 concentration was found to occur around 9–11 am. The data showed two daily peaks for both CO and NO_2 concentrations, indicating vehicle-derived pollution. A rapid reaction of NO_2 and VOCs with strong sunlight led to the photochemical formation of O_3, which declined to lower levels at night. Two daily peaks for PM_{10} concentrations

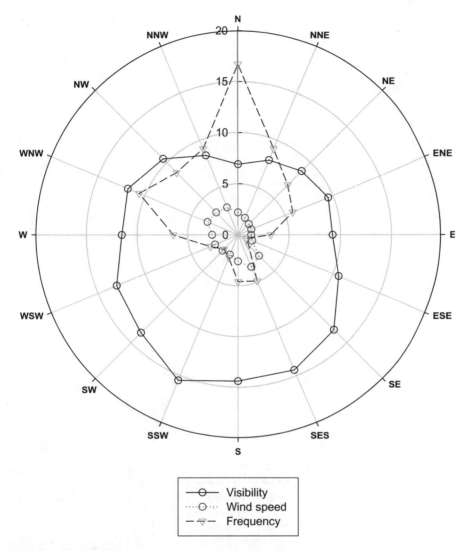

Fig. 4 Mean visibility and wind speed based on wind direction for entire year during the period of 1979–2004 in urban Kaohsiung

occurred at both midday and midnight. The midday peak was probably attributed to the formation of secondary aerosols derived from photochemical formation into the atmosphere, while the midnight peak was probably triggered by a worse atmospheric dispersion associated owing to lower wind speed and higher relative humidity.

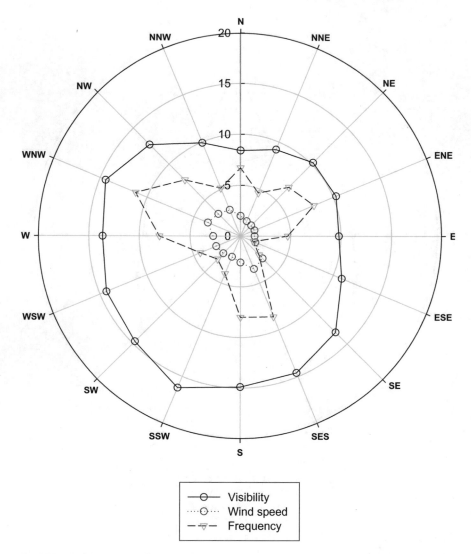

Fig. 5 Mean visibility and wind speed based on wind direction for clear half year during the period of 1979–2004 in urban Kaohsiung

Table 1 Average air pollutant levels in urban Kaohsiung during the period of 1999–2004		SO$_2$	CO	O$_3$	PM$_{10}$	NO$_2$
		(ppbv)	(ppmv)	(ppbv)	(μg m^{-3})	(ppbv)
	Mean	6.4	0.8	28.8	70.8	23.1
	Std. dev.	6.6	0.4	27.3	48.7	14.3

Table 2 Statistical frequency of occurrence of annual air pollution episodes in urban Kaohsiung during 1999–2004

Year	Pollutant	Occurrence (days)	Percentage (%)
1999	PM_{10}	11	37
	O_3	19	63
2000	PM_{10}	19	68
	O_3	9	32
2001	PM_{10}	18	47
	O_3	20	53
2002	PM_{10}	2	5
	O_3	41	95
2003	PM_{10}	2	9
	O_3	20	91
2004	PM_{10}	7	35
	O_3	13	65

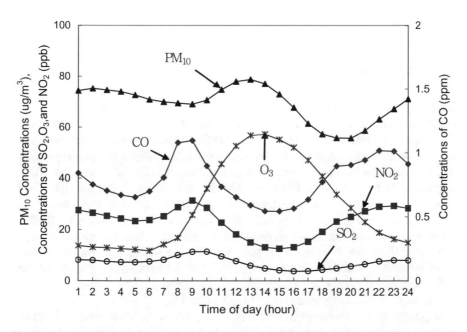

Fig. 6 Diurnal variation of criteria air pollutants in urban Kaohsiung during the period of 1999–2004

Table 3 Correlation matrix of monthly visibility, meteorological parameters, and air pollutants in urban Kaohsiung over the period of 1999–2004

	Vis	Press	Temp	RH	WS	SO$_2$	CO	O$_3$	NO	NO$_2$	PM$_{10}$
Vis[a]	1.00										
Press[b]	−0.80	1.00									
Temp[c]	0.76	−0.89	1.00								
RH[d]	−0.64	−0.79	0.59	1.00							
WS[e]	0.44	−0.25	0.08	0.13	1.00						
SO$_2$	−0.49	0.48	−0.55	−0.34	−0.13	1.00					
CO	−0.76	0.70	−0.76	−0.42	−0.24	0.67	1.00				
O$_3$	−0.74	0.61	−0.37	−0.67	−0.33	0.25	0.49	1.00			
NO	−0.07	0.14	−0.38	0.24	0.05	0.48	0.41	−0.39	1.00		
NO$_2$	−0.87	0.87	−0.85	−0.66	−0.25	0.62	0.87	0.57	0.33	1.00	
PM$_{10}$	−0.88	0.83	−0.86	−0.60	−0.17	0.63	0.83	0.54	0.31	0.93	1.00

[a] Visibility
[b] Atmospheric pressure
[c] Temperature
[d] Relative humidity
[e] Wind speed

4   Conclusions

Table 3 presents a correlation matrix of monthly average visibility, meteorological parameters, and criteria air pollutants in urban Kaohsiung over the period of 1999–2004. Correlation analysis revealed that the visibility increased with both temperature and wind speed, but decreased with relative humidity, atmospheric pressure, and also had relatively high correlation with all the air pollutants except for NO. A negative correlation between PM$_{10}$ and visibility was observed with the highest correlation coefficient of −0.88. Furthermore, a significant correlation between PM$_{10}$ and NO$_2$, and CO and SO$_2$ implicated that these air pollutants might result from the same or at least similar urban or industrial sources, which explained the negative association with visibility. NO$_2$ and CO were demonstrated as the second most significant air pollutants which deteriorated visibility. Despite an insignificant correlation between atmospheric pressure and NO, there was a relatively high correlation for atmospheric pressure with NO$_2$, PM$_{10}$, CO, and O$_3$.

Figure 7 shows a markedly inverse relationship between the mean monthly PM10 concentration and mean monthly visibility over the period of 1999–2004. Higher PM10 concentration resulted in lower visibility and vice versa. Mean PM10 concentration was the highest in November–January of next year, followed by February, October, March, April, May, September, and June–August. It was clear that the lowest PM10 concentration and the highest visibility were found in the summer months. Owing to the dominant role played by PM2.5 in visibility impairment, a targeted

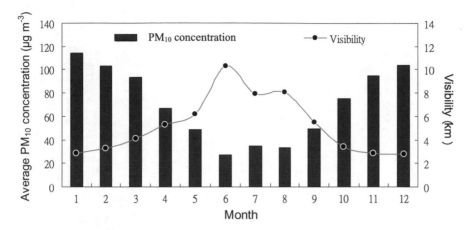

Fig. 7 Monthly variations of visibility with PM_{10} concentration in urban Kaohsiung during the period of 1999–2004

reduction in PM10 concentration might not completely improve the visual range. However, there is still a strong correlation between presence of PM10 and PM2.5 in most urban areas of the world [2]. It indicated that long-term visibility trends were highly correlated to PM10 concentrations. It is clear that the decreased visibility was related mostly to the increase in PM10. Consequently, this study focused on investigating the influences of particulate matter characteristics on visibility impairment in urban Kaohsiung.

References

1. Tsai, Y. I., Lin, Y. H., Lee, S. Z.: Visibility variation with air qualities in the metropolitan area in Southern Taiwan. Water Air Soil Pollut. **144**, 19–40 (2003)
2. Chan, Y. C., Simpson, R. W., Mctainsh, G. H., Vowles, P. D., Cohen, D. D., Bailey, G. M.: Source apportionment of visibility degradation problems in Brisbane (Australia) – using the multiple linear regression techniques. Atmos. Environ. **33**, 3239–3250 (1999)

Implementation of Data Collection in Smart Living Application

Yuh-Chung Lin, Yu-Chun Liu, Chun-Cheng Wei and Xiaogang Wang

Abstract The technology of AI also applies to the smart living application. However, before using AI technology to the application of our daily life, we have to solve the problem about data collections. In order to implement smart living application, the deployment of sensors is necessary around the house. Based on these sensing data, smart living control system can control all the electronic devices in the house to make the living place a comfortable environment. In this paper, we try to integrate the scattered documents and to describe the complete procedures for collecting data. It can be a reference for students who would like to learn how to build a data collection system by themselves.

Keywords Arduino UNO · ESP8266 · IoT · Smart home

1 Introduction

With the advancement of computer technologies, the AI is one of the most popular topics in either academy or industries. Since the AlphaGO beats the best human go chess player (Lee Sedol), AI technology quickly enters our daily life. A lot of companies claim that their products have a built-in AI functions to make a better performance. Why AI technologies make such a big progress? In addition to the substantial increase in hardware computing power, many new efficient algorithms also have been proposed. Another hero is the big data. A large amount of data provides

Y.-C. Lin (✉)
College of Information Science and Engineering Fujian University of Technology,
Fuzhou, China
e-mail: yuhchung@gmail.com

Y.-C. Liu · C.-C. Wei
Department of Information Application and Management, Tajen University,
Pingtung, Taiwan

X. Wang
School of Software, Changzhou College of Information Technology, Changzhou, China

© Springer Nature Singapore Pte Ltd. 2019
J.-S. Pan et al. (eds.), *Genetic and Evolutionary Computing*,
Advances in Intelligent Systems and Computing 834,
https://doi.org/10.1007/978-981-13-5841-8_39

a good training data set in different application areas. AlphaGo is just one of the most famous examples. The analysis of the election of democratic politics not only uses the traditional telephone poll, but also analyzes the big data in the Internet, which can get more accurate analysis results.

The technology of AI also applies to the smart living application. However, before using AI technology to the application of our daily life, we have to solve the problem about data collections. Recently, the development of AI is based on the accumulation of large amount of data. Years and years development of big data technology makes the AI to come into our daily life, not only running in the academic lab. In order to implement smart living application, the deployment of sensors is necessary around the house. Based on these sensing data, smart living control system can control all the electronic devices in the house to make the living place a comfortable environment. With the assistance of AI system, all the processes are automatically executed according to the sensing data, the behavior of people who lived in this place, etc.

We would like to implement a data collection system using Arduino and ESP8266 module, which is for the control system in a smart living application. During the implementation, we encountered a lot of difficulties. It is hard to find a complete document to explain how to integrate Arduino and ESP8266 module to be a sensing node. Another issue is how to transmit the sensing data to a cloud storage for a further process. There is no document to describe the processes from making a sensing node to transmitting sensing data to the cloud database. In this paper, we try to integrate the scattered documents and to describe the complete procedures for collecting data. It can be a reference for students who would like to learn how to build a data collection system by themselves.

The rest of the paper is organized as follows. Section 2 introduces some related works on the directed diffusion for a wireless sensor network. Section 3 presents the load-balance directed diffusion scheme for route selection and maintenance. Some simulation results have been shown in Sect. 4. Finally, a short conclusion is presented in Sect. 5.

2 Related Works

2.1 Arduino

"Arduino is an open-source computer hardware and software company, project, and user community that designs and manufactures single-board microcontrollers and microcontroller kits for building digital devices and interactive objects that can sense and control objects in the physical and digital world" [1]. The Arduino project's products are distributed as open-source hardware and software, which are licensed under the GNU Lesser General Public License (LGPL) or the GNU General Public License (GPL). The goal of Arduino project was to create simple, low-cost tools for creating digital projects by non-engineers. Everyone can manufacture the Arduino boards and software distribution without license fee. Developers can select a board from the Arduino board family which is suitable for their projects. Many makers around the world choose the Arduino board to create the prototype of their ideas. In education, Arduino board is also the first option for students to implement their digital projects, because of its low cost and as it is easy to learn.

2.2 Communication Modules

In the smart home application, data collection is a very important part which will integrate some sensors with a communication module to transmit data to the server. Following this, we will introduce some communication modules that are commonly utilized in IoT applications.

The ESP8266 [2] is produced by Shanghai-based Chinese manufacturer Espressif Systems which is a low-cost Wi-Fi microchip with full TCP/IP stack and microcontroller capability. In August 2014, the ESP-01 module which is made by a third party (Ai-Thinker) got the attention of Western makers. This small module without many external components allows microcontrollers to connect to a Wi-Fi network and make simple TCP/IP connections using Hayes style commands. Due to very low cost, many hackers were attracted to explore the module, chip, and the software on it.

Espressif Systems released a software development kit (SDK) for programming the chip directly. Usually, ESP8266 is only a communication module which has to be integrated with other microcontrollers, such as Arduino. The limitation is removed by the new released SDK. It is more convenient for developers. There are two versions of SKD maintained by Espressif—one is based on FreeRTOS and the other is based on callbacks. An alternative to the official SDK is the open-source ESP-Open-SDK which is based on the GCC toolchain. There are also many different modules made by Espressif, Ai-Thinker, and other producers. With earlier ESP-xx modules, developers have to purchase the USB-to-serial adapter and the regulator separately which have to be wired into the ESP-xx circuit. Some other boards are created which integrate an onboard USB-to-UART bridge and a micro-USB connector, coupled with a 3.3-volt regulator to provide both power to the board and connectivity to the software development console on the computer.

There are other options for the communication module, such as ESP32 [3] and LinkIt 7697 [4]. ESP32 is a successor to the ESP8266 microcontroller which is a series of low-cost, low-power system on chip microcontrollers with integrated Wi-Fi and dual-mode Bluetooth. It can perform as a complete standalone system or as a slave device to a host. LinkIt 7697 development board is developed by MediaTek which is a Combo development board that integrates Wi-Fi and low-power Bluetooth (BLE) communication technology.

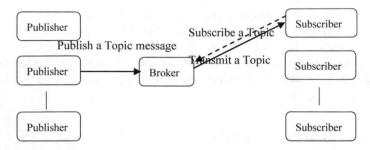

Fig. 1 MQTT publish/subscribe message transmission mechanism

2.3 *Message Queuing Telemetry Transport (MQTT)*

MQTT [5] is an extremely lightweight publish/subscribe message transport protocol which is designed for constrained devices and low-bandwidth, high-latency, or unreliable networks. The design principles are to minimize network bandwidth and device resource requirements which make it ideal for a machine-to-machine (M2 M)/"Internet of Things" connectivity.

The MQTT protocol includes four components which are the Publisher, the Subscriber, the Topic, and the Broker. The Publisher is the source of message which includes information with the "Topic". It transmits the message to the Broker. The Subscriber can subscribe to the Broker with a "Topic" which the subscriber would like to receive. Figure 1 shows the MQTT transmission mechanism.

There are three different transmission QoS levels as follows:

- QoS 0: at most once,
- QoS 1: at least once, and
- QoS 2: exactly once.

The application can set the QoS level according to the importance of message.

3 System Architecture

In a smart home, the automation system will control lighting, climate, entertainment systems, and appliances. It may also include an access control and alarm systems. Figure 2 shows a simple architecture of smart home application. There are many sensors to monitor the home environment such as temperature, humidity, brightness, etc. All home devices are connected with the Internet which are the important constituents of the Internet of Things.

The sensors will be deployed around the house to monitor the weather conditions, and the time of sunrise and sunset. All the data will use MQTT protocol to transmit to the MQTT Broker. The smart home application server subscribes all Topics to gather

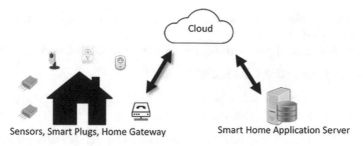

Fig. 2 System architecture of smart home

all the related information and store into database waiting for further analysis. The smart home system can utilize machine-learning technology to analyze the weather data to find out a schema to turn on air conditioner. It can automatically adjust the temperature of the house to make it a comfortable living place. Also, it can record the lifestyle of people who live in there, or track the position of people, to make a decision on how to adjust the house temperature. There are many application scenarios. But, first of all, data collection is the first step to implement such an automatic smart home management system. In next section, we will focus on the implementation of data collection and describe how to implement a data collection system.

4 Implementations

In this section, we will describe how to implement the data collection system step by step. A sensor node was built to collect temperature and humidity data and transmit to the Broker by MQTT. A procedure run at the Smart Home Application Server takes the responsibility to subscribe and store the received data to database. Following is the detail.

4.1 Sensor Node

The sensor node is composed of a sensing module and a communication module. The sensing module is made by Arduino UNO board and DHT11 sensor. The communication module is ESP8266 ESP-12E UART WIFI Wireless Shield Development Board. Following steps are the procedure to set up the communication module of sensor node.

1. Download the new firmware and ESP8266 flasher tool.

   The download websites are listed in the following for your reference (or you can search on the Internet):

https://raw.githubusercontent.com/sleemanj/ESP8266_Simple/master/firmware/ai-thinker-v1.1.1-115200.bin

https://github.com/JhonControl/ESP8266-Flasher/tree/master/esp8266_flasher.

2. Connect Arduino UNO and ESP8266 shield.

In this step, we use the debug port of ESP8266 shield to connect to the Arduino UNO. The pin connection is shown in Table 1.

3. Set the DIP of ESP8266 shield to (DOWN, DOWN, UP, UP) which is in flashing mode.
4. Test the connection between Arduino UNO and ESP8266 shield using AT command.

    i. Execute Arduino IDE.
    ii. Open a new empty project and upload to UNO. At this time, the DIP must be set to all DOWN.
    iii. Open serial monitor window and change the baud rate to 115200.
    iv. Input "AT" command in the transmission window. Check the response. If the response is "OK", the connection is good.

5. Update the firmware by ESP8266 flash tool as Fig. 3.
6. Install WiFiEsp library in Arduino IDE. We are ready to program the sensor node.

The next step is to install a sensor. In our implementation, we choose temperature and humidity sensing as an example. DHT11 is first our choice because it is easy to obtain and inexpensive. In Arduino IDE, the DHT library has to be installed before we can program. It is very simple to install the DHT11 sensor to the Arduino UNO. Therefore, we are not going further to describe the details [6].

Table 1 The connection mapping between ESP8266 shield and Arduino UNO	ESP8266 shield	Arduino UNO
	Debug Port TX	UNO Pin 1 (TX)
	Debug Port RX	UNO Pin 0 (RX)
	Debug Port 5 V	UNO 5 V
	Debug Port GND	UNO GND

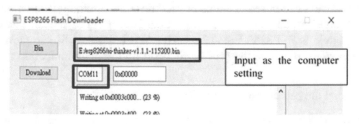

Fig. 3 Update firmware of ESP8266 shield

Fig. 4 The information for connecting to MQTT broker

Instance info

Server	m14.cloudmqtt.com
User	bbjbxecj
Password	xLFR1Lz_14Lm
Port	15675
SSL Port	25675
Websockets Port (TLS only)	35675
Connection limit	5

C Restart

C Rotate

4.2 MQTT Server

The sensor node will publish the sensing data to the MQTT broker. How to establish an MQTT broker? The simple way is to use the MQTT broker in public domain, such as cloudmqtt [7], RabbitMQ, and HiveMQ. The other way is to establish your own MQTT server.

First, we introduce how to use cloudmqtt as a broker. It is easy to set up an MQTT broker. The following is a detailed description.

1. Register an account or simply use Google account to login cloudmqtt.
2. Create an MQTT instance.
3. Follow the step-by-step instruction and fill in the relevant information for creating MQTT broker.
4. After creating an instance, click on the name of instance. The information for connecting to the MQTT broker is presented, as shown in Fig. 4.

For greater flexibility, you may choose to build your own MQTT broker. First, you must build a Linux server and install the mosquito MQTT broker. The installation instruction for Ubuntu is listed below:

```
$ sudo apt-add-repository ppa:mosquito-dev/mosquito-ppa
$ sudo apt-get update
$ sudo apt-get install mosquitto
$ sudo apt-get install mosquitto-clients
```

To stop and start the service, following instructions are needed to be used:

```
$ sudo service stop mosquitto
$ sudo service start mosquitto
```

4.3 Sample Code

There are three parts of the code: Wi-Fi connection setup, MQTT broker connection, and data sensing and transmitting. Because limited by space, we only describe how to connect to the Wi-Fi and MQTT broker.

4.3.1   Wi-Fi Connection

```
#include "WiFiEsp.h"
#include <PubSubClient.h>

char ssid[] = " your network SSID (name)";
char pass[] = " your network password ";

// initialize serial for ESP module, the communication baud rate set to 9600
  Serial1.begin(9600);
  // initialize ESP module
  WiFi.init(&Serial1);

  // check for the presence of the ESP8266 shield
  if (WiFi.status() == WL_NO_SHIELD) {
    Serial.println("WiFi shield not present");
    // don't continue
    while (true);
  }

  // attempt to connect to WiFi network
  while ( status != WL_CONNECTED) {
    Serial.print("Attempting to connect to WPA SSID: ");
    Serial.println(ssid);
    // Connect to WPA/WPA2 network
    status = WiFi.begin(ssid, pass);
  }
```

4.3.2   MQTT Broker Connection

Once the node connects to Wi-Fi successfully, the connection to the MQTT broker must be set up as following code. If using the cloudmqtt as a broker, the needed information can be found in the website of cloudmqtt as shown in Fig. 4.

```
const char* mqttServer = "m14.cloudmqtt.com";
const int mqttPort = 15675;
const char* mqttUser = "bbjbxecj";
const char* mqttPassword = "xLFR1Lz_14Lm";
WiFiEspClient espClient;
PubSubClient client(espClient);

client.setServer(mqttServer, mqttPort);
client.setCallback(callback);

void callback(char* topic, byte* payload, unsigned int length) {
  Serial.print("Message arrived in topic: ");
  Serial.println(topic);
  Serial.print("Message:");
  for (int i = 0; i < length; i++) {
    Serial.print((char)payload[i]);
  }
}
```

5   Conclusion

In this paper, we describe the implementation of data collection that is part of a smart home application. Many resources could be found on the Internet. The purpose of this paper is to integrate all the necessary resources needed to build a data collection system. In our experiment, the implemented system could work perfectly to gather the sensing data.

In the future, by referring to the machine learning and gathering more data (including people behavior), we would like to build a real smart home application system.

Acknowledgements This work is supported by Top-notch Academic Programs Project of Jiangsu Higher Education Institutions (TAPP) under Grant No. PPZY2015A090.

References

1. https://en.wikipedia.org/wiki/Arduino
2. https://en.wikipedia.org/wiki/ESP8266
3. https://www.espressif.com/en/products/hardware/esp32/overview
4. https://labs.mediatek.com/en/platform/linkit-7697
5. Singh, M., Rajan, M.A., Shivraj, V.L., Balamuralidhar, P.: In: Secure MQTT for Internet of Things, 2015 Fifth International Conference on Communication Systems and Network Technologies, Gwalior, India
6. http://www.circuitbasics.com/how-to-set-up-the-dht11-humidity-sensor-on-an-arduino/
7. https://www.cloudmqtt.com/

Generalized Benford's Distribution for Data Defined on Irregular Grid

Li-Fang Shi, Bin Yan, Jeng-Shyang Pan and Xiao-Hong Sun

Abstract In forensic analysis, such as forensic auditing, multimedia forensic, and financial fraud detection, the auditor needs to detect data tempering to find clue for possible fraud. First digit distribution such as Benford's law is proved to be an efficient tool and is used by many auditing companies to preprocess the data before the actual auditing. However, when the range of the data is limited, the first digit distribution usually does not conform to Benford's law. Using temperature data from a sensor network, we show that if the data can be modeled by a graph signal model, then after the graph Fourier transformation, the distribution of first digits conforms to a generalized Benford's law. In addition, a graphic model based on historical data provides better fit to the Benford's model than that based on geodesic distance. This model is evaluated for simulated data and temperature sensor network. This finding may help to build models for forensic analysis of accounting data and sensor network data for fraud detection.

Keywords Fraud detection · Benford's law · Graph signal · Complex network · Graph Fourier transform

L.-F. Shi · X.-H. Sun
Department of Auditing, Shandong University of Science and Technology,
Qingdao, China
e-mail: 13793299153@163.com

B. Yan (✉)
College of Electronics Communication and Physics, Shandong University of Science
and Technology, Qingdao, China
e-mail: yanbinhit@hotmail.com

J.-S. Pan
College of Computer Science and Engineering, Shandong University of Science
and Technology, Qingdao, China

© Springer Nature Singapore Pte Ltd. 2019
J.-S. Pan et al. (eds.), *Genetic and Evolutionary Computing*,
Advances in Intelligent Systems and Computing 834,
https://doi.org/10.1007/978-981-13-5841-8_40

1   Introduction

Benford's law for the distribution of first digits, or most significant digits (MSD), is a powerful tool for data forensic. Since its discovery by Benford [2], it has found many applications in forensic accounting [3], financial fraud detection [12], multimedia forensic [6, 10], etc. Recently, it is extended to forensic analysis of social networks [5]. For these applications, if the empirical distribution of the MSD violates the Benford's law, then it is a flag of possible data tempering. This flag may lead the auditor or researcher to focus on a smaller amount of data within a large amount of data. So, it is very effective for reviewing a large amount of data, such as data from sensor network or general ledger. For large listed companies, the size of accounting data facing the auditors may exceed tens of gigabytes each year. The application of Benford's law for forensic analysis relies on the premise that the data without manipulation follows the Benford's law. This is not always ensured for many realistic data, such as image and temperature. For these data, samples come from a single source and the range of data is limited. Take, for example, temperature. For most places in the world, the temperature is from -30 to $40\,°C$. So, most MSDs are limited to 1, 2, 3, and 4. Fortunately, it is found that, by proper transforming of the data, the MSD follows Benford's law or generalized Benford's law [6, 11]. Jolion found that by taking the gradient of pixels, the MSD of the magnitude of gradient follows the Benford's law very well [6]. Pérez-González et. al found that if the data follows a generalized Gaussian distribution, then the MSD follows a generalized Benford's distribution [11]. This result is employed for detecting hidden watermark or double compression [4].

The data studied in transformation-based Benford's analysis is limited to data defined on regular grid. For example, the image signal studied in [6, 11] is defined on regular two-dimensional (2D) grid. So various transformations are available, such as 2D discrete cosine transformation (DCT), 2D discrete wavelet transformation (DWT), and 2D Fourier transformation. For data lacking in this regular domain, it is not clear how to transform the data. For example, the temperature, rainfall data are defined on irregular locations determined by the longitude and latitude. In 3D mesh model and point cloud model, the domain of the data is also irregular in 3D space. For these examples, *the range of the data is limited and the domain of the data is irregular*. How to properly transform the data to find a useful distribution of the MSD? This is still an undressed problem.

The research problem of this paper can be stated as extending the transformation-based approach to data defined on irregular grid, trying to find the distribution of the MSD. To this end, we model the domain of the data using a weighted graph, and model the data as a graph signal. We found that, by transforming the data using *graph Fourier transformation* (GFT), the MSD of the transformed data follows a generalized Benford's law.

In Sect. 2, we review briefly the graph signal model and distribution of MSD. Then, the distribution of MSD in graph Fourier transform domain is presented in

Sect. 3. Using temperature data, the experimental results are discussed in Sect. 4. Finally, we conclude the paper in Sect. 5.

2 Graph Signal Model for Temperature Data and Benford's Law

This section briefly reviews the graph signal model and Benford distribution. A recent and comprehensive review of graph signal processing (GSP) can be found in [9].

The domain of the irregular data, such as temperature network, can be modeled as a weighted graph $\mathcal{G} = (V, E, \mathbf{W})$, where V is the set of vertices $V = \{\mathbf{v}_1, \ldots, \mathbf{v}_N\}$. The set E contains all the edges of the graph, and each such edge is recorded as a pair of vertices $(\mathbf{v}_i, \mathbf{v}_j)$. The matrix $\mathbf{W}$ is a weighting matrix with its component $w_{i,j}$ denoting the weight of the edge connecting vertex $\mathbf{v}_i$ and vertex $\mathbf{v}_j$.

Graph signal is a mapping f defined on a graph, i.e., $f : \mathcal{G} \to \mathbb{R}$, where the range of the signal is the set of real numbers. For the problem studied here, the range is a closed interval of $\mathbb{R}$. Given a fixed indexing of vertices $\mathbf{v}_i$, the graph signal can also be denoted by a vector $\mathbf{f} = (f_1, f_2, \ldots, f_N)$. Figure 1 shows a graph signal model for the temperature of 160 cities in China for December, 2000.

Benford's law gives the distribution of MSD of a lot of naturally occurring data:

$$P(d) = \log_{10}(1 + 1/d), \quad d = 1, 2, \ldots, 9, \tag{1}$$

where d is the MSD and $P(d)$ is the probability of d. However, for range-limited data, such as temperature, the MSD does not conform to the Benford's distribution. Figure 2 shows the comparison between distribution of MSD for temperature data and the Benford's distribution. It is evident that the MSD for temperature data does not conform to the Benford's distribution. It is shown in next section that by transforming the temperature data using graph Fourier transform, the MSD of the data conforms to a generalized Benford's law.

Fig. 1 Example of graph signal of temperature for 160 cities in China. The filled circles correspond to location of cities, and the vertical bars with filled squares are the temperature values

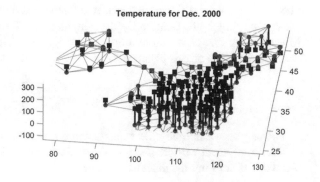

Fig. 2 Distribution of MSD for temperature data

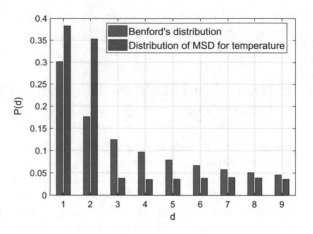

3 First Digit Distribution in Graph Fourier Transform Domain

First, we show that after graph Fourier transform, the coefficients follow the generalized Gaussian distribution. Then, according to the result in [11], the MSD of these coefficients follows a generalized Benford's law.

Just like the Fourier transform uses the characteristic function of the Laplacian operator as basis function to expand a given signal, the graph Fourier transform (GFT) uses the eigenvectors of graph Laplacian as basis functions. The graph Laplacian is defined as $\mathbf{L} = \mathbf{D} - \mathbf{W}$, where $\mathbf{D}$ is the diagonal degree matrix $\mathbf{D} = \text{diag}\left(D_{1,1}, D_{2,2}, \ldots, D_{N,N}\right)$ and $D_{i,i} = \sum_{j=1}^{N} w_{i,j}$.

The weights $w_{i,j}$ are specified using truncated distance between feature vector on the vertices $\mathbf{v}_i$ and $\mathbf{v}_j$:

$$d(i, j) = \begin{cases} d(i, j), & \text{if } d(i, j) \leq \tau; \\ \infty, & \text{otherwise.} \end{cases} \tag{2}$$

where $d(i, j)$ is the distance between two feature vectors on vertices $\mathbf{v}_i$ and $\mathbf{v}_j$ and τ is a fixed threshold. Using a fixed τ, the number of neighbors for each vertex is different. One can also fix the number of neighbors. The weights are assigned as Gaussian function of this distance, i.e.,

$$w_{i,j} = \exp\left[-\frac{d^2(i, j)}{2\sigma^2}\right]. \tag{3}$$

The purpose of defining the weights and distances is to produce a smooth graph signal. So, adjacent vertices should have similar values. For this purpose, we explore two ways of defining the distance measures.

Fig. 3 Graph for temperature data using difference between historical data as distance measure

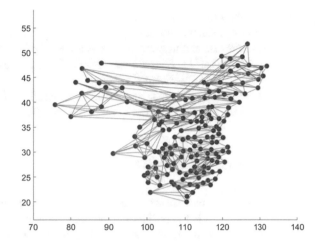

- *Using the Geodesic distance*. Given the location of each city $\mathbf{v}_i = (a_i, o_i)$ specified by latitude a_i and longitude o_i, the geodesic distance between two vertices is the arc length between them, where the arc is part of a great circle dividing the earth. The graph in Fig. 1 is build using this distance measure.
- *Using feature distance*. If the graph is dynamic with time, such as temperature data and dynamic point cloud, then the historical data can be employed to design the weights such that they reflect the similarity between the vertices. Let the available data be collected in a matrix $\mathbf{F} = [\mathbf{f}_1, \ldots, \mathbf{f}_N] \in \mathbb{R}^{K \times N}$, where N is the number of vertices and K is the number of time instants. Then, the distance is defined as the Euclidean distance: $d(i, j) = \left\| \mathbf{f}_i - \mathbf{f}_j \right\|_2$. The constructed graph using six neighbors is shown in Fig. 3. Compared to the geodesic distance result in Fig. 1, we see that, for a given city, it prefers neighbors with similar latitudes. This reflects the fact that temperatures of the same latitude are similar.

The basis for the graph $\mathcal{G}$ is the eigenvectors of the Laplacian matrix $\mathbf{L}$. Let the eigendecomposition of $\mathbf{L}$ be $\mathbf{L} = \mathbf{U} \Lambda \mathbf{U}^T$, where $\mathbf{U} = [\mathbf{u}_1, \ldots, \mathbf{u}_N]$ is the matrix of eigenvectors, and $\Lambda = \mathrm{diag}\,(\lambda_1, \ldots, \lambda_N)$ is a diagonal matrix of eigenvalues. Using these eigenvectors as an orthogonal basis, we obtain the transformed signal $\hat{\mathbf{f}} = \mathbf{U}^T \mathbf{f}$. Since the graph is constructed to produce a smooth graph signal so that the signal values between connected vertices are similar, it is expected that the transformed coefficients $\hat{\mathbf{f}}$ are sparse. Such a sparse distribution can be fitted by generalized Gaussian distribution. According to [11], the MSD for a generalized Gaussian random numbers follows a generalized Benford's distribution.

To find the generalized Benford's distribution for MSD, we need to first fit the data to generalized Gaussian distribution. We propose to use the mixture of generalized Gaussian (MGG) distribution to fit the empirical distribution of the data:

$$p(f) = \sum_{k=1}^{K} \omega_k p_k(f) = \sum_{k=1}^{K} \omega_k \frac{\beta_k}{2\alpha_k \Gamma\,(1/\beta_k)} \exp\left\{ -\left(\frac{|f - \mu_k|}{\alpha_k} \right)^{\beta_k} \right\}, \quad (4)$$

where ω_k is the proportion of component density $p_k(f)$ in the mixture. μ_k, β_k, $and \alpha_k$ are the location, shape, and scale parameters for component density k, respectively.

To estimate the parameters for generalized Gaussian distribution, we use the method described in [1]. The parameters for MGG can be estimated by an expectation–maximization (EM) algorithm similar to [7].

It is shown in Sect. 4 that, compared to generalized Gaussian model, the MGG model provides a better fit to the coefficients in GFT domain.

4 Experimental Results

This section provides the experimental results for the generalized Benford's distribution. The data used is the publicly available temperature data of 160 cities in China. This data consists of monthly averaged temperature from the year 1951 to 2000, totally 600 samples for each city. The graph $\mathcal{G}$ is constructed as described in Sect. 3. The basis $\mathbf{v}_1$ and $\mathbf{v}_2$ are shown in Fig. 4, and the resulting graphs are shown in Fig. 1 and Fig. 3, respectively.

To fit the MGG distribution to coefficients $\hat{\mathbf{f}}_i$, the EM algorithm is employed and the initial values for ω_k, μ_k, β_k, $and \alpha_k$ are estimated using parameter estimation for mixture of Gaussian density. We note that, since the EM algorithm strongly depends on the initial value, we may need to initialize the parameters several times before getting a good starting point.

The histogram of $\hat{\mathbf{f}}_i$ is nonsymmetric, which poses difficulty in estimating the parameters of MGG. Considering the fact that the first digit will not change if the polarity of a number is changed, so we multiply +1 and -1 randomly on each coefficient before parameter estimation. To better fit to the histogram, we also limit the data of $\hat{\mathbf{f}}_i$ to the range $[-30, 30]$. This helps in the convergence of the EM algorithm and is shown to not seriously affect the conformity of the actual data to generalized Benford's law.

The fitting results for $\hat{\mathbf{f}}_i$ are shown in Fig. 5. The result for graph using geodesic distance is shown in Fig. 5a and the result for graph using feature distance is shown

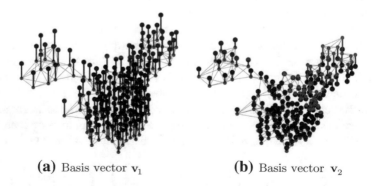

(a) Basis vector $\mathbf{v}_1$ **(b)** Basis vector $\mathbf{v}_2$

Fig. 4 Two basis vectors for the graph Fourier transform

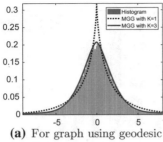

(a) For graph using geodesic distance.

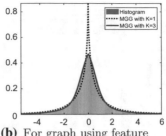

(b) For graph using feature distance.

Fig. 5 Fitting the histogram of transformed coefficients $\hat{\mathbf{f}}$

Table 1 p-value for Kolmogorov–Smirnov test

	Geodesic distance	Feature distance
$K = 1$	1.2072^{-35}	2.3952×10^{-34}
$K = 3$	4.6695×10^{-4}	0.55

in Fig. 5b. In each case, we compare the fitting performance using $K = 1$ and $K = 3$, where $K = 1$ corresponds to using the generalized Gaussian and $K = 3$ corresponds to using the mixture model. Obviously, using mixture model provides better fit in both cases. For the graph constructed from feature distance, the mixture model provides the best fit. We also use the two-sample Kolmogorov–Smirnov goodness-of-fit test to evaluating the fitting. Only the fitting using $K = 3$ for graph using feature distance passes the hypothesis testing under the significance level 0.05. The p−values are listed in Table 1.

Next, we present the result for MSD distribution using the generalized Benford distribution. The generalized Benford distribution is calculated using the estimated MGG models. The histogram for the data is calculated for all samples including those outside of $[−30, 30]$.

Note that the generalized Benford's distributions are different for the two graph models since different parameters are estimated. Obviously, using the feature distance, we get better fitting. To measure the fitting qualitatively, the mean absolute difference (MAD) is used to evaluate if the empirical distribution conforms to the generalized Benford's distribution. Nigrini proposed the following threshold for test of conformity using MAD, as shown in Table 2 [8]. The MADs for Fig. 6a, b are 0.0045 and 0.0013, respectively, which are "Close conformity" according to Table 2.

Table 2 Threshold for MAD [8]

Range of MAD	Conclusion
0.000–0.006	Close conformity
0.006–0.012	Acceptable conformity
0.012–0.015	Marginally acceptable conformity
Above 0.015	Nonconformity

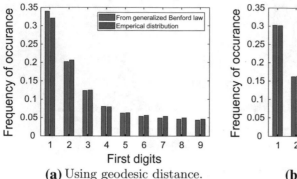
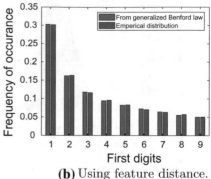

(a) Using geodesic distance.   **(b)** Using feature distance.

Fig. 6 Comparing the MSD distribution and generalized Benford law using estimated mixture model

5 Conclusion

Using temperature data, we show that by modeling the data as a graph signal and transform the data using graph Fourier transformation, the transformed coefficients follow a generalized Benford's law from fitting the coefficients using mixture of generalized Gaussian distribution. Using the distance between feature vectors to construct the graph, we get better fitting than using the geodesic distance. Furthermore, for both models, we get close conformity using the MAD measure. This generalized Benford's distribution can be used for fraud detection and multimedia forensic analysis.

Acknowledgements This work is supported by Ministry of Education in China (MOE) Projects of Humanities and Social Science (No. 15YJC790087, 18YJAZH110), National Statistics Science Project (No. 2015LZ59).

References

1. Beelen, T., Dohmen, J.: Parameter estimation for a generalized Gaussian distribution. CASA-report, Technische Universiteit Eindhoven (12) (2015)
2. Benford, F.: The law of anomalous numbers. Proc. Am. Philos. Soc. **78**(4), 551–572 (1938)
3. Bhattacharya, S., Kumar, K.: Forensic accounting and Benford's law. IEEE Signal Process. Mag. **25**(2), 150–152 (2008)
4. Comesana, P., Pérez-González, F.: The optimal attack to histogram-based forensic detectors is simple(x). In: IEEE International Workshop on Information Forensics and Security, pp. 137–142 (2015)
5. Golbeck, J.: Benfords law applies to online social networks. Plos One **10**(8) (2015)
6. Jolion, J.: Images and Benford's law. J. Math. Imaging Vis. **14**(1), 73–81 (2001)
7. Mohamed, O.M.M., Jaidane-Saidane, M.: Generalized gaussian mixture model. In: Signal Processing Conference, 2009 European, pp. 2273–2277 (2009)

8. Nigrini, M.J.: Benford's Law: Applications for Forensic Accounting, Auditing, and Fraud Detection. Wiley, Hoboken, NJ, USA (2012)
9. Ortega, A., Frossard, P., Kovačević, J., Moura, J.M.F., Vandergheynst, P.: Graph signal processing: overview, challenges, and applications. Proc. IEEE **106**(5), 808–828 (2018)
10. Pasquini, C., Comesana-Alfaro, P., Perez-Gonzalez, F., Boato, G.: Transportation-theoretic image counterforensics to first significant digit histogram forensics. In: IEEE International Conference on Acoustics, Speech and Signal Processing, pp. 2699–2703 (2014)
11. Pérez-González, F., Heileman, G.L., Abdallah, C.T.: A generalization of Benford's law and its application to images. In: Proceedings of the European Control Conference, pp. 3613–3619 (2007)
12. Shi, J., Ausloos, M., Zhu, T.: Benfords law first significant digit and distribution distances for testing the reliability of financial reports in developing countries. Physica Stat. Mech. Appl. **492**, 878–888 (2018)

Application of Bluetooth Low Energy in Smart Connected Lighting

Kai Yang

Abstract Smart connected lighting is beginning to generate serious momentum. Lighting is a powerful environmental factor, heavily influencing how we go about our daily lives. In this paper, we develop the testbed and software of Bluetooth Low Energy for smart lighting. Afterward, the overall scheme was sophisticatedly designed for system performance testing. The experiment results show that the Bluetooth Low Energy system can carry out a comprehensive controlling of lighting, which affords opportunities for users to gain convenience and effectiveness.

Keywords Bluetooth Low Energy · Smart connected lighting · CC2640R2F

1 Introduction

Recent years have witnessed the increasing adoption of smart lighting in commercial applications like retail and health care, as well as municipal use like streetlights. For example, smart buildings and factories, where connected, energy-efficient LED lighting can generate substantial energy savings, help personalize work spaces and optimize manufacturing by leveraging the right multiple wireless protocols, including ZigBee, Thread, and Bluetooth Low Energy with SoCs and certified wireless stacks. Smart lighting can help make municipalities safer, reduce energy costs, and encourage community engagement. While the smart city includes smart buildings, it is much more than that. It brings together the complexity of not only sheer size but also multiple vendors that must be managed effectively and efficiently. As we mentioned above, there are three primary benefits that are driving the adoption of smart connected lighting: convenience, intelligence, and data analytics [1].

K. Yang (✉)
School of Software, Changzhou College of Information Technology, Changzhou, China
e-mail: yangkai@ccit.js.cn

© Springer Nature Singapore Pte Ltd. 2019
J.-S. Pan et al. (eds.), *Genetic and Evolutionary Computing*,
Advances in Intelligent Systems and Computing 834,
https://doi.org/10.1007/978-981-13-5841-8_41

Bluetooth is one of the most widespread wireless technologies in the world. It is been in existence since 2000 and has found its way into billions of devices. Since then, Bluetooth has been carefully and systematically improved, enabling it to keep pace with market requirements, continuing to support and inspire innovation. As a wireless communication standard of low power consumption and low rate, Bluetooth Low Energy (LE) is optimized to use as little energy as possible, able to operate and communicate wirelessly, powered by only a coin-sized battery which could often last for years. It is hard to find a smartphone or tablet that does not support Bluetooth LE nowadays [2, 3].

The structure of this paper is as follows: (1) System design: the scheme of the whole system and its implementation principles are proposed. (2) Testbed design: it provides descriptions of the system's testbed compositions including CC2640R2F module and working principles. (3) Software design: it introduces the details of embedded software design and programming with tool of Bluetooth Developer Studio (BDS) to run properly and robustly on the testbed. (5) Experimental results and evaluation: deployment, environment, and procedures of experiments are illustrated, and two metrics are used to evaluate the system performance. (6) Conclusion: according to the experimental results of the system and our related work, some important conclusions can be drawn.

2 System Design

Bluetooth Low Energy supports multiple topology options to best meet the unique wireless connectivity needs of a diverse, global developer population [4]. The point-to-point topology available on Bluetooth Low Energy is optimized for data transfer and is well suited for connected device products, such as fitness trackers, health monitors, and PC peripherals and accessories. Accounting for this characteristic, the point-to-point topology is adopted for our system.

As shown in Fig. 1, the system is made up of two parts. The left part is a smartphone, which must support Android 4.3 or iOS 4.0 and above. The app is running on the smartphone and is responsible for communicating with the testbed in smart lighting by means of Bluetooth Low Energy. On the right side, there is a smart lighting, which can be an ordinary LED bulb embedded with the testbed. On the one hand, the testbed receives the controlling command from the smartphone. On the other hand, it parses the command packet and then decides whether to switch on or off the bulb. Obviously, the core part of the system lies in the application of Bluetooth Low Energy.

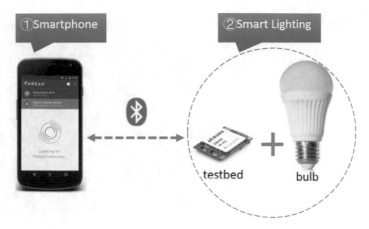

Fig. 1 System diagram

3 Hardware Components

The hardware part of the system is the testbed. It is not only designed for the smart controlling of the bulb but also for the other engineer's project. The hardware diagram is shown in Fig. 2.

The MCU for the testbed is the CC2640R2F, which is a wireless microcontroller with 2.4-GHz RF Transceiver Compatible. With Bluetooth Low Energy (BLE) 4.2 and 5 Specifications, CC2640R2F device contains a 32-bit ARM Cortex-M3 core that runs at 48 MHz as the main processor. The actual module in the testbed is shown in Fig. 3.

Benefiting from its small size and endurance, CR2032 is chosen as the power supply for the CC2640R2F module. Generally, it can be used for 2–3 years.

Accounting for the limited capability of the output current of CC2640R2F, we choose a relay module as an interface for CC2640R2F and bulb.

Fig. 2 Hardware diagram

Fig. 3 CC2640R2F module
in the testbed

4 Software Programming

As we all know, every protocol is used for its own specific situations [5]. For Bluetooth
Low Energy, data exchange occurs over the air according to the Attribute Protocol
(ATT). Benefiting from its small data characteristic, ATT is not complicated. ATT
allows a device to expose some data, known as attributes, to another device. The
device's exposing attribute is called the Server, and the peer device needing them
is called the Client. Obviously, in our system, the smartphone serves as the Client
whereas the smart lighting is the Server as shown in Fig. 4. By the way, the "switch"
is an attribute which can be read or written by Client.

4.1 Introduction to Basics of Bluetooth Low Energy

From a BLE application point of view, however, data is exchanged using the Generic
Attribute Protocol (GATT) which can be viewed as a meta-layer on top of ATT. An
Attribute is the smallest addressable unit of data used by ATT and GATT [6]. Several
of these attributes are needed to define a Characteristic. A Service is a collection
of characteristics. A Profile defines a collection of one or more services and how
services can be used to enable an application or use case. The hierarchy of Profile
→ Service → Characteristic → Attribute is shown in Fig. 5. As shown in Fig. 6,

Fig. 4 Client and server

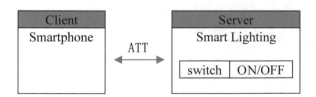

Fig. 5 The hierarchy of server

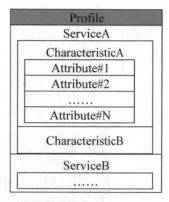

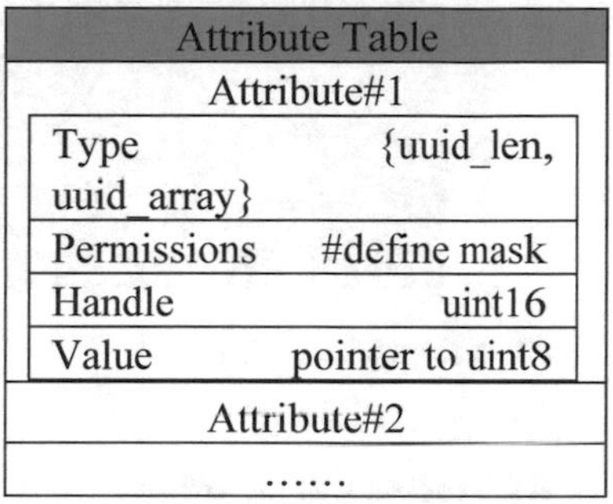

Fig. 6 Attribute table

an attribute is consisted of type, permissions, handle, and value. Permissions is not visible externally, but is a signal to the GATT Server.

4.2 Testbed Programming

Traditionally, in order to develop executable BLE software for this system, we may have to do the following works: (1) Create the files (switchService.c and switchService.h) for a new Project named SmartControl. (2) Register the service in initialization procedure and verify it. (3) Add a characteristic called switchValue to the attribute table and implement the handling of read request, write command. (4) Add API and callback handling to tie remote actions on the switch service to callbacks in our implementation.

However, SIG's Bluetooth Developer Studio (BDS), which is a graphical user interface for designing Bluetooth profiles and generating buildable source code which implements the services and characteristics that have been designed, is preferred in this paper. In the process of using the tool, some essential parameters have been configured as in Table 1.

Table 1 Settings of parameters

Profile name	SmartControl	Characteristic	① Name: switch
Service name	Lighting		② Read: mandatory
GAP settings	① Discovery mode: general		③ Write: mandatory
	② BR/EDR not supported		④ Field: name/state
	③ Include in packet: device name, flags		Format: unit8

Fig. 7 A snapshot of BDS

As expected, a snapshot of our BDS Profile looks like Fig. 7. Finally, we can generate code for the server, selecting the Texas Instruments plugin CC26XX BLE SDK.[1]

4.3 Smartphone Programming

Bluetooth Developer Studio cannot only generate code for the testbed, but also for the Android. The procedure is generally the same as testbed programming. One thing we should be cautious about is that the smartphone serves as the client as shown in Fig. 3. Hence, the code should generate for clients with the Android client plugin.[2]

5 Experimental Results and Evaluation

In order to verify the actual effect of the system, experiments were, respectively, carried out with Meizu (Android Platform) and iPhone 6S (iOS Platform) smartphone

---

[1]http://software-dl.ti.com/lprf/bds/ti-bds-plugin.html#bluetooth-developer-studio-plugin-downloads.

[2]https://www.bluetooth.com/develop-with-bluetooth/build/developer-kits/bluetooth-starter-kit.

Fig. 8 The success rate of the system

Fig. 9 The latency time of the system

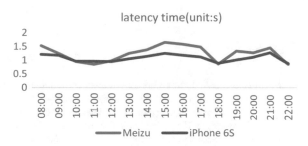

[7] at SA318, an ordinary classroom in school of software in CCIT. We use the smartphone to switch on the light and instantly off it every hour from 8:00 am to 10:00 pm lasting for one month. In the procedure, two metrics (success rate and latency time) were recorded in order to evaluate the performance [8, 9].

The success rate of controlling the bulb by Meizu and iPhone 6S smartphone is shown as Fig. 8. The average success rate by iPhone 6S is 96% surpassing 94% achieved by Meizu. On the whole, the success rate of the system is accepted.

The latency time of controlling the bulb by Meizu and iPhone 6S smartphone is shown in Fig. 9. The average latency time by iPhone 6S is 1.08 s, less than 1.24 s achieved by Meizu. Meanwhile, the latency time during lunch or dinner time is less than the rest. That is because when people go to dinner, the Wi-Fi indoor is less active. Thus, the BLE signal suffers less interference from Wi-Fi than the rest time, and the less interference is corresponding to less latency time.

6 Conclusion

This paper builds a smart connected lighting system based on Bluetooth Low Energy through reliable hardware structures and software designs. It adopts state-of-the-art CC2640R2F as Bluetooth LE module to realize the wireless real-time controlling of the lighting. The software programming of the system features the use of a graphical programming tool instead of writing much source code by yourself. According to the experimental data, it can accurately control the connected lighting and realize small latency time effectively. In addition, by using smartphones with Android or iOS, the system performance is slightly different. Above all, with strong applicability and high stability of the system, it plays a positive role in the emerging smart connected lighting market.

Acknowledgements This work is supported by Top-notch Academic Programs Project of Jiangsu Higher Education Institutions (TAPP) under Grant No. PPZY2015A090 and the Scientific Research Project of Changzhou College of Information Technology under Grant No. CXZK201701Y and "Laboratory of Application and Research of Artificial Intelligence Technology" under Grant No.PYPT201805Z.

References

1. Smart-connected lighting.https://www.silabs.com/documents/referenced/white-papers/smart-connected-lighting.pdf http://www.ncbi.nlm.nih.gov
2. Waterman.: Bluetooth history and the future of Bluetooth mesh networking. **107**, 145–147 (2017)
3. Raza, S., Misra, P., He, Z., Voigt, T.: Building the internet of things with bluetooth smart. Ad Hoc Netw. http://www.sciencedirect.com/science/article/pii/S1570870516302050 (2016)
4. Honkanen, M., Lappetelainen, A., Kivekas, K.: Low end extension for bluetooth. In: Proceedings of 2004 IEEE Radio and Wireless Conference, pp. 199–202
5. Gomez, C., Paradells, J.: Wireless home automation networks: a survey of architectures and technologies. IEEE Commun. Mag. **48**, 92–101 (2010)
6. Ko, J., Terzis, A., Dawson-Haggerty, S., Culler, D.E., Hui, J.W., Levis, P.: Connecting low-power and lossy networks to the internet. IEEE Commun. Mag. **49**(96–101), 13 (2011)
7. Gomez, C., Demirkol, I., Paradells, J.: Modeling the maximum throughput of Bluetooth low energy in an error-prone link. IEEE Commun. Lett. **15**, 1187–1189 (2011)
8. Silva, I., Guedes, L.A., Portugal, P., Vasques, F.: Reliability and availability evaluation of wireless sensor networks for industrial applications. Sensors **12**, 806–838 (2012)
9. Reis, C., Mahajan, R., Rodrig, M.: Measurement-based models of delivery and interference in static wireless networks. In: Proceedings of the 2006 Conference on Applications, Technologies, Architectures, and Protocols for Computer Communications (2006)

Part X
Technologies for Next-Generation Network Environments

Applying Wireless Sensors to Monitor the Air Conditioning Management System

Tung-Ming Liang, Yi-Nung Chung and Chao-Hsing Hsu

Abstract Because the fossil fuel is the major energy in this world, which let the global greenhouse effect problem become very serious. Therefore, how to save energy and have more efficient energy management technology play an important role. The demand for air conditioning has risen in accordance with the daily living condition of modern day life. The energy consumption of air conditioning is huge in daily life. In order to enhance the efficiency of air conditioning and save energy, a new technology which applies ice storage design and wireless monitoring unit is developed in this paper. Based on the experimental test, the sensor system and new air conditioning structure developed in this paper have better quality, higher efficiency, lower consumption, and save a lot of energy.

Keywords Global greenhouse effect · Air conditioning · Ice storage design · Wireless monitoring unit

1 Introduction

The demand for air conditioning has risen in accordance with the daily living condition. In addition, constant rising demands from the high technology and traditional manufacture industries have continued to increase the need for the cooling system and dramatic energy usage. Especially during the summertime, the air condition-

T.-M. Liang · Y.-N. Chung (✉)
Department of Electrical Engineering, National Changhua University of Education, Changhua 500, Taiwan
e-mail: ynchung@cc.ncue.edu.tw

T.-M. Liang
e-mail: v36625@gmail.com

C.-H. Hsu
Department of Information and Network Communications, Chienkuo Technology University, Changhua 500, Taiwan
e-mail: hsu@cc.ctu.edu.tw

© Springer Nature Singapore Pte Ltd. 2019
J.-S. Pan et al. (eds.), *Genetic and Evolutionary Computing*,
Advances in Intelligent Systems and Computing 834,
https://doi.org/10.1007/978-981-13-5841-8_42

ing system will require over one-third of the energy requirement in a typical power system. The natural energy resource is short in Taiwan. But the environment of Taiwan is subtropical high temperature, humid environment, nonstop industrial, and commercial growth, which forces Taiwan to import 98% of its energy requirement. If there are obstacles to the imported energy sources, it may drastically impact Taiwan's economy, and even the operation of its national security. Yet, renewable energy sources [1–3] such as solar power, wind power, and geothermal power are not able to provide a stable and specific beneficial power source at a large scale for the time being. So in this condition of constant inflating fuel and electricity cost, the aim is to lower operating and production cost, while not decreasing the production line and yield rate. In this study, a monitoring system is applied to monitor the power consumption during peak power period to avoid unnecessary energy waste.

In this paper, an effective energy saving design for air conditioning is developed. First, an ice storage system [4, 5] is designed for the air conditioning system. In the ice storage system, the chiller makes ice by using cheaper nighttime electricity and let the ice become as cold energy. The storage of ice type energy can be released to the air supply system during the daytime when it is necessary. In this study, a ZigBee sensor unit [6, 7] is also applied to monitor the conditions of space and air conditioning system. And then the management algorithm can adjust the control procedure to have the best arrangement to save the energy.

ZigBee is a wireless unit, which is used for short distance wireless transmission. ZigBee has its own unique wireless standard, so it can transmit signals among many different sensors and communicate with one another. Because it is short distance signal transmission equipment, if it needs to monitor longer distance place, it should use the deliver way to get to the final computer for analysis and access. This paper applies ZigBee wireless sensor network to monitor various environment information and to control the air conditioning situations. Based on the experiment test, this design has higher efficiency and it can save a lot of energy.

2 System Design

The air conditioning system [5, 8] is composed of the exchange process by the production of chiller through exchanging heat to water, and the exchange from water to the heat by the refrigerant. Structurally, it can be divided into the water side system and air side system. The air conditioner's water side system can further be divided into the chilling side system and the water cooling side system. Chilling side system utilizes the chiller pump and area pump [9, 10] to provide the power needed for transporting the chiller. The chiller is transported to each of the designated air conditioning areas to absorb the heat from the load side. The water cooling side system uses the cooling water pump as the source of power to transmit the heated chiller to the cooling tower for heat dissipation. The chilling side system is basically composed of the main unit, water pump, expansion tank, terminal device, chiller transport duct, etc. Primary components of the water cooling side system include

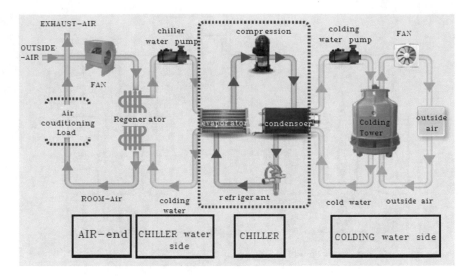

Fig. 1 Structure of air conditioning system

the cooling tower and cooling water pump. The air conditioning system is shown as Fig. 1, which is generally divided into four parts. They are chiller, pump and cooling tower, chilling system, and air conditioning box. In this system, there are a chiller recycling pump, which can carry the ice water to air conditioning box in the specific area.

The system has two kinds of developments which are primary-only system and primary/secondary side chilling system. Primary-only system is further divided into direct return system and reverse return water system. The primary-only system uses the chiller pump to circulate the chiller between the chiller main unit and the terminal devices in the system. Due to the difference in channel pipe length between the terminal heat exchange devices in the direct return system, the terminal pipes lose pressure greatly along different parts of the system. There is a lower volume flow for the pipes near the chiller source's terminal device as opposed to other parts. To balance the pressure within the channels and equalize the flow, a balance valve can be added to each pipe to regulate the amount of water or pressure depending on the needs. The reverse return water system partially extends the pipes to equalize the channel length in the system, by eliminating the pressure difference between the terminals due to pipe length. This system has a successful application in a case study in regards to a system that could not keep balance.

In a large air conditioning system, the chilling system shown as Fig. 2 is used. In this structure, the system is composed of primary side pump and secondary side pump. The primary side pump provides the chiller main unit's recycling water, and the secondary side pump provides the load side of the cycling water volume. There is a connecting channel duct between the two sides also.

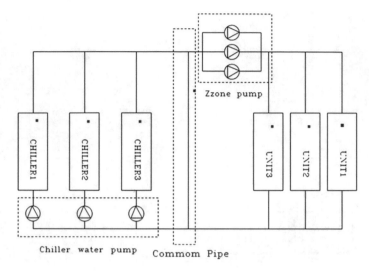

Fig. 2 Structure of the primary/secondary side chilling system

In air conditioning application, there is a common duct between the primary and secondary side of the chilling system. When the primary side is designated with a fixed volume, the secondary side will have variable streaming volume. The flow direction and volume within the common duct will vary in accordance with the secondary side's flow. When the load varying occurs, there are three possible scenarios within the common duct. The first scenario, when the primary side pump's chiller volume equals the required volume of the secondary side's flow volume, the water will flow from the primary side to the secondary side. The second scenario, when water volume provided by the primary side is greater than the secondary side, the chiller not only flows towards the secondary side but will also directly utilize the common duct to flow back toward the primary side. Thus, the chiller from the primary side will mix with the heated recycled water at the common duct allowing the water that will enter the chiller unit to decrease its temperature. The third scenario, when the primary side flow is lower than the secondary side. Thus, secondary side's reverted water flow will make intake of the heated recycled water to replenish the inadequate volume. So when the heated recycled water from the common duct flow back towards the secondary side's supply, it will cause the supply's temperature to rise, resulting in inadequacy of its cooling ability. This scenario should be controlled to avoid this type of occurrence.

In this study, the ice storage style air conditioning system is developed. The key advantage of this system is to use off-peak hours or night time to generate ice by using the ice chiller maker unit and keep the ice-cold energy in adequate storage. After the formation process from water to ice, a large amount of the potential cold energy is stored and can be later released during peak hour usage to address the system's need. This system can transfer the peak hour power usage time to the off-peak time to fully benefit from the cheaper power cost during those hours. Therefore, the user not only reduces the common power usage quota to generate cost saving from the usage transfer time to discounted hour but also increases air conditioning system reliability whenever there might be an energy shortage.

The ice storage style air conditioning system has two kinds of developments. One is full ice storage style system which is started by storing the ice energy for all the air conditioning's required energy for the daytime into the ice storage unit during the night time. Then the next day, it releases the stored ice-cold energy through melting during air conditioning hour to absorb all the heat energy in the room. The other is the part ice storage style system that is to store ice for part of the air conditioning capacity needs during the night or off-peak time. During the next day, air conditioning first uses the main chiller unit to reach the desired cold temperature, and then supplement the remaining required air conditioning by melting the stored ice. Vice versa, one can also first utilize the stored ice energy for cooling, and then activate the main chiller unit to reach the desired cooling temperature level.

In this study, the part ice storage style system is chosen. Under this mode, the main unit can operate at full loading during the daytime if it is necessary. During the air conditioning hours, the cold energy released from the melting ice in the ice storage tank will be the primary cold energy source. When the required cold capacity for the room is greater than the cold energy released by the ice storage, the main unit can then supplement the rest required air conditioning. However, if we like to save the most energy, the chiller main unit and cooling tower operating should be adjusted optimally. The relationship between the water cooling tower and the chiller main unit's energy consumption is relative. To maintain the lower temperature for the cooling water and reducing the main unit's energy consumption, the cooling tower fan must consume more energy to maintain the lower temperature for the water. There exists an optimal operating point between the chiller unit and cooling tower. This optimal point consists of the lowest value in both the cooling tower fan's air volume energy consumption increase rate and the chiller unit's energy consumption reduction rate.

However, the optimal point must be based on the main unit, cooling tower's performance line curve, and partial load rate. The system utilizes the data from the monthly outdoor average humidity and temperature according to each area, and the chiller unit and cooling tower performance curve. The best temperature adjustment setting can be set for both chiller and cooling water source to gain better energy saving effect. In order to detect the outdoor humidity and temperature for a long term, a wireless sensor denoted ZigBee unit is applied to this system. The ZigBee unit can monitor and record the data of outside humidity and temperature. The data of humidity and temperature detected by sensors can be transmitted to the receiving terminal device through wireless transmission. This information is also can transmit to a computer and present to users. Moreover, the system can use the software to access this information and to the control functions of entire system.

3   Experimental Test

In this study, the part ice storage style system is chosen. During the air conditioning hours, the cold energy released from the melting ice in the ice storage tank will be the primary source. When the required cold capacity for the room is greater than the cold energy released by the ice storage, the main unit can then supplement the rest required air conditioning. In order to set the best temperature adjustment point, a ZigBee wireless sensor unit is applied to monitor the humidity and temperature in this system. The data of humidity and temperature detected by sensors can be transmitted to the control system. The system uses the software to access the best set point and let the system operate under optimal condition.

Another experiment test is applied to the sensor network for cold and heat energy recycle utilization which is a total heat exchanger energy saving technique. For example, the heat enthalpy for the inside air and the outside air has a very large difference. A design of full heat exchanger is shown in Fig. 3. It induces the two air current source to engage in heat exchange during fresh air intake and vent, this could lead to conserving a large majority of the external air loading. In this system, a heat exchanger is used to absorb the air humidity and the heat of the external air. It allows the lower temperature and less humid external air to go in and the venting air through the total heat exchanger will carry the humidity and heat to the outside. It could save 30% of the energy consumption for the external air transformation requirement. During the fall and winter seasons, when the external temperature is lower than the set room temperature, one can consider intaking more external air

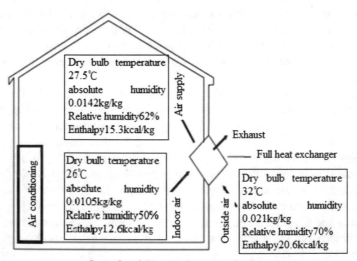

Cross-flow full heat exchanger application

Fig. 3   A design of full heat exchanger

to decrease the air conditioning loading. The temperature should not be the only external air factor to be considered, but also the value of enthalpy.

The concept of this study also can be applied to the building energy management system which is a centralized monitoring and management for a building's control system. This system provides the monitoring status, energy saving management, and the decentralized controls for a building, such as power distribution, illumination, elevator, air conditioning, heating, and plumbing component.

4 Conclusion

In this study, a sensor system and a new air conditioning structure are developed. The advantages of this design are using the monitoring performance data from the utilizing space, air conditioning box and cooling tower to increase energy conservation, optimal operating efficiency, and system maintenance. It also uses an energy monitor system to control each area's air conditioning temperature and power switching to avoid unnecessary energy waste. This research aims to provide users and energy management to have an effective control as well as good energy conservation concept. It also aims to promote energy conservation consensus among the general public.

Acknowledgements This work was supported by the Ministry of Science and Technology under Grant MOST 106-2221-E-018-028-.

References

1. Ji, J., He, H., Chow, T.: Distributed dynamic modeling and experimental study of PV evaporator in a PV/T solar-assisted heat pump. Int. J. Heat Mass Transf. **52**(5–6), 1365–1373 (2009)
2. Pei, G., Fu, H.D., Ji, J., Chow, T.T., Zhang, T.: Annual analysis of heat pipe PV/T systems for domestic hot water and electricity production. Energy Convers. Manag. **56**, 8–21 (2012)
3. Chang, L.-Y., Chang, J.-H., Chao, K.-H., Chung, Y.-N.: A low-cost high-performance inter-leaved inductor-coupled boost converter for fuel cells. MDPI Ener. **9**(10), 792–813 (2016)
4. Braun, J.E., Klein, S.A., Mitchell Beckman J.W., Beckman, W.A.: Applications of optimal control to chilled water system without storage ASHRAE Trans. **95**(1) (1989)
5. Halderman, J.D.: Automotive heating and air conditioning. In: The Amazon Book Review 1 Jan 2017
6. Ijaz, F., Adeel, A., Siddiqui, B.K.I., Lee, C.: Remote management and control system for LED based plant factory using ZigBee and internet. In: Advanced Communication Technology (ICACT) (2012)
7. Hsu, S.-P., Chung, Y.-N., Hsu, Y.-C.: Applying power management and wireless sensing system to plant factory. J. Inf. Hiding Multimed. Signal Process. **7**(2), 419–424 (2016)
8. Taylor, S.T.: Degrading chilled water plant delta-t causes and mitigation, ASHRAE Trans. **108**(1), (2002)
9. Fiorino, D.P.: How to raise chilled water temperature differentials, ASHRAE Trans. **108**(1), (2002)
10. Dewalt HVAC code reference: based on the 2015 international mechanical code, spiral bound version (DEWALT Series) (2015)

Research on the Method of Solving Danger Level of Cascading Trips in Power System

Hui-Qiong Deng, Bo-Lan Yang and Chao-Gang Li

Abstract A new method for calculating danger distance of cascading trips is proposed in this paper. The method is based on enhanced particle swarm optimization. An example on IEEE 39-system proves the effectiveness of the method.

Keywords Danger level · Cascading trips · Power system · Enhanced particle swarm optimization

1 Introduction

Electric power system is considered as critical infrastructure requiring the highest levels of reliability. However, large-scale blackout resulting from cascading trips imposes massive loss to the world [1]. Blackouts in Taiwan on August 25, 2017 caused varying degrees of traffic disruption that made many people to get trapped in elevators and affected 6.68 million households [2]. On March 4, 2011, a major blackout occurred in eastern Brazil. This event affected eight states in the northeast which is equivalent to an economic loss of approximately 60 million dollars [3]. On August 14, 2003, the biggest blackout called took place in the US and Canada of all time. In the same year, Italy suffered a 9/28 blackout [4, 5].

A lot of experts who research the difficulty mainly focus on two perspectives which are complex network theory and complex systems theory. Researches on relevant phenomena and problems of cascading failure have achieved many valuable research results [6].

H.-Q. Deng · B.-L. Yang (✉) · C.-G. Li
Fujian University of Technology, Fuzhou 350118, China
e-mail: 282289192@qq.com

H.-Q. Deng
e-mail: 1943387146@qq.com

C.-G. Li
e-mail: leechaogang@qq.com

© Springer Nature Singapore Pte Ltd. 2019
J.-S. Pan et al. (eds.), *Genetic and Evolutionary Computing*,
Advances in Intelligent Systems and Computing 834,
https://doi.org/10.1007/978-981-13-5841-8_43

411

Particle swarm optimization (PSO) is simple, fast, and easy to implement. It is suitable for solving complex optimization problems that are difficult for classical optimization algorithms. PSO has been applied in fields of power system. These fields include reactive power optimization of wind farm grid-connected power system based on improved quantum, reactive power optimization of power system, and the unit commitment problem [7–9].

In this paper, enhanced particle swarm optimization (EPSO) is used to calculate danger distance. Danger distance is the minimum Euclidean distance between the current state after initial fault and the critical state [10]. Based on nodal injection, danger distance will be obtained by EPSO and IEEE 39-node network will validate the method.

2 Brief Introduction About the Model

A two-node network that is supposed is shown in Fig. 1. The circle indicates the set of critical state [10] of power system. Point A in the circle is supposed the current state after initial failure. Then, point B, C, and D are in the critical state. n_1 and n_2 are the horizontal longitudinal coordinates which represent the nodal injection of the two nodes in the system. Active power is only considered, and the unit of which is kW. Among the three points on the circle, the distance between A and B is the shortest. Therefore, the distance between the current state and the critical state can be expressed as

$$\min D = \sqrt{(P_1 - P_1')^2 + (P_2 - P_2')^2},\qquad(1)$$

where $\min D$ is the danger distance in the two-node system. Both P_1, P_1' are nodal injection of n_1, the former one under critical state and the later one at the current state. P_2, P_2' are similar to the above.

Fig. 1 A two-node network [1]

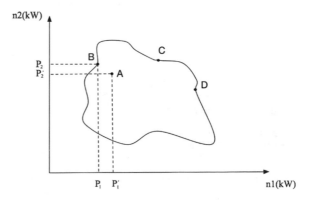

Considered reactive power, danger distance in multi-node network can be further extended to

$$\min D = \sqrt{\sum_{m=1}^{N} (P_m - P'_m)^2 + (Q_m - Q'_m)^2}, \tag{2}$$

where m is the number of branches and N is the total number of branches except the initial fault one in a multi-node system [10].

Except for the initial fault line, whether any other branches are disconnected can be inferred from [11]

$$\boldsymbol{J} = \mathrm{diag}(z_{1 \cdot dist}, \ldots, z_{j \cdot dist}, \ldots z_{l \cdot dist}), \mathbf{j} \neq \mathbf{i}, \mathbf{j} = 1, 2, \ldots, 1, \tag{3}$$

where i is the number of initial fault line and l is the total number of branches [11]. Once $|\boldsymbol{J}| = 0$, cascading trips occur.

$$z_{1 \cdot dist} = \left| z_{j \cdot lim} \right| - \left| z_j \right|. \tag{4}$$

In Eq. (4), $z_{1 \cdot dist} < 0$ in the normal operating state. Therefore, the critical condition is as follows:

$$\begin{cases} |\boldsymbol{J}| = 0 \\ z_{1 \cdot dist} \leq 0, i \neq j, j = 1, 2, \ldots, l \end{cases}. \tag{5}$$

In normal operating state, equality and inequality constraints can be summarized below:

$$\begin{cases} h^i(x^i, u^i) = 0 \\ h^0(x^0, u^0) = 0, \\ g^0(x^0, u^0) \leq 0 \end{cases} \tag{6}$$

where the first equality constraint is in the state after initial fault. The second equality and the inequality constraints are in normal operating state. Combining Eqs. (2)–(6), the model for solving danger distance can be constructed as follows:

$$\begin{cases} \min D = \sqrt{\sum_{m=1}^{N} (P_m - P'_m)^2 + (Q_m - Q'_m)^2} \\ |\boldsymbol{J}| = 0 \\ h^i(x^i, u^i) = 0 \\ h^0(x^0, u^0) = 0 \\ g^0(x^0, u^0) \leq 0 \\ z_{j \cdot dist} \leq 0, j \neq i, 1, 2, \ldots, l \end{cases}. \tag{7}$$

3 Method for Solving the Model

In this part, the method based on enhanced particle swarm optimization (EPSO) is applied for solving the model mentioned in Part 2. The settings of penalty function and fitness function are similar to Ref. [10], so they will not be reiterated here.

3.1 The Description of EPSO

Compared with particle swarm optimization, the accuracy and global searching ability of EPSO will improve. Flow migration technique is applied in EPSO. The technique is used to determine the value of p_r which is compared with r set in advance. When $p_r \geq r$, the particles are migrated in a wide range and the position is adjusted by v_{rand}. When $p_r < r$, the particles are migrated in a small range and the position is adjusted by $v_{rand} + v_{rand}^\varepsilon$ [12].

The iterative equation is as follows [12]:

$$\begin{cases} v^{k+1} = wv^k + c_1 r_1 \left(p_{best} - x^k \right) + c_2 r_2 \left(g_{best} - x^k \right) \\ \quad x^{k+1} = \begin{cases} \tilde{x}^{k+1}, r\left(p_r \& x^{k+1}\right) x^k + v^{k+1} \\ \quad x^k + v^{k+1}, others \end{cases} \end{cases} . \tag{8}$$

In Eq. (8), x and v, respectively, represent the position and velocity of the particle. The implication of parameters which is not mentioned in Ref. [10] only is explained. r is a random number between 0 and 1. k is the number of iterations. $\tilde{x}^{k+1}$ is defined below:

$$\tilde{x}^{k+1} = \max\left\{x^k + v_{rand}, x^k + v_{rand} + v_{rand}^\varepsilon\right\}, \tag{9}$$

where v_{rand} is a random number between v_{min} and v_{max}. And, v_{rand}^ε is a random number between εv_{min} and εv_{max}. The value of ε is between 0.01 and 0.1. p_r is defined as

$$p_r = p_0(k_{max} - k)/k_{max}. \tag{10}$$

In Eq. (10), p_r represents the mobility of the kth generation particles. p_0 is the initial value of p_r.k_{max} is the maximum number of iteration. The setting of p_0 is a random number bigger than 0 and no bigger than 0.5.

3.2   The Method Based on EPSO for Solving the Model

The overview of EPSO has been written in 3.1, and then it is specifically introduced to solve the model. The specific steps are as follows:

1. Generate new ground-state data based on original ground-state data. Monte Carlo method is randomly used to generate new ground-state data which includes load power and power supply of PQ node and load power of PV node.
2. Test the convergence of the new generated ground-state data, or regenerate it if power flow is unsolvable. The test is divided into two steps. First, testing whether the power flow is of convergence or not. Second, judging power flow is of convergence after the initial failure. The program will proceed if the new ground-state data passes the tests, otherwise assign the original ground-state data as the new one.
3. Generate particles swarms for optimization solution. Nodal injection of PQ node and PV node satisfied condition of convergence is stored in matrix.
4. Initialize randomly the velocity of each particle.
5. Establish a fitness function [10].

$$
\begin{aligned}
fitness(x, u) = 1/\{ & D(x, u) + M_1[|min(0, V - V_{min})|^2 + |min(0, V_{max} - V)|^2]+ \\
& M_2[|min(0, P - P_{min})|^2 + |min(0, P_{max} - P)|^2]+ \\
& M_3[|min(0, Q - Q_{min})|^2 + |min(0, Q_{max} - Q)|^2]+ \\
& M_4[|min(0, LP - LP_{min})|^2 + |min(0, LP_{max} - LP)|^2]+ \\
& M_5|J|^2 + M_6 \sum_{j=1}^{n-1} |min(0, w_{j \cdot dist})|^2 \}.
\end{aligned}
\tag{11}
$$

Once the value of fitness takes the maximum, danger distance will be obtained.

6. Evaluate the fitness of each particle.
7. Update constantly velocity and position according to Eq. (8). In position updating, retaining the maximum fitness one.
8. Check new particles and form formal particle swarms.
9. Stop the loop and output the result after finishing the iteration. But if it does not, return to step 6 keeping on finding out the optimal solution.

4   Result

In this part, IEEE 39-system is used to verify the validation and effectiveness of the model and the method. The connection diagram is shown in Fig. 2. Initial fault branch is set L_{16-24}. In this procedure, the per-unit which uses virtual data is applied. And, base capacity is 100 MVA and reference voltage is 115 kV.

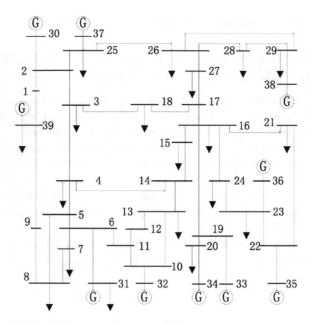

Fig. 2 The connection diagram of the IEEE 39-system

After running the procedure, the graph results will be obtained. The graphs are shown as follows:

Figure 3 takes the number of iteration as the abscissa and the value of fitness as the ordinates. Figure 4 indicates the process of min*D* and the abscissa is the same as above (Table 1).

The table shows the danger distance is 1380.2 and the corresponding fitness is 2.16×10^{-75}. In this program, there is no better optimum after running many times. Compared with PSO, EPSO's ability of accuracy and global research are stronger and the optimum is better. The example can prove the rationality and effectiveness of the method.

5 Conclusion

In nutshell, the method based for danger level of cascading trips was successfully designed. The Euclidean distance of the model can be optimized by EPSO. In this paper, another method is proposed to solve the model. And, the method can be advised to the dispatcher. Therefore, the relevant measures can be taken to prevent the occurrence of cascading trips.

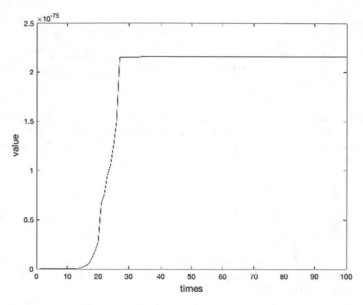

Fig. 3 The process to generate *fitness*

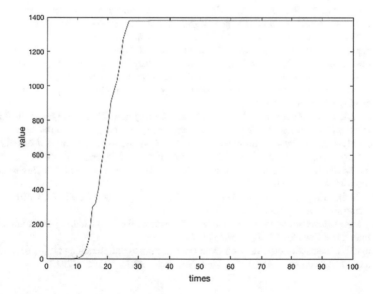

Fig. 4 The process to generate min*D*

Table 1 The running result
of EPSO

	The running results of PSO	
	min$D$	*Fitness*
Value	1380.2	2.16×10^{-75}

Acknowledgements This research was financially supported by Fujian Provincial Natural Science Foundation of China under the grant 2015J01630, Doctoral Research Foundation of Fujian University of Technology under the grant GY-Z13104, and Scientific Research and Development Foundation of Fujian University of Technology under the grant GY-Z17149.

References

1. Guo, L.Q., Liang, C.: Monotonicity properties and spectral characterization of power redistribution in cascading failure. In: Fifty-Fifth Annual Allerton Conference Allerton House, UIUC, Illinois, pp. 918–925 (2017)
2. Wang, P., Wang, Y.: Who is the One to Take Responsibility for Blackout in Taiwan, Relations Across Taiwan Straits, vol. 36(16) (2017)
3. Lin, W., Tang, Y., Sun, H., et al.: Blackout in Brazil power grid on February 4 2011 and inspirations for stable operation of power grid. Autom. Electr. Power Syst. **35**(9), 1–5 (2011)
4. Veremyev, A., Sorokin, A., Boginski, V., et al.: Minimum vertex cover problem for coupled interdependent networks with cascading failures. Eur. J. Oper. Res. **232**(3), 499–511 (2014)
5. Rosato, V., Issacharoff, L., Tiriticc, F., et al.: Modelling interdependent infrastructures using interacting dynamical models. Int. J. Critical Infrastruct. **4**(1–2), 63–79 (2008)
6. Hu, P., Mei, T., Fan, W.L.: Cascading failure forecast of complex power grids: a review. Sci. Sinica **47**(4), 355–363 (2017)
7. Guo, H.: Reactive Power Optimization of Power System with Wind Farm based on Quantum-Particle Swarm Optimization. Nanjing University of Posts and Telecommunications
8. Que, S.M.: Reactive Power Optimization of Power System Based on Improved Particle Swarm Optimization Algorithm. Nanjing University of Posts and Telecommunications
9. Wu, H.H.: Study of The Unit Commitment Problem Based on Improved Particle Swarm Optimization. Hunan University
10. Deng, H.Q., Yang, B.L.: A New Method of Calculating Danger Level of Cascading Failure Occurring in Power System
11. Deng, H.Q.: Research on the safety margin of power network considering cascading tripping. J. Fujian Univ. Technol. **14**(3), 255–261 (2016)
12. Huang, P., Zhang, Y., Zeng, H.: Improved particle swarm optimization solving economic dispatch of electric power. J. Huazhong Univ. Sci. Tech. **38**(1), 121–124(2010)

Service Migration Based on Software-Defined Networking

Chin-Shiuh Shieh, Cheng-Ta Lu, Yu-Chun Liu and Yuh-Chung Lin

Abstract Mobile wireless networking is a definite trend for communication networks. Mobility support, such as mobile IP, at client side is well developed. However, there is limited effort paid for the server side. Quality of Service (QoS) degrades as a mobile client moves farther away from its original server. To this problem, a new service paradigm, named service migration, is proposed in this article. The idea is to migrate an ongoing service from distant server to a new server in proximity. Both individual user and the entire network environment will benefit from the service migration. Service migration can be implemented at different layers. In this study, we investigate the possibility of implementing service migration with Software-Defined Networking (SDN). SDN has been distinguished for its capability in traffic monitoring and flexibility in network administration. A fully functional prototype of service migration based on SDN for streaming service is implemented. Such a prototype validates the feasibility of the proposed service framework.

Keywords Service migration · Software-defined networking

1 Introduction

With the advancement of the technology for mobile wireless communication, more and more users nowadays access the Internet via mobile devices. Traditionally, the mobile client's request is directed to the nearest server in the beginning. When the

C.-S. Shieh (✉) · C.-T. Lu
Department of Electronic Engineering, National Kaohsiung
University of Applied Sciences, Kaohsiung, Taiwan
e-mail: csshieh@cckuas.edu.tw

Y.-C. Liu
Department of Information Application and Management, Tajen University,
Pingtung, Taiwan

Y.-C. Lin
College of Information Science and Engineering, Fujian
University of Technology, Fuzhou, China

© Springer Nature Singapore Pte Ltd. 2019
J.-S. Pan et al. (eds.), *Genetic and Evolutionary Computing*,
Advances in Intelligent Systems and Computing 834,
https://doi.org/10.1007/978-981-13-5841-8_44

client roams into different network domains with the mobile user moving around, mobility support, such as mobile IP [1] technology, helps to maintain the connection, as well as the service, with the original server. However, as the client moves farther away from the original server, the connection will have longer "weighted network distance" and results in degradation in QoS.

In conventional network environment, the network topology is decided at the time of deployment and remains fixed. However, the fixed configuration of network topology may fail to accommodate dynamic traffic changes since the network environment is a highly time-variant system. Furthermore, the available bandwidth for different routes varies significantly from time to time depending on traffic loads. For better provision of QoS, it is highly desirable to have more control over the network topology as well as the packet routing. Then, we can locate, for an ongoing service, a route with higher bandwidth and lower delay.

To these problems, we pioneer the concept of service migration [2] with software-defined networking [3] in this study. The proposed architecture takes the advantage of SDN's flexibility in network administration. It tries to locate a best route for an ongoing service and helps an ongoing service migrate from the distant server to the proximity server. With service migration, every individual user can enjoy supreme QoS. The entire network environment can also benefit from service migration in terms of better resource utilization.

We implement a functional prototype for streaming service to validate the feasibility of the proposed architecture. In our implementation, we adopt OpenFlow [4] as our SDN environment and Ryu [5] as the SDN controller. The system is designed to deliver streaming service with transparent service migration to client ends. It monitors the traffics of the entire environment and conducts bandwidth estimation upon the detection of client site handover. It invokes service migration in due situation and updates the routing paths accordingly to facilitate the service migration.

2 Software-Defined Network

Software-defined networking, as a new technology for network virtualization, has received considerable attention since 2014. As the name implies, in SDN, the network topology and packet routing are governed by software. According to Open Networking Foundation [6], the fundamental idea of SDN is the separation of control plane and data plane, as shown in Fig. 1.

In traditional networks, routers and switches operate according to their own individual software and hardware configurations. Control planes and data planes are mixed and fused together resulting in integrating and coordinating routing devices extremely difficult. On the other hand, in SDN, control plane is separated from data plane and SDN controller is introduced to control the forwarding behaviors of all switches in the same domain. Switches are now merely responsible for the packet forwarding. With SDN, when network topology changed, there will be no more tangling of network cables. The administrator simply commands the controller to send adequate forwarding rules to individual switches. It is not necessary to configure all

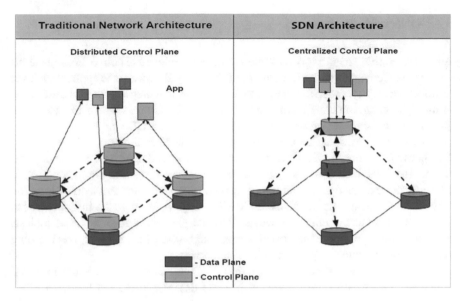

Fig. 1 Separation of control and data plans in SDN

the routers and switches one by one. As a result, high level of flexibility and quick deployment can be achieved by SDN.

In SDN, all network devices are controlled by the SDN controller in a centralized manner. SDN controller is responsible for the maintenance of network structure and provides useful Application Programming Interface (APIs) to upper layer. With these APIs, upper layer applications can monitor and control the entire network, as illustrated in Fig. 2. Various applications become possible, such as network security, virtual segmentation, load balancing, QoS support, and so forth.

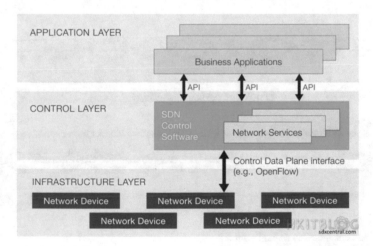

Fig. 2 Layered structure of SDN

3 Service Migration

According to IETF RFC 3466, multiple servers are deployed to deliver homogeneous content in a Content Distribution Network (CDN) [7]. By way of the application layer protocol, servers are grouped together to form a CDN in which a content is distributed to multiple, geographically distinct nodes. Once the end users would like to access the content, the connection will be directed to the nearest CDN node.

As shown in Fig. 3, at first the mobile client connects to the nearest server. When time goes by, the mobile client may move to other areas and far away from the original server. The "weighted network distance" becomes larger, resulting in lower QoS. In addition, the extended connection will also compete with the cross-flow along the way. To solve the problem, one of the feasible ways is to move the initial server to a location close to the client. However, it is not physically feasible. Nonetheless, this observation prompted the idea of service migration which is to migrate ongoing services from remote servers to neighboring servers.

In the framework of service migration [8], three main modules are included which are the Proximity Management Module (PMM), Migration Decision Mod-

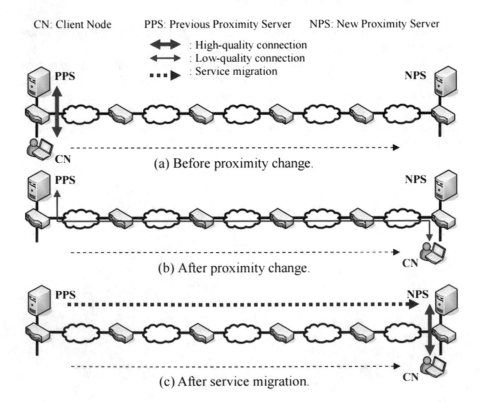

Fig. 3 The concept of service migration

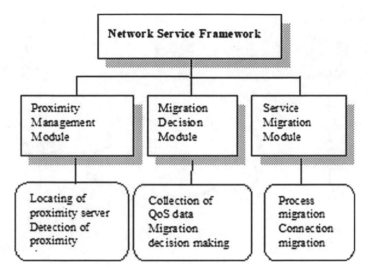

Fig. 4 Functional blocks of the new network service framework

ule (MDM), and Service Migration Module (SMM). They will work closely with others to contribute to the tentative goals. Each module has its responsibility, as shown in Fig. 4. First, the proximity management module is responsible for locating the location of the proximity server and detecting the variation during approaching to the target server. The anycast technology or a portal node of the content distribution system can be utilized to locate the proximity server. Two most common reasons which cause the proximity changes are network topology changes and client mobility. To detect network topology changes, we can measure the round-trip time of the packets and the number of hops they traveled which are the main indicators for detecting changes in proximity. For the latter, Mobile IP is able to detect changes in its access domain. The migration decision module is responsible for monitoring the quality of service and determining server migration based on the data reported from PMM. Following parameters should be considered which are network bandwidth, network latency, server capacity, service type, service remaining time, and connection fees. The service migration module is designed to perform the actual process and connection moves, which will be explained in detail in the subsequent sections.

4 System Design

Figure 5 is a typical scenario intended by the proposed service framework. A mobile client is initially connected to the Streaming Server 1. As it moves to a new domain covered by Base Station 2, the QoS degrades and then service migration is invoked.

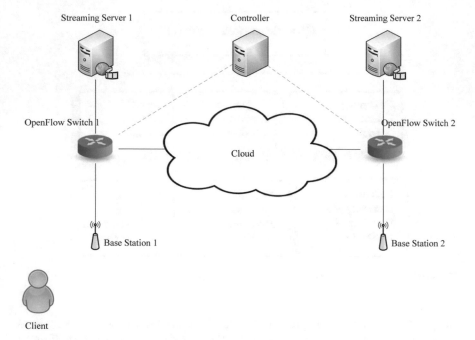

Fig. 5 A typical scenario for service migration

The Streaming Server 2 will take over and continue the delivery of streaming service to the client.

Figure 6 is the signaling diagram. Functional modules implemented on the controller cooperate closely to achieve the design objective. For PMM, as a client approaches a base station, switch will report to the controller and let it keep track of client's location. Controller passes client's location with information of its nearest server to MDM for migration decision-making.

To make a judicious decision, MDM need to know available bandwidth on all links. Fortunately, SDN has a built-in mechanism for traffic monitoring. As shown in Fig. 6, the controller obtains the statistics information of incoming and outgoing packets on each port for each flow entry by querying switches periodically. Therefore, the MDM can be aware of the loading of each link, and has a global view on the entire network. SMM performs the actual migration, as shown in Fig. 6. As Switch 2 received a packet from newly arrived client, it sends the packet info to the controller. The MDM will choose a best route for the client based on its knowledge on the network environment. Once service migration is confirmed, the controller will add a new flow entry on Switch 2 to direct packets from Streaming Server 2 to the client. It then commands Switch 1 to delete the connection between the Streaming Server 1 and the client.

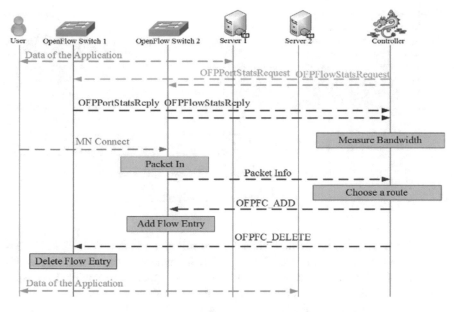

Fig. 6 Signaling diagram for bandwidth estimation and service migration

5 Implementation and Measurement

The hardware and software used to implement the topology in Fig. 5 are listed in Table 1. As a mobile client moves farther away from its original server, the user will experience increased delay and decreased bandwidth. To create a scenario with insufficient bandwidth, we limit the bandwidth of the link from the Switch 1 to the Switch 2—150 KB using Linux's built-in traffic control [9]. In addition, to simulate third-party traffic, we inject around 70 KB traffic into the link from the Switch 1 to the Switch 2. The streaming service and the third-party traffic will compete with each other for the available bandwidth.

Table 1 Table captions should be placed above the tables

Item	OS	Hardware/software
OpenFlow controller	Ubuntu 14.04	Ryu 3.19
OpenFlow switch	Ubuntu 14.04	OpenvSwitch 2.0.0 [11]
Streaming server	Ubuntu 14.04	VLC
Mobile device	Windows 7	Asus x450v
Test clip		AVI (>90 KBps)

Fig. 7 Data flows on ports of Switch 1 without service migration

First, we examine the results when service migration is disabled. Figure 7 represents the data flow on ports of Switch 1. Before handover, the streaming packets are transmitted on the path: Streaming Server 1 → OpenFlow Switch 1 → Base Station 1 → Client. There is third-party traffic of 70 KB/s from OpenFlow Switch 1 to OpenFlow Switch 2. The handover takes place at around 133 s. After handover, the streaming packets are transmitted on the path: Streaming Server 1 → OpenFlow Switch 1 → OpenFlow Switch 2 → Base Station 2 → Client.

Apparently, without service migration, once handover happened, all traffics flow on the link between Switch 1 and Switch 2. It may cause an overload on the link and packet drops. QoS cannot be guaranteed.

Figure 8 presents data flow of switches' ports with service migration enabled. Before handover, the streaming packets are transmitted on the path: Streaming Server 1 → OpenFlow Switch 1 → Base Station 1 → Client. The handover takes place at around 122 s and the service migration is invoked. After handover and service migration, the streaming packets are transmitted on the path: Streaming Server 2 → OpenFlow Switch 2 → Base Station 2 → Client. Streaming Server 2 takes over Streaming Server 1 and continues to deliver the service to the Client. Both streaming service and third party enjoy supreme QoS, and no longer compete with each other for limited bandwidth. This is a good example demonstrating the main appeal of service migration.

Figure 9 shows the data flows at the client with service migration enabled. It can be seen that, before handover and service migration, packets are delivered from Streaming Server 1 and from Streaming Server 2 afterward. There is a small interval within which the client receives packets from both servers, as shown in Fig. 9. In average, the service migration process takes only 75.45 ms. However, due to the time stamp and sequence number mechanisms in Real-time Transport Protocol (RTP) [10], player takes 3–5 s to reorganize incoming packets and continues the playback.

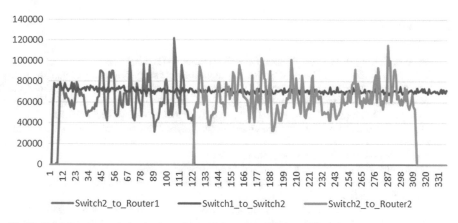

Fig. 8 Data flows at switches' ports when service migration is enabled

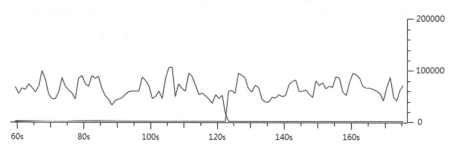

Fig. 9 Data flows at the client when service migration is enabled

6 Conclusions

Mobile wireless networking is well deployed recently. Users nowadays are used to accessing the Internet anywhere, anytime, and on the move. Content distribution is one of the most important applications of the Internet. CDN have offered certain solutions to this problem. However, in the context of mobile networking, moving away from the initial server induces QoS degradation. To address this problem, we propose a new service paradigm, named service migration. The idea is to migrate an ongoing service from distant server to a new server in proximity. Taking advantages of SDN's capability in traffic monitoring and flexibility in network administration, service can be achieved within SDN's environment, as reported in this article.

Acknowledgements We are grateful to the support from the Ministry of Science and Technology, Taiwan, ROC, under the grant number MOST 106-2221-E-151-007.

428 C.-S. Shieh et al.

References

1. Perkins, C. (ed.) IP Mobility Support for IPv4, RFC-3344 (2002)
2. Lai, W.K., Shieh, C.-S., Hsu, C.-W.: Service migration—a new paradigm for content distribution systems. In: Proceedings of the Third International Conference on Communications and Networking in China (ChinaCom-2008), pp. 34–38, Hangzhou, China, 25–27 Aug 2008
3. Nadeau, T.D., Gray, K.: SDN: Software Defined Networks. O'Reilly Media (2013)
4. Hu, F.: Network Innovation Through OpenFlow and SDN: Principles and Design. CRC Press (2014)
5. Ryu SDN Framework. https://osrg.github.io/ryu
6. Open Networking Foundation. https://www.opennetworking.org
7. Kurose, J.F., Ross, K.W.: Computer Networking: A Top-Down Approach Featuring the Internet, 6th ed. Pearson (2012)
8. Shieh, C.-S., Gu, H.-X., Yan, J.-Y., Horng, M.-F.: SCTP-based service migration for VOD service. In: Proceedings of the Twelfth International Conference on Digital Information Management, pp. 165–168, Fukuoka, Japan, 12–14 Sept 2017
9. Panagiotis Vouzis, How to Use the Linux Traffic Control. https://netbeez.net/blog/how-to-use-the-linux-traffic-control
10. Schulzrinne, H., Casner, S., Frederick, R., Jacobson, V.: RTP: A Transport Protocol for Real-Time Applications, RFC-3550 (2003)
11. Open vSwitch. https://www.openvswitch.org

Discussions About Construction of Digital Campus in the Development of Higher Vocational Colleges

Haiying Zhou

Abstract In higher vocational colleges, construction of digital campus system, which includes teaching, scientific researches, management, service, and many other functions, is one of the most important issues. We should focus on four aspects which are creating a new concept of construction, innovating a new mode of educating people, bringing birth a smart management mode and carrying out a development mechanism, and regard them not only as goals, but also as mainly means of achieving goals to achieve a new kind of smart digital campus.

Keywords Discussions · Digital campus · Construction · Mode and mechanism

1 Introduction

With the rapid development of higher vocational education and the popularization of information technology, the construction of digital campus has become a basic project in higher vocational colleges. It has become an effective guarantee for improving the quality of teaching and management efficiency [1], and also it has been already an important part of the training of talents in higher vocational colleges. The level of the digital campus construction directly reflects the degree of school education information, and also reflects the level of the school decision-makers for the development trend of modern higher vocational education. It is also one of the important indicators to measure the overall level of teaching and scientific research and the ability of running a school. The scientific construction of digital campus system has very important practical significance for higher vocational colleges to realize education information, information systematization, resource sharing, process standardization, personnel reduction, and so on.

H. Zhou (✉)
Changzhou College of Information Technology, Changzhou 213164, China
e-mail: 80557277@qq.com

© Springer Nature Singapore Pte Ltd. 2019
J.-S. Pan et al. (eds.), *Genetic and Evolutionary Computing*,
Advances in Intelligent Systems and Computing 834,
https://doi.org/10.1007/978-981-13-5841-8_45

2   Why We Need to Build Digital Campus

2.1   It Can Effectively Solve the Problem of Insufficient Resources and Single Way in Curriculum Teaching

First, construction of digital campus can create a teaching resource center. The digital resource should include online open courses, micro-courses, professional teaching resource database, virtual simulation training center, digital teaching materials, and so on. Second, construction of digital campus can create a network-teaching platform. We can improve the growth space of teachers and students, realize the connection between local learning platform and cloud-based courses [2], set up mobile learning client, and realize the coverage of smart education services on mobile phones and tablets, so as to provide more for students with convenient, flexible, and personalized learning conditions. Third, construction of digital campus can optimize teachers' teaching management. Through homework, interaction, questionnaires, data statistics, and other functions, we can enable teachers to understand the real-time learning situation, improve or reduce the difficulty of the course in time, adjust the progress of learning, and also we can give students a relaxed learning environment. Students can also directly send difficult points or doubts to the teacher and feedback information.

2.2   It Can Effectively Solve the Problem of Unclear Orientation and Unclear Direction in the Development of Teachers [3]

First, the digital campus can show teachers' achievements. Through the collection of basic data, the extraction of process data, and the application of data mining technology, the construction of digital campus can clearly show teachers' annual work performance, and provide guidance and auxiliary decision-making for teachers' development needs. Second, the digital campus can build the space for teachers to communicate. By creating a space to show the content of teachers' life and promote the interaction among all the teachers, we can make teachers have a strong sense of belonging both at work and in life. Teachers can consult their predecessors and communicate with their peers on their career development at any time. Third, the digital campus can set up a teacher-training platform. The construction of digital campus can provide a series of online training application channels for teachers by improving their professional level, teaching ability, scientific research ability, and thesis writing ability [4].

2.3 It Can Effectively Solve the Problems of Vague Responsibility and Low Efficiency in Teaching Management

First, the platform can effectively sort out key businesses. By rebuilding the business process, drawing the business flow chart, and visualizing, the digital campus construction can help to solve the problem of unclear workflow. Second, information supervision can be carried out by means of SMS and business schedule. The applicants can view the progress of business process in real time. Digital campus construction can help to solve the problem of opaque progress and greatly improve the satisfaction of teachers and students. Third, the platform can set up online business hall. Through the digital campus construction, each application module is interconnected and interoperable, which can form an online and off-line collaborative office system with instructions, multiple system responses, and multiple task function modules simultaneously [5], thus greatly improving the efficiency of the management department.

3 Main Contents of Digital Campus Construction in Higher Vocational Colleges

The digital campus system of higher vocational colleges includes teaching, scientific researches, management, service, and many other functions. Absolutely all the various systems are interconnected and interact with each other. Therefore, the construction of the digital campus system is mainly to realize the full digitization of the four fields: digitalization of the field of education and teaching, the digitalization of scientific research and technology, the digitalization of the field of organization and management, the digitalization of the social service field, the understanding of the relationship between the four systems, and the final realization of the four systems.

3.1 Digitization in the Field of Education and Teaching

The information platform is integrated with the digitalization of the field of education and teaching. The core of the construction of the digital campus is the education and teaching project based on the network technology [6]. The main body of education and teaching is all the students in the school and the students who receive the network education. The construction of the digital campus system is to create an ideal learning environment suitable for most students under the network, and to build an ideal learning atmosphere. This environment and atmosphere cannot be restricted by time and not restricted by the region. It is based on the learner's willingness to learn and enrich the diversity of interactive parties. With information platform, all the teachers, staffs, and students could share all kinds of resources in varied ways of interaction, and in quick and flexible access. This way of education, especially the network education system of multimedia teaching, can fully explore the main role of the learners and give full play to their independent creativity and exploration. The learners can make full use of the digital library and various network resources, search for the information they need and study, solve the related problems to achieve the learning goals, and the learners can also communicate with other schools, teachers, and scholars in other schools and other fields through the Internet [7]. And so on, a wide range of cooperative exploration and discussion exchange.

3.2 Digitalization in the Field of Organization and Management

The school is a small concentrated society. The effective and efficient operation of the school requires a high level of organization and management system. The development of the digital campus system needs to establish a set of high-end and perfect organizational management system, including institutions, personnel, finance, assets, publicity, enrollment, employment, foreign affairs, and other aspects, in order to ensure the normal operation of the school. The realization of digitalization in the field of organization and management will effectively solve the problem of information island. It cannot only achieve effective sharing of information, change the segmentation of functional departments, but also unifies the process of business, realizes a more simple and convenient coordination, gradually reduces the burden of management personnel, and improves the efficiency of management personnel. The realization of digital in the field of organization and management will fully realize office automation and paperless. The teachers and students in higher vocational colleges no longer have busy telephone communication, transfer notifications, and files. They can complete the notification and arrange the meeting through the network. The circulation documents, the publication documents, the material and so on, the traditional paper notice, documents, and so on will be replaced by the network information publishing system, the circulate system, the office mail, and so on. They will be replaced by the network information system, the circulate system, and the office mail. The digitalization of the field of organization and management can be more efficient and convenient for the teaching and research service of the school, and save more manpower, material, and financial resources for the school, and make less staff make more things in a shorter time and achieve twice the result with half the effort.

3.3 Digitalization in the Field of Science and Scientific Research

The mode and mechanism of scientific research in higher vocational colleges will be directly influenced by the construction of digital campus. It is the requirement of the information age to realize the sharing of scientific research resources between colleges and universities through the Internet. On the one hand, resources sharing can make scientific research information spread faster, facilitate interscholar and international academic exchanges and academic communication, and make effective research on online cooperation among colleges and universities; on the other hand, the higher vocational colleges can make use of the network platform to promote them. The mutual transformation between scientific and technological achievements and the field of education and teaching, so as to achieve the industrialization and marketization of scientific and technological achievements, and to improve the innovation level and radiation scope of scientific research and technology. The digitalization of internal scientific research and technological fields in higher vocational colleges is mostly reflected in the establishment of digital libraries and the establishment of professional websites and discussion platforms.

3.4 Digitalization in the Field of Social Service

The service field of higher vocational colleges mainly refers to the public service system and the campus community service system. The public service system of higher vocational colleges refers to a series of complete supporting service system provided for the school scientific research and production. It mainly includes network and education technology service, student work system service, museum service, library service, art museum service, and so on. The public service system is fully digitized, which fully indicates that the public service of the school has achieved a greater degree of computerization, networking, and intellectualization, whether it is the service content or the service form. Campus community service refers to the logistics system services such as life, leisure and entertainment, public infrastructure equipment maintenance and management service, campus security system service, financial management service, and stylistic entertainment services [8]. The campus community service system has the intervention and promotion of the computer network, its service will be more comprehensive, more convenient, and more efficient, and the digitalization of the service system will further promote the socialization of the campus community service.

4 How to Build the Smart Digital Campus

4.1 We Should Focus on Creating a New Concept of Construction

Based on cloud computing and large data technology, we should make the overall plan from three aspects, the business architecture, technical architecture, and data architecture. We should set up the development of unified data standards and application interface specifications. On the information facilities, we should get the original infrastructure and various types of business systems together to integrate computing, storage, and bandwidth, to carry out the cloud distribution of all the resources. On the platform building, we should make the comprehensive realization of teaching and scientific research, management services and cultural construction of digital, networked, intelligent together, and get one kind of "Smart Campus" platform.

4.2 We Should Focus on Innovating a Mode of Educating People

We can focus on building a network-learning platform, a network communication platform, and a networking meeting platform to innovate a mode of educating people. The network-learning platform should include integrates specialty, curriculum, and teaching resources. The network communication platform could allow teachers and students to share their feelings at anytime, anywhere. The network-meeting platform should provide information sharing and real-time feedback between home and school. We should construct a whole-course training mode for students before and after graduation, and a cooperative education mode between school and parents. Students provide convenient and independent channels for knowledge learning and self-demonstration, create boundless areas between teachers and students, home and school, and effectively improve the quality of personnel training.

4.3 We Should Focus on Bringing to Birth a New Management Mode

We can build all kinds of applications to form a kind of "one-stop office", in which we can use procedure, standardization, data, and information technology to achieve the real sunshine operation for administrative power, public service, and online information. And, we should make the best use of the smart digital campus system, to truly facilitate all the teachers and students, to reduce the cost of management, to improve efficiency, and to truly simplify complex things.

4.4 We Should Focus on Carrying Out a New Development Mechanism

With the construction of digital campus as a link and the development of platform application module as a carrier, the teachers and staff can attract information-based enterprises and colleges to cooperate and establish institutes, enterprise studios, and other institutions. Teachers of relevant specialties cooperate with enterprises to jointly research and develop application modules, so as to promote teachers' practical teaching ability and scientific research service level.

5 Concluding Remarks

At present, the development of higher vocational colleges needs to accelerate the construction of digital campus. The construction of digital campus is a necessary condition for the scientific development of higher vocational colleges. The field of digital campus construction mainly includes the field of education, scientific research and technology, organization and management, and social service [9]. Only these four fields have realized the digitalization and the combination of each other so as to realize the high efficiency and wisdom of the campus digitalization.

Acknowledgements This work is supported by Top-notch Academic Programs Project of Jiangsu Higher Education Institutions (TAPP) under Grant No. PPZY2015A090, and Jiangsu top six talent summit project under Grant No. XXRJ-011.

References

1. Fengliang, X., Binbin, S.: Current situation and exploration of digital campus construction in Colleges and universities. China Modern Educ. Equip. **15**, 23–25 (2009)
2. Chen, J., Ma, J.: Research on the mode of effective use of digital campus. Reform Open. Up (12), 180–266 2009
3. Li, P., Chen, Y., Zheng J.: The construction of digital campus in Chinese colleges and universities and thinking of. Inf. Sci. (3), 356–362 (2004)
4. Wang, L.: Several common problems in the construction of digital campus in higher vocational college. Sci. Technol. Inf. (7), 104–105 (2009)
5. Zhou, B., Hua, M.W.: Application of WCF in the construction of digital campus. J. Zhongshan Univ. (Natural Science Edition) (48), 335–337 (2009)
6. Zhao P.: Analysis of "smart campus" based on cloud computing. Journal (2), 8–10 (2012)
7. Bao, W., Zhou, H., Xie, F.: An optimal design of web based online English teaching by using struts framework. Bol. Tecnico/Technical Bul. **55**(10), 58–63 (2017)
8. Bao, W., Zhou, H., Wang, L., Xie, F.: The system of knowledge management using web based learning and education technology. Comput. Syst. Sci. Eng. **31**(6), 469–473 (2016)
9. Bao, W.: Research on college English teaching based on intercultural communication in engineering universities. World Trans. Eng. Technol. Educ. **12**(3), 191–195 (2014)

Part XI
Sensor and RFID Techniques

Parking Information Acquisition System Based on LoRa and Geomagnetic Sensor

Zhai Wenzheng

Abstract Aiming at the problems of poor management and inefficiency of the exiting outdoor parking system, a set of parking information acquisition system based on LoRa wireless network and geomagnetic sensor technology is proposed. The design of front-end acquisition equipment and LoRa gateway is introduced in detail. Combined with the preprocessing of the collection data and the dynamic threshold value correction algorithm, the vehicle detection sensor effectively reduces the false and the missed detection rate. The experimental results show that the system has high detection accuracy and reliability.

Keywords Data acquisition · Geomagnetic sensor · LoRa wireless
communication · Internet of things (IoT) · Parking system

1 Introduction

With the rapid development of urbanization, the number of vehicles increases every year, causing difficulties in parking and increasing road congestion. As a core component of intelligent traffic, parking guidance system can provide real-time urban parking information, improve parking utilization rate of parking facilities, and solve urban parking difficulties, which has become a hot and difficult point in recent years.

Parking information acquisition plays an important role in a parking guidance system. The common parking detectors make use of induction coil, ultrasonic, infrared, and video detection [1, 2], but they are difficult in installation and maintenance, expensive and easy to be affected by the environmental factors such as wind, rain, snow, or fog. Magnetoresistive sensor has been proposed for vehicle detection

Z. Wenzheng (✉)
Software School, Changzhou College of Information Technology, Changzhou,
Jiangsu 213164, China
e-mail: zhwenzheng@shu.edu.cn; 493333151@qq.com

Changzhou Changgong Electronic Technology Co. Ltd., Changzhou,
Jiangsu 213031, China

© Springer Nature Singapore Pte Ltd. 2019
J.-S. Pan et al. (eds.), *Genetic and Evolutionary Computing*,
Advances in Intelligent Systems and Computing 834,
https://doi.org/10.1007/978-981-13-5841-8_46

because it is quite sensitive, small, and more immune to environment [3]. Many research works have been carried out to improve the detection accuracy [4, 5].

LoRa is the representative radio technology in the Low-Power Wide Area Network (LPWAN), and it used narrow bandwidth and spread spectrum modulation to support various Internet of Things (IoT) services with the aim of providing low cost and scalability.

This paper presents the design and implementation of a set of parking information acquisition system based on LoRa wireless network and geomagnetic sensor technology. The specific purposes of this study are to (1) describe the core hardware and software design of the acquisition system and illustrate its working, (2) propose the algorithm for output signal processing and vehicle states recognition, and (3) validate the approach with actual experiments.

The remainder of this paper is organized as follows: Sect. 2 describes the overall system architecture, including the parking space detection system composition and the operating principle. Sections 3 and 4 introduce the implementation of a prototype system in hardware and software development. Section 5 conducts an experiment to evaluate the system and present the results with a discussion. Section 6 makes a brief conclusion.

2 Principle and System Architecture

The mathematical model to describe the magnetic signature of a vehicle is a magnetic point dipole [6], with a magnetic moment M centered in the vehicle and parallel to the Earth's field.

$$B = \frac{\mu_0}{4\pi d^5}\left[3(M \cdot d)d - d^2 M\right], \tag{1}$$

where μ is the permeability of free space and d is the distance from the dipole to the observation point.

Figure 1 shows working principle of the magnetic sensor. In order to obtaining the changes in the magnetic field, the sensor decomposes the Earth's magnetic field and outputs individual components over the x-, y-, and z-axes relative to the local coordinate frame of a compass.

Fig. 1 Working principle of the magnetic sensor

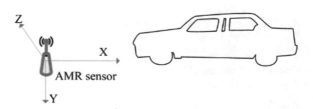

The field components B_{ix}, B_{iy}, B_{iz} produced by M are

$$
\begin{cases}
B_{ix} = \frac{\mu_0}{4\pi} \frac{M_x(2x^2-y^2-z^2)+3M_y xy+3M_z xz}{d^5} \\
B_{iy} = \frac{\mu_0}{4\pi} \frac{3M_x xy+3M_y(2y^2-x^2-z^2)+3M_z yz}{d^5} \\
B_{iz} = \frac{\mu_0}{4\pi} \frac{3M_x xz+3M_y yz+3M_z(2z^2-x^2-y^2)}{d^5}
\end{cases} ,
$$

where $\mu_0 = \mu$ (when in the air), M_x, M_y, and M_z are the magnetic dipole moments in x-, y-, and z-directions, respectively, and $d = \sqrt{x^2 + y^2 + z^2}$.

The large amount of ferrous materials in a car body means that changes in the magnetic field are easily affected by vehicle entering or leaving, but also by nearby cars. A magnetic sensor close or below a vehicle will detect this change in the Earth's magnetic field. *Because changes of B_{iz} is behind B_{ix} and B_{iy}*, a magnetic z-axis sensor can sense the vertical component and detect its change when a vehicle passes or is parked over the sensor.

The proposed parking information acquisition system is shown in Fig. 2, which is composed of the geomagnetic sensors, LoRa gateways, and cloud server.

The geomagnetic sensor is designed to detect the real-time intensity of the parking space magnetic field and to periodically acquire the signals of X, Y, and Z-axes of the variation in the geomagnetic field and to data process for recognition of vehicle states, including signal filtering and geomagnetic baseline correction. It is installed under the road parking space. LoRa gateway receives the parking information from the geomagnetic sensors and transmits cloud server to analyze and display the relevant information.

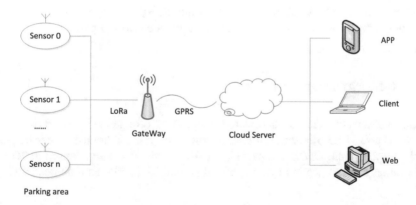

Fig. 2 System composition diagram

3 Implementation of Hardware

3.1 LoRa End Node Hardware Design

Parking data acquisition, analysis, and transmission process are working based on the hardware components. Figure 3 shows the parking information acquisition hardware structure design, which is composed of AMR sensor, micro control processor unit, battery supply, and a LoRa data communication module.

Geomagnetic sensor HMC5983L and HMC5983R can detect the presence of vehicle and analyze the state of vehicles that is passing or parking. The detection mode of two sensors greatly enriches the variation information of geomagnetic field extracted.

When a vehicle passes, disturbances will be generated on the Earth's magnetic field around it. HMC5983 sends the geomagnetic data to STM32L MCU, after smart algorithm processing and judgment, MCU transmits parking state data to SX1278 LoRa module.

The Honeywell HMC5983 is a temperature compensation three-axis integrated circuit magnetometer. Anisotropic Magnetoresistive (AMR) technology is the symbol of this chip. It contains a 12-bit ADC that enables 1° to 2° compass heading accuracy, and can achieve the resolution with 5 m Gauss in the magnetic field of ±8 Gauss. The IIC or SPI serial bus allows for easy interface. These characteristics enable it to be of low cost and with extreme sensitivity.

STM32 is an ARM cortex-M3 kernel specially designed for high-performance, ultra-low-power, and low-cost embedded applications. It has abundant internal resources and standard interface. In this design, STM32L151 is adopted to collect data of geomagnetic sensor in real time and judge the states of parking space.

3.2 Gateway Hardware Design

Figure 4 shows the gateway hardware design. The LoRa gateway uses a SIM900A General Packet Radio Service (GPRS) module as a communication device to transmit parking space data. GPRS is a packet-oriented mobile data service used in 2G and 3G cellular communication systems global system for mobile communications.

Fig. 3 Parking information acquisition hardware structure design

Fig. 4 Gateway hardware design

The ARM Cortex-M4-based STM32F407 MCU series leverages ST's NVM technology and ART Accelerator to reach the industry's highest benchmark scores for Cortex-M-based microcontrollers with up to 225 DMIPS/608 CoreMark executing from flash memory at up to 180 MHz operating frequency.

4 Implementation of Software

4.1 Signal Filtering

The core of the vehicle detection algorithm is to extract effective vehicle information from the collected three-axial magnetometer sensor signals. Owing to the interference of high-frequency signals, it is necessary to filter and denoise before analyzing it. The sliding average filter has a good inhibiting effect on high-frequency interference and is highly smooth and easy to handle, this method is adopted in this paper to preprocess the signal. The calculation of the sliding average filtering is as follows:

$$m_i(k) = \begin{cases} \frac{m(k)+m(k-1)+\cdots+m(1)}{k} & (k < N) \\ \frac{m(k)+m(k-1)+\cdots+m(k-N+1)}{N} & (k \geq N) \end{cases}, \quad (3)$$

where m(k) is the signal collected by the magnetoresistive sensor, and $m_i(k)$ is the signal after sliding average filtering, and N is the window length of the sliding average filter.

4.2 Geomagnetic Baseline Correction

The Earth's magnetic field varies with location and over time. Also, the change of temperature and the traffic of surrounding vehicles can have an effect on the geomagnetic signals. Therefore, baseline tracking must be carried out on the basis of sliding filtering to adapt to the changing baseline values. The main design idea of baseline tracking is to update the baseline values when there is no parking, while the baseline values are consistent with the previous state when there is a large disturbance.

The baseline adaptive update is achieved by the following weighting function:

$$m_{base}(k) = \begin{cases} m_{base}(k-1) \times \alpha + m_i(k) \times (1-\alpha) & (No\ parking) \\ m_{base}(k-1) & (Occupation) \end{cases}, \quad (4)$$

where $m_{base}(k)$ is the baseline value, α is the weighting coefficient, and $m_i(k)$ is a filtered magnetic signal.

Obviously, the larger the weighting coefficient α, the larger the baseline adjustment amplitude, and the faster the tracking speed. After repeated experiments, the weighting coefficient is equal to 0.05, and has good effect on baseline tracking.

4.3   LoRa Communication Protocol Design

The geomagnetic sensor node encapsulates the transmitted content into frames according to a certain format and sends it to the LoRa gateway. After that, the LoRa gateway analyzes the received data package according to the frame format and delivers it to GPRS by TCP/IP protocol.

LoRa frame format is shown in Table 1. The geomagnetic sensor uses a 20-byte frame to send information. Because every LoRa module is assigned a unique address and has 32 alternative channels, the destination address is assigned 15 bytes and the destination channel is assigned 1 byte. Forward Error Correction (FEC) is redundant information, which can enhance LoRa communication anti-interference ability.

The frame load is the effective information to be transmitted, which is composed of the parking number and the parking status. The parking number uses the address of LoRa module itself and the parking status value in 0/1 means "Vacant" or "Occupation".

In addition, since the entire LoRa network uses the same channel, multiple terminal nodes send data to LoRa gateway at the same time, resulting in data conflict. To avoid this situation, the sending/reply mechanism is adopted. That is to say, after receiving valid information, the gateway replies to the end node. If the end node does not receive the reply frame, it will randomly delay a period of time to send again until the transmission is successful.

4.4   Software Workflow of End Nodes

The magnetic sensor is controlled by stm32L microcomputer to measure the magnetic field. To keep the real-time detection and transmission, optimize the control workflow including data collecting, information processing, and data communication is required.

As shown in Fig. 5, the sleep–wake–sleep programming model makes the power consumption minimum. When the time slice is out, end nodes start geomagnetic acquisition and compare the sampling value with the baseline value signal. Once the

Table 1 LoRa frame format

Destination	Channel	Load	FEC

difference exceeds the specified threshold value, the final judgment is made and the parking state data will be uploaded to the LoRa gateway. Otherwise, the end nodes continue sleeping until the next wake. (In this design, the time slice is set to 1 s.)

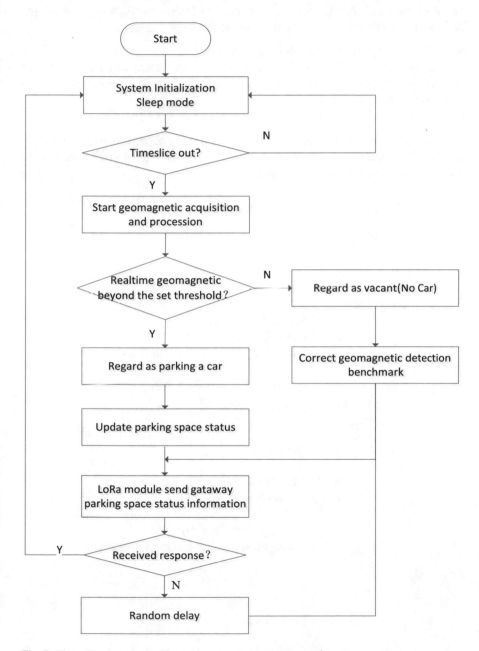

Fig. 5 Time-slice-based algorithm to detect vehicle presence or absence

4.5 Software Workflow of Gateway

The gateway node runs the free RTOS operating system, which realizes the simultaneous execution of multiple tasks and ensures the long-term stable operation of the system. The workflow of gateway is shown in the following steps:

Step 1 System initialize when device powered on.

Step 2 Program come into RTOS process to create four tasks, which aim to receive and parsing the LoRa data package, upload data regularly in intervals, and wait to reply data request timely.

Step 3 Each task runs perpetually under the workflow logic control.

The task 4 is responsible for receives and parses the LoRa end nodes data package. Its code fragment is as shown below. LoRa module is initialized to RX mode. After then, the task goes into an endless loop.

```
void StartTask04(void const * argument)
{
      SX1278Init ();
      RFRxMode ();
   for (;;)
   {
   delay(500);
                  app_loraRecvNode ();
                  app_loraSendCtrl();
                  if(a_timerT.t2_gprsFlag && a_usartT.gpReptFlag)
                  {
                          a_timerT.t2_gprsFlag = 0;
                          app_createCjson();

                  }
                  app_parseCjson();

                  a_keyT.keyName = key_scan(1);
                  if(a_keyT.keyName == KEY0_PRES)
                  {
                          a_keyT.keyName = 0;
                          app_createCjsonAutoinfo(&a_cjsonT);
                  }
   }
}
```

4.6   Data Analysis

The system adopts Ali cloud server, which runs a socket monitor program to receive the data uploaded by GPRS and store it in the database.

The cloud server abstracts the packet forwarder UDP protocol running on LoRa gateways into JSON over MQTT. As is shown in Fig. 6, the information of the database has been presented in hexadecimal notation after the decoding of base64. It contains International Mobile Equipment Identity (IMEI), parking state data, current time, etc.

5   Experiment and Discussion

As shown in Fig. 7, a prototype system of parking space state information acquisition end node was designed and developed. The magnetic sensor is installed on the ground of parking space.

Table 2 is the data set sampled at free parking spaces and Table 3 is the data set sampled at occupied parking spaces.

It can be seen from the data in the table that when there is a car or no car in a parking space, the reading changes of X-axis and Y-axis are not particularly obvious, but the reading changes of Z-axis. It is only necessary to judge whether there is a car in a parking space based on the data of Z-axis.

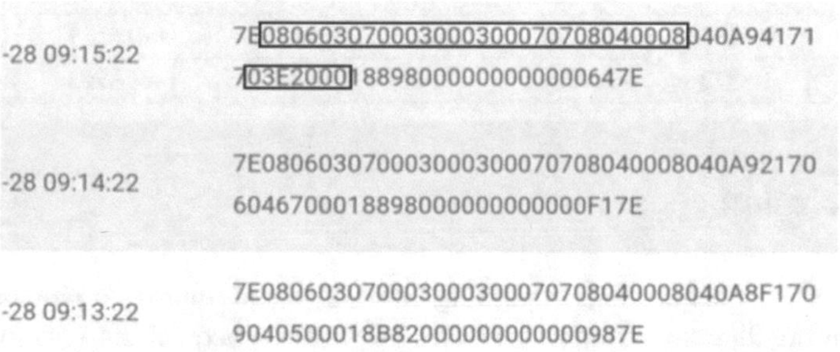

-28 09:15:22

7E 080603070003000300070708040008 040A94171
7 03E2000 18898000000000000647E

-28 09:14:22

7E08060307000300030007070804000804 0A92170
60467000188980000000000000F17E

-28 09:13:22

7E08060307000300030007070804000804 0A8F170
9040500018B82000000000000987E

Fig. 6   The information of the database

Fig. 7 Prototype system

 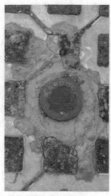

(a) PCB design of LoRa
end node

(b) The layout of
magnetic sensor

Table 2 Data set sampled at free parking spaces

Times	1	2	3	4	5	6
X-axis	+0.6705	+0.6727	+0.6727	+0.6727	+0.6795	+0.6750
Y-axis	−0.3705	−0.3705	−0.3773	−0.3750	−0.3727	−0.3659
Z-axis	−0.8386	−0.8386	−0.8409	−0.8409	−0.8364	−0.8409

Table 3 Data set sampled at occupied parking spaces

Times	1	2	3	4	5	6
X-axis	0.8159	0.8136	0.8159	0.8182	0.8159	0.8136
Y-axis	−0.4182	−0.4136	−0.4159	−0.4205	−0.4159	−0.4136
Z-axis	−1.3773	−1.3818	−1.3841	−1.3795	−1.3795	−1.3841

6 Conclusion

In order to solve the questions of parking difficulties and improve urban traffic intelligence, an intelligent parking space states information acquisition system based on LoRa network and geomagnetic sensors are designed which are integrated with a variety of GPRS, Lora WAN, and Embedded system technologies. The actual installation test results show that the system has high detection accuracy and reliability. The system has a very important practical significance.

Acknowledgements This work is supported by Blue Project of Jiangsu Province, Top-notch Academic Programs Project of Jiangsu Higher Education Institutions (TAPP) under Grant No. PPZY2015A090 and Laboratory of Application and Research of Artificial Intelligence Technology under Grant No. PYPT201805Z.

References

1. Zhu, H., Yu, F.: A cross-correlation technique for vehicle detections in wireless magnetic sensor network. IEEE Sens. J. **16**(11), 4484–4494 (2016)
2. Bagenda, D.N., Parulian, C.: Online information of parking area using ultrasonic sensor through wifi data acquisition. In: Proceedings of the International Conference on Industrial Technology for Sustainable Development, Bandung, Indonesia (2018)
3. Mazhar, R.M., Anand, P., Awais, A., Gwanggi, J.: IoT-based big data: From smart city towards next generation super city planning. Int. J. Sem. Web Inf. Syst. **13**(1), 28–47 (2017)
4. Ji, Y., Fu, P., Blythe, P.T., et al.: An examination of the factors that influence drivers' willingness to use the parking guidance information. KSCE J. Civil Eng. **19**(7), 2098–2107 (2015)
5. Nguyen, T.T., Dao, T.K., Pan, J.-S.: An improving data compression capability in sensor node to support sensor ML-compatible for internet-of-things. J. Netw. Intell. **3**(2), 74–90 (2018)
6. Ma, S., Xu, C., Bao, X., et al.: Reliable wireless vehicle detection using magnetic sensor and distance sensor. Int. J. Digit. Content Technol. Appl. **8**(1), 112–121 (2014)

Design of Library Reader Deployment and Management System Based on RFID Technology

Lijuan Shi and Yafei Wang

Abstract In order to solve the problem of library's increasingly serious bookshelf, loss, and concealment, a UHF reader was designed by combining RFID technology, Wi-Fi technology, and Arduino open platform, and a book positioning and management system was established through the location deployment of the reader and upper computer software. The experimental results showed that the system could effectively complete the functions of automatic book borrowing, book positioning, and book counting. The system met the design requirements preliminarily.

Keywords RFID · Reader · Deployment · Management

1 Introduction

With the rapid development of society, the library has made greater progress in automation, digitalization, and networking than before. However, the management of books and documents in the collection, such as books and periodicals borrowing, book positioning, and bookshelf, still adopts the traditional manual method. The traditional book management mode is tedious in operation procedures, low in efficiency, heavy in workload, and high in error rate. At the same time, the problems of bookshelf, loss, and concealment often occur, which bring many difficulties to the query, update, and maintenance of a large amount of data. In recent years, the greater the reader's demand for knowledge has led to a great increase in the number of books, literature, and capacity of the library compared with a decade ago,

L. Shi (✉)
School of Electrical and Electronic Engineering, Changzhou College of Information Technology, Changzhou, Jiangsu, China
e-mail: 346600457@qq.com; shilijuan@ccit.js.cn

Y. Wang
Department of Library, Changzhou College of Information Technology, Changzhou, Jiangsu, China

© Springer Nature Singapore Pte Ltd. 2019
J.-S. Pan et al. (eds.), *Genetic and Evolutionary Computing*,
Advances in Intelligent Systems and Computing 834,
https://doi.org/10.1007/978-981-13-5841-8_47

which gradually aggravates these problems. Therefore, it is urgent to explore new and efficient intelligent library management methods.

In view of the above problems, many domestic and foreign libraries have begun to adopt the model of closed book storage, and transfer books to closed book storage system, and conduct automatic management of books through precise instruments and equipment. A variety of solutions are proposed. Some of them adopt the Automated Storage and Retrieval System (ASRS), which is widely used in industry, and is applied to the library to borrow and return books automatically. Some use Radio Frequency Identification (RFID) technology to track, identify, and scan books [1].

RFID technology mainly realizes short distance identification. Wi-Fi is a wireless network communication technology, which can carry out LAN data transmission. The integration of RFID technology and wireless communication can realize object recognition and transmission and meet the needs of many application fields [2].

Considering book number, the borrower's personal experience, efficiency of library, library management costs, etc, with the demand of the current library intelligent management, this paper constructed a library reader–writer deployment and management system based on RFID to solve the problems of self-borrowing books, book consolidation, book positioning, and so on. This paper focuses on system scheme design, hardware design, reader deployment, and upper computer program development [3].

2 System Design

The application environment of this thesis was university library. The system was mainly composed of passive UHF electronic tags, readers, servers, terminal devices, and PC software systems. Each book was labeled with a UHF electronic label with a unique serial number. Wireless readers should be placed at intervals on each shelf according to the distribution of library shelves. The reader recognized and collected the unique label information and location information of the book, and sent the data to the server through Wi-Fi [4]. Users can conduct online book enquiries, appointments, etc. through the book management system. At the same time, they can borrow books at any time and any place without staff intervention. The system scheme structure frame based on RFID was shown in Fig. 1.

3 RFID Reader Hardware Design and Deployment

3.1 RFID Reader Design

Considering the application environment, reading range of reader, collision between reader devices, and metal interference, this paper selected RFID UHF band for system design. Wi-Fi module was configured on the reader structure to realize short distance

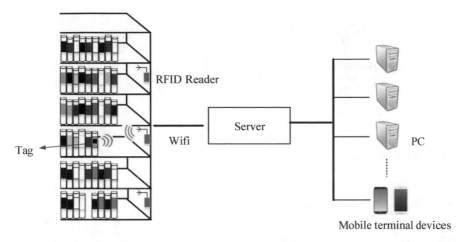

Fig. 1 System scheme structure diagram based on RFID

recognition and long-distance transmission of book information. The reader adopted the open-source Arduino + radio frequency module + Wi-Fi module. By designing the reader in this way, the cost can be saved and the system development cycle can be reduced. The reader used a circularly polarized antenna. The schematic diagram of the hardware circuit of the reader was shown in Fig. 2. The structure diagram of the reader was shown in Fig. 3 [5].

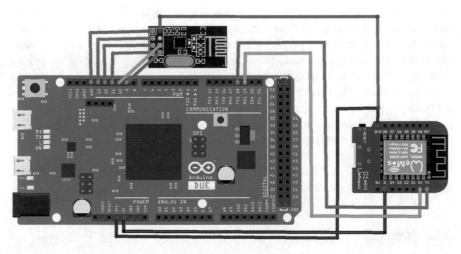

Fig. 2 Schematic diagram of reader hardware circuit

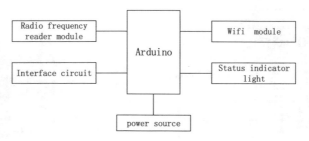

Fig. 3 Reader structure

3.2 Reader Deployment

3.2.1 The Problem Description

The reader deployment plays an important role in this system. In the system, library books have been placed on fixed shelves according to the book classification number. Readers can be designed as fixed readers and mobile readers. If the system uses mobile readers, the number of readers can be reduced, but it can cause collisions between readers. If the system uses fixed readers, it can avoid collision between readers, and at the same time it can locate books in a small range, but it needs a number of readers. Considering the cost and difficulty of system development, this paper used a stationary reader to recognize books on the shelves. In the complex communication environment, cross-coverage, and inevitable interference between readers, the following aspects should be considered when deploying readers nodes to ensure the full coverage of the system and the full connectivity of the network: the network hardware as low as possible, the communication quality as good as possible, the number of readers as small as possible, the collision as small as possible, and the coverage area as large as possible [6].

In order to simplify the network model, it assumed that the network area was a two-dimensional rectangular plane S. The reader and test labels in library were discretized into finite point sets. The network consisted of M reader nodes and N test tags, assuming that the set of readers was R and the set of labels was T. Each reader was assumed to have the same reading range as a circle with radius R_0. V_{RSSI} was the minimum signal strength threshold value, that is, the radio frequency signal strength of the reader received by each test tag was greater than V_{RSSI}.

3.2.2 Coverage

The reader-to-tag coverage performance indicator was represented by f_1:

$$f_1 = \sum_{j=1}^{N} c_j. \tag{1}$$

In formula 1, c_j represented the coverage of the jth test tag. $c_j =$
$\begin{cases} 1, V_{ij} > V_{RSSI} \\ 0, V_{ij} \leq V_{RSSI} \end{cases}$. V_{ij} represented the signal value of the ith reader node received by
the jth test tag. The larger the value of the target function f_1, the greater the coverage
of the reader to the test tags.

3.2.3 Load Balancing

This system was a multi-target recognition system where both readers and test tags
were relatively dense. The communication between readers and tags were basically
a many-to-many relationship. In order to prevent collision between different devices,
indicators should be set to ensure that the number of tags read by each reader was
relatively close, so as to achieve the load-balancing effect. The system used f_2 to
represent the load balance of the network.

$$f_2 = \prod_{j}^{N} \frac{1}{d_j}. \tag{2}$$

In formula 2, d_j represented the number of readers who read the jth test tag. The
function f_2 represented the reciprocal of the number of readers who read each test
tag, making the load of each reader relatively balanced. The larger the value of f_2
function, the more balanced the load distribution.

3.2.4 Objective Function

In order to make the network not only meet the requirement of tag coverage, but also
make the reader balance load well, the optimization model of reader network was
set to f:

$$f = w_1 f_1 + w_2 f_2 = w_1 \sum_{j=1}^{N} c_j + w_2 \prod_{j}^{N} \frac{1}{d_j}. \tag{3}$$

In formula 3, w_1 was the weight of coverage, and w_2 was the weight of load
balancing. w_1 and w_2 needed to meet $w_1 + w_2 = 1$. The larger w_1 can guarantee
the maximum coverage of tags by reader. The larger w_2 can minimize interference
between readers. In order to ensure the maximum coverage of the tags by the readers
and minimize the interference between the readers, in this paper, we had set $w_1 =$
0.9, $w_2 = 0.1$.

3.2.5   Crossover and Mutation

Good solutions in the population can be passed on to the next generation through cross-processing, and better new individuals can be produced. The crossover probability P_c was the ratio of the number of individuals in a population to the number of subalgebras produced by crossover in each generation. When P_c was larger, the space of solution could be improved and the probability of selecting local optimal solution individual could be reduced. However, if P_c was too high, the search speed of the algorithm would be reduced.

The probability of variation was P_v. P_v was the ratio of the number of mutants in the population to the number of individuals in the entire population, which affected the proportion of new individuals entering the population. If P_v was small, new individuals had difficulty entering the population. However, if P_v was too high, the search speed of the algorithm would be reduced.

3.2.6   Reader Deployment

We assumed that there were 20 rows of bookshelves in a reading room of the library. Each row of bookshelves had 6 layers. Each shelf was separated by multiple partitions, and each shelf can hold 20 books. The system randomly placed 120 books with test labels on a 6-storey shelf (8 m * 1.8 m), as shown in Fig. 4. In Fig. 4, a solid circle represented a reader, a rectangle represented a test tag, and a triangle represented an auxiliary tag. In order to guarantee the full coverage and connectivity of the network and reduce the system cost so as to minimize the number of readers, it was assumed that N readers were required to collect test tags data in local network. Data transmission between tag and reader was done by inductively coupled mode. The working frequency was 2.4 GHz and the radio frequency radius of the reader was 1.5 m. Other parameters were set as follows: $w_1 = 0.9$, $w_2 = 0.1$, $P_c = 0.9$, and $P_v = 0.05$. The readers were deployed as shown in Fig. 4 finally.

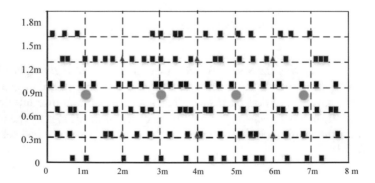

Fig. 4 Reader deployment

4 Upper Computer Software Design

The upper computer software system was divided into access control system, alarm system, self-borrowing and returning, book inquiry, book positioning, book inventory, personal information management, and other modules. Considering that the current general library system can realize book inquiry, personal information management, access control system, and alarm system, we mainly carried out self-help loan, book location, and automatic book counting function design.

How do you find a book you need from a large number of books? Or we need a book in the library, but the book has been maliciously hidden or misplaced. General traditional library management system can search books and show the room where the book is located. However, the traditional library management system only roughly positions the books according to the agreed book classification, but not accurately positions the books in real time.

This system used the deployment of readers, test tags, and auxiliary tags to realize the recognition and positioning of different books. Each reader had a fixed location and a fixed code. The serial numbers of the books in the corresponding range of recognition had been written in the memory of different readers. If the reader recognized another tag, the tag was "illegal," which should be reported to the server. The server can calculate the position based on the reader's location and auxiliary tags. A hand-held reader can also be used to locate misplaced books accurately. In the library, a number of self-service loan terminal devices were installed, and campus cards were used for identity authentication. It was simple and quick to use the labeled labels to borrow books. Through multiple readers, the system counted books quickly, which brought great convenience to book management.

This paper used LabVIEW software and database to design upper computer software. The book location search software interface was shown in Fig. 5. If the book was misplaced or was intentionally hidden, the system displayed the information as shown in Fig. 6.

5 Conclusion

The system used RFID technology, Arduino platform, and Wi-Fi technology to design UHF readers. The system recognized and located books with the help of reader deployment. Combined with the existing library management mode, it realized the intelligent management of library self-borrowing, book counting, and book positioning. In order to realize intelligent library better, we can use RFID technology and automation technology to design automatic sorting function for returned books, which is also one of our future research directions.

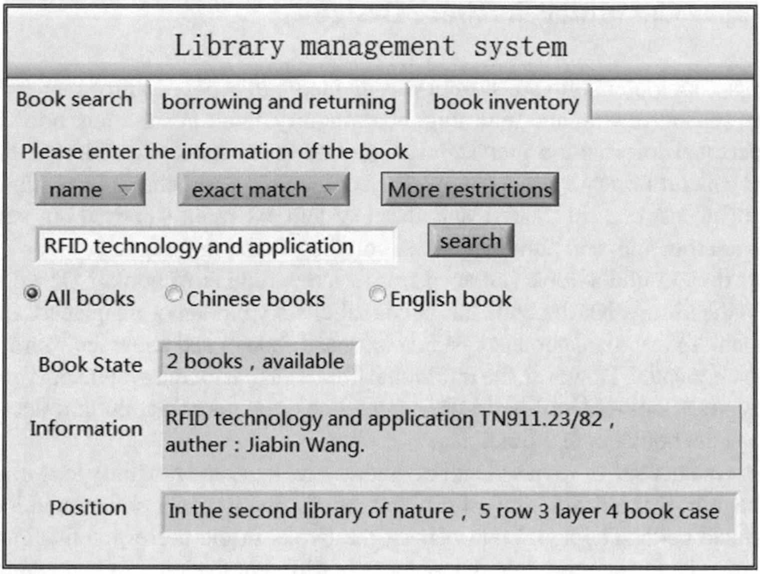

Fig. 5 Book search display

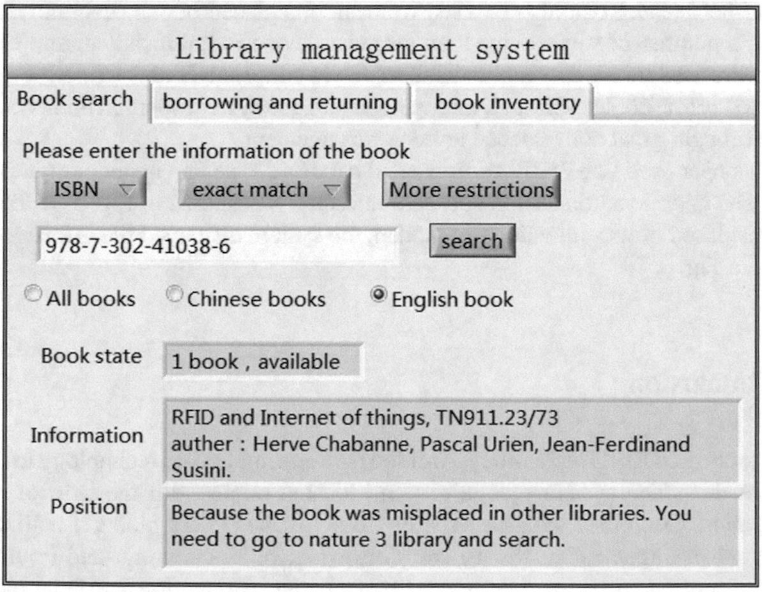

Fig. 6 Misplaced book search display

Acknowledgements This work is supported by the Changzhou University higher vocational education research project under grant CDGZ2018047, Teaching reform of higher vocational education of CCIT under grant 2018CXJG10, University philosophy social science research fund project of Jiangsu Province under grant 2017SJB1822.

References

1. Fujisaki, K.: An RFID-based system for library management and its performance evaluation. In: International Conference on Complex, pp. 105–109. IEEE (2015)
2. Cheng, H., Huang, L., Xu, H., et al.: Design and implementation of library books search and management system using RFID technology, pp. 392–397. In: INCoS (2016)
3. da Silva, C.E.A., de Souza Tavares, J.J.-P.Z., Ferreira, M.V.M.: Arduino library developed for petri net inserted into RFID database and variants. In: Petri Nets, pp. 396–405 (2018)
4. Deepika, K., Usha, J.: Design & development of location identification using RFID with WiFi positioning systems. In: International Conference on Ubiquitous and Future Networks, pp. 488–493 (2017)
5. Sun, X., Ma, L.: Application of nRF2401 in the network management of chemical industry. In: International Conference on Electronic and Mechanical Engineering and Information Technology, pp. 3365–3367. IEEE (2011)
6. Wang, Y.C., Liu, S.J., Wang, Y.C., et al.: Minimum-cost deployment of adjustable readers to provide complete coverage of tags in RFID systems. J. Syst. Softw., 134 (2017)

Reliable and Efficient Task Allocation Method in Wireless Sensor Networks

Da-Xiu Zhang, Chia-Cheng Hu, Xiao-Juan Zhu and Rui-Bin Xu

Abstract In this paper, a task allocation algorithm is proposed for reliable and efficient transmission in wireless sensor networks. The task allocation problem is formulated as a minimum energy consumption problem under multiple constraints. The feasible solution of the formulated problem is obtained by the proposed algorithm. The task with large resource demand is mapped to the multi-node cooperation by using the principle of proximity. Finally, the proposed algorithm is evaluated on MATLAB platform. The results show that its reliability is guaranteed in the case of low energy consumption.

Keywords Wireless sensor networks · Task mapping · DPSO algorithm · Reliability

1 Introduction

In recent years, wireless sensor networks (WSNs) have attracted a lot of attention from researchers and been widely used in many applications, such as agriculture, environmental monitoring, and perception, as well as military, medical, and disaster hazards [1]. In WSNs, one of the main challenges is how to efficiently assign the tasks of monitoring ambient conditions for the applications to the sensors for the following purposes. First, the overall power consumption is used efficiently. Second, the tasks

D.-X. Zhang · C.-C. Hu (✉) · R.-B. Xu
Quanzhou University of Information Engineering, Quanzhou, Fujian 362000, China
e-mail: 2958761515@qq.com

D.-X. Zhang
e-mail: 1451358113@qq.com

R.-B. Xu
e-mail: 4771597@qq.com

X.-J. Zhu
Anhui University of Science and Technology, Huainan, Anhui 232001, China
e-mail: xjzhu@aust.edu.cn

© Springer Nature Singapore Pte Ltd. 2019
J.-S. Pan et al. (eds.), *Genetic and Evolutionary Computing*,
Advances in Intelligent Systems and Computing 834,
https://doi.org/10.1007/978-981-13-5841-8_48

are finished by the assigned sensors before their deadlines. Third, the network lifetime is extended. Fourth, the high reliable transmission of the monitoring ambient data by the sensor nodes in WSNs can be obtained. To support reliable transmission in WSNs is important and necessary when WSN is applied to the applications in gathering the monitoring data. Without doubt, the WSN performance closely depends on the task allocation.

In this paper, we propose a task allocation algorithm for the above four purposes. In the proposed algorithm, the problem of task allocation in WSNs is first to be converted to a task mapping problem. A formal definition of the task mapping problem in WSNs is given and formulated as a mapping energy minimization problem under multiple constraints, such as reliability and scheduling length. Then, a discrete particle swarm optimization mapping algorithm is proposed for solving the minimization problem by using the discrete particle swarm optimization (DPSO) mapping algorithm to iteratively calculate the optimal mapping the tasks to the sensors. The inertia factor of the traditional PSO algorithm is improved to decrease linearly with the iteration number. When tasks with large resource requirements need to be mapped to multiple sensors to complete cooperatively, the mapping principle of optimal node proximity is added to the (DPSO) mapping to reduce the communication energy consumption between tasks. Finally, the task mapping algorithm based on ant colony is compared with that based on MATLAB platform in terms of energy consumption. The results show that the task mapping algorithm in this paper consumes less energy. Then, the reliability of task mapping under different energy consumptions is analyzed. The results show that the reliability of task mapping can be guaranteed under the condition of less energy consumption.

2 Related Works

In [2], the authors solve the problem of the task allocation by proposing a bi-graph matching algorithm, in which two sets are used to represent the tasks and nodes. The proposed algorithm aims to determine a bi-graph, which is a feasible solution to the problem, between the two sets. In [3], Kobrosly proposes a quickly and efficiently heuristic algorithm to obtain the feasible solution of the problem. In [4], Lin proposes a dynamic probability algorithm, which is an enhancement of the heuristic algorithm in [3], to solve the problem.

On the other hand, many algorithms of task allocation have been proposed for WSNs. In [5], Yu W proposes a distributed optimal online task allocation algorithm by taking into account the energy cost of communication, calculation, perception, and sleep activities. In [6], Lei M develops an effective decomposition-based approach to achieve the optimal solution of the problem in WSNs, in order to achieve the goal of real-time task mapping. In [7], Zeng et al. obtain an approximate optimal solution of the problem in WSNs by proposing an energy-balanced directed acyclic graph task scheduling algorithm and a genetic algorithm combined with chromosome coding.

In [8], Abdelhak proposes a task allocation algorithm in order to extend the life of a WSN. In [9], Morteza proposes a bidirectional task allocation technology based on queuing theory by considering the energy and task completion time in WSNs. In [10], Fan W proposes an energy-efficient flow allocation algorithm for multi-home WSN gateways with limited resources. The proposed algorithm aims to minimize the power consumption of the gateways. In [11], Zhu J proposes an optimization model of balancing network reliability and maximizing network life on the basis of network utility maximization. They solve the model through the hop-by-hop retransmission mechanism for ensuring the reliability of network resource allocation, improving the network life, and ensuring the reliable transmission of tasks in the network.

Most of the studies [5–11] mainly aim to facilitate energy consumption, network life cycle, and task execution time. However, since the network reliability is not considered, they cannot achieve reliable transmission of gathering the monitoring data in WSNs.

3 Problem Definition

In this paper, a directed acyclic graph $G = (V, E, W, C)$ and a set X are used as the input and output of the problem, respectively. In G, the element $m_i \in V$ is a task i to be performed, the edge $e_{i,j} \in E$ is the finished time dependency between tasks i and j where tasks i is needed to be finished before task j, the element $w_i \in W$ is the computation overhead of the task i, and the edge $c_{i,j} \in E$ is the communication load between the tasks i and j. In X, the element $x_{i,j}^t \in X$ is the mapping map() as shown in (1) for assigning task i to sensor j at time t.

$$\forall_{m_i} \in M, map(m_i \rightarrow n_k), n_k \in N \tag{1}$$

In this paper, we aim to determine the optimal mapping from the tasks to sensors for the purpose of executing the tasks with minimum energy consumption under the constraints of reliability, communication relationship, and deadline. Here, n_k represents the kth node. Therefore, the above mapping problem is defined as follows:

$$\min E_{map} = w_1 \cdot c_1(map) + w_2 \cdot c_2(map)) \tag{2}$$

$$s.t \ x_{i,k}^t \in \{0, 1\}, \forall n_k \in N, \forall m_i \in M, \forall t \in T \tag{3}$$

$$\sum_{t \in T} \sum_{n_k \in N} x_{i,k}^t = 1, \forall m_i \in M \tag{4}$$

$$\sum_{m_i \in M} x_{i,k}^t \leq 1, \forall n_k \in N, \forall t \in T \tag{5}$$

$$lengtn(H) \leq DL \tag{6}$$

$$R_{map} \geq \varepsilon \tag{7}$$

The objective function (2) is to minimize the total energy consumption. The energy consumption for a mapping *map* includes the computing energy consumption c_1 (*map*) for executing the tasks, and the communication energy consumption c_2 (*map*) for transmitting among the assigned sensors, where w_1 and w_2 are two weight parameters for the energy consumption. Constraint (3) represents the mapping relationship of the task mapping; 1 indicates that task m_i is mapped on node n_k in time slot t. Constraint (4) indicates that each task must be mapped to a node. Constraint (5) means that up to one task can be assigned to a sensor node in a time slot. Constraint (6) constrains the task to be completed before the deadline, that is, for the effective task mapping and scheduling solution H, H is the scheduling length, and DL is the task processing deadline. Constraint (7) constrains the reliability of the task map, indicating that if the reliability of the task map is greater than ε, the mapping is successful; otherwise, it fails. Here, ε represents the reliability threshold, which is set to 0.7 as different reliability under different circumstances.

The reliability of task mapping includes the reliability of node and communication links, where $e^{\lambda_k t}$ is the reliability of any node k in time slot t, and λ_k is the failure rate of node k. $\sum_{i=1}^{m} T_i$ represents the time at which m tasks are performed on node k, and T_i represents the execution time of the ith task. Further, m and n represent the numbers of tasks and nodes, respectively. So the reliability of node k is

$$R_k = e^{-\lambda_k \sum\limits_{i=1}^{m} T_i}, \forall i \in (1, 2, \ldots m), \forall k \in (1, 2, \ldots, n) \tag{8}$$

The reliability of the communication links is

$$R_{path_{ij}} = e^{-\mu_{ks} \sum\limits_{e_{ks} \in path_{ij}} \frac{tc_{ij}}{p_{ks}}}, \forall i, j \in (1, 2, \ldots, m), \forall k, s \in (1, 2, \ldots, n) \tag{9}$$

where $tc_{i,j}$ represents the data traffic between tasks i, j; $\mu_{k,s}$ represents the failure rate of the link between nodes k, s; and p_{ks} represents the transmission rate between nodes k and s. The total reliability of the task mapping *map* is

$$R_{map} = \prod_{k=1}^{n} e^{-\lambda_k \sum\limits_{i=1}^{m} T_i} \prod_{\forall e_{ks} \in path_{ij}} e^{-\mu_{ks} \sum\limits_{e_{ks} \in path_{ij}} \frac{tc_{ij}}{p_{ks}}}, \forall i, j \in (1, 2, \ldots, m),$$
$$\forall k, s \in (1, 2, \ldots, n) \tag{10}$$

The task is scheduled according to the earliest execution time (EST) and the latest execution time (LST). The (DPSO) algorithm is used to search for the node with the minimum total energy consumption under multiple constraints such as reliability and scheduling length, and then the tasks are mapped to the nodes. The task mapping

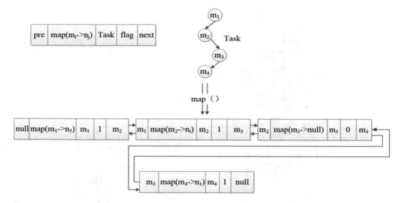

Fig. 1 The process of task mapping

process is shown in Fig. 1, where each five-dimensional linked list represents a feasible solution to the target function (the task is mapped to the node).

4 Proposed Algorithms

The steps of the proposed algorithm are as follows and the flowchart is given in Fig. 2:

Step 1. The task request arrives.

Step 2. The scheduling list of tasks is obtained based on the earliest execution time (EST) and the latest execution time (LST).

Step 3. Calculate θ. If $\theta < 1$, you only need to map to an optimal node, and then perform all the steps except step 12; otherwise, continue to perform step 4 to step 13.

Step 4. Initialize the solution set, the mapping task is performed according to the task schedule, and randomly obtain the global optimal node.

Step 5. Update the position and speed of the task according to the update formula of the position and speed.

Step 6. Calculate the fitness value of the task mapping.

Step 7. Determine whether the fitness value is less than the value of the last iteration, if it is true, perform step 8, otherwise perform step 11.

Step 8. Update the local optimal node pbest and the global optimal node gbest.

Step 9. Iter ++.

Step 10. Determine if the number of iterations equals 200. If true, it is stored in the pre-reserved linked list, and return the global optimal node gbest, otherwise go to step 5.

Step 11. Obtain a global optimal node gbest with the smallest adaptation value.

Step 12. Map the task to the node near the global optimal node gbest.

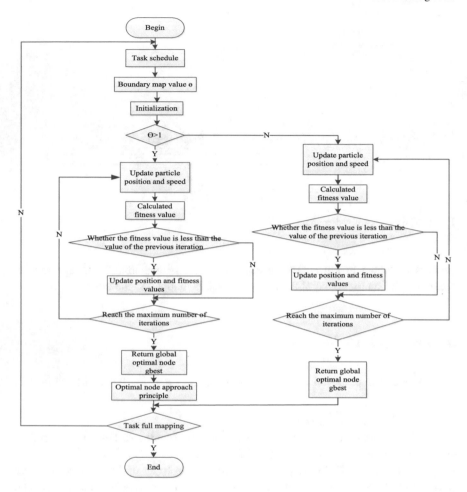

Fig. 2 Task mapping flowchart

Step 13. Determine if all tasks have been mapped, and if true, end; otherwise return to step 2.

In Step 3, the boundary mapping value θ is calculated in formula (12), where r_i and r_j are the required resource of task i and the available resource of sensor j, respectively. If $\theta > 1$, the tasks will be mapped to multiple sensors for collaborative completion by using the proximity principle. Otherwise, every task will be assigned to only one sensor.

$$\theta = \frac{\max_{m_i \in M} r_i}{\max_{n_j \in N} r_j} \tag{11}$$

In Steps 4 and 12, some operations are performed by using the DPSO algorithm as follows:

1. The particle position is represented by the vector $P_i(t) = (x_{i,1}, x_{i,2}, x_{i,3}, ..., x_{i,s})$, which represents the node to which the ith task may be mapped.
2. Particle velocity is used to guide the task to map to the optimal node, represented by $V_i(t) = (v_{i,1}, v_{i,2}, ..., v_{i,j}, ..., v_{i,s})$. In the process of adjusting the task map, the update formula of the position and velocity of the particles is as follows:

$$P_i(t + 1) = P_i(t) + V_i(t + 1) \tag{12}$$

$$V_i(t + 1) = \omega V_i(t) + c_1 r_1 (pbest_i - P_i(t)) + c_2 r_2 (gbest_i - P_i(t)) \tag{13}$$

where ω, $c_1 r_1$, $c_2 r_2$ indicate different probabilities of particle flight, $c_1 r_1$ and $c_2 r_2$ take values between 0 and 1, and $\omega + c_1 r_1 + c_2 r_2 = 1$. The pbest$_i$ and gbest$_i$, respectively, represent the local optimal node and the global optimal node of the ith task map. The subtraction indicates the difference between the two mappings in $(pbest_i - P_i(t))$ and $(gbest_i - P_i(t))$. If the two mapping results are the same, the result is 0 (i.e., no difference), otherwise is 1. In order to maintain better search results for particles, this paper sets 1 as a dynamic parameter that decreases linearly as the number of iterations increases. The formula is (15), where $iter$ is the number of iterations, and $iter_{max}$ is the maximum number of iterations. This paper sets it to 200.

$$w = 0.96 - \frac{iter}{iter_{max}} \tag{14}$$

The fitness value function is to minimize the EC of the task map, including computational and communication EC.

$$fitness = w_1 \cdot \min(c(m_i)) + w_2 \cdot \min(E(m_i)) \tag{15}$$

5 Simulations

In order to evaluate the performance of the proposed algorithm, the experimental analysis will be conducted from two aspects: energy consumption and reliability. There are 50 sensor nodes randomly deployed in the area of 100 m * 100 m. Then, the simulation parameters of DPSO algorithm are set as shown in Table 1. In each iteration, the DPSO algorithm obtains the fitness value comparison between the global optimal node and the global optimal node of the last iteration, and retains the node of the iteration with smaller adaptation value.

The first set of experiments is used to analyze the EC of the task map as shown in Fig. 3, which compares the EC of DPSO mapping algorithm and ant colony (ACO)-based task mapping algorithm when dealing with different numbers of task mappings.

Table 1 Parameter table for task mapping

Parameter	Value
Node deployment area	100 * 100 m
Population size	50
Nodes number	50
Initially ready tasks	10
Max iterations	200
Learning factor c_1	2
Learning factor c_2	2
Inertia factor ω	$0.96 - \frac{iter}{200}$

Fig. 3 Energy consumption of task mapping

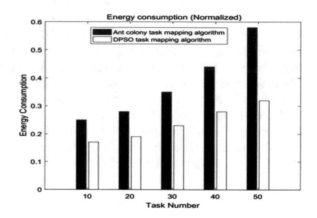

The results show that the EC by the DPSO mapping algorithm is almost lower than the EC obtained by the ACO-based task mapping algorithm. This is mainly because the DPSO algorithm adopts the linear decreasing inertia factor, which makes the particle start to obtain the optimal position within a larger area quickly and accurately before local search, to avoid the local optimal stagnation phenomenon and reduces the large EC due to the stagnation. DPSO has a sharing mechanism, and the task does not generate preemption and congestion. However, the task mapping algorithm based on ACO can only perceive local information and cannot directly use global information, and the task will consume energy until the global optimal node is obtained. Therefore, the DPSO mapping algorithm consumes less energy than the task mapping algorithm based on ACO.

The second set of experiments evaluates the reliability of task mapping. Figure 4 shows the relationship between EC and reliability in the task mapping process under different task mapping methods. The results show that the DPSO task mapping method can guarantee the reliability of task mapping in the case of low EC. This is mainly due to the inclusion of reliability constraints in the process of task mapping, which can only be mapped if the reliability is greater than 0.7. However, the ant colony (ACO)-based task mapping algorithm does not consider reliability when task

Fig. 4 Reliability of task mapping

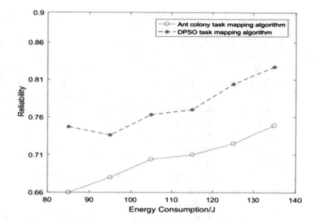

mapping, but updates and searches through the pheromone feedback ant, and maps it to the node. The reliability of the ant colony (ACO)-based task mapping algorithm is calculated by the calculation method of this paper. The results show that under the same energy consumption, the reliability of the task mapping in this paper is significantly higher than that of the ant colony (ACO)-based task mapping algorithm.

6 Conclusions

In order to achieve reliable and efficient task mapping, this paper transforms the task mapping into the minimum EC problem under multiple constraints such as reliability and uses the improved DPSO algorithm to solve the optimal solution of the problem. The simulation results show that our proposed task mapping algorithm consumes less energy. Then, the reliability of task mapping under different energy consumptions is analyzed. The results show that the reliability of task mapping can be guaranteed under the condition of less energy consumption.

Acknowledgements This work was supported in part by the Fujian Provincial Key Laboratory of Cloud Computing and Internet-of-Thing Technology, China. And this work is jointly supported by National Natural Science Foundation of China (Grant No. 51504010) and Key Projects of Anhui Province University Outstanding Youth Talent Support Program (Grant No. gxyqZD2016083).

References

1. Kocakulak, M., Butun, I.: An overview of wireless sensor networks towards internet of things. In: Comput and Commun Workshop and Conference 2017, pp. 1–6. IEEE (2017)
2. Yao, H.: Research on mobile Cloud Computing Environment Task Allocation. Nanjing University of Posts and Telecommunications, Nanjing (2014)

3. Kobrosly, W.: Implementation of an optimal task allocation strategy. In: Proceedings of the IEEE Southern Tier Technical Conference 1988, Binghamton, pp. 212–219 (1988)
4. Zeng, H., Cui, W.: Heuristic construction strategy for initial solution of Lin-Kernighan algorithm. J. Shandong Univ. (Eng. Ed.) **42**(2), 30–35 (2012)
5. Yu, W., Huang, Y., Garcia-Ortiz, A.: Distributed optimal on-line task allocation algorithm for wireless sensor networks. Sens. J. IEEE **18**(1), 446–458 (2017)
6. Lei, M., Kritikakou, A., Sentieys, O.: Decomposed task mapping to maximize QoS in energy-constrained real-time multicores. In: International Conference on Computer Design 2017, ICCD, IEEE Computer Society, pp. 493–500 (2017)
7. Zeng, Z.: A highly efficient DAG task scheduling algorithm for wireless sensor networks. In: Proceedings of the 9th International Conference for Young Computer Scientists 2008. IEEE, pp. 570–575 (2008)
8. Abdelhak, S.: Energy-balancing task allocation on wireless sensor networks for extending the lifetime. In: Proceedings of the 53rd IEEE International Midwest Symposium on Circuits and Systems 2010, IEEE, pp. 781–784 (2010)
9. Morteza, O., Hossein, M.: Task allocation to actors in wireless sensor actor networks: an energy and time aware technique. Procedia Comput. Sci. **3**(1), 484–490 (2011)
10. Fan, W., Tang, B., Liu, Y.: Energy efficient traffic allocation for resource-constrained multi-homed WSN gateway. Int. J. Smart Home **9**(10), 77–86 (2015)
11. Zhu, J.: Rate-lifetime tradeoff for reliable communication in wireless sensor networks. Comput. Netw. **52**(1), 25–43 (2008)

Neural Network and Deep Learning

Automatic Training Data Generation Method for Pixel-Level Road Lane Segmentation

Xun Pan, Yutian Wu and Harutoshi Ogai

Abstract Lane detection or road detection is one of the key features of autonomous driving. By using deep convolutional neural network based semantic segmentation, we can build models with high accuracy and robustness. However, training a pixel-level semantic segmentation needs very fine-labeled training data, which requires large amount of labor. In this paper, we propose an automatic training data generating method, which can significantly reduce the effort of the training phase. Experiments prove that our method can generate high-quality training data for lane segmentation task.

Keywords Lane detection · Semantic segmentation · Training data generation

1 Introduction

Drivable area detection, or more specifically, road detection, and lane detection have always been a key part of autonomous driving. In the past few years, the revival of convolutional neuron networks [1] (CNN) has led to a breakthrough in computer vision, especially in semantic understanding, where road and lane detections can significantly benefit from. A lot of CNN-based algorithms are proposed in autonomous driving field and the performance is largely improved compared with traditional non-CNN-based methods.

X. Pan (✉) · Y. Wu · H. Ogai
The Graduate School of Information, Production and Systems, Waseda University,
2-7 Hibikino Wakamatsu-ku, Kitakyushu, Fukuoka, Japan
e-mail: xunpan@ruri.waseda.jp
URL: https://www.waseda.jp/fsci/gips/

Y. Wu
e-mail: wuyutian@fuji.waseda.jp

H. Ogai
e-mail: ogai@waseda.jp

© Springer Nature Singapore Pte Ltd. 2019
J.-S. Pan et al. (eds.), *Genetic and Evolutionary Computing*,
Advances in Intelligent Systems and Computing 834,
https://doi.org/10.1007/978-981-13-5841-8_49

473

In this paper, we define the lane detection task as a pixel-level image segmentation problem. The lane marking pixels are defined as one class and all other pixels are defined as background class. Unlike other CNN-based tasks such as image classification or object detection, pixel-level image segmentation requires pixel-level finely labeled ground truth as training data. To solve the tedious image labeling task for generating training data, we propose an automatic training image labeling method based on Hough transform [2] and learning-based image matting [3]. Experiments show that our method can greatly improve the efficiency of training data generation for road lane segmentation.

The main contribution of this paper is that we propose an automatic training data generating method for pixel-level lane segmentation, which overcomes one of bottlenecks for applying CNN-based method—generating training data. The rest of the paper is organized as follows: In Sect. 2, we introduce about CNN and a pixel-level lane segmentation method. In Sect. 3, we give a brief introduction about online open dataset for image-related tasks and explain about why we need an efficient training data generating method. In Sect. 4, we introduce our training label generating algorithm which saves people from manually labeling the training images. In Sect. 5, we show the performance of our method with experiments. Finally, we draw conclusions in Sect. 6.

2 Pixel-Level Lane Segmentation

Fully convolutional networks [4] (FCN) is a popular pixel-wise segmentation DCNN model. By upsampling the output of convolutional layers, FCN can classify each pixel in the original image, which is, in fact, semantic image segmentation. DeepLab [5] successfully improves the accuracy of FCN using fully connected conditional random field (CRF) to refine the segmentation result. DeepLab structure takes several seconds to process one image, to decrease the inference time SegNet [6] is proposed, which reduces the inference time to about 400 ms. Hengshuang Z. et al. propose ICNet [7] to further reduce the inference time to about 30 ms, which is applicable in real-time system.

Since the main theme of this paper is how to generate training data for the segmentation model, we will not talk too much about the model structure, which is beyond the scope of this paper. We only briefly show that what the input and output of the model are, and what kind of training data we need to train such a model. Figure 1 shows an example of road scene segmentation. The first row shows the input image and ground truth (i.e., the training image), the second row shows the output from different segmentation models. From the figure, we can see that the input of the model is the image captured by the camera, the training image is of the same size and resolution as the input image, and for every pixel in the input image, we need to assign a class (marked as one certain color) to it at the corresponding position in training image. As for our specific task, what we want to segment out is only the

Fig. 1 Examples of different segmentation results in road scene. First row shows the original road scene image and ground truth. Second row shows segmentation results of FCN, DeepLab, and SegNet

boundaries of the road lane, the ground truth can be simplified as a binary mask, where white represents the lane boundaries and black represents others.

3 Online Open Datasets of Images

There are a variety of datasets or benchmarks that offer training data for image-related tasks. For example, PASCAL VOC [8] is a very famous benchmark. It contains data for main subjects in computer vision, i.e., classification, detection, and segmentation. However, it focuses more on compact objects—like persons, different kinds of animals, vehicles, indoor objects, and so on—than infrastructures, such as roads and lanes. Another widely used general-purpose dataset is COCO [9] which is built by Microsoft. COCO has more classes than PASCAL VOC, but it does not contain road data either.

There are also some road scene focused datasets. The KITTI [10] benchmark is a widely used road scene benchmark. It contains various kinds of data about the road scene like color images, stereo images, and laser point information. But it does not offer training data for lane markings. Cityscapes [11] is famous for its high-resolution and fine-labeled training images, but similar to KITTI, it does not contain lane markings either. CamVid [12] has labeled lane markings in its dataset, but it treats all kinds of lane markings as the same class, including lane boundaries, arrows, and text on the road. Figure 2 shows some examples in existing datasets.

As described above, the lack of proper training data for lane boundaries drives us generating the training data by ourselves. Usually, to get training data that is

Fig. 2 Examples of segmentation labels in different road scene datasets. The left shows examples in Cityscapes and KITTI datasets, labels are directly marked on the images. The right shows an example in CamVid dataset, top is original image, bottom is labeled image

not contained in open dataset, one can either rely on commercial data company or generate the data by oneself. For classification problems, the training labels are very easy to produce. We only need to assign one class (usually mapped as one index number) for one whole image. For detection problems, labeling becomes complicated since there can be multiple objects to be detected in one image. For each object, we need to draw a rectangle around it. Training labels for pixel-wise segmentation problems are the most complicated. We need to do the labeling pixel-wisely, and usually the boundaries between different classes are irregular. So it is very useful to develop an algorithm that can do the labeling automatically.

4 Generation of Training Data

Detecting the road in "normal" road conditions can be relatively simple even for non-CNN methods. The difficult part is how to deal with unexpected noises, such as shadow, reflection, traffic, changing illumination, and so on. Fortunately, for generating training data for CNN, we do not need a perfect algorithm that can always detect the lane boundaries correctly, we only need an algorithm that can do the detection in majority of the cases. And for cases that algorithm cannot do well, we do it manually.

Now, we can make some assumptions that hold for majority of the road scenes as given below:

1. The lane boundaries are straight or nearly straight.
2. The color of the lane boundaries is uniform and distinguishable from the surroundings.
3. Lane boundaries are the most prominent edges inside the road area.

With these three assumptions, we can detect the lane boundaries with two main steps. First, we use Hough transform [2] to detect edges of the lane boundaries. Second, we use image alpha matting method to calculate the boundaries area. The details are described below.

4.1　Edge Extraction

First, we use Canny edge detector [13] to get the edges of a road scene image. Then, with the Assumption 2 above, we can filter out irrelevant edges by color. Concretely, we first convert the road scene image from RGB color space to HSV color space, then generate a mask by specifying a range of Hue channel value which corresponds to the color (usually white or yellow) of the lane boundaries. Then, we multiply the edge image by the mask. Figure 3 shows an example of the above procedure.

After getting the edge image, we can use Hough transform, which is a classic straight lines detection algorithm in image processing area, to detect the position of the lane boundaries. Since we assume that the lane boundaries are straight, we are supposed to detect the boundaries correctly, as shown in Fig. 3.

4.2　Filling Up Lane Boundaries

Now that we have calculated the edges of the lane boundaries, there is one more step to get the final training data. Since we define lane boundaries detection as a pixel-wise image segmentation problem, the training data we need is pixel-wise

Fig. 3 An example of boundary edge extraction. Top left is the original image, top middle is the full edge image, top right is the image in HSV color space. Bottom left is the color mask calculated by the HSV image, bottom middle is the masked edge image, bottom right is the result of Hough transform on masked edge image

Fig. 4 An example of filling up lane boundaries by image matting. Top left is the original image; top right is the result from the edge extraction procedure introduced in last subsection. Bottom left is the tri-map for image matting, white area represents the foreground, black area represents the background, gray area represents unknown part; bottom right is the matting result

label image, which marks out every lane boundary pixel rather than the edge or skeleton of the boundary. So we need to fill up the lane boundary area. We choose an image alpha matting method [3] to do this. Alpha matting refers to the problem of extracting the foreground from an image, given a tri-map which specifies known foreground/background and unknown pixels. Here, we use the boundary edges we calculated in last subsection as the known foreground, and use image dilation to generate an area around the edges as the unknown part, and set all other parts of the image as the background. Figure 4 shows an example of filling up lane boundaries by image matting.

5 Experiments

To evaluate the performance of our training data generating algorithm, we compare the algorithm-generated training labels and manually generated ground truth. The metrics we use are similar to those in [4], which are pixel accuracy, mean accuracy, and mean IU (region intersection over union). Let n_{ij} be the number of class i predicted to class j, n_{cl} be the number of classes, and $t_i = \sum_j n_{ij}$ be the total number of pixels of class i. The metrics are defined below:

- pixel accuracy: $\sum_i n_{ii} / \sum_i t_i$,
- mean accuracy: $(1/n_{cl}) \sum_i n_{ii}/t_i$,
- mean IU: $(1/n_{cl}) \sum_i n_{ii}/(t_i + \sum_j n_{ji} - n_{ii})$,
- frequency weighted IU:
  $(\sum_k t_k)^{-1} \sum_i t_i n_{ii}/(t_i + \sum_j n_{ji} - n_{ii})$.

Since we only have two classes—the lane boundary and non-boundary—we can define lane boundary as positive class and non-boundary as the negative class, then the above definition can be simplified as

- pixel accuracy: $(TP + TN)/(P_{all} + N_{all}))$,
- mean accuracy: $(TP/P_{all} + TN/N_{all})/2$,
- mean IU: $(TP/(P_{all} + FP) + TN/(N_{all} + FN))/2$,
- frequency weighted IU:
  $(TP \times P_{all}/(P_{all} + FP) + TN \times N_{all}/(N_{all} + FN))/(P_{all} + N_{all})$.

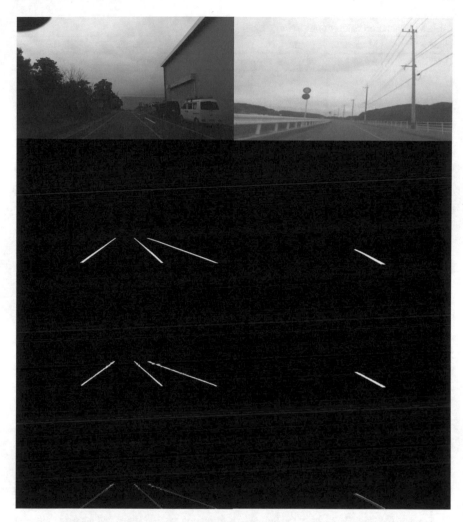

Fig. 5 An example of evaluation of training label generation. The first row shows original images, the second and third rows are manually marked ground truth and algorithm calculated label (white represents lane boundary and black represents non-boundary). The last row shows comparison between ground truth and calculated label, black represents true negative, red represents true positive, blue represents false negative, and green represents false positive

where TP, TN, P_{all}, and N_{all} represent true positive, true negative, all positive, and all negative, separately. We evaluate our label generating algorithm on 48 images (since this evaluation requires manually made ground truth, we only tested limited images). Figure 5 shows some examples of the evaluation in different road scenes. Considering the fact that lane boundary only occupies a small portion of the image, most area of a road scene image is non-boundary part, calculating accuracy of lane boundary class is more meaningful than calculating the average accuracy between

Table 1 Evaluation result of our training label generation

Pixel acc.	Mean acc.	Mean IU	f.w. IU	Lane acc.	Lane IU
0.996	0.883	0.821	0.991	0.757	0.643

both lane boundary part and non-boundary part. So we add two more metrics as shown below:

- lane accuracy: TP/P_{all},
- lane IU: $TP/(P_{all} + FP)$.

Table 1 shows the evaluation result. According to the table, we can see that our algorithm can generate training labels for the CNN well.

6 Conclusion

In this paper, we propose a non-CNN-based method to generate training data for pixel-level lane boundary segmentation, which is based on edge extraction and image alpha matting. This algorithm can significantly save the effort when preparing training data. Experiments show that we can get reliable training data from various road scenes. By combining the current state-of-the-art DCNN method and traditional non-CNN method together, we can build a reliable lane detection system efficiently.

References

1. LeCun, Y., Boser, B., Denker, J.S., Henderson, D., Howard, R.E., Hubbard, W., Jackel, L.D.: Backpropagation applied to handwritten zip code recognition. Neural Comput. **1**(4), 541–551 (1989)
2. Duda, R.O., Hart, P.E.: Use of the hough transformation to detect lines and curves in pictures. Commun. ACM **15**, 11C15 (1972)
3. Zheng, Y., Kambhamettu, C.: Learning based digital matting. In: Proceedings of the 20th IEEE International Conference on Computer Vision, pp. 889–896 (2009)
4. Long, J., Shelhamer, E., Darrell, T.: Fully convolutional networks for semantic segmentation (2014). arXiv:1411.4038
5. Chen, L.-C., Papandreou, G., Kokkinos, I., Murphy, K., Yuille, A.L.: Semantic image segmentation with deep convolutional nets and fully connected CRFs (2014). arXiv:1412.7062
6. Badrinarayanan, V., Kendall, A., Cipolla, R.: SegNet: a deep convolutional encoder-decoder architecture for image segmentation (2015). arXiv:1511.00561
7. Zhao, H., Qi, X., Shen, X., Shi, J., Jia, J.: ICNet for real-time semantic segmentation on high-resolution images (2017). arXiv:1704.08545
8. Everingham, M., Gool, L., Williams, C.K., Winn, J., Zisserman, A.: The Pascal visual object classes (VOC) challenge. Int. J. Comput. Vis. **88**, 303–338 (2010)
9. Lin, T.-Y., Maire, M., Belongie, S., Bourdev, L., Girshick, R., Hays, J., Perona, P., Ramanan, D., Zitnick, C.L., Dollar, P.: Microsoft COCO: common objects in context (2015). arXiv:1405.0312v3

10. Fritsch, J., Kuehnl, T., Geiger, A.: A new performance measure and evaluation benchmark for road detection algorithms (2013)
11. Cordts, M., Omran, M., Ramos, S., Rehfeld, T., Enzweiler, M., Benenson, R., Franke, U., Roth, S., Schiele, B.: The cityscapes dataset for semantic urban scene understanding. In: 29th IEEE Conference on Computer Vision and Pattern Recognition, pp. 3213–3223 (2016)
12. Brostow, G.J., Fauqueur, J., Cipolla, R.: Semantic object classes in video: a high-definition ground truth database. Pattern Recognit. Lett. **30**, 88–97 (2009)
13. Canny, J.: A Computational approach to edge detection. IEEE Trans. Pattern Anal. Mach. Intell. **6**, 679–698 (1986)

Breaking CNN-Based Steganalysis

Zhenxing Qian, Cheng Huang, Zichi Wang and Xinpeng Zhang

Abstract With the rapid development of deep learning, a lot of CNN-based steganalyzers have emerged. This kind of steganalyzer uses statistical learning to investigate the properties caused by steganography, which is the most efficient approaches for breaking information hiding. However, we find a vulnerability of CNN-based steganalyzer that it can be defeated by dual operations. In this paper, we propose an easy yet effective algorithm to perturb the stego images against neural network, which can evade CNN-based steganalyzer with high probabilities. We elaborated on the theoretical basis of the method we proposed and proved the feasibility of this method through experiments.

Keywords Steganography · Steganalysis · Deep learning

1 Introduction

Steganography is a technology of hiding secret data into a digital media by slightly modifying cover data without drawing suspicion [1]. It is one of the hot topics in information security and has drawn lots of attention in recent years. Steganography is often used in secret communications. Especially in the fast-growing social networks, there is an abundance of images and videos, which provide more opportunities and challenges for steganography. Therefore, the design of a secure steganography scheme is of critical importance. The STC (Syndrome Trellis Coding) [2] based embedding is the most effective steganography scheme currently, which achieves minimal additive distortion between cover and stego image with a user-defined distortion function.

Z. Qian (✉) · C. Huang · X. Zhang
School of Computer Science, Fudan University, Shanghai 200444, China
e-mail: zxqian@shu.edu.cn

Z. Wang
School of Communication and Information Engineering, Shanghai University, Shanghai 200444, China

© Springer Nature Singapore Pte Ltd. 2019   483
J.-S. Pan et al. (eds.), *Genetic and Evolutionary Computing*,
Advances in Intelligent Systems and Computing 834,
https://doi.org/10.1007/978-981-13-5841-8_50

There are numbers of STC-based embedding methods, e.g., WOW [3], UNIWARD [4], HILL [5], and MiPOD [6].

Contrarily, steganalysis which aims to disclose the covert transmission has also been developed rapidly meanwhile. With the development of machine learning, most steganalytic methods use supervised machine learning to investigate the models of the covers and the stegos, e.g., spatial rich model (SRM) [7]. Recently, deep learning is employed in image steganalysis. It emerges a large group of excellent steganalytic methods which rival traditional method based on machine learning. GNCNN [8] is the earliest approach of CNN-based steganalysis, which achieves a comparable performance with SRM. Different from the traditional CNN networks [9], the images are processed by KV convolution kernel before inputted into CNN. Another difference is that the Gaussian activation function is used in the entire model. In 2016, Xu et al. [10] further improve the structure of convolutional network, including ABS activation function, batch normalization, and tanh activation function. The detectability of this network surpasses SRM.

At present, the most powerful tool of CNN-based steganalysis is the "deep learning hierarchical representations (DLHR)" [11]. Comparing with traditional CNN nets, the DLHR net contains a special convolutional layer, which is used to simulate the residual computation. As investigated in SRM, the residual computation can improve the SNR (stego signal to image content), which is vital for image steganalysis. Inspired by this, the DLHR structure uses all the 30 high-pass filter kernels in SRM, instead of random weights. This construction also accelerates the convergence of the net. Experimental results indicate that DLHR achieves satisfactory performance.

Although the CNN-based methods have made a success in steganalysis, it is easily being attacked as same as the traditional CNN networks. Szegedy et al. [12] discovered that the traditional neural networks like CNN are vulnerable to adversarial samples due to some inherent characteristics of neural networks. This vulnerability of neural network has become a critical topic [13, 14]. Szegedy et al. [12] first disclose that deep neural network may misclassify an image by applying a certain hardly perceptible perturbation. They attribute this phenomenon to the discontinuous of the input–output mappings learned by neural networks. Goodfellow et al. [15] further explained the reasons for this phenomenon. They show that a simple linear model can be easily attacked when it is inputted with sufficient dimensionality. Recently, Su et al. [16] proposed an effective method by changing only one pixel to fool deep neural networks, which achieves well performance.

These methods are not designed to fool steganalytic network. In this paper, we improve the existing attack methods and propose a scheme to attack steganalysis network by adding perturbations. Meanwhile, the secret data can be extracted from the perturbed image. Using the proposed method, the CNN-based steganalysis misclassify a stego image with high probabilities.

The rest of the paper is organized as follows. In Sect. 2, we present a brief description of our proposed method. Structure and implementation are shown in Sect. 3, which is followed by the experimental results and analysis in Sect. 4. Finally, the concluding remarks are drawn in Sect. 5.

2 Method Description

The framework of steganalytic network consists of three parts, i.e., noise residual computation, feature extraction, and binary classification. For CNN-based steganalysis methods, these parts are merged into the structure of CNN. Obviously, a more unified structure helps improve the performance of steganalysis. Compared to the traditional SRM method, this new CNN-based steganalysis methods can achieve better results by optimizing the network structure and updating network parameters. Feature extraction is the critical step in steganalysis. In the traditional method, feature extraction is a very difficult task, which takes a lot of manpower and time. Contrarily, it is simulated with convolutional layers in CNN-based steganalysis methods. A well-established convolutional structure could obtain more discriminative features, which results in the improvement of the detection accuracy.

However, the feature map obtained by convolutional operation could not truly reflect the characteristics of the cover images sometimes. It is due to the limitation of precision in convolutional operation. Feature map could obtain information correctly in MSB (Most Significant Bits), but most information in LSB (Least Significant Bits) is neglected oppositely. Although the steps of residual computation in traditional steganalysis method can improve SNR to avoid this drawback to some extent, for CNN-based steganalysis methods, it is still difficult to obtain all information of an image. Accordingly, it is possible to add some small perturbations into cover image to fool steganalysis method.

The added perturbation should be neglected by the convolutional operations of CNN. Meanwhile, considering that the added perturbation should be as small as possible, we need to limit the size of the perturbation so that the addition of the perturbation does not have a big impact on the original convolution operation. Let δ be the perturbation and τ be the minimum precision that convolutional operation will neglect. We limit the maximum norm constraint by $\|\delta\|_{\infty} < \tau$. It is worth noting that although the added perturbation is small, it is continuously amplified by the convolution operation, which is enough to affect the result of the final output of steganalysis.

To fool the CNN-based steganalysis method, it is needed to limit the direction of perturbation. In this paper, we use the direction of gradient to guide the adding of perturbation. Nowadays, the gradient descent algorithm is used in the training of CNN networks. The gradient of each parameter is calculated through back-propagation algorithm. Then each parameter is updated according to the gradient. By iterating the gradient descent algorithm, the network finally reaches a convergence state. We simulate the gradient descent algorithm to let the steganalysis network judge the stego images as cover images. First, assign the label of cover images to the target stego images. Then keep the parameters unchanged, and add a perturbation according to the gradient of stego images. We describe the perturbation as

$$\delta = \tau \, \text{sign}(\nabla_x L(\boldsymbol{\theta}, \mathbf{X}_\text{s}, y_\text{c})) \tag{1}$$

Fig. 1 Sketch of the adversarial method

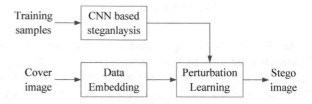

where τ is the minimum precision that convolutional operation can neglect, $L(\cdot)$ is the cost function we used, $\mathbf{X}_s$ is the stego images we input, and y_c is the label of cover images.

3 Structure and Implementation

In Fig. 1, we propose a framework to reveal the vulnerability of CNN-based steganalysis. Given a cover image, we generate an initial stego image using a popular steganography approach. With training samples, we first construct a CNN-based steganalyzer for simulation. According to the simulation network, a learning algorithm of adding perturbation into the initial stego is proposed, and the final stego is generated to evade the CNN-based steganalyzer.

As long as the adversary finds that the CNN-based steganalysis is being used by the guard, he can also build a CNN-based network for simulation by himself. After generating the labeled examples, the adversary trains a CNN-based steganalysis model. Let $f(\cdot)$ be the judgment function, θ the parameters of a model, $\mathbf{X}$ the input image, y the target label for $\mathbf{X}$, $Loss(\cdot)$ the cost function, and ε the learning rate. Then the training algorithm is described in Algorithm 1.

Once the iterative training is done, the adversary can identify an approximate network using the parameters θ. As long as the training examples belong to the widely used images, the parameters would be close to the real network mastered by the guard. With the fixed parameters θ, the adversary owns a simulation steganalysis network.

Algorithm 1 Network Training

---
for number of training iterations steps **do**

$\quad \widetilde{y} = f(\mathbf{X})$

$\quad\quad L(\theta, \mathbf{X}, y) = Loss(\widetilde{y}, y)$

Update θ

$\quad\quad \theta = \theta - \varepsilon \cdot \nabla_\theta L(\theta, \mathbf{X}, y)$

end for

Next, he modifies the stego by perturbation learning. Let $\mathbf{X}_c$ be the clean image, $\mathbf{X}_s$ the stego image, y_c the label of $\mathbf{X}_c$, y_s the label of $\mathbf{X}_s$, $\widetilde{y}$ the output, and τ the minimum precision that convolutional operation can neglect. Algorithm 2 describes the adversarial perturbation operations.

Algorithm 2 Adding perturbations

$$\tilde{y} = f(\mathbf{X}_s)$$
$$L(\mathbf{\theta}, \mathbf{X}_s, y_c) = Loss(\tilde{y}, y_c)$$

Update $\mathbf{X}_s$

$$\mathbf{X}_s = \mathbf{X}_s - \tau \cdot \text{sign}(\nabla_x L(\mathbf{\theta}, \mathbf{X}_s, y_c))$$

In this procedure, the adversary no longer modifies $\mathbf{\theta}$. Rather, he uses the gradient descent to learn the steps of modifying the stego image. Therefore, perturbation learning is dual operation to the network training.

In Algorithm 2, the parameter τ is used to control the extent of perturbation. Generally, a small τ is enough to defeat the CNN-based steganalysis. To guarantee an exact data extraction, τ should be properly chosen. For example, in ternary STC [2], we can use $\tau = 4$ ensure the correctness of data extraction. In this algorithm, the adversary updates $\mathbf{X}_s$ only once, which is different from the iteration ways in training process. It is because that a single disturbance is enough to defeat the CNN-based steganalysis.

4 Experimental Results

4.1 Experiment Setup

The CNN-based steganalysis method DLHR [11] is employed in our experiments. In order to contrast with the experimental result of DLHR, we use the same dataset as DLHR, which is the image dataset BOSSbase ver. 1.01 [17], containing 10000 uncompressed grayscale images with each sized 512×512. All the images are resized into 256×256 as the settings in [11]. The popular algorithms of WOW and HILL are used for embedding. All embedding tasks are done by ternary STC. The payloads are specified as 0.05, 0.1, 0.2, 0.3, 0.4, and 0.5 bpp (bit per pixel), respectively. Let the adversarial results for WOW and HILL be WOW-P and HILL-P, respectively. All experiments are implemented in Pytorch.

4.2 Visual Quality of Perturbed Stego Image

Figure 2 provides an example of perturbation adding. Figure 2a shows an initial stego image produced from data set BOSSbase ver. 1.01. Figure 2b is the perturbation which is added in the initial stego image. After perturbation added, the obtained perturbed stego image is shown in Fig. 2c.

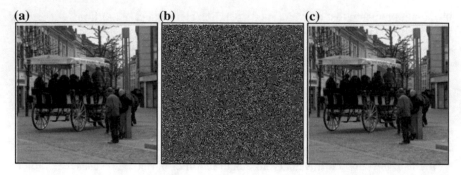

Fig. 2 Perturbation adding. **a** Initial stego image, **b** the added perturbation, **c** perturbed stego image

Table 1 Data extraction error rate

Number of images	10	50	100	1000
Extraction error	0%	0%	0%	0%

We can see that there is no evident distortion in the perturbed stego image. In spite of this, this perturbed stego image could be mistakenly judged as a clear image by the DLHR steganalyzer. Although DLHR added a preprocessing operation to perform residuals calculation, and improve SNR to detect noise signal, it is still difficult for DLHR to distinguish the added perturbation.

4.3   Correctness of Data Extraction

We conducted some experiments to verify the correctness of data extraction after the perturbation is added. After secret data is embedded, we add perturbation into a certain amount of stego images. The final extraction error rate is shown in Table 1. It shows that the secret data can be extracted correctly.

For ternary STC, only the lowest two bits of the pixel will be modified. The added perturbation will not affect the extraction of secret data when τ is as large as 4. This is also the reason that why we set τ as 4. This value can ensure the correctness of data extraction with the minimal distortion.

4.4   Undetectability Against Steganalysis

We compare the perturbed stego image with the original stego images to verify the performance of our method. The missed detection rates of the two kinds of images are tested. We have not test the false alarm rate since the perturbation is added into the stego images and did not change the cover images.

Fig. 3 Comparisons of
missed detection using
WOW

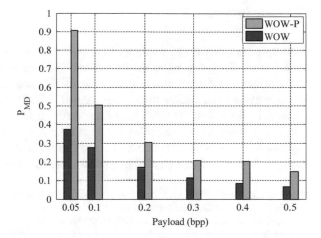

Fig. 4 Comparisons of
missed detection using HILL

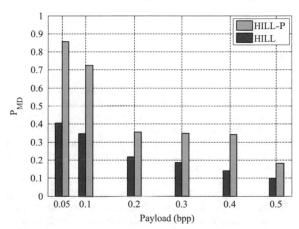

Figures 3 and 4 show comparisons results of the missed detection rates using
WOW, HILL, and the perturbed versions, in which P_{MD} is the average value of
missed detection rate by ten times averaging Ye et al. DLHR method. The results
indicate that the missed detection rates clearly increased after the perturbations. In
other words, more stego images are regarded as clear images after the adversarial
implementations. In addition, the values of P_{MD} are around 0.9 for the payload 0.05
bpp, which means that a perturbed stego with a small payload is undetectable using
DLHR.

5 Conclusion

Although the CNN-based methods have made a success in steganalysis, it is easily being attacked as same as the traditional CNN networks. This paper proposes a perturbation-based algorithm to defeat the CNN-based steganalyzer and proves the effectiveness of this method through experiments. The algorithm is easy to be implemented. It shows that the CNN-based steganalyzer is not stable against adversarial operations, while the vulnerability does not exist in the traditional supervised learning based steganalysis tools like SRM. For further study, it is significant to develop rival perturbation adding algorithm against these traditional steganalyzers.

Acknowledgements This work was supported by the Natural Science Foundation of China (U1736213, U1536108, 61572308, 61103181, U1636206, 61373151, and 61525203), the Natural Science Foundation of Shanghai (18ZR1427500), the Shanghai Dawn Scholar Plan (14SG36) and the Shanghai Excellent Academic Leader Plan (16XD1401200).

References

1. Fridrich, J.: Steganography in Digital Media: Principles, Algorithms, and Applications [M]. Cambridge University Press, New York (2009)
2. Filler, T., Judas, J., Fridrich, J.: Minimizing additive distortion in steganography using syndrome-trellis codes. IEEE Trans. Inf. Forensics Secur. **6**(3), 920–935 (2011)
3. Holub, V., Fridrich, J.: Designing steganographic distortion using directional filters. In: Proceedings of IEEE International Workshop on Information Forensics and Security, Binghamton, NY, USA, pp. 234–239 (2012)
4. Holub, V., Fridrich, J., Denemark, T.: Universal distortion function for steganography in an arbitrary domain. EURASIP J. Inf. Secur. **2014**(1), 1–13 (2014)
5. Li, B., Wang, M., Huang, J., Li, X.: A new cost function for spatial image steganography. In: Proceedings of the IEEE International Conference on Image Processing, Paris, France, pp. 4206–4210 (2014)
6. Sedighi, V., Cogranne, R., Fridrich, J.: Content-adaptive steganography by minimizing statistical detectability. IEEE Trans. Inf. Forensics Secur. **11**(2), 221–234 (2016)
7. Fridrich, J., Kodovský, J.: Rich models for steganalysis of digital images. IEEE Trans. Inf. Forensics Secur. **7**(3), 868–882 (2012)
8. Qian, Y., Dong, J., Wang, W., Tan, T.: Deep learning for steganalysis via convolutional neural networks. Proc. SPIE **9409**, 94090J (2015)
9. LeCun, Y., Bottou, L., Bengio, Y., Haffner, P.: Gradient-based learning applied to document recognition. Proc. IEEE **86**(11), 2278–2324 (1998)
10. Xu, G., Wu, H.-Z., Shi, Y.Q.: Structural design of convolutional neural networks for steganalysis. IEEE Signal Process Lett. **23**(5), 708–712 (2016)
11. Ye, J., Ni, J., Yi, Y.: Deep learning hierarchical representations for image steganalysis. IEEE Trans. Inf. Forensics Secur. **12**(11), 2545–2557 (2017)
12. Szegedy, C., Zaremba, W., Sutskever, I., Bruna, J., Erhan, D., Goodfellow, I.J., Fergus, R.: Intriguing properties of neural networks. ICLR (2014). arXiv:1312.6199
13. Barreno, M., Nelson, B., Sears, R., Joseph, A.D., Tygar, J.D.: Can machine learning be secure? In: Proceedings of the 2006 ACM Symposium on Information, Computer and Communications Security, pp. 16–25 (2006). ACM
14. Barreno, M., Nelson, B., Joseph, A.D., Tygar, J.: The security of machine learning. Mach. Learn. **81**(2), 121–148 (2010)

15. Goodfellow, I.J., Shlens, J., Szegedy, C.: Explaining and harnessing adversarial examples (2014). arXiv:1412.6572
16. Su, J., Vargas, D.V., Kouichi, S.: One pixel attack for fooling deep neural networks (2017)
17. Bas, P., Filler, T., Pevný, T.: Break our steganographic system: the ins and outs of organizing BOSS. In: Proceedings of the 13th International Conference on Information Hiding, Prague, Czech Republic, pp. 59–70 (2011)

Meteorite Detection and Tracing with Deep Learning on FPGA Platform

Kuo-Kun Tseng, Jiangrui Lin, Haichuan Sun, K. L. Yung and W. H. Ip

Abstract At present, the strength of space exploration represents the strength of a country. Meteorite exploration is also part of the space field. The traditional meteorite detection and tracking technology are slow and not accurate. With the development of deep learning, computer detection technology becomes more and more accurate and efficient, which makes it possible to improve the accuracy and speed of meteorite detection. In this paper, the deep learning algorithm implemented by FPGA is applied to the meteorite detection, and the popular tracking algorithm is applied to the meteorite tracking, so that the structure of the meteorite detection and tracking system can meet the practical requirements.

Keywords Meteorite detect · Tracking · Deep learning · FPGA

1 Introduction

Meteorite detection is an important part of space detection. Meteorites vary in size, and there are space stations, satellites, and lots of space junk floating in space. How to avoid the collision between meteorites and artificial satellites and keep satellites safe by accurately identifying meteorites, detecting and tracking them, predicting the track and speed of their operation becomes a hot topic in planetary exploration.

K.-K. Tseng (✉) · J. Lin · H. Sun
School of Computer Science and Technology, Harbin Institute of Technology (Shenzhen),
Shenzhen, China
e-mail: kktseng@hit.edu.cn

K. L. Yung · W. H. Ip
The Department of Industrial and Systems Engineering, The Hong Kong Polytechnic University,
Hung Hom, Hong Kong

W. H. Ip
The Department of Mechanical Engineering, University of Saskatchewan,
Saskatoon, SK S7N 5A9, Canada

© Springer Nature Singapore Pte Ltd. 2019
J.-S. Pan et al. (eds.), *Genetic and Evolutionary Computing*,
Advances in Intelligent Systems and Computing 834,
https://doi.org/10.1007/978-981-13-5841-8_51

493

The traditional machine learning method has some shortcomings such as low applicability and poor accuracy in image, so it is difficult to apply to practice. In recent years, with the rise of deep learning and the development of computer hardware, the field of computer vision has made explosive progress. Prior to this, because of the low accuracy and poor real-time performance of the image algorithm, the research results of machine vision can only be applied to simple scenes. With the development of science and technology, computer vision is more pressing to cope with complex scenes. With the breakthrough development of deep learning, artificial intelligence, and other fields, many excellent large data sets of computer vision have emerged, which indicates that machine vision has entered a new era. Because of its low cost, large amount of information, and so on, the development potential is self-evident.

With the progress of deep learning, we believe that meteorite detection will face a chance to improve its technology in the speed and accuracy of detection. In the paper, the related research, algorithm, and FPGA implementation have been studied as the following sections.

2 Related Research

2.1 Detection Work

Traditional object detection methods usually only aim at one object, such as popular and mature face detection, pedestrian detection, and fingerprint recognition. Feature extraction uses some human-experience-based features such as HOG [1] features or SIFT [2] features, and the classifiers are usually SVM classifiers or AdaBoost classifiers. In general, the strategy of exhaustion is used in detection. It extracts all possible target boxes, traverses these target boxes, and classifies each window.

With the development of convolutional neural networks, the advancement of hardware technology, and the rise of machine learning, people begin to study large-scale object detection. The emergence of excellent data sets is also important. Li Feifei, a professor at Stanford University, led to the establishment of a large data set, ImageNet. In addition, the common datasets are PASCAL VOC, Microsoft COCO. Some initial research work based on convolution neural network adopted sliding window method and used CNN to classify different sizes and positions of windows. There are many regions generated by exhaustive sliding windows, and these methods often take a long time to train and test, so they cannot be used in the above data sets. AlexNet's [3] success in ILSVRC 2012 has not only affected the direction of image classification, but also attracted the attention of other researchers in the field of computer vision. At that time, object detection was still carried out by traditional algorithms, but no major breakthroughs have been made in the standard test data sets for object detection such as PASCAL VOC. Therefore, Cirshick [4] applies convolution neural network to object detection, and proposes the R-CNN model. In the R-CNN model, about 2000 candidate regions are generated for each image, and for each image, all candidate regions are extracted separately, which makes the time consumed in feature extraction become the bottleneck of the total test time. The research team of Microsoft Asia Research Institute [5] applies SPP-Net to object detection, and improves the defect of R-CNN, but there are still some shortcomings. In the process of training and testing, candidate regions are proposed, image features are extracted, and classified according to the features. The three processes form a multi-stage pipeline. A direct result of this is that additional space is needed to store the extracted features for use by the classifier. So, Girshick [6], one of the designers of R-CNN, proposed a further improvement on R-CNN, called Fast R-CNN. DeepID-Net [7] is also an influential model in object detection. The model is further improved on the basis of R-CNN training process, and the model pretraining method is improved. All of these models are based on candidate regions.

In addition to the candidate-region-based approach, YOLO introduced by Redmon [8] and others adopted the idea of regression. Then, the SSD algorithm proposed by Liu et al. [9] is faster than the previous fastest YOLO algorithm in speed, and can be compared with Faster R-CNN in detection accuracy.

2.2 Tracking Work

As a key technology of intelligent video surveillance, target tracking has important application value in many fields and has been a hot research topic in related academic fields. After decades of exploration and development, researchers at home and abroad have realized the change and optimization of tracking algorithm. According to different classification criteria, visual target tracking has the following categories [10]:

1. According to whether the tracking target belongs to rigid object, it can be divided into rigid body tracking and non-rigid tracking.
2. According to the number of cameras used to collect information in the tracking scene, it can be divided into monocular vision tracking and multi-vision vision tracking.
3. According to the different representations of the target, it can be divided into region-based, contour-based, and feature-based tracking.

3 Algorithms

3.1 Object Detection

In order to ensure the accuracy of meteorite detection algorithm, the popular depth-learning algorithm is used as the benchmark algorithm, and the algorithm is optimized and improved on the basis of this algorithm.

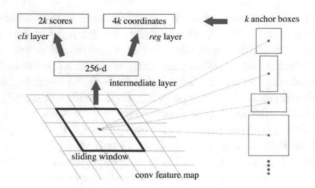

Fig. 1 The whole graph is obtained by CNN and the characteristic graph is obtained. After the convolution of kernel 3 * 3 * 2563 * 3 * 256, it is predicted that K anchor boxes are objects at each point, and the position of anchor boxes is fine-tuned. Extracts the object frame and classifies it in the same way as Fast R-CNN. Shares a CNN network with the categories [6]

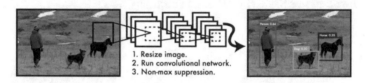

Fig. 2 Partitioned the image into an S * SS * S cell. Each cell outputs B rectangular boxes (redundant designs) containing the location information of the box (x, y, w, h) and Confidence of the object. Each cell then outputs C conditional P (Class|Object) P (Class|Object). The final output layer should have S * S × (B * 5 + C) S × S × (B 5 + C) unit [8]

R-CNN:

In the early days, window scanning was used for object recognition. R-CNN [4] removes window scanning and uses clustering method to segment and group images to get hierarchical groups of multiple candidate frames (Fig. 1).

Faster R-CNN:

Extracting candidate box is running on CPU, which is inefficient. Faster R-CNN uses CNN to predict candidate boxes. Its innovations are shown in Fig. 2.

YOLO:

Fast R-CNN needs 20 K anchor boxes to determine whether the object is, and then object recognition. It has two steps. YOLO combines the selection and recognition of object boxes, and outputs them one step at a time.

3.2 Comparison

Comparing the performance of traditional methods, region-proposal-based algorithm, and regression-based algorithm on PASCAL VOC data sets, the results are

Table 1 Contrast of different detection algorithms in PASCAL VOC

Algorithm	Training set	mAP	FPS
30HzDPM	2007	26.1	30
YOLO	2007 + 2012	63.4	45
Fast R-CNN	2007 + 2012	70.0	0.5
Faster R-CNN	2007 + 2012	73.2	7

Fig. 3 Sample of a circulant matrix

5	6	4	4	5	6	6	4	5
8	9	7	7	8	9	9	7	8
2	3	1	1	2	3	3	1	2
2	3	1	1	2	3	3	1	2
5	6	4	4	5	6	6	4	5
8	9	7	7	8	9	9	7	8
8	9	7	7	8	9	9	7	8
2	3	1	1	2	3	3	1	2
5	6	4	4	5	6	6	4	5

shown in Table 1. Experiments show that the accuracy of deep learning method is much higher than that of traditional computer vision method, and Yolo can meet the real-time requirement with relatively high accuracy (Fig. 3).

3.3 Object Tracking

This paper uses the popular tracking algorithm KCF [11] algorithm, which is based on correlation filter tracking algorithm, the effect is very good, high speed, and ideas and implementation are very simple. Among them, the method of fast computation using circulant matrix is worth learning.

The innovation of the algorithm is that the circulant matrix is used to represent the image block, which greatly increases the speed of operation.

During training, a series of displacement sampling around the current position can be represented by a two-dimensional block circulant matrix X, and the ij block represents the result of moving down the i row and right j column of the original image. Similarly, when testing, a series of displacement sampling near the result of the preceding frame can also be represented by X. Such X can quickly accomplish many linear operations by Fourier transform.

Because the tracking algorithm requires fast real time, how to track the target quickly and accurately is a very important problem. In the KCF algorithm, besides using the cyclic matrix for fast calculation, linear regression training and kernel function method are also used to speed up the algorithm. The experimental results are shown below.

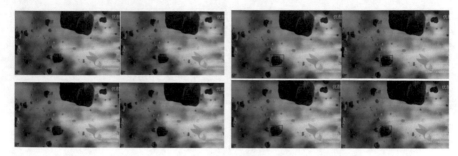

Fig. 4 Original map (left) and impression drawing (right) (http://www.youku.com/)

In the actual detection, the tracking speed reaches about 40 frames per second, which fully meets the requirements of real time. At the same time, the tracking target is not lost under the occlusion condition (Fig. 4).

4 Hardware Architecture Used FPGA

In order to save cost and meet the real-time requirement of target detection, this project intends to use FPGA instead of GPU and CPU as computing hardware resources. The advantage is that FPGA is cheaper than GPU and the computation speed is much better than that of CPU.

4.1 Object Detect with PYNQ-Z1 FPGA

The Low Power Target Detection System Design Challenge, sponsored by DAC 2018, the Top Conference on Electronic Automation Design, ended June 28, 2008 in San Francisco, California. The competition aims to design a high-precision and energy-efficient object detection system for UAVs to meet the needs of real complex scenes. In group FPGA, third teams iSmart2 use the following architecture to detect object (Fig. 5).

In addition, Xilinx provides an example of pynq running CNN. Pynq is equipped with CIFAR10 and LeNet5 networks, respectively. The CIFAR10 network has a test time of 3.3333906849999977 s and an accuracy of 73.2% at 500 batch size. The LeNet5 network has a test time of 1.1257555660000023 s at 600 batch size and an accuracy of 98.3%. (https://github.com/Xilinx/PYNQ-ComputerVision).

Overall System Diagram

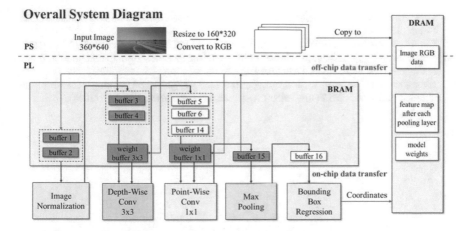

Fig. 5 The team iSmart2 adopted Mobilenet-based lightweight network design, with 12 layers. In terms of hardware implementation, the team utilizes PYNQ-Z1 and uses a module-based (IP) reuse architecture, allowing the same kind of network layer to reuse the same module to save hardware resources (https://baijiahao.baidu.com/s?id=1606045093668063557&wfr=spider&for=pc)

5 Conclusion

The deep learning architecture based on FPGA runs fast and has high accuracy. With the popular target-tracking algorithm, it can effectively solve the meteorite detection and tracking tasks. Believe that it can help the development of space industry in the future.

References

1. Dalal, N., Triggs, B.: Histograms of oriented gradients for human detection. In: IEEE Computer Society Conference on Computer Vision and Pattern Recognition 2005 San Diego, CA, USA. https://doi.org/10.1109/cvpr.2005.177
2. Lowe, D.G.: Object Recognition from local scale-invariant features. IEEE **2**(3), 1150–1157 (1999)
3. Krizhevsky, A., Sutskever, I., Hinton, G.E.: ImageNet classification with deep convolutional neural networks. In: International Conference on Neural Information Processing Systems, pp. 1097–1105. Curran Associates Inc. (2012)
4. Girshick, R., Donahue, J., Darrell, T., et al.: Rich feature hierarchies for accurate object detection and semantic segmentation. In: IEEE Conference on Computer Vision and Pattern Recognition, pp. 580–587. IEEE Computer Society (2014)
5. He, K., Zhang, X., Ren, S., et al.: Delving deep into rectifiers: surpassing human-level performance on ImageNet classification. In: IEEE International Conference on Computer Vision, pp. 1026–1034. IEEE (2016)
6. Girshick, R.: Fast R-CNN. Comput. Sci. (2015)
7. Ouyang W., Zeng, X., Wang, X. et al.: DeepID-Net: deformable deep convolutional neural networks for object detection. IEEE Trans. Pattern Anal. Mach. Intell. (99), 1–1 (2016)

8. Redmon, J., Divvala, S., Girshick, R., et al.: You Only Look Once: Unified, Real-Time Object Detect. 779–788 (2015)
9. Liu, W., Anguelov, D., Erhan, D., et al.: SSD: Single Shot MultiBox Detect. 21–37 (2015)
10. Huang, K., Reng, W., Tieniu, T.: Survey of image object classification and detection algorithms. J. Comput. Sci. **37**(6), 1225–1240 (2014)
11. Henriques, J.F., Caseiro, R., Martins, P., Batista, J.: High-Speed tracking with Kernelized correlation filters. IEEE Trans. Pattern Anal. Mach. Intell. **37**(3), 583–596 (2015)

Multi-local Feature Target Detection Method Based on Deep Neural Network

Guojie Li, Wenxue Wei and Wen Sun

Abstract In application of video surveillance system, the algorithm of object detection is affected by occlusion easily, and the results on small object are not satisfying. This paper develops a multi-local feature object detection method based on deep neural network. The image is used as input to calculate the position and category probability of the object through a single network calculation, which improves the operating efficiency. The core of the method is to extract multiple local features of the target for detection. When the target is partially occluded, it can identify the target by the unoccluded patch. In addition, the high-level and low-level features in the convolutional network integrate to improve the detection effect on small targets. Experimental results show that the proposed method has a good effect on the detection of occluded targets and small objects.

Keywords Deep learning · Convolutional neural network · Local feature · Feature fusion · Object detection

1 Introduction

Surveillance video usually serves as a necessity when the police are investigating cases, and tracking specific objects in the crime scene. It is of great significance for improving the detection rate and shortening the detection time. In order to ensure that a large amount of video information can be processed in a timely and effective manner, manual screening only is not favorable and equals much difficulty. How to quickly and accurately identify the interested object from surveillance video is a very big challenge for police. Now the accuracy of face recognition has reached a very high level, if the suspect's facial information can be collected in the live video, it will help the police to locate the suspect object in the surveillance system. However,

G. Li · W. Wei (✉) · W. Sun
College of Computer Science and Engineering, Shandong University of Science and Technology, Qingdao, China
e-mail: wwxjyh@163.com

© Springer Nature Singapore Pte Ltd. 2019   501
J.-S. Pan et al. (eds.), *Genetic and Evolutionary Computing*,
Advances in Intelligent Systems and Computing 834,
https://doi.org/10.1007/978-981-13-5841-8_52

in actual situations, the suspect usually uses a hat, a mask, etc. to occlude his or her face, so face recognition cannot achieve best results in practical applications.

Traditional object detection algorithm mainly contains two parts, namely, feature extraction and classifier design. The features of the object are usually divided by color, texture, and geometry [1]. Common classifiers include Bayes classifiers [2], neural network classifiers [3], Support Vector Machine (SVM), AdaBoost [4], etc.

In recent years, researchers have made a huge breakthrough in object detection using the deep learning method. Based on the object detection of region proposal, Ross Girshick proposed the object detection framework combining region proposal and CNN classification (including R-CNN [5], SPP-NET, Fast R-CNN, etc.), and R-CNN is the representation of the basic idea of converting object detection into regression problems, Joseph Redmon et al. raised You Only Look Once (YOLO) [6], a object detection system based on a single neural network; Wei Liu put forward Single Shot Detector (SSD) [7], which is based on the infusion of high-level features and low-level features, Tsung-Yi Lin came up with Feature Pyramid Network (FPN) [8], which shows obvious optimization in small object detection.

The analysis disclosed that the traditional object classification detection based on color, texture, and geometric features tend to be affected by factors such as illumination and background. Therefore, it is not easy to design a robust feature, and the selection of features directly affects the accuracy of classification [9]. The deep learning method can directly use the original images as input to automatically learn the expression of the features from the training sample, possessing strong robustness and fault tolerance. This paper proposes a method of detecting the object using multiple local features based on deep neural network. In detail, it extracts multiple local features from the suspect in the live video, and using these features to search for the object suspect in the monitoring system. Though some features of the object suspect have changed, the method can still detect the object through other features, so that the detection rate of the target is highly raised.

2   Related Work

2.1   Convolutional Neural Network

The two core operations of CNN are convolution and pooling. Convolution: mainly to extract features and also have certain dimensionality reduction. Pooling: mainly for dimensionality reduction, the commonly used pooling methods include maximum pooling and average pooling.

$$f(x) = \sum_{i=1}^{n} \sum_{j=1}^{n} \theta_i x_{ij} \tag{1}$$

The convolution operation is shown in Eq. (1). Where, θ_i represents the element in the convolution kernel, x_{ij} represents the element input into the corresponding location, and f(x) represents the convolution result.

2.2 Feature Pyramid

Feature pyramid is an important method in multi-scale object detection. As the depth increases, the receptive field of the convolutional neural network becomes larger and larger, and its abstraction ability and resistance to scale changes become stronger. The high-level features have low resolution and high semantic information, and only using high-level features in the classification can achieve favorable effects. However, in the object detection, this will greatly weaken the effect of small object detection and the accuracy of the bounding box. In response to such a problem, Tsung-Yi Lin proposed Feature Pyramid Network (FPN), which can modify directly on the original network, and integrates high-level features and low-level features, not only maintaining certain scale invariance, but also working well in handling multi-scale issues of object.

This paper also draws on this method to merge the characteristics of different convolution layers, which has produced satisfying results in small object detection and response to scale changes.

3 Network Design

3.1 Definition of Output Representation

The network directly takes the original images as input, and the data of the output layer is the detection result of the network. The input layer divides the input images into $9 * 9$ grids, and each grid is responsible for detecting objects at the center of the object falling into the grid.

Each grid is responsible for predicting two bounding boxes and their confidence. Each bounding box contains five predicted values: (x, y) represents the central coordinate of the bounding box, w(width) and h(height) indicate the width and height of the bounding box, and confidence shows the confidence rate of the object contained in the bounding box.

Each grid also predicts the probability that the grid belongs to a certain type of object. The number of object categories is C, which is determined according to the specific data and labeling.

Therefore, the output layer is a $9 * 9 * (2 * 5 + C)$ tensor, each grid correspondingly outputs a $1 * (2 * 5 + C)$ vector. The structure is shown in Fig. 1.

Fig. 1 Each grid corresponds to output layer vector structure

3.2 Definition of Loss Function

This paper uses mean square error as the loss function, as shown in Eq. (2)

$$loss = coordError + iouError + classError, \tag{2}$$

where coordError is the object coordinate error, which represents the error between the predicted coordinate value of object center point (x, y) and the actual value($\hat{x}$, $\hat{y}$), plus the error between the predicted width and height value of predicted frame (w, h) and the actual value ($\hat{w}$, $\hat{h}$); iouError is position confidence error, indicating the error between the confidence at the predicted position of the object and the actual value. The actual value of 1 indicates the existence of an object in the location, and 0 means no object; classError is the classification error, which shows the error between the predicted probability value and actual value of a certain category. The specific calculation formulas of the three errors are shown in (3)–(5).

$$coordError = \sum_{i=0}^{S^2}\sum_{j=0}^{B} 1_{ij}^{obj}[(x_i - \hat{x}_i)^2 + (y_i - \hat{y}_i)^2]$$

$$+ \sum_{i=0}^{S^2}\sum_{j=0}^{B} 1_{ij}^{obj}[(\sqrt{w_i} - \sqrt{\hat{w}_i})^2 + (\sqrt{h_i} - \sqrt{\hat{h}_i})^2], \tag{3}$$

$$iouError = \sum_{i=0}^{S^2}\sum_{j=0}^{B} 1_{ij}^{obj}(C_i - \hat{C}_i)^2] + \sum_{i=0}^{S^2}\sum_{j=0}^{B} 1_{ij}^{noobj}(C_i - \hat{C}_i)^2], \tag{4}$$

$$iouError = \sum_{i=0}^{S^2} 1_{ij}^{obj} \sum_{c \in classes}[p_i(c) - \hat{p}_i(c)], \tag{5}$$

where x, y, w, h, C, and p are the predicted network values, $\hat{x}$, $\hat{y}$, $\hat{w}$, $\hat{h}$, $\hat{C}$, and $\hat{p}$ are the labeled values, indicates that the object fell into the 1_i^{obj} border of grid i, means that the object did not fall into the 1_i^{noobj} border of grid i, shows that the object fell into grid i.

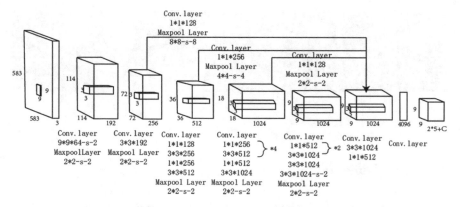

Fig. 2 Network structure

3.3 Network Structure

Based on convolutional neural network and deep learning theory, the network structure of deep convolutional neural network is designed as shown in Fig. 2.

Based on the idea of regression, this network implements a single neural network to predict the probability of the position and category of the predicted object by one calculation from the original input images. It contains 27 convolution layers and two fully connected layers. Different form GoogLeNet's inception model, the network adopts a structure similar to You Only Look Once (YOLO), adding a 1 * 1 convolution layer after the 3 * 3 convolution layer, and the 3 * 3 convolution layer is used to extract the feature map, 1 * 1 convolution layer is to reduce the feature space and calculating amount.

Considering that high-level features have low resolution and high semantic information, while low-level features have high resolution and low semantic information, the network design merges the features of high and low levels, respectively, connecting the 5, 14, and 20th convolution layers to the previous layer of the fully connected layers to combine with the features of the upper layer, which improves the detection of small object to a certain level.

4 Experimental Analysis

Four sets of data are used in this paper, three of which are Crossing, Girl2, and Basketball in TB-100 sequences, and another set is the data collected from the authors simulated scene (Fig. 3).

(1)   (2)   (3)   (4)

Fig. 3 Experimental data

4.1 Data Labeling

Use the open-source tool labelImg to label the data. Label the object of interest at the crime scene, and try to divide the object into multiple categories. For instance, in the video sequence Girl2, the little girl's tops, pants, and bicycle are labeled as three small objects, respectively. In this way, an alarm can be triggered as long as an object is detected. This will inevitably reduce the accuracy to a certain extent, but it also greatly improves the detection rate of the object, substantially avoiding the incapability of detecting the object. Moreover, we have adopted different training strategies as per the features of data, improving the object recall rate as much as possible on the basis of ensuring a certain accuracy rate.

4.2 Experimental Results

(1) The experimental results for dataset 1 are shown in Fig. 4 and Table 1.

The dataset contains a total of 120 frames of images. Pedestrians are crossing the road along the zebra crossing. The camera is fixed and the object moves in a shooting area. There are two pedestrians in the frames, one of them is selected as the object and the other as an interference. All the 120 frames contain the object and

Fig. 4 Result 1

Table 1 Result 1

Features	Recall	Accuracy rate
Object	0.8	0.83
Non-object	0.83	0.88

Fig. 5 Result 2

Table 2 Result 2

Features	Recall	Accuracy rate
Girl's jacket	0.76	0.77
Girl's pants	0.72	0.72
Girl's scooter	0.66	0.63
Man's bicycle	0.60	0.67
Man's cup	0.55	0.73

108 frames contain the interference. 90 frames of them are used as training samples and 30 frames as test samples.

Recall rate: the number of object detected/the number of object in the sample containing the object; accuracy rate: the number of object detected/the total number of object detected. Since the scene is relatively singular, the trained network can well detect the object and non-object.

(2) The experimental results for dataset 2 are shown in Fig. 5 and Table 2.

This dataset contains 1500 frames of images and the camera follows the little girl moving from one scene to another. Some features on the object will be occluded and deformed during the movement. 1200 frames are used as training samples and 300 frames as test samples.

The results of the network on dataset 2 show the advantages of dividing the object into local features for detection. In data labeling, we mark the target girl's tops, pants, and bicycle as the features to be detected. The target features are relatively obvious and we try to raise the accuracy of each feature as much as possible. In the subsequent scenes, because the targets partial features are occluded with some dressing changes, etc., the recall rate of single feature decreases to a certain level, but the recall rate of the entire object reaches nearly 100%, largely reducing the possibility of missing detection.

Fig. 6 Result 3

Fig. 7 Result 4

(3) The experimental results for dataset 3 are shown in Fig. 6.

This dataset contains 725 frames of images, the camera follows the athletes' fast movement, and there are multiple similar moving object in this scene. Reduce the accuracy during training to improve the object recall rate. The final recall rate is 89%, and the accuracy rate is 53%.

The results indicate that although we have labeled the No. 9 athlete wearing a green jersey, the reduced accuracy leads to multiple detected object in each frame, the No. 9 player included. Although the detecting accuracy is decreased, it effectively avoids missing detection, which facilitates the police in spotting the object and detecting the case.

(4) The experimental results for dataset 4 are shown in Fig. 7 and Table 3.

The dataset 4 contains 200 frames of images. It simulates the scene that the suspect leaves the video information on the scene after committing the crime, and the suspect in the car is monitored after passing a certain road. 160 frames are used as training samples and 40 frames are used as test samples.

The experimental results show that the labeled feature 3 is the suspect's blouse. When the suspect is sitting in the car, affected by the illumination change and occlusion, feature 3 completely loses the function of detecting the object. Although features

Table 3 Result 4

Features	Recall	Accuracy rate
Orange hat	0.73	0.77
White hat	0.77	0.75
Man's jacket	0.3	0
Man's mask	0.76	0.65

1, 2, and 4 are interfered by windshield, but well-trained networks can still detect, and the recall rate of the final suspect can reach 0.95 above. The object occupies a small proportion in multiple frames, but after the integration of high- and low-level features, the recall rate of features remains at a high level.

5   Conclusion

This paper proposes a method based on deep neural network to detect object using multiple local features. The method uses a single network as the foundation with the original images as input, and the position and category of the object as output resulting in an improvement of operating efficiency. On the one hand, considering that the object is occluded or the dressing is partially changed in the video, the paper proposes a multi-local feature detection method that divides the object into multiple parts, which can well adapt to these situations; on the other hand, most of the objects are small in the surveillance video, the paper put forward the idea of integrating the high- and low-level features of the convolutional network to enhance the detection effect of small object. The experimental results show that the object recall rate can be improved at the expense of certain accuracy rate, which can better avoid the probability of missing object and is more in line with the requirements of detection, and it proves the advantages of integrating multi-feature detection with high- and low-level features. When the object is small and some features change, a high recall rate of features can still remain.

References

1. Niblack, C.W., et al.: QBIC project: querying images by content, using color, texture, and shape. Storage and Retrieval for Image and Video Databases, vol. 1908. International Society for Optics and Photonics (1993)
2. Simonyan, K., Vedaldi, A., Zisserman, A.: Deep inside convolutional networks: visualising image classification models and saliency maps (2013). arXiv:1312.6034
3. Cireşan, D.C., et al.: Mitosis detection in breast cancer histology images with deep neural networks. In: International Conference on Medical Image Computing and Computer-Assisted Intervention. Springer, Berlin (2013)
4. Yamada, K., et al.: Image recognition for automatic traveling wheelchair. In: 2016 IEEE 5th Global Conference on Consumer Electronics. IEEE (2016)

5. Girshick, R., et al.: Rich feature hierarchies for accurate object detection and semantic segmentation. In: Proceedings of the IEEE Conference on Computer Vision and Pattern Recognition (2014)
6. Redmon, J., et al.: You only look once: unified, real-time object detection. In: Proceedings of the IEEE Conference on Computer Vision and Pattern Recognition (2016)
7. Liu, W., et al.: Ssd: single shot multibox detector. In: European Conference on Computer Vision. Springer, Cham (2016)
8. Lin, T.-Y., et al.: Feature pyramid networks for object detection. In: CVPR, vol. 1. No. 2. (2017)
9. Armanfard, N., Komeili, M., Kabir, E.: TED: a texture-edge descriptor for pedestrian detection in video sequences. Pattern Recognit. **45**(3), 983–992 (2012)

Medical Image Diagnosis with Deep Learning on FPGA Platform

Kuo-Kun Tseng, Ran Zhang and Lantian Wang

Abstract In recent, cancer has become a major public health problem and the leading cause of death. Therefore, early screening and accurate diagnosis of cancer are extremely important. With the development of medical image equipment, image diagnosis has made great contributions to the improvement of the medical standard. In this paper, the deep learning algorithm is applied to the computer-aided detection, so that the structure of the medical image detection can meet the practical requirements. Moreover, we have tried to implement the deep learning algorithm on FPGA, which can improve the performance effectively with compared to GPU and CPU architectures.

Keywords Medical image · Diagnosis · Deep learning · FPGA

1 Introduction

Medical Image Diagnosis (MID) based on deep learning and image processing technology has gradually become a research hotspot in the medical field [1], and related research has made some progress. MID based on machine learning mainly includes four aspects: (1) image preprocessing (2) region of interest (ROI) segmentation (3) feature extraction, selection and classification (4) tumor region recognition (classification or segmentation) [2]. Among them, efficient feature extraction is particularly critical [3].

At present, MID systems based on traditional machine learning structure are highly dependent on the features selected by human beings and the integration of features by classifiers. Moreover, because the traditional machine learning structure cannot meet the requirements of complex function modeling in practical applications, it is difficult to distinguish the relationship between high-dimensional features, which

K.-K. Tseng (✉) · R. Zhang · L. Wang
School of Computer Science and Technology, Harbin Institute of Technology (Shenzhen), Shenzhen, China
e-mail: kktseng@hit.edu.cn

© Springer Nature Singapore Pte Ltd. 2019
J.-S. Pan et al. (eds.), *Genetic and Evolutionary Computing*,
Advances in Intelligent Systems and Computing 834,
https://doi.org/10.1007/978-981-13-5841-8_53

usually requires dimensionality reduction. Therefore, we need to simplify and optimize the process of feature selection in MID technology to improve the accuracy of the MID system for auxiliary diagnosis.

In recent years, as a kind of multilayer neural network learning algorithm, deep learning technology [4] can learn features through nonlinear deep network structure, and form more abstract deep representation (attribute class or feature) by combining low-level features, so that complex function approximation can be realized essential feature of data set can be learned.

Therefore, the deep learning algorithm applied to MID system has the following advantages: first, as a data-driven automatic feature learning algorithm, it can extract features directly from training data, thus workload of feature extraction and the impact of manual intervention are greatly reduced. Second, the intrinsic deep structure of the neural network can represent the interaction and hierarchical structure between features, thus revealing the relationship between high-dimensional features. Third, feature extraction, feature selection, and feature classification can be achieved in the optimization of the same deep structure. Therefore, deep learning is expected to solve the MID problem based on traditional machine learning, greatly improving the ability of auxiliary diagnosis.

Generally speaking, deep learning algorithms can be divided into four categories: model based on restricted Boltzmann machine (RBM), convolutional neural network (CNN), model based on autoencoder, and model based on sparse coding [5]. At present, deep learning networks for cancer medical image analysis mainly include RBM-based DBN, CNN, self-coding-based SAE, DAE, and their improved algorithms.

As the deep learning model can learn the hidden high-level features from the image directly, it can achieve better results than other learning algorithms based on manual extraction of low-level features (such as color, shape, texture, and edge). Generally, the training time of deep learning model is longer than that of various structure algorithms, but it depends on the model design and the selection of training set. For different types of data sets, the performance of different deep learning models is different. In addition, the hybrid model which is composed of the deep learning model and other learning algorithms may improve the accuracy compared with the method using only the basic deep learning model, while the improved deep learning model may improve the segmentation accuracy and reduce the running time.

In tumor classification, the deep learning models can greatly improve the accuracy, sensitivity, and specificity of the MID system. Generally speaking, large-scale image data sets can greatly improve the classification accuracy of deep learning model [6]. Therefore, in order to overcome the problem of small sample size in medical image data sets, the migration learning method is proposed, and to a certain extent, the accuracy of tumor classification is improved. Therefore, for different types of tumors and images, it is necessary to consider when choosing a deep learning model and designing an algorithm framework, so as to achieve better performance.

2 Related Research

Medical image diagnosis models based on deep learning are mostly based on the convolutional neural network model, so we mainly divide them into three aspects: network structure, learning algorithm, and hardware system. The network structure is mainly the input layer, the hidden layer and the number of CNN. We mainly improve the hidden layer, from the initial progressive convolution network to multiscale, multichannel convolution network [7], and then to an adaptive cascaded convolution neural network. The learning algorithm is mainly optimized algorithm and combined with other methods. The optimization algorithm is embodied in the dropout, Relu, sparse representation, Fisher criterion and so on. Here, we can combine PCA [8], SVM [9], Softmax classifier [10] and other methods. The hardware system develops from the initial use of CPU computing, to the CUDA-based GPU computing, and then to the FPGA-based computing method.

In the 1990s, the Convolutional Neural Network (CNN) proposed by Lecun et al. [11] combines the two-dimensional discrete convolution operation in image processing with the multilayer neural network. Since G.E. Hinton [12] put forward the concept of deep learning in 2006, deep learning has been successful in many fields, showing unparalleled performance advantages. In 2007, when NVIDIA launched the GPU software interface of CUDA, its speed was 70 times faster than that of the traditional dual-core CPU. With the continuous updating of NVIDIA GPU, it provides great help for deep learning in hardware.

In 2012, Hinton's student Krizhevsky proposed the deep convolution neural network model AlexNet [13], and successfully applied ReLU, Dropout, and LRN in CNN for the first time. The error rate of top-5 in ILSVRC 2012 was reduced to 16.4%. In 2014, Google introduced the GoogleNet [14] with the top 5 error rates of 6.67% in the ILSVRC 2014 competition. In addition, Karen Simonyan and Andrew Zisserman have established a 19-tier deep network VGGNet [15], which also yields good results, ranking first and second in ILSVRC 2012. In 2015, Carneiro et al. [16] based on mammographic mammography (including bilateral mammary axis and bilateral oblique mammary gland) and segmented microcalcifications and mass areas, used ImageNet pretrained CNN model to classify the masses, thereby assessing the risk of breast cancer development in patients. In 2016, Andrew Beck of the BIDMC used deep learning to develop a classification model [17], which examines lymph node images to determine whether breast cancer can be shown, with a diagnostic accuracy of 92%. When they combined the pathologist's diagnosis with our automated method of diagnosis, the accuracy of the diagnosis increased to 99.5%. Kawahar et al. [18] proposed an improved CNN model BrainNetCNN, which was used to predict the neural development of brain networks in premature infants. Unlike the traditional image-based CNN (spatial local convolution), BrainNet CNN is a deep learning framework specially designed for brain network data. It utilizes the topological location of the structured brain network and has a unique design of convolution kernel from edge to edge, from edge to node, and from node to graph. The predicted cognitive and motor development results show that the BrainNetCNN framework is superior to other similar methods.

3   Related Technology

Caffe is an open source software framework (Fig. 1), which provides a basic programming framework to implement deep convolutional neural networks, Deep Learning and other algorithms under the GPU parallel architecture.

Since the data formats supported by Caffe are LMDB and LevelDB, we need to convert the medical image data in the database into the formats supported by Caffe. Here, we choose the format that supports better LevelDB in Windows to build our data layer. In addition, in order to speed up the training and improve the diagnostic accuracy, we will average all training samples and save them as a mean file.

The training process can be divided into two stages: forward and backward. First, we use several convolution kernels of different sizes to convolute the sample set in the data layer, and obtain conv Layer, and extract local features. Then, in order to reduce the computational complexity and data dimension, we take the maximum value of the region as the sampling value, and get the pooling layer (Pooling Layer), which greatly reduces the generation of over-fitting. In downsampling, we first add weights and biases to the features obtained by convolution, and then get the feature map by a Relu activation function. Among them, the Relu activation function is

$$f(x) = \max(0, x) \tag{1}$$

Then, multiple convolutions and maximum pooling operations are performed, respectively, to get the depth features. The probability likelihood value of each class is obtained by softmax classifier, and loss is obtained. Among them, the softmax formula is

$$f(z_j) = \frac{e^{z_j}}{\sum_{i=1}^{n} e^{z_j}} \tag{2}$$

The loss function of softmax uses logarithmic loss function:

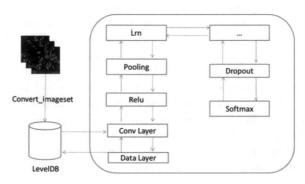

Fig. 1 Universal disease diagnosis framework

$$\text{loss} = -\log f(z_j) \tag{3}$$

Among them, j is the label of the sample (that is, the correct output of the sample). Usually, we have more than one batch of pictures for training, such as batch size = 32, then

$$\text{loss} = \sum_{i=0}^{batch\,size} -\log f(z_j) \tag{4}$$

Finally, from the softmax level, the backpropagation is carried out. Random gradient descent algorithm is used to calculate the gradient of each layer according to loss function, and then adjust the weight to update the loss, so as to optimize the network structure.

Next, we can classify the pictures in the test set according to the network model after training, and calculate the correct rate and the final loss.

4 Hardware Architecture Used FPGA

In order to save cost and meet the real-time requirement of disease diagnosis, this paper intends to use FPGA instead of GPU and CPU as computing hardware resources. The advantage is that FPGA is cheaper than GPU and the computation speed is much better than that of CPU.

4.1 Python Integration on PYNQ

To leverage the use of hardware acceleration units and simplify the use of advanced programming frameworks, we developed the required APIs to allow transparent deployment of accelerators. Specifically, we developed the libraries needed to instantiate the kernel from a high-level language such as python, which is widely used for machine learning tasks. The whole process is done using Pynq Board, a prototype of Digilent that comes with a Linux image containing Python libraries that help designers use the kernel from Python scripts. The whole process is described as follows:

We created a bit stream for our IP matching new device (PYNQ), using Vivado.

Then, the software part of the algorithm is written by Python, using efficient libraries (NUMPY, SISPY).

Finally, by using the Linux image library of PYNQ, suitable software drivers are created for software and hardware communication.

In the last step of the above process, we must manually perform the automatic execution of the SDSOC framework. The Python library is a C language wrapper for interprocess communication. This wrapper is accomplished by using a library

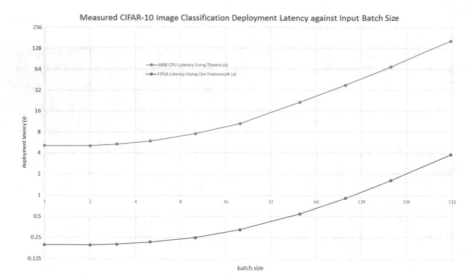

Fig. 2 Plot of deployment latency for CPU and FPGA against input batch size

called cffi, which allows Python scripts to execute precompiled C code as source code. This means that this integration can happen on any platform, not Pynq. In the experiment, FPGA has outperformance than ARM processor against any input batch size for CIFAR-10 image classification (Fig. 2).

5 Conclusion

With the development of hardware equipment, the computing ability will be improved, to some extent, it can help to solve the efficiency problem of deep learning technology. In summary, on the basis of the combination of medical imaging technology and image processing technology with deep learning, it can be predicted that MID technology is promising to improve the efficiency and accuracy of clinical diagnosis.

References

1. Doi, K.: Computer-aided diagnosis in medical imaging: historical review, current status and future potential. Comput. Med. Imaging Graph. **31**(4), 198–211 (2007)
2. Xu, Y., Jia, Z., Ai, Y., et al.: Deep convolutional activation features for large scale brain tumor histopathology image classification and segmentation. In: IEEE International Conference on Acoustics, Speech and Signal Processing, pp. 947–951. IEEE (2015)

3. Cheng, J.Z., Ni, D., Chou, Y.H., et al.: Computer-aided diagnosis with deep learning architecture: applications to breast lesions in US images and pulmonary nodules in CT scans. Sci. Rep. **6**, 24454 (2016)
4. Lecun, Y., Bengio, Y., Hinton, G.: Deep learning. Nature **521**(7553), 436 (2015)
5. Guo, Y., Oerlemans, A., Oerlemans, A., et al.: Deep learning for visual understanding. Neurocomputing **187**(C), 27–48 (2016)
6. Cho, J., Lee, K., Shin, E., et al.: How much data is needed to train a medical image deep learning system to achieve necessary high accuracy?. Comput. Sci. (2016)
7. Eigen, D., Puhrsch, C., Fergus, R.: Depth map prediction from a single image using a multi-scale deep network. In: International Conference on Neural Information Processing Systems, pp. 2366–2374. MIT Press (2014)
8. Yang, J., Zhang, D., Frangi, A.F., et al.: Two-dimensional PCA: a new approach to appearance-based face representation and recognition. IEEE Trans. Pattern Anal. Mach. Intell. **26**(1), 131–137 (2004)
9. Chen, P., Lin, C., Bernhard, S.: A tutorial on ν-support vector machines. Appl. Stoch. Model. Bus. Ind. **21**(2), 111–136 (2005)
10. Liu, W., Wen, Y., Yu, Z., et al.: Large-margin softmax loss for convolutional neural networks. In: International Conference on International Conference on Machine Learning. JMLR.org, pp. 507–516 (2016)
11. Cun, Y.L., Boser, B., Denker, J.S., et al.: Hand written digit recognition with a back-propagation network. Adv. Neural. Inf. Process. Syst. **2**(2), 396–404 (1990)
12. Hinton, G.E., Salakhutdinov, R.R.: Supporting online material for reducing the dimensionality of data with neural networks. Methods 504 (2006)
13. Krizhevsky, A., Sutskever, I., Hinton, G.E.: ImageNet classification with deep convolutional neural networks. In: International Conference on Neural Information Processing Systems, pp. 1097–1105. Curran Associates Inc. (2012)
14. Szegedy, C., Liu, W., Jia, Y., et al.: Going deeper with convolutions. 1–9 (2014)
15. Simonyan, K., Zisserman, A.: Very deep convolutional networks for large-scale image recognition. Comput. Sci. (2014)
16. Carneiro, G., Nascimento, J., Bradley, A.P.: Unregistered multiview mammogram analysis with pre-trained deep learning models. In: International Conference on Medical Image Computing and Computer-Assisted Intervention, pp. 652–660. Springer, Cham (2015)
17. Wang, D., Khosla, A., Gargeya, R., et al.: Deep learning for identifying metastatic breast cancer (2016)
18. Kawahara, J., Brown, C.J., Miller, S.P., et al.: BrainNetCNN: convolutional neural networks for brain networks; towards predicting neurodevelopment. Neuroimage **146**, 1038–1049 (2016)

Part XIII
Artificial Intelligence Applications

Innovative Financial Prediction Model Based on Lasso and Keras Neural Network

Xiaogang Wang and Shichen Zhai

Abstract Data analysis has been widely used for mining valuable information. This paper first studies the influence factors for the data of city revenue, and then mines the hidden relationship utilizing the Lasso regression analysis of selected key features. Finally, we construct our innovative financial prediction model based on Lasso and Keras neural network (PMBLanK). By comparing the performance of our PMBLanK model with other widely used prediction models, we can see that our PMBLanK model has higher performance over existing models with lower MSE, which can be used to predict fiscal revenue of coming years effectively.

Keywords Data analysis · Feature engineering · Neural networks · Model prediction

1 Introduction

With the rapid economic development and rapid growth of GDP, financial revenue is also increasing. Based on the historical data of fiscal revenue, it is very important for the city government to make scientific fiscal policy by analyzing factors affecting revenue and predicting effective fiscal income. Considering the availability of the data, we use the data obtained from the statistical yearbook, mainly including 1994–2013 fiscal revenue and related factors [1].

Nowadays, many domestic scholars are studying the influence factors of fiscal revenue by establishing multivariate linear model between the fiscal revenue and the undetermined factors of regression model. Furthermore, they use the least squares

X. Wang (✉)
School of Software, Changzhou College of Information Technology, Changzhou, P.R. China
e-mail: wangxiaogang@ccit.js.cn

S. Zhai
Computer and Software Engineering Department, Anhui Institute of Information Engineering, Wuhu, P.R. China
e-mail: sczhai@iflytek.com

© Springer Nature Singapore Pte Ltd. 2019
J.-S. Pan et al. (eds.), *Genetic and Evolutionary Computing*,
Advances in Intelligent Systems and Computing 834,
https://doi.org/10.1007/978-981-13-5841-8_54

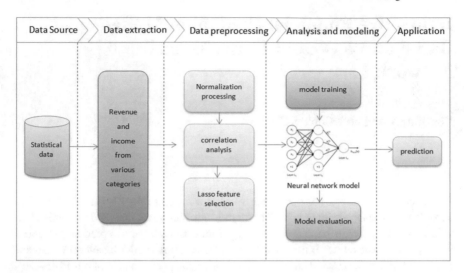

Fig. 1 Analysis process of PMBLanK model

method to estimate the coefficients of the regression model. Such coefficients can be used to test the relationship between various factors. However, the results of the model are highly dependent on the data, and the algorithm is often used to solve the local optimal solution. The subsequent verification may lose its due significance.

Therefore, in order to improve the forecasting model, we first study the correlation between the factors of the impact of fiscal revenue using Pearson algorithm, and then use Lasso [2] regression analysis method of key feature selection. Finally, we use the history data to establish a training model based on Keras [3] (i.e., neural network), since such a neural network has the practicability and good fault-tolerant ability. In this way, more accurate prediction results can be obtained.

The overall analysis process of our PMBLanK model is illustrated in Fig. 1.

2 Relative Algorithms

2.1 Pearson Correlation

Correlation analysis refers to the analysis of two or more feature elements with correlation, so as to measure the degree of correlation between feature factors. Statistically, Pearson correlation coefficients are commonly used to measure the linear correlation between two features X and Y in the range of $[-1,1]$.

If we have two vectors $X = (x_1, x_2, \ldots x_n)$ and $Y = (y_1, y_2, \ldots y_n)$, the Pearson correlation coefficients between them are as follows:

$$COR(X, Y) = \frac{\sum_1^n (X_i - \bar{X})(Y_i - \bar{Y})}{\sqrt{\sum_1^n (X_i - \bar{X})^2 \sum_1^n (Y_i - \bar{Y})^2}} \quad (1)$$

where (X_i, Y_i) means sample point, $\bar{X}$ and $\bar{Y}$ stand for the average. If the coefficient value belongs to $[0,1]$, we can say that positive relation exists between the two vectors. If the value belongs to $[-1,0]$, then we can say the negative correlation. If the value is 0, then there is no relation between the two vectors.

2.2 Lasso Regression Method

Lasso regression is constructed by adding regularization parameters for compression based on ordinary least squares linear regression. Such regression model includes a penalty function. In order to reduce the feature set, model features can be compressed and the regression coefficient is set to 0. Such a method of feature selection can be widely used for model selection and improvement. We can first construct the cost function using ordinary least squares linear regression.

$$RSS = \sum_{i=1}^n \left(y_i - \beta_0 - \sum_{j=1}^p \beta_j x_{ij} \right) \quad (2)$$

where RSS can be minimized by fitting coefficient beta. The method is simple. Partial derivative can be solved by linear algebra. According to the theory of linear algebra, as long as the sample size is suitable, we can obtain a unique optimal solution of the model. The RSS reaches a minimum, however, all characteristics concerned are equally important degrees causing overfitting.

The expression of the cost function used in Lasso regression is as follows:

$$\sum_{i=1}^n \left(y_i - \beta_0 - \sum_{j=1}^p \beta_j x_{ij} \right) + \lambda \sum_{j=1}^p |\beta_j| \quad (3)$$

where Lambda is a nonnegative parameter. If Lambda is 0, it is consistent with the RSS, did not play any role for punishment. When Lambda is approaching infinity, the penalty is infinite. In order to make the minimum cost function, we can only compress coefficient (β) and set its value close to 0.

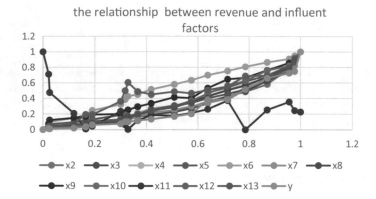

Fig. 2 Relationship between financial revenue (y) and the independent variable

2.3 Keras Neural Network

Keras is such a kind of neural network based on Theano and TensorFlow neural network [4]. Keras can be quickly constructed and deployed in Windows and MacOS or Linux environment. Keras neural network shown in Fig. 2 consists of a unit input layer, an output layer, and a hidden layer. Keras is a deep learning framework implemented by pure Python Theano and TensorFlow. Keras is a high-level API neural network, supporting rapid test. It has some advantages, such as user-friendliness, good modularity, easy extensibility, and supporting Python programming.

3 Data Analysis

3.1 Data Exploration and Analysis

There are many factors affecting financial revenue(y). Here, we choose following factors: the number of employees in the society (x1), total wages of workers on duty (x2), total retail sales of consumer goods (x3), per capita disposable income of urban residents (x4), per capita consumption expenditure of urban residents (x5), total population at the end of the year (x6), the investment in fixed assets of the whole society (x7), regional gross product (x8), the output value of the first industry (x9), tax (x10), consumer price index (x11), the output value ratio of tertiary industry to secondary industry (x12), and consumption level of residents (x13).

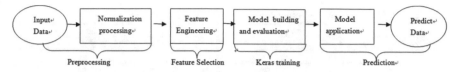

Fig. 3 Process for Keras prediction model

3.2 Correlation Analysis

In order to analyze the correlation between dependent variables and independent variables, we first calculate the Pearson correlation coefficient between variables to determine whether there is a correlation.

From the table, a correlation coefficient can be found in the consumer price index and (x_{11}) showed a negative correlation with the linear relationship between fiscal revenue and fiscal revenue, the remaining variables were positively correlated.

This form does not seem intuitive. In order to more accurately analyze the relation between fiscal revenue and the relationship between the variables, we first normalize values by making the data belonging to [0,1]. The standardization of data (Normalization) is proportional to zoom in a specific small interval. The data will be transformed into pure numerical dimensionless without considering different units or the magnitude of the index.

Figure 2 shows the relationship between the strain y (fiscal revenue) and the independent variables. We can see that the overall decline trend of X11 is a negative correlation with y, and other dependent variables are positively correlated with y.

4 Keras Prediction Model

Based on the above analysis, we can select the features of the sample data and construct the evaluation and prediction model. The basic process is shown in Fig. 3.

4.1 Feature Selection Model Based on Lasso

We first propose the Lasso variable selection algorithm as follows:

```
#Lasso variable selection
From sklearn.linear_model import Lasso
Model = Lasso (10) # calls the Lasso () function and sets the value of lambda to 10.
To improve accuracy, the penalty factor for loss needs to be adjusted
Model.fit (data.iloc[:, 0:13], data['y'])
q = model.coef_ #The coefficients of each characteristic
```

Table 1 Relationship between influence factors and characteristic coefficients

X1	X2	X3	X4	X5	X6	X7
-0.000	-0.3155	0.4329	-0.0316	0.0758	0.0004	0.2413

X8	X9	X10	X11	X12	X13
-0.037	-2.5545	0.4414	0.0000	0.000	-0.0399

q = pd. DataFrame (q, index = ['x1', 'x2', 'x3', 'x4', 'x5', 'x6', 'x7', 'x8', 'x9', 'x10', 'x11', 'x12', 'x13')).Tnp.round (q, 4).to_csv ('dataLasso.csv')

The result by executing the above Lasso variable selection algorithm is shown in Table 1.

Obviously, the result shows that X_1, X_{11}, and X_{12} are 0, i.e., these factors have a very small influence on financial income and are negligible.

4.2 Model Building and Evaluation with Keras

We can then construct the grey prediction and neural network prediction model by utilizing Lasso variable selection method, where the parameter is set to 0.0000001 accuracy, learning times is 10000, the number of the input neurons is 12 (except X12). After several tests, we found that the input layer node is 12, the hidden layer node is 12, and the output node is 1 for the neural network model.

The model evaluation index for Keras model can be shown in Table 2.

The comparison of the real value y and the predicted value y_pred of fiscal revenue is shown in Fig. 4.

We can see that the various indicators have reached the optimal by comparing the true value and predicted values. Obviously, the fitting effect is very good. Our proposed prediction model can be used to predict financial revenue.

Based on the evaluation index model and comparison analysis, we can see that our PMBLanK has a higher performance with higher accuracy rate.

Table 2 Model evaluation index for Keras model

Index name	Index result
Mean absolute error	0.28508953857421704
Mean square error (MSE)	0.13034514592594496
Absolute error of median	0.25281188964842727
Interpretable value of variance	0.9999997121396272
R square value	0.9999996303644205

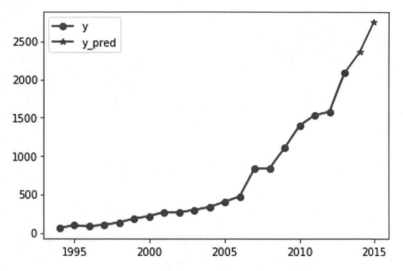

Fig. 4 Comparison of the real value y and the predicted value y_pred of fiscal revenue

5 Summary

On the basis of studying the influence factors for the data of city revenue, we propose an innovative financial prediction model based on Lasso and Keras neural network (PMBLanK). Estimation and experimental data show that our proposed MBLanK model can well fit the trend of fiscal revenue based, has high precision, can be used to guide the practical work.

Acknowledgements This work is supported by Top-notch Academic Programs Project of Jiangsu Higher Education Institutions (TAPP) under Grant No. PPZY2015A090, and Jiangsu top six talent summit project under Grant No. XXRJ-011.

References

1. Hongmei, H., Liangjun, Z.: Python data analysis and application [M]. People's Post and Telecommunications Publishing House (2018) (3)
2. Tibshirani, R.: Regression shrinkage selection via the LASSO [J]. **73**(3), 273–282 (2011)
3. Bickel, P.J., Ritov, Y., Tsybakov, A.B.: Simultaneous analysis of Lasso and Dantzig selector [J]. Ann. Stat. **37**(4), 1705–1732 (2009)
4. Liping, H., Xinyue, H., Li, L.: Artificial neural network model for credit risk analysis of commercial banks [J]. Syst. Eng. Theory Pract. **21**(5), 62–69 (2001)

Innovative Text Extraction Algorithm Based on TensorFlow

Shichen Zhai, Xiaogang Wang, Di Xiao and Zhiwen Li

Abstract Extracting business registration information exploiting graphic recognition algorithms on the Internet nowadays is vital to e-commercial business. However, business registration information is usually presented in graphics and existing graphic recognition systems have been hindered because of their slow detection speed, low accuracy, and complex operations. Thereby, we propose an innovative text extraction algorithm based on TensorFlow (TEAT). We first utilize the web crawler to obtain the data source, and then extract the character information by using our TEAT based on TensorFlow framework recognition technology. Our TEAT algorithm can extract business registration information efficiently and effectively. Comparing with existing text extraction algorithm based on Tess4j framework for extracting Tmall shop business license picture information, our TEAT has obvious advantages over Tess4j framework with higher accuracy and efficiency.

Keywords OpenCV · Convolutional neural network · Network crawler · Multithread distributed

1 Introduction

With the rapid development of the Internet, electronic commerce has been widely deployed, requiring more and more information processing. However, existing image recognition technology needs to be improved due to the low effectiveness and efficiency of image information processing. Therefore, we propose an innovative text extraction algorithm based on TensorFlow, combining with a web crawler and mul-

X. Wang (✉)
School of Software, Changzhou College of Information Technology, Changzhou, P.R. China
e-mail: wangxiaogang@ccit.js.cn

S. Zhai · D. Xiao · Z. Li
Computer and Software Engineering Department, Anhui Institute of Information Engineering, Wuhu, P.R. China
e-mail: sczhai@iflytek.com

© Springer Nature Singapore Pte Ltd. 2019
J.-S. Pan et al. (eds.), *Genetic and Evolutionary Computing*,
Advances in Intelligent Systems and Computing 834,
https://doi.org/10.1007/978-981-13-5841-8_55

529

tithreading convolutional neural network. Comparing with existing text extraction algorithm based on the tess4j framework, our TEAT is simpler and can effectively optimize the graphic analytical performance.

2 Key Technologies

2.1 CNN Convolution Neural Network Algorithm

Convolutional Neural Network (CNN) is an important algorithm for deep learning. We use CNN to calculate the propagation output value, to adjust the weights and bias of backpropagation. The structure of CNN is different from BP [1], where adjacent layers of neural cells are not fully connected. That is, the perception of unit area in CNN comes from parts of the upper nerve unit, not from all the nerve units like BP.

CNN consists of four main processes. Image convolution processing is used to obtain image eigenvalues. Image pooling can be used to reduce the size of feature mapping and the number of feature values. A multilayer convolution layer, pooling layer, is constructed to express the picture information with fewer eigenvalues, thereafter reducing the amount of calculation and speeding up the operation. Full connection layer processing is used to extract eigenvalue information for image classification by eigenvalue information.

2.2 TensorFlow

TensorFlow [1] is an open source software library used for numerical calculation by using a data flow diagram (data flow graphs). It can also be called the second generation of artificial intelligence learning system based on DistBelief or Google research. By obtaining enough data crawling under the premise, TensorFlow uses a convolutional neural network (CNN) to train the crawler to obtain the picture, and then calls the training model for image recognition. So it is often used in machine learning and deep neural network research.

3 Text Extraction Algorithm Based on TensorFlow

Our TEAT consists of three steps including data acquisition and preprocessing, business feature extraction and recognition model evaluation.

3.1 Data Acquisition and Preprocessing

The business license pictures of online stores are usually uploaded by e-commercial websites. We use Spyder and Selenium to simulate the browser and shortcut keys for crawling data. Since e-commercial websites usually deploy the protection mechanism for the access to business license pictures, we are required to simulate user inputs over the analog input box and make confirmation by clicking the OK button. If the verification code is not correct, we need to download the code image again until the model can get the correct recognition results. Successfully downloading a business license image will visit the next business license image URL link. The working process of data acquisition is as shown in Fig. 1.

After obtaining the shop business license pictures, we separate them into two different folders by manual removal of duplication and separation of the background, so as to guarantee the validity of the model when training images. We can obtain comprehensive business license pictures to increase the number of Chinese characters training set, thus improve the recognition accuracy.

Since any picture on a web page is coded into a binary file and stored in the BLOB field on the Cloud database, we first use nonblocking thread pool mode for multithreaded batch image processing. We then use the third party package based on Python for image preprocessing. We modify the Alpha (transparency) channel value for RGBA image to 255 (i.e. non-transparent), so as to effectively remove the watermark through threshold processing later.

We first use OpenCV to convert the image into a correlation matrix which is further processed by gray-scale processing, so as to convert the picture into a black

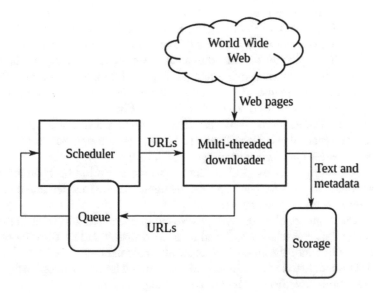

Fig. 1 Working process of data acquisition

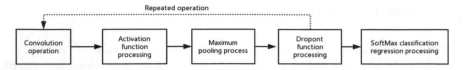

Fig. 2 Feature value extraction flowchart

and white picture. A black and white picture can reflect the image features better, reducing the original picture data and increasing processing efficiency. We then use the image threshold to remove the watermark in the image. We can get the blank position coordinates between text characters by looking for valley points over the projection histogram. According to the position coordinates, we can obtain a picture containing only a single character. Finally, each image is converted into the same pixel size picture and stored in the list through the graphic transformation. Such single-character pictures in the list can be effectively used to facilitate the convolution neural network training.

3.2 Feature Extraction

Our TEAT extracts the image information using the convolution neural network. We first store the single-character pictures accompanied with annotation tag sets by manual into a batch. Each batch contains 100 text images and the corresponding labels for batch processing so as to speed up the training of the model. Since the Chinese labels cannot be properly identified, we convert Chinese characters into Unicode encoding from 0x4E00 to 0x9FA5, where each code represents a Chinese character. We use one-hot markers, i.e. a bit represents a Chinese character.

In our TEAT, all pictures contained in a batch are processed using multiple convolutions, activation function, maximum pooling, and dropout operations. Finally, the image features are extracted by a full connection layer and are used to classify images. The process flow for our TEAT is shown in Fig. 2.

Step1: Multiple convoluting—We use convolution kernel size of 3 * 3, a step size of 1 to map the eigenvalues of text images so as to extract the eigenvalues of a picture. An image of 10 * 10 pixels is convoluted to an image of 8 * 8 pixels.

Step2: Activating—We use ReLU function as the activation function to process nonlinear equations so as to speed up the convergence of the function and reduce the time required for training the model.

Step 3: Maximum pooling operating—We output the maximum value of the window by using the 2 * 2 window size and 2 step size so as to reduce the dimension of eigenvalues and reduce the amount of eigenvalue calculation.

Step 4: Dropout function operating—We use dropout function operation to prevent the final function curve from over-fitting by adding noise points.

Step 5: Repeat convolution to dropout operating—Such an operation aims to the repetition of feature extraction and dimension reduction with fewer coefficients to represent images, thus reducing the time required to calculate the similarity value of feature recognition process and speeding up the running speed.

Step 6: Fully connected layer processing—This process is to extract the features of the given images by fully connected layer. We can represent the values of the parameters used to calculate similarity information in image recognition.

Step 7: Calculating eigenvalues by using softmax function—The calculation of the eigenvalues and the probability of a similar Chinese characters characteristic value help to achieve better image classification performance.

3.3 Model Construction and Evaluation

Our TEAT first constructs the model and then evaluates the performance.

(1) Construction

The feature values extracted by feature extraction are further processed by a neural network. Our single neuron network formula is presented as follows:

$$Z = W * X + B \tag{1}$$

where Z is the output, w means weight, b means the Paranoid value, and X stands for the input representing size, distance, space, curvature.

The best w-weights and b-paranoid values can be obtained through model training. We can obtain the Chinese character with the maximum probability prediction value by the softmax function (Fig. 3).

(2) Assessment

We evaluate the performance of our TEAT model by calculating the correct recognition rate of the test pictures. First of all, we load the trained convolutional neural network model, and obtain a batch for image feature extraction operation through the same image segmentation. Then, eigenvalues are loaded into the fitted model to find the most similar eigenvalues and return the corresponding vector values. This process is the inverse process, which converts the vector value into the corresponding text. This process is called vector to text. The loss value is calculated by cross entropy, and the smaller the loss value is, the better the actual effect of the model is.

We evaluate our TEAT after iteration of 50 times. When the training set is large enough, the accuracy of our model can exceed 99.7%.

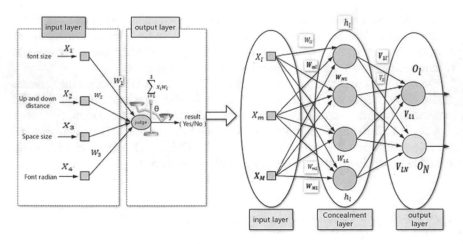

Fig. 3 CNN convolution neural network diagram

4 Experiments and Application Analysis

4.1 Data Source Acquisition

The business license information pictures of online stores are obtained by using web crawler technology in our experiment. Such information obtained automatically is effective and accurate.

4.2 Experimental Method

We use tess4j technology and deep learning based on TensorFlow technology respectively to extract the business license information over the same set of test pictures. Under the same condition, we choose 10, 50, 100, 300, 500 pieces, respectively, and calculated the recognition results in statistics.

4.3 Evaluation Criteria

Three evaluation criteria are adopted in our experiment.

(1) Recognition speed is calculated by the number of pictures over recognition time.
(2) Recognition accuracy is divided into character recognition accuracy and industrial and commercial number recognition accuracy.

Fig. 4 Efficiency comparison between TensorFlow and tess4j

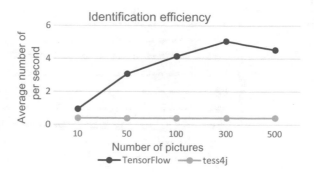

Character Recognition Accuracy is calculated by the number of correct characters recognized over the number of characters recognized.

Industry and Commerce Identification Accuracy is calculated by the number of correct identification of industry and commerce numbers over the number of industry and commerce numbers.

(3) The operation complexity includes the complexity of manual picture processing and recognition, image recognition and other operations.

4.4 Experimental Results

Based on the data source recognition results, statistical analysis and calculation results are presented in Figs. 4 and 5.

Our TEAT can recognize more than 320 pictures in a minute, whereas tess4j technology can only recognize 24 pictures.

5 Conclusion

We propose our TEAT algorithm to extract business registration information utilizing TensorFlow framework and deep learning recognition algorithms. We first utilize the web crawler to obtain the data source, and then extract the character information by using TensorFlow framework recognition technology. Comparing with existing text extraction algorithm based on the tess4j framework when identifying Tmall shop business license picture information, our TEAT has obvious advantages.

Since we deploy only on the same kind of servers, the evaluation in various environmental dimensions has not yet verified. In the future, we will deploy our TEAT in more complicated environment and verify its performance further.

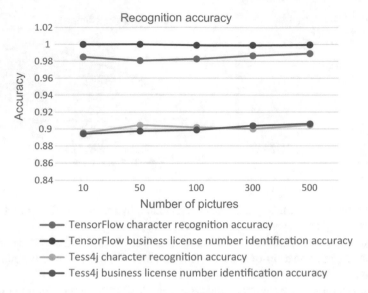

Fig. 5 Accuracy comparison between TensorFlow and tess4j recognition

Acknowledgements This work is supported by Top-notch Academic Programs Project of Jiangsu Higher Education Institutions (TAPP) under Grant No. PPZY2015A090, and Jiangsu top six talent summit project under Grant No. XXRJ-011.

References

1. Pan, F.: Research on the Construction Method of Adaptive Convolution Neural Network Based on Mutual Information. Beijing University of Technology (2015)
2. Abadi, M., Barham, P., Chen, J., et al.: TensorFlow: a system for large-scale machine learning (2016)
3. Chen, F.: Verification Code Image Recognition for Web Agents. Nanjing University of Technology (2007)

Convolutional Neural Networks Implemented by TensorFlow for the Segmentation of Left Ventricular MRI Images

Chih-Yu Hsu and Kuo-Kun Tseng

Abstract Convolutional neural networks applied to medical image segmentation with low-compute power computers are investigated. A convolutional neural network architecture is implemented by TensorFlow for a synthetic image and left ventricular MRI image segmentation. Finding parameters of a trained model is an optimization problem. A flowchart is proposed to solve the compute resource problem for the training procedure. The main point is to train a model by updating the checkpoint step by step. With the limitation of the low compute power, experiments can also have good enough results. The experiment results in image segmentation are good enough. The accuracy of both of them is above 95%.

Keywords Image segmentation · Deep learning algorithm · MRI image · Convolutional neural network · TensorFlow

1 Introduction

Alan Turing [1] proposed Turing test with several criteria to assess whether a machine could be said be intelligent. Walter Pitts and Warren McCulloch [2] proposed the first mathematical model of a neural network, called a McCulloch-Pitts neuron, mimicked a neuron but without an ability to learn.

Frank Rosenblatt's "Perceptron" [3] known as an electronic device was constructed in accordance with biological principles and showed an ability to learn. The perceptron would be the foundation of the modern neural networks lead to true AI with a learning procedure that solution and could recognize letters and num-

C.-Y. Hsu · K.-K. Tseng (✉)
Department of Information and Communication Engineering, ChaoYang University
of Technology, Taichung, Taiwan
e-mail: kktseng@foxmail.com

C.-Y. Hsu · K.-K. Tseng
Department of Computer Science and Technology, Shenzhen Graduate School, Harbin
Institute of Technology, HIT Campus of ShenZhen University Town, Xili, Shenzhen, China

© Springer Nature Singapore Pte Ltd. 2019
J.-S. Pan et al. (eds.), *Genetic and Evolutionary Computing*,
Advances in Intelligent Systems and Computing 834,
https://doi.org/10.1007/978-981-13-5841-8_56

bers. Marvin Minsky showed the perceptron was incapable of learning the simple exclusive or (XOR) function in his book entitled "Perceptrons" [4]. Geoff Hinton [5, 6] demonstrated the use of generalized backpropagation algorithm for training multilayer neural nets that have many hidden layers could be effectively trained by a relatively simple procedure. The weakness of the perceptron was solved by adding hidden layers endowed the network with the ability to learn nonlinear functions. Networks had the ability to learn any function that is demonstrated as a universal approximation theorem. He is now a leading figure in the deep learning community known as the "Godfather of Deep Learning" [7–11]. A team from the University of Toronto in Canada proposed an algorithm called SuperVision, that is, a deep convolutional neural network had won the ImageNet challenge competition 2012 [12]. Machines are now almost as good as humans at object recognition. The turning point event occurred in 2012 revolutionized the field of computer vision [13]. Due to the advancements in GPU processing power promotes the success of deep learning. In 2016, fast real-time detection and classification on images (YOLO) [14] are driven by both developments in convolutional neural networks (CNNs). However, everyone wants to study deep learning but may not have a computer with high compute power. The aim of this paper is to show a method to segment the slice images of left ventricle of MRI with low compute power. A cardiac MRI left ventricle segmentation challenge took place in 2009 [15]. Several new algorithms were proposed to find the best segmentation of the left ventricle. The magnetic resonance images and the data have been provided by Sunnybrook Health Sciences [16]. Tran [17] first trains a CNN on a left ventricle dataset to recognize left ventricle contours. They then use the LV network's weights as a starting point to train a network to segment the right ventricle. Tran [17] uses transfer learning to improve the performance of a network on the right ventricle segmentation task in cardiac MRIs.

TensorFlow library and deep neural networks are used to recognize the area of the left ventricle which shows the image segmentation process by deep learning technology. The performance and hardware issues of low compute power are considered in the training procedure.

2 Dataset

Our synthetic image is created by self-written code, and it is a gray image. The values of gray levels are different for background and foreground. The foreground is a square region with 28×28 pixels in a image size 256×256 pixels. The Sunnybrook Cardiac Data provided by the Cardiac Atlas Project [18]. The data were exported by the DICOMS as PNGs and converted the list of coordinates of the contours to PNGs as well, by Matthew Shun-Shin [19]. The MRIs show the heart through different perspectives. The same images also exist in their segmented form which shows the left ventricle white on black as a binary image. The image shown in Fig. 1a is a portable network graphics (png) format and 256×256 pixels with 32 bits per pixel. The image shown in Fig. 1b is the left ventricle white on black as a binary image. In

(a)   (b)

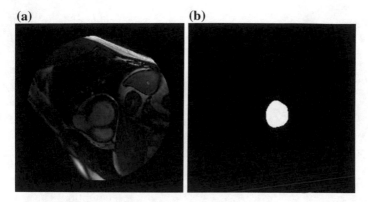

Fig. 1 **a** Medical image and **b** label image as the ground truth

total, there are 805 images and 805 labels. Images are divided into a training set of 526 images and a testing set of 279 images.

3 Proposed Method

The proposed method is training a CNN for segmentation of images pixel by pixel. Image segmentation problem is transformed into the classification problem of patch images. Patch images are sampling from the original image to be segmented. The classification problem of patch images is solved by the convolutional neural network (CNN).

3.1 Architecture of the Convolutional Neural Network

The architecture of the convolutional neural network is shown in Fig. 2. The layers which were used were an input layer, two convolution layers, each with its own pooling layer, a fully connected layer, and a output layer.

3.2 Image Patch Sampling

The purpose is to train the CNN for segmentation of the image with the size 256 by 256 pixels. Patch with the size 28 by 28 pixels is generated for training CNN. The center pixel of a patch is checked for the class label. If the center pixel mapping to the pixel of the label images and the pixel of the label images is 1, then the patch image

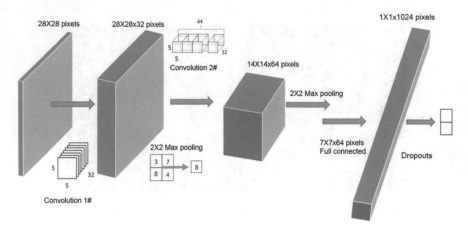

Fig. 2 Architecture of the CNNs for image segmentation

is class label 1 and put it into the folder 1. On the other hand, if the center pixels mapping to the pixel of the label images and the pixel of the label images is 0, then the patch is class label 1 and put it into the folder 0 as shown in Fig. 4. In one image with the size 256 by 256 pixels, a mask with size 28 by 28 pixels sweep with Z-scan to generate $(256 - 27)x(256 - 27) = 52441$ patch images. The balance problem between the class 0 and class 1 will be found by the way to sample patch. It is better to decrease the frequency of sampling proportional to the distance to the region of interest (ROI). Each patch has to be converted into a 2D numpy array and saved in the right format for being appended to the right MNIST file format. Sampling patches out of every image was a time-consuming activity. If the RAM is not enough, the MNIST file should not include too many patch images to avoid memory errors. For train CNN model for prediction of the segmentation results of medical images, it is necessary to build the model step by step. The flowchart to train a model is as shown in Fig. 3. The first step is preparing MNIST files for the training and test sets and setting the initial random values for parameters of CNN. At the first step, there is no checkpoint file that can be loaded to obtain pertained parameters. The CNN model is trained with the initial parameters and save all values of parameters into one checkpoint file. If there are some MNIST files that are unselected, continue the loop.

3.3 Loss Function

The loss function is used for training CNN. The cross-entropy function is a loss function defined as follows:

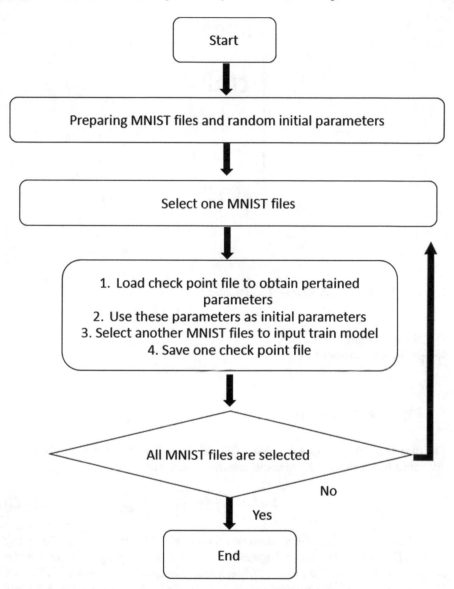

Fig. 3 The flowchart to train a model

$$H_{y'}(y) := -\sum_{i=1}^{M} y'_i log(y_i),\qquad(1)$$

M—Number of classes
log—the natural log

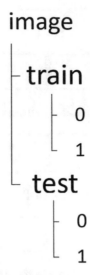

Fig. 4 The patch images stored in the folder directory

y_i'—binary indicator (0 or 1)
y_i—predicted probability.

4 Experimental Results

The dice similarity coefficient (DSC) [20, 21] is a formula used to calculate the difference between the created result and the ground truth.

$$DSC := \frac{2|A \cap B|}{|A| + |B|} \qquad (2)$$

where A is the set of the created result and B is the set of the ground truth. The image analysis [22] is based on image segmentation. The synthetic image to be segmented can be seen in Fig. 5a. The ground truth is a binary image that can be seen in Fig. 5b. The segmented result is a binary image that can be seen in Fig. 5c. The dice similarity coefficient is 0.9979 comparing Fig. 5c with Fig. 5.

The medical image to be segmented can be seen in Fig. 6a. The ground truth is a binary image that can be seen in Fig. 6b. The segmented result is a binary image that can be seen in Fig. 6c. The dice similarity coefficient is 0.9650, comparing Fig. 6c with Fig. 6b. It is a success that trained CNNs were used to predict the class of patch images and finish the image segmentation process.

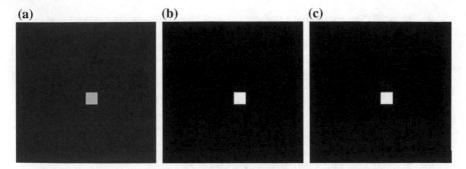

Fig. 5 **a** Synthetic image **b** ground truth and **c** segmented result

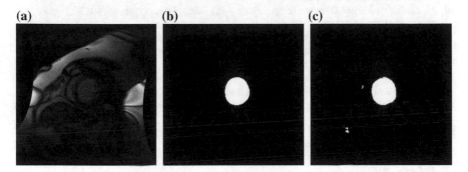

Fig. 6 **a** Medical image **b** ground truth and **c** segmented result

5   Conclusions

If you wanted to inspect the image patch samples, the whole operating system froze, because it could not handle those huge amounts of files. Because a large of image data to process the normal CPUs were unable to cope with the huge amounts of files, the limitation of compute power makes preprocessing a dataset with considering reasonable size. Sampling every picture and creating 65.535 patches out of every image was a time-consuming activity.

Acknowledgements Thanks to the Department of Mathematics and Computer Science at OTH Regensburg in Germany invited to give a lecture titled Deep Learning for Image Classification and Segmentation. Some ideas of the paper were initiated from questions asked by students.

References

1. Turing, A.M.: Computing machinery and intelligence. Mind New Ser. **59**(236), 433–460 (1950)
2. McCulloch, W., Pitts, W.: A logical calculus of ideas immanent in nervous activity. Bull. Math. Biophys. **5**, 115–133 (1943)

3. Rosenblatt, F.: The perceptron: a probabilistic model for information storage and organization in the brain. Psychol. Rev. **65**(6), 386–408 (1958)
4. Minsky, M., Papert, S.: Perceptrons, An Introduction to Computational Geometry. MIT Press, Cambridge, MA, USA (1969)
5. Rumelhart, D.E., Hinton, G.E., Williams, R.J.: Learning representations by back-propagating errors. Nature **323**, 533–536 (1986)
6. Hinton, G.E.: Relaxation and its role in vision. lib.ed.ac.uk (PhD thesis). University of Edinburgh. hdl:1842/8121. OCLC 18656113. EThOS uk.bl.ethos.482889 (1977)
7. Geoffrey Hinton was briefly a Google intern in 2012 because of bureaucracy-TechCrunch. https://techcrunch.com
8. Somers, J.: Progress in AI seems like it's accelerating, but here's why it could be plateauing. MIT Technol. Rev. https://www.technologyreview.com/
9. How U of T's 'godfather' of deep learning is reimagining AI. University of Toronto News. https://www.utoronto.ca/news
10. Godfather' of deep learning is reimagining AI. https://phys.org/news/2017-11-godfather-deep-reimagining-ai.html
11. Hinton, G.: The 'godfather' of deep learning, on AlphaGo. Macleans.ca, 18 Mar 2016. https://www.macleans.ca/society/science/the-meaning-of-alphago-the-ai-program-that-beat-a-go-champ/
12. Krizhevsky, A., Sutskever, I., Hinton, G.E.: ImageNet Classification with Deep Convolutional Neural Networks, pp. 1097–1105. Curran Associates Inc. (2012)
13. How a Toronto professor's research revolutionized artificial intelligence I Toronto Star. https://www.thestar.com
14. Abbott, R., Del Rincon, J. M., Connor, B., Robertson, N.: Deep object classification in low resolution LWIR imagery via transfer learning, Deep object classification in low resolution LWIR imagery via transfer learning. In: Proceedings of the 5th IMA Conference on Mathematics in Defence (2017)
15. A Cardiac MRI left ventricle segmentation challenge took place in 2009. http://www.cardiacatlas.org/challenges/lv-segmentation-challenge/
16. Cardiac atlas project. http://www.cardiacatlas.org
17. Tran, P.V.: A Fully Convolutional Neural Network for Cardiac Segmentation in Short-Axis MRI 2016. https://arxiv.org/pdf/1604.00494.pdf
18. Radau, P., Lu, Y., Connelly, K., Paul, G., Dick, A.J., Wright, G.A.: Evaluation framework for algorithms segmenting short axis cardiac MRI. The MIDAS J. Card. Left Ventricle Segm. Chall. http://hdl.handle.net/10380/3070
19. Matthew Shun-Shin, 01 Jun 2018. https://github.com/mshunshin/SegNetCMR
20. Srensen, T.: A method of establishing groups of equal amplitude in plant sociology based on similarity of species and its application to analyses of the vegetation on Danish commons. K. Dan. Vidensk. Selskab. **5**(4), 1–34 (1948)
21. Dice, L.R.: Measures of the amount of ecologic association between species. Ecology **26**(3), 297–302 (1945)
22. Tian, D.: Research on PLSA model based semantic image analysis: a systematic review. J. Inf. Hiding Multimed. Signal Process. **9**(5), 1099–1113 (2018)

Financial Analysis with Deep Learning

Kuo-Kun Tseng, Chiye Ou, Ao Huang, Regina Fang-Ying Lin
and Xiangmin Guo

Abstract Nowadays, there are massive data in the financial industry that can provide a lot of information to activities of capital supply and economic operation. Artificial Neural Networks (ANN) allow input characteristics of high dimensional, and have the ability to describe the complex nonlinear relationships. Therefore, deep learning methods in the financial sector have great potential. Further, deep learning has extensive application in computer vision and speech recognition, due to which many measure approaches are able to be applied to the financial sector. In this work, we predicted the stock price by using Recurrent Neural Networks (RNN), indicating that data characteristics of the financial sector are suitable for using deep learning method, and the result has certain practical value.

Keywords Financial analysis · Artificial neural network · Deep learning algorithm

1 Introduction

When it comes to applying deep learning in the financial sector, many people may immediately think of the prediction of stock price movements. This is indeed a high-profile research direction, but the practical application of deep learning in the financial sector is far more extensive.

The ability of the sustainable development and profit are important for investors. Tam and Kiang [1] applied artificial neural networks to predict bank bankruptcy.

K.-K. Tseng · C. Ou
School of Computer Science and Technology, Harbin Institute of Technology (Shenzhen), Shenzhen, China

A. Huang · R. F.-Y. Lin (✉)
School of Economics and Management, Harbin Institute of Technology (Shenzhen), Shenzhen, China
e-mail: r.lin@hit.edu.cn

X. Guo
School of Architecture, Harbin Institute of Technology (Shenzhen), Shenzhen, China

© Springer Nature Singapore Pte Ltd. 2019   545
J.-S. Pan et al. (eds.), *Genetic and Evolutionary Computing*,
Advances in Intelligent Systems and Computing 834,
https://doi.org/10.1007/978-981-13-5841-8_57

They used a back propagation (BP) neural network with single hidden layer to predict bankruptcy for Texas banks between 1985 and 1987. The results showed that such a simple neural network is more accurate and stable than multivariate discriminant analysis model, logistic model, K-nearest neighbors (KNN) model, and decision tree model.

Hutchinson et al. [2] also used ANN for analyzing option prices and hedging S&P 500 futures options from 1987 to 1991. It is verified the result is better than the Black-Scholes-Merton option pricing model, which is the classic option pricing model in the financial studies.

Due to the limitations of the linear regressions, the exchange rate has been considered unpredictable in the past period of time, but the ability of the neural network to describe the nonlinear relationship opens up a new way to predict the exchange rate. Lee and Chen [3] used RNN to predict the trend of the five currencies against the US dollar exchange rate, and demonstrated significant market synchronization ability in the forecast of the yen against the US dollar and the British pound against the US dollar. It reveals the potential of neural networks to describe nonlinear relationships.

In 2006, Hinton [4] published a paper in *Science* that proposed the Deep Belief Network (DBN), breaking the bottleneck of gradient disappearance in multilayer network training, and letting the concept of deep learning enter people's field of vision. The improvement of computer hardware performance also makes the training of multilayer networks more efficient. Therefore, more and more researchers use deep learning methods to solve the complex problems that shallow neural networks are invalid, which includes the problems of financial analysis.

Heaton et al. [5] gave a brief overview of the development of deep learning in the finance, summarizing the three models of deep learning, namely Autoencoder, Rectified Neural Networks, and Long Short-Term Memory (LTSM). They also introduced two ways to avoid over-fitting, namely regularization and dropout. They then used the Autoencoder model as an example, and selected a small number of stocks to build two training sets (One training set is the 10 stocks most similar to the S&P500 trend, another set based on the former, adding 10 stocks that are the least similar to the S&P500 trend), trained the S&P500 stock index between 2014 and 2015. The results showed that as long as the samples are diverse enough, a small number of samples can approximate the real value with almost any precision by deep learning.

Nowadays, countless researchers have devoted a lot of energy to study deep neural networks, and have harvested many excellent achievements. These studies allow us to stand on the shoulders of giants, so we have a strong basis on financial analysis using deep learning.

2 Related Research

Deep learning has developed to a certain scale, with a knowledge system containing many models and optimization methods. Deep Belief Network (DBN) is a classic deep learning model proposed by Hinton [4]. It is based on the Restricted Boltzmann Machine (RBM), and the entire network can be regarded as a stack of several RBMs. In DBN, the layer-by-layer unsupervised training method is used to gradually optimize the parameters of each RBM structure, which is called "pre-training". After pre-training, the parameters close to the optimal solution can be obtained, and the problem of gradient disappearance when training multilayer neural networks is solved to some extent. Then, the whole neural network is trained by BP algorithm, which is called "fine tuning". The DBM (Deep Boltzmann machine) model is another RBM-based model proposed by Salakhutdinov and Hinton [6], which has been proven to perform well in handwritten digit recognition and object recognition. The difference between DBM and RBM are the multiple hidden layers (RBM has only one hidden layer), and the connection between all adjacent layers is undirected (in the DBM structure, only the last two layers are undirected connections).

Convolutional Neural Network (CNN) is a forward artificial neural network model proposed by Yann LeCun et al. in 1998. Its typical network structure includes convolutional layer, pooled layer, and fully connected layer. It has been widely used in computer vision. The operation process is that we give a picture (a training sample) as an input, and sequentially scan the input picture pixel through multiple convolution operators. The scan result is activated by the activation function to obtain the feature map. Then we use the pooling operator to get the downsample of the feature map and output the result as the input to the next layer. After all the convolution and pooling layers, we use the fully connected neural network for further operations. Finally, the result is given by the output layer.

Recurrent Neural Network (RNN) is a type of neural network efficient for processing time series data. In the traditional neural network, the signals of neurons in each layer can only be propagated to the next layer, and the data processing is independent at each moment. However, the RNN memorizes the previous information and applies it to current data processing. That is, the hidden layer not only receives the output from upper layer, but also receives the output made by itself from the previous time. Saad et al. [7] applied Time-Delay Neural Network, Probabilistic Neural Network, and RNN models to predict the closing price of the stock. The experimental results show that all three models have predictive abilities. However, the gradient disappearance occurred on the RNN, which means that the gradient generated at a certain moment disappears after several propagations on the time axis, and the long-range information cannot be effectively memorized. Long Short-Term Memory (LSTM) is developed to solve this weakness of RNN. LSTM introduces the concept of cell state, which allows add or forget information, and also form a closed loop through the gates, thereby overcoming the weakness that RNN cannot effectively memorize long-range information. Chen et al. [8] tried to use LSTM to predict stock returns in China's stock market. Experiments show that compared with the stochastic forecasting method, the LSTM model's prediction accuracy for stock returns increased from 14.3 to 27.2%.

Autoencoder is a neural network that reproduces the input signal as much as possible. Its learning process is first to encode the input and then decode the reconstructed input as an output. This approach is often used for feature extraction, denoising, and so on. Autoencoder generally refers to a network model with only one hidden layer, while Deep Autoencoder is a network model with multiple hidden layers. The training of Deep Autoencoder is similar to DBM. Pre-training is performed between every two layers using RBM, and then the parameters are fine-tuned by the BP algorithm.

Because of the large number of layers in deep learning, the generated models are usually very complex. Over-fitting is prone to occur when the training sample set is not large enough. Therefore, some techniques are needed to control the training process, and to reduce or even prevent serious over-fitting. Commonly used techniques include early stopping, regularization, and drop out.

Early stopping is a method to prevent over-fitting by controlling the epochs of trainings. It stops iteration at the right timing to prevent over-fitting. Prechelt [9] made a deep study in this method. The regularization method refers to adding a regular term of the weights to the loss function, avoiding the overly complicated weights lead to an over-fitting model. The main idea of drop out is to randomly remove certain connections between the hidden layers in each training epoch to reduce the complexity of the model. The papers by Heaton et al. [5] introduce practical applications of regularization and drop out.

3   Case Study: Short-Term Prediction of the Shanghai Composite Index Using RNN

In this example, the RNN will be applied to make short-term prediction of the Shanghai Stock Index. The prediction method is to use the opening price, the highest price, the lowest price, the closing price and the trading volume of the previous 10 days to predict the highest price of the next day. The input characteristic vector is

$$X = (Opening\ price, highest\ price, lowest\ price, closing\ price, trading\ volume)$$
(1)

Then the time series inputted is

$$(X_1, X_2, X_3, X_4, X_5, X_6, X_7, X_8, X_9, X_{10})$$
(2)

The expected output is

$$Y_{11}^* = (highest\ price)_{t=11}$$
(3)

3.1   Data Collection and Processing

First, we collected transaction data of the Shanghai Stock Exchange Index from January 4, 2005 to June 30, 2017 for a total of 3,034 trading days, including date, opening price, highest price, lowest price, closing price, and trading volume. The data from January 4, 2005 to December 31, 2015, a total of 2,671 transaction days were used as training sets. The data from January 4, 2016 to June 30, 2017, a total of 363 trading days were used as test sets.

As input, the transaction data of the previous 10 days are processed as a moving time series, and the highest price of the 11th day is taken as the expected output, namely:

$$(X_1, X_2, X_3, X_4, X_5, X_6, X_7, X_8, X_9, X_{10}) => \left(Y_{11}^*\right) \tag{4}$$

$$(X_2, X_3, X_4, X_5, X_6, X_7, X_8, X_9, X_{10}, X_{11}) => \left(Y_{12}^*\right) \tag{5}$$

After building the moving time series, the scales of input data were normalized into [0, 1]. So far the data processing is completed and they were prepared to input to the model.

3.2 Model Structure and Parameters

The neural network has an RNN layer as its hidden layer, and this hidden layer contains 32 neurons, which are itcratively trained in 10 time steps. The hidden layer activation function is tanh. Because it is a regression analysis problem, the output layer does not perform nonlinear transformation, that is, the linear unit y = x is directly used. The mean square error (MSE) is selected as the loss function, namely:

$$MSE = \frac{1}{n} \sum_{i=1}^{n} \left(y^* - y\right)^2 \tag{6}$$

where n is the total number of training samples, y* is the expected output, and y is the label of the model, namely real highest price of the 11th day.

The optimization strategy was gradient descent, and the learning rate was set to 0.01. The early stopping technique was used to prevent over-fitting. When the value of loss function did not reduce during continuous 5 iterations, the training would stop, the preset iteration time is 500 epochs, and the actual iteration time is 407 epochs.

3.3 Analysis of Results

The comparison of the predicted curve of the model on the test set with the real curve is shown in Fig. 1.

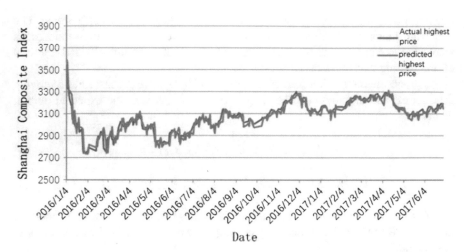

Fig. 1 The predicted curve of the highest price of the Shanghai Composite Index generated by RNN

Table 1 Statistics on the daily error rate of Shanghai Composite Index	Daily error rate	Number of days	Proportion (%)
	DER ≤ 1%	224	61.71
	1% < DER ≤ 2%	95	26.17
	2% < DER ≤ 5%	40	11.02
	DER > 5%	4	1.10

It can be seen that the predicted curve is basically consistent with the actual curve in the trend, though it still has some local deviations. The daily error rate is used to further measure the accuracy of the prediction. The calculation formula is

$$DER = \frac{ABS(y^* - y)}{y*} * 100\% \tag{7}$$

The ABS is absolute value function, and the daily error rate of the 363 trading day predictions of the test set is shown in Table 1.

In the 363 trading days of the test set, the number of days with predicted daily error rates less than or equal to 1% were 224, accounting for 61.71%. The predicted daily error rates ranged from 1 to 2% were 95 days, accounting for 26.17%. In general, the generalization performance on the test set was good and the model can be considered to have certain practical value. We analyzed the 4 days with the highest predicted error rate and found that it appeared on January 5, 12, 27, and February 26, 2016. These 4 days' Shanghai Composite Index has experienced a sharp decline compared to other days. According to the securities news, China's stock market experienced a serious crash in 2015. So the circuit breaker was officially implemented on January 4, 2016. However, the circuit breaker was repeatedly triggered by the continuous

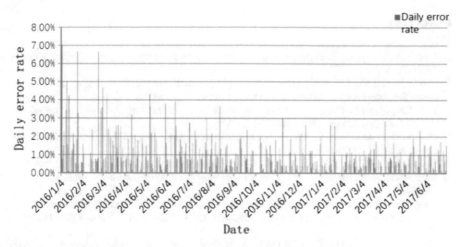

Fig. 2 Daily error rate histogram of predicted Shanghai Composite Index

falling of the stock price. This mechanism was implemented for only 4 days. Even so, in January 2016, it was the biggest drop in a single month in the past 8 years (last time is the global financial crisis in 2008). This explains the high deviations and shows that RNN's prediction of stock index as a technical analysis method has the disadvantage of being insensitive to bad news in the market (Fig. 2).

4   Conclusion

This paper studies the applications of deep learning in finance and focuses on applying a recurrent neural network to stock price prediction. The experimental result suggests that in the context of big data, deep learning is very suitable for financial analysis and can achieve high performance. Therefore, it is very promising to apply deep learning to solve more financial problem.

Acknowledgements The authors would like to thanks funding from 1. The Science Research Project of Guangdong Province, "Research on the Technical Platform of Rural Cultural Tourism Planning Based on Digital Media" (No: 2017A020220011). 2. The Natural Science Funding of Guangdong Province, "Research on the Sponge River Design Based on Multi-disciplines: a Case Study of Shenzhen" (No: 2016A030313659).

References

1. Tam. K.Y., Kiang, M.: Predicting Bank Failures: A Neural Network Approach. Taylor & Francis, Inc. (1990)
2. Hutchinson, J.M., Lo, A.W., Poggio, T.: A nonparametric approach to pricing and hedging derivative securities via learning networks. J. Financ. **49**(3), 851–889 (2012)
3. Lee, T.S., Chen, I.F.: Forecasting exchange rates using feedforward and recurrent neural networks. J. Appl. Econom. **10**(4), 347–364 (1995)
4. Hinton, G.E., Osindero, S., Teh, Y.W.: A fast learning algorithm for deep belief nets. Neural Comput. **18**(7), 1527–1554 (2006)
5. Heaton, J.B., Polson, N.G., Witte, J.H.: Deep learning in finance (2016). arXiv:1602.06561
6. Salakhutdinov, R., Hinton, G.: Deep Boltzmann machines. J. Mach. Learn. Res. **5**(2), 1967–2006 (2009)
7. Saad, E.W., Prokhorov, D.V., Nd, W.D.: Comparative study of stock trend prediction using time delay, recurrent and probabilistic neural networks. IEEE Trans. Neural Netw. **9**(6), 1456–1470 (1998)
8. Chen, K., Zhou, Y., Dai, F.: A LSTM-based method for stock returns prediction: a case study of China stock market. In: IEEE International Conference on Big Data, pp. 2823–2824. IEEE (2015)
9. Prechelt, L.: Early stopping—but when? Neural Netw. Tricks Trade **1524**(1), 55–69 (1998)

An LSTM-Based Topic Flow Pattern Learning Algorithm and Its Application in Deceptive Review Detection

Na Liu, Lu-yu Dong and Shu-Juan Ji

Abstract Great progress has been made in detecting deceptive reviews based on traditional machine learning. However, existing classification methods for detecting deceptive reviews mainly focus on combining word-embedding with the traditional machine learning algorithm, while neglecting the latent semantic meaning and its temporal relation of the topic in a review. In this paper, we propose a deceptive review detection model based on the learning of topic flow pattern. First, this paper analyzes the topic of reviews and considers the temporal relations of topics simultaneously to construct discriminative characteristics of reviews. Second, the Long Short-Term Memory (LSTM) neural network is used to classify reviews that contain information on topic series. Experimental results on three domains of datasets show that our proposed method is superior to benchmark methods.

Keywords Deceptive review detection · Neural language processing · Topic flow pattern · Long short-term memory

1 Introduction

Due to its convenience, shopping online has gradually become a popular way. E-commerce websites allow consumers to comment on purchased products, so followers can decide their purchase behavior through referencing reviews. To increase the number of transactions, many malicious sellers hire buyers to write fake reviews, which can mislead consumers purchase behavior and bring negative social impacts. Thus, the research in detecting deceptive reviews is an urgent and meaningful task.

In general, detecting deceptive reviews can be seen as a text classification problem. Most existing methods follows Ott et al. [1] and adopts machine learning algorithms to build classifiers. However, the weaknesses of the standard supervised learning

N. Liu · L. Dong · S.-J. Ji (✉)
Key Laboratory for Wisdom Mine Information Technology of Shandong Province, Shandong University of Science and Technology, Qingdao, China
e-mail: jane_ji2003@aliyun.com

© Springer Nature Singapore Pte Ltd. 2019
J.-S. Pan et al. (eds.), *Genetic and Evolutionary Computing*,
Advances in Intelligent Systems and Computing 834,
https://doi.org/10.1007/978-981-13-5841-8_58

algorithm are obvious. First, they usually do not provide explicit probabilistic prediction about whether a review is truthful or not. Secondly, it is difficult to find the semantic information implied in a review.

To solve the above problems, researchers have done a lot of research mainly on feature extraction and classification algorithms. In terms of feature extraction, researchers mainly used topic models to learn semantic information of reviews [2]. Till now, almost all topic models adopted the assumption of the bag of word model, which considers there is no order between words in a document. In terms of the classifier, deep learning method based on the neural network has been widely applied. For example, CNNs are widely used for semantic combination and automatically capturing of n-gram information [3, 4]. Furthermore, research based on fake review has mostly used time-agnostic models which fail to capture temporal relationship information within a review, so the classification accuracy is relatively low. Differently, sequential models such as LSTM can process reviews with temporal relationships very well.

In this paper, we propose a novel method based on topic flow pattern and LSTM (named TFNN) to learn document representation of review and detect fake reviews. The learning of document representation based on topic flow can capture sequential relationship and implicit semantic features. It is vital that the topic flow pattern is first used jointly in deceptive review detection. Experiments indicate that our method is superior to baselines and improve the semantic expression of the document.

In the following parts, Sect. 2 introduces related work. Section 3 focuses on the introduction of the proposed detecting algorithm. Section 4 demonstrates the experimental setup and analyzes results. Section 5 summarizes this paper.

2 Related Work

As we all know, the feature-based models have high computational efficiency. Moreover, as a typical statistical model, the topic model is good at exploring implicit topics in documents with high computational performance [5]. In the implementation process, the topic model considers that each document is composed of several topics and each topic consists of several words. Generally, topic models can be divided into two categories, i.e., unsupervised ones and supervised ones. The unsupervised ones mainly include Probabilistic Latent Semantic Analysis [6], Latent Dirichlet Allocation [2], while the supervised ones include Labeled Latent Dirichlet Allocation [7] and Hierarchically Supervised Latent Dirichlet Allocation [8]. Given the good scalability of LDA model, researchers often combine more knowledge (e.g., sentiment, label) into LDA to learn different levels (e.g., document, sentence) of text. For example, the HSLDA model is designed based on the basic LDA model by assuming that a document was only relevant to the hierarchical label corresponding to the document [8].

In recent years, neural network based algorithms have been widely used in natural language processing to shorten the time in analyzing data features. Some researchers (e.g., the authors in [9]) believed that the representation learning is a transformation of learning data that can make it easier to extract useful information when designing classifiers. Therefore, in order to learn a continuous representation of review, they use representation learning methods in different granularities, such as words, sentences, paragraphs, and documents. Besides, sequential models are often used for recurrent semantic combinations [10]. Till now, there are many approaches to learn the expressions from distributed word representations and large text segments. For example, Le and Mikolov [11] presented a paragraph vector to learn the representation of text. Socher et al. [12] used a variety of recursive neural networks to learn the semantic information of sentence levels. Li et al. [13] proposed a sentence weighted neural network model to learn the representation of text in document level and detect fake opinions. Ren et al. [14] explored a neural network model to learn document-level representation for detecting deceptive review. Topic models are good at mining hidden semantic information from documents. In contrast, neural network models have the advantages of learning different levels of document representation. So far, we did not find research that combining topic model with a neural network to improve classification performance in deceptive reviews detection field.

3 The TFNN Model

In this section, we present an LSTM-based topic flow pattern learning algorithm for deceptive review detection. It mainly includes two stages: topic flow feature extraction and deep learning neural network classification. In the first stage, it contains text data preprocessing and topic flow feature generation. In the second stage, a topic flow feature is input into a deep neural network for training through the embedding layer, and then the classification result is obtained. Table 1 introduces the terms used in this work.

Table 1 Definition of characters in TFNN algorithm

Character	Definition
K	The number of topics in the review document
α, β	Hyper-parameters
o	The output of our algorithm, 0 stands for truthful, 1 stands for deceptive
W	The whole review $W = [w_1, w_2, \ldots, w_m]$, where w_i represents the ith review
Y	The label vector $Y = [y_1, y_2, \ldots, y_m]$, where y_i is the label of the ith review
X_i	For w_i, generate topic flow vector of each review $X_i = [x_1, x_2, \ldots, x_n]$, where x_i is the topic of each word in the review
X	For W, generate the topic flow matrix $X = [X_1, X_2, \ldots, X_m]$

Table 2 TFNN algorithm pseudo code

Algorithm: TFNN
Input: W, K, α and β.
Output: o
Process:
1:        **Text preprocessing**
2:        **LDA training**
3:          For each review in review document:
4:            Get topic flow vector of each review X_i
5:          End for
6:        **Deep neural network classification**
7:          Input variable-length topic flow X
8:          While true
9:            LSTM training
10:        End while

Detailed steps of our algorithm are as follows (see Table 2). First, each review in each document is preprocessed to remove special symbols and stop words, then LDA is used to train model, the preprocessed document W is generated by the trained model to generate the topic vector X_i with a temporal relationship. At last, the topic and label vector (X, Y) of each document is obtained, which forms the feature matrix representation of the whole review, as shown in Formula (1). This matrix is selected as the input of the LSTM model, then this deep neural network model is trained, finally, the classification results can be obtained.

$$(X, Y) = \begin{bmatrix} x_{11} & x_{12} & \cdots & x_{1n} & y_1 \\ x_{21} & x_{12} & \cdots & x_{2n} & y_2 \\ \vdots & \vdots & \cdots & \vdots & \vdots \\ x_{m1} & x_{m2} & \cdots & x_{mn} & y_m \end{bmatrix} \tag{1}$$

4 Experiments

To verify the effectiveness of our proposed model and to compare it with typical benchmark models for deceptive review detection. We design three experiments. In the first experiment, we study how the number of topics influences the performance of classification. The purpose of the second experiment is to verify the classification performance over three domains of datasets. The third experiment statistically analyzes the result derived from the classification model.

Domain	Turker	Employee	Customer
Hotel	400/400	140/140	400/400
Restaurant	200/0	0	200/0
Doctor	356/0	0	200/0

Table 3 Statistics of three domain dataset

4.1 Data Description and Evaluation Criteria

We use the standard datasets over three domains (e.g., Doctor, Hotel, and Restaurant) that are used in Li et al. [15], which include truthful and deceptive reviews. In each domain, there are three different types, i.e., "Turker", "Employee", "Customer". "Customers" means that the reviews are truthful ones that come from reviewers' real consumption experiences. The "Employees" and "Turkers" represent the reviews are deceptive ones that come from dishonest reviewers with expert-level domain knowledge and the automatic review robot, respectively. Table 3 shows these datasets. In the experiment, we adopt accuracy (A), precision (P), recall (R), and Macro-F1 (F1) as evaluation metrics to verify the performance of our method and the benchmarks.

4.2 Result and Analysis

Results of three domain data under different topic numbers. Since the TFNN model is based on the topic flow, the final performance of it is related to the number of topics. To verify the performance of this model, we must first select an appropriate number of topics. So, this paper first chooses 5–50 topics as training parameters and compares the experimental results in Doctor, Hotel, and Restaurant domains.

Figure 1 shows the results we get from the first experiment. From Fig. 1a, we can see that, with the increase of the number of topics, the values of the four metrics (i.e., accuracy (A), precision (P), recall(R), and Macro-F1 (F1)) increase first and then decrease. From Fig. 1b, it can be seen that the curves of four metrics have local optimal results. Especially, when the number of topics is 15, the value of these metrics is extremely similar to ones that are gotten with 30 topics. However, in Fig. 1c, we can see that the curves of A, P, R, and F1 has large fluctuations, which mainly because the reviews in the restaurant domain are relatively too few. From the overall analysis, we can conclude that, when the number of topics range from 10 to 30, good results can be achieved.

Comprehensively considering the values under these four metrics, we can conclude that our TFNN model achieves the highest performance when the number of topics is assigned to 15. To further state the good performance of this model, in the following second experiment, we only list the results when topic is 15 and compare it with benchmarks by taking accuracy and F1 as evaluation metrics, which is similar to the literature [15].

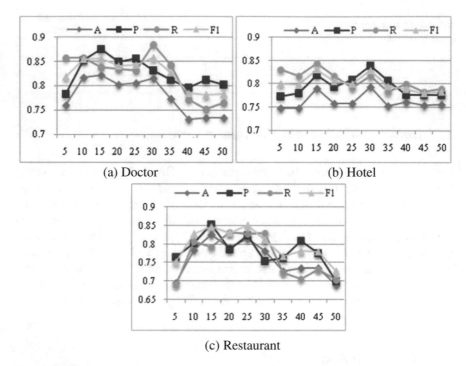

(a) Doctor   (b) Hotel

(c) Restaurant

Fig. 1 Performance of different topics for the model in three domains

Comparison results about TFNN and the benchmarks. In the second experiment, we compare the performance of our model with benchmark models (i.e., the SVM-based model proposed by Li et al. [15], the two models proposed by Ren et al. [14]).

Table 4 illustrates the results. For the Hotel domain, the classification accuracy and F1 values of TFNN model are 79.0% and 83.1%, respectively, which is higher than other models. The accuracy on the Restaurant domain is higher than that of discrete models, which is slightly lower than the neural model, but the F1 value obtains its optimal value. Because there are fewer reviews in the restaurant domain, the accuracy of the model is lower than that of the other three models. The results over the Doctor domain is very similar to those we got on the Hotel domain, where the TFNN model is obviously superior to other models.

Statistical analysis of topic distribution over three domain datasets. To fully explain the above conclusions, this paper further statistically analyzes the number of model topics that obtain the best performances in three domains.

As shown in Fig. 2a, through qualitative analysis, the curve of truthful reviews in the Doctor domain is above the curve of deceptive ones when the number of topics is less than 8. The opposite phenomenon can be observed when the number of topics greater than 8, which indicates that the topic distribution in the Doctor domain is

Table 4 Performance comparison of the models over three domain dataset

Domain	Methods	Accuracy (%)	Macro-F1 (%)
Hotel	Li et al.	66.4	67.3
	Ren's Logistic	65.9	66.7
	Ren's Neural	78.3	74.0
	TFNN	**79.0**	**83.1**
Restaurant	Li et al.	81.7	82.2
	Ren's Logistic	82.1	82.3
	Ren's Neural	**84.4**	84.6
	TFNN	82.6	**84.8**
Doctor	Li et al.	74.5	73.5
	Ren's Logistic	73.7	72.4
	Ren's Neural	74.6	72.8
	TFNN	**82.2**	**85.6**

consistent with human common sense. That is, the deceptive reviews contain more topics. For Hotel domain (see Fig. 2b), the data distribution is opposite to the Doctor domain. That indicates that the data distribution in this domain is contrary to human common sense exactly, which leads to slightly lower classification accuracy. From Fig. 2c, we can see that people are more inclined to review comprehensively and express their opinions fully according to their own experience. These statistical analysis results indicate that our model has good interpretability. Generally, the proposed model in this paper achieves good classification performance in different domains when detecting deceptive reviews.

5 Conclusion

This paper proposes a novel model named TFNN for detecting deceptive spam reviews, which is designed based on the learning of topic flow pattern. It mainly includes two aspects. First, the model proposes a topic flow pattern, which transforms text into a topic flow representation. Secondly, the LSTM network can capture time-series relationship information with a long time span so that it can improve the classification accuracy of the detection model. The experimental results show that the proposed model in this paper has good performance and outperforms the benchmark ones. Although this paper first utilizes the topic flow pattern as a feature

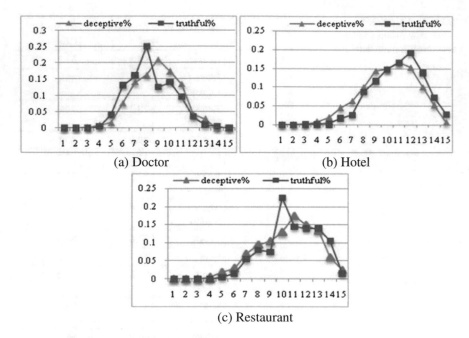

(a) Doctor   (b) Hotel

(c) Restaurant

Fig. 2 Topic distribution of truthful and deceptive reviews in three domains

vector to improve classification accuracy, it does not consider emotion information of reviews. Therefore, in the future work, we will study whether the combination of topic flow pattern and emotion can further enhance the efficiency of classification in terms of deceptive review detection.

Acknowledgements This work is supported by the Natural Science Foundation of China (No. 71772107, 71403151, 61502281, 61433012), Qingdao social science planning project (QDSKL1801138), the National Key R&D Plan (2018YFC0831002), Humanity and Social Science Fund of the Ministry of Education (18YJAZH136), the Key R&D Plan of Shandong Province (2018GGX101045), the Natural Science Foundation of Shandong Province (Nos. ZR2018BF013, ZR2013FM023, ZR2014FP011), Shandong Education Quality Improvement Plan for Postgraduate, the Leading talent development program of Shandong University of Science and Technology and Special funding for Taishan scholar construction project.

References

1. Ott, M., Choi, Y., Cardie, C., et al.: Finding deceptive opinion spam by any stretch of the imagination. In: Meeting of the Association for Computational Linguistics: Human Language Technologies, pp. 309–319. Association for Computational Linguistics (2011)
2. Blei, D.M., Ng, A.Y., Jordan, M.I.: Latent Dirichlet allocation. J. Mach. Learn. Res. Archive **3**, 993–1022 (2003)

3. Johnson, R., Zhang, T.: Effective use of word order for text categorization with convolutional neural networks. EprintArxiv (2014)
4. Kalchbrenner, N., Grefenstette, E., Blunsom, P.: A convolutional neural network for modelling sentences. EprintArxiv 1 (2014)
5. Pennebaker, J.W., Facchin, F., Margola, D.: What our words say about us: the effects of writing and language. Neuropsychol. Clin. Psychol. (2010)
6. Hoffman, T.: Probabilistic latent semantic analysis. Uncertain. Artif. Intell. **15**(6), 289–296 (1999)
7. Ramage, D., Hall, D., Nallapati, R., et al.: Labeled LDA: a supervised topic model for credit attribution in multi-labeled corpora. In: Conference on Empirical Methods in Natural Language Processing, pp. 248–256. Association for Computational Linguistics (2009)
8. Perotte, A.J., Wood, F., Elhadad, N., et al.: Hierarchically supervised latent Dirichlet allocation. In: International Conference on Neural Information Processing Systems. Curran Associates Inc. (2011)
9. Bengio, Y., Courville, A., Vincent, P.: Representation learning: a review and new perspectives. IEEE Trans. Pattern Anal. Mach. Intell. **35**(8), 1798–1828 (2012)
10. Li, J., Luong, M.T., Dan, J., et al.: When are tree structures necessary for deep learning of representations? Comput. Sci. (2015)
11. Le, Q.V., Mikolov, T.: Distributed representations of sentences and documents. **4**, II-1188 (2014)
12. Socher, R., Perelygin, A., Wu, J.Y., et al.: Recursive deep models for semantic compositionality over a sentiment treebank (2013)
13. Li, L., Qin, B., Ren, W., et al.: Document representation and feature combination for deceptive spam review detection. Neurocomputing (2017)
14. Ren, Y., Ji, D.: Neural networks for deceptive opinion spam detection: an empirical study. Inf. Sci. **385–386**, 213–224 (2017)
15. Li, J., Ott, M., Cardie, C., et al.: Towards a general rule for identifying deceptive opinion spam. In: Meeting of the Association for Computational Linguistics, pp. 1566–1576 (2014)

Deep Learning for Image Denoising: A Survey

Chunwei Tian, Yong Xu, Lunke Fei and Ke Yan

Abstract Since the proposal of big data analysis and Graphic Processing Unit (GPU), the deep learning technique has received a great deal of attention and has been widely applied in the field of imaging processing. In this paper, we have an aim to completely review and summarize the deep learning technologies for image denoising in recent years. Moreover, we systematically analyze the conventional machine learning methods for image denoising. Finally, we point out some research directions for the deep learning technologies in image denoising.

Keywords Deep learning · Convolutional neural networks · GPU · Image denoising

1 Introduction

Image processing has numerous applications including image segmentation [28], image classification [12, 25, 32, 38], object detection [13], video tracking [35], image restoration [46] and action recognition [34]. Especially, the image denoising

C. Tian · Y. Xu (✉) · K. Yan
Bio-Computing Research Center, Harbin Institute of Technology, Shenzhen,
Shenzhen, Guangdong 518055, China
e-mail: yongxu@ymail.com

C. Tian
e-mail: chunweitian@163.com

K. Yan
e-mail: yanke401@163.com

C. Tian · Y. Xu · K. Yan
Shenzhen Medical Biometrics Perception and Analysis Engineering Laboratory,
Harbin Institute of Technology, Shenzhen, Shenzhen, Guangdong 518055, China

L. Fei
School of Computers, Guangdong University of Technology, Guangzhou 510006, China
e-mail: flksxm@126.com

© Springer Nature Singapore Pte Ltd. 2019
J.-S. Pan et al. (eds.), *Genetic and Evolutionary Computing*,
Advances in Intelligent Systems and Computing 834,
https://doi.org/10.1007/978-981-13-5841-8_59

technique is one of the most important branches of image processing technologies and is used as an example to show the development of the image processing technologies in last 20 years [42]. Buades et al. [5] proposed a non-local algorithm method to deal with image denoising. Lan et al. [19] fused the belief propagation inference method and Markov random fields (MRFs) to address image denoising problem. Dabov et al. [9] proposed to transform grouping similar two-dimensional image fragments into three-dimensional data arrays to improve sparsity for image denoising. These feature selection and extraction methods have amazing performance in image denoising. However, the conventional methods have the following two challenges [45]. Firstly these methods are non-convex, which need to manually set parameters. Secondly these methods refer to a complex optimization problem for the test stage, which results in high computational cost.

In recent years, researches have shown that deep learning technologies can reply to deeper architecture to automatically learn and find more suitable image features rather than manually setting parameters, which effectively address drawbacks of traditional methods mentioned above [18]. Especially, big data and GPU are also essential for deep learning technologies to improve the learning ability [16]. The learning ability of deep learning is finished by model (also referred to as network) and the model consists of many layers including the convolutional layer, pooling layer, batch normalization layer, and full connection layer. In other words, deep learning technologies can convert input data (e.g., images, speech, and video) into outputs (e.g., object category, password unlocking, and traffic information) by the model [24]. Specifically, convolutional neural network (CNN) is one of the most typical and successful deep learning networks for image processing [20]. CNN originated LeNet from 1998 and it was successfully used in handwritten digit recognition, achieving excellent performance [21]. However, convolutional neural networks (CNNs) have not been widely used in other real applications before the rise of GPU and big data. The real success of CNNs attributed to ImageNet Large Scale Visual Recognition Challenge 2012 (ILSVRC 2012), where new CNN was proposed, named AlexNet and became a world champion in this ILSVRC 2012 [18, 43].

In subsequent years, deeper neural networks have becoming popular and obtain promising performance for image processing [29]. Karen Simonyan et al. [29] increased the depth of neural networks to 16–19 weighted layers and convolution filter size of each layer was 3×3 for image recognition. Christian Szegedy et al. [30] provided a mechanism by using sparsely connected layer [2] instead of fully connected layers to increase the width and depth of the neural networks for image classification, named as Inception V1. Inception V1 effectively prevented to over-fit from enlarged size (width) of network and reduced the computing resource from increased depth of network. Previous researches show that the deeper networks essentially use an end-to-end fashion to fuse plug-in components of CNN (i.e., convolution and activation function) [17] and classifiers, where the extracted features are more robust by increasing the number of depth in networks. Despite deeper networks have obtained successfully applications for image processing [27], they can generate vanishing gradient or exploding gradient [4] with increased network depth. That makes network hamper convergence, which can be solved by normalized initialization [39].

However, deeper neural networks get to converge, it is saturated and degrade quickly with increasing depth of networks. The appearance of residual network effectively dealt with problems above for image recognition [15]. ResNeXt method is tested to be very effective for image classification [40]. Spatial–temporal attention (SPA) method is very competitive for visual tracking [50]. Residual dense network (RDN) is also an effective tool for image super-resolution [49]. Furthermore, DiracNets [44], IndRNN [23] and variational U-Net [11] also provide us with many competitive methods for image processing. These deeper networks are also widely applied in image denoising. For example, the combination of kernel-prediction net and CNN is used to obtain denoised image [3]. BMCNN utilizes NSS and CNN to deal with image denoising [1]. GAN is utilized to remove noise from noisy image [33].

Although the researches above expose that deep learning technologies have obtained enormous success in the applications of image denoising, to own knowledge, there is no comparative study of deep learning technologies for image denoising. Deep learning technologies refer to properties of image denoising to propose wise solution methods, which are embedded in multiple hidden layers with end-to-end connection to better tackle them. Therefore, a survey is important and necessary to review the principles, performance, difference, merits, shortcomings, and technical potential for image processing. Deeper CNNs (e.g., AlexNet, GoogLeNet, VGG, and ResNet), which can show the ideas of deep learning technologies and successful reasons for image denoising. To better show the robustness of deep learning denoising, the performance of deep learning for image denoising is shown. The potential challenges and directions of deep learning technologies for image denoising in the future are also offered in this paper.

The remainder of this paper is organized as follows. Section 2 overviews of typical deep learning methods. Section 3 provides deep learning techniques for image denoising. Section 4 points out some potential research directions. Section 5 presents the conclusion of this paper.

2 Typical Deeper Network

Nowadays, the most widely used model is trained with end-to-end in a supervised fashion, which is easy and simple to implement for training models. The popular network architecture is CNNs i.e., ResNet. This network is widely used to deal with applications of image processing and obtain enormous success. The following sections will show the popular deep learning technology and discuss the merits and differences of the method in Sect. 2.

2.1 ResNet

Deep CNNs have resulted in a lot of breakthroughs for image recognition. Especially, deeper network plays an important role on image classification [30]. Many

Fig. 1 Residual network: a
building block

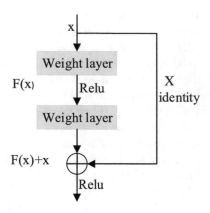

other visual recognition applications are beneficial from deeper networks. However, deeper network can have vanishing/exploding gradients [30]. This problem has been effectively solved by normalized initialization [33], which makes the network converge. When this network starts converging, it's performance gets degraded. For example, the depth of this network is increased, and the errors in the training model are increasing. The problem is effectively addressed by ResNet [15]. The ideas of ResNet are that outputs of each two layers and their inputs are added as the new input. ResNet includes many blocks, and the block is shown in Fig. 1, where x and f respectively denote input and activation functions. A residual block is obtained by $f(x) + x$. The ResNet is popular based on the following reasons. Firstly ResNet is deep rather than width, which effectively controls the number of parameters and overcomes the overfitting problem. Secondly it uses less pooling layers and more downsampling operations to improve transmission efficiency. Thirdly it uses BN and average pooling for regularization, which accelerates the speed of training model. Finally, it uses 3×3 filters of each convolutional layer to train model, which is faster than the combination of 3×3 and 1×1 filters. As a result, ResNet takes the first place in ILSVRC 2015 and reduces 3.57% error on the ImageNet test set.

In addition, deformation networks of residual network are popular and have been widely used in image classification, image denoising [41], and image resolution [31].

3 Image Denoising

Image denoising is topic application for image processing. We take image denoising as an example to show the performance and principle for deep learning technologies in image applications.

The aim of image denoising is to obtain clean image x from a noisy image y, which is explained by $y = x + n$. Especially, n denotes the additive white Gaussian noise (AWGN) with variance σ^2. From the machine learning knowledge, it is seen that the image prior is an important for image denoising. In the past 10 years, a

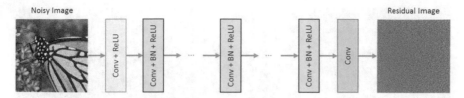

Fig. 2 The architecture of DnCNN

lot of methods are proposed for model with image priors, such as Markov random field (MRF) method [19], BM3D [9], NCSR [10], and NSS [6]. Although these methods perform well for image denoising, they have two drawbacks. First, these methods need to be optimized which increases computational cost. Second, these methods are non-convex, which needs manual settings to improve performance. To address the problems, some discriminative learning schemes were proposed. A trainable nonlinear reaction–diffusion method was used to learn image prior [26]. A cascade of shrinkage fields fuses the random field-based model and half-quadratic algorithm into a single architecture [47]. Despite the existing methods improve the performances for image denoising, they are limited to the specified forms of prior. Another shortcoming is that these methods cannot use a model to deal with blind denoising.

Deep learning technologies can effectively handle problems above. And deep learning technologies are chosen for image denoising based on the following three-fold. First, they have deep architectures which can learn more extractions. Second, BN and ReLU are added to deep architectures, which can accelerate the training speed. Third, networks of deep learning methods can run on GPU, which can train more samples and improve the efficiency. For example, the proposed DnCNN [45] uses BN and ResNet to perform image denoising. It not only deals with blind denoising, but also addresses image super-resolution task and JPEG image deblocking. Its architecture is as shown in Fig. 2. Specifically, it obtains the residual image from the model and it needs to use $y = x + n$ to obtain clean image in the test phase. It obtained PSNR of 29.13, which is higher than the state-of-the-art BM3D method of 28.57 for BSD68 dataset with $\sigma = 25$.

FFDNet [47] uses noise level map and noisy image as input for different noise levels. This method exploits a single model to deal with multiple noise levels. It is also faster than BM3D on GPU and CPU. As shown in Fig. 3, performance of FFDNet outperforms the CBM3D [19] method in image denoising. IRCNN [48] fuses the model-based optimization method and CNN to address image denoising problem, which can deal with multiple tasks via single mode. In addition, it adds dilated convolution into network, which improves the performance for denoising. Its architecture is shown in Fig. 4.

In addition, many other methods also obtain well performance for image denoising. For example, fusion of the dilated convolution and ResNet is used for image denoising [36]. It is a good choice for combing disparate sources of experts for image denoising [8]. Universal denoising networks [22] for image denoising and

(a) Noisy ($\sigma = 35$)   (b) CBM3D (29.90dB)   (c) FFDNet (30.51dB)

Fig. 3 Results of CBM3D and FFDNet for color image denoising

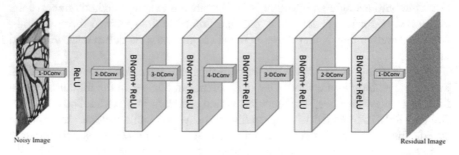

Fig. 4 The architecture of IRCNN

deep CNN denoiser prior to eliminate multiplicative noise [37] are also effective for image denoising. As shown in Table 1, deep learning methods are superior to the conventional methods. And the DnCNN method obtains excellent performance for image denoising.

Table 1 Comparisons of different methods with $\sigma = 25$ for image denoising.

Methods	PSNR	Dataset
BM3D [9]	28.57	BSD68
WNNM [14]	28.83	BSD68
TNRD [7]	28.92	BSD68
DnCNN [45]	29.23	BSD68
FFDNet [47]	29.19	BSD68
IRCNN [48]	29.15	BSD68
DDRN [36]	29.18	BSD68

4 Research Directions

4.1 The Challenges of Deep Learning Technologies in Image Denoising

According to existing researches, deep learning technologies achieve promising results in image denoising. However, these technologies also suffer from some challenges as follows. (1) Current deep learning denoising methods only deal with AWGN, which is not effective for real noisy images, such as low-light images. (2) They cannot use a model to deal with multi low-level vision tasks, such as image denoising, image super-resolution, image blurring, and image deblocking. (3) They cannot use a model to address blind denoising.

4.2 Some Potential Directions of Deep Learning Technologies for Image Denoising

According to the previous researches, deep learning technologies have the following changes for image denoising above. First, deep learning technologies design different network architectures to blind denoising. Second, they can fuse the optimization and discrimination methods to make a tradeoff between performance and efficiency. Third, they can deal with real noisy. Fourth, they can use a model to address multi low-level tasks.

5 Conclusion

This paper first comprehensively introduces the development of deep learning technologies on image processing applications and then shows the implementations of typical CNNs. After that, image denoising is illustrated in detail, which concludes the differences and ideas of different methods for image denoising in real world. Finally, this paper shows the challenges of deep learning methods for image processing applications and offers solutions. This review offers important cues on deep learning technologies for image processing applications. We believe that this paper could provide researchers with a useful guideline working in the related fields, especially for the beginners worked in deep learning.

Acknowledgements This paper was supported in part by Shenzhen Municipal Science and Technology Innovation Council under Grant no. JCYJ20170811155725434, in part by the National Natural Science Foundation under Grant no. 61876051.

References

1. Ahn, B., Cho, N.I.: Block-matching convolutional neural network for image denoising (2017). arXiv:1704.00524
2. Arora, S., Bhaskara, A., Ge, R., Ma, T.: Provable bounds for learning some deep representations. In: International Conference on Machine Learning, pp. 584–592 (2014)
3. Bako, S., Vogels, T., McWilliams, B., Meyer, M., Novák, J., Harvill, A., Sen, P., Derose, T., Rousselle, F.: Kernel-predicting convolutional networks for denoising monte carlo renderings. ACM Trans. Graph 36(4), 97 (2017)
4. Bengio, Y., Simard, P., Frasconi, P.: Learning long-term dependencies with gradient descent is difficult. IEEE Trans. Neural Netw. 5(2), 157–166 (1994)
5. Buades, A., Coll, B., Morel, J.-M.: A non-local algorithm for image denoising. In: CVPR 2005. IEEE Computer Society Conference on Computer Vision and Pattern Recognition, 2005, vol. 2, pp. 60–65. IEEE (2005)
6. Buades, A., Coll, B., Morel, J.-M.: Nonlocal image and movie denoising. Int. J. Comput. Vis. 76(2), 123–139 (2008)
7. Chen, Y., Pock, T.: Trainable nonlinear reaction diffusion: a flexible framework for fast and effective image restoration. IEEE Trans. Pattern Anal. Mach. Intell. 39(6), 1256–1272 (2017)
8. Choi, J.H., Elgendy, O., Chan, S.H.: Integrating disparate sources of experts for robust image denoising (2017). arXiv:1711.06712
9. Dabov, K., Foi, A., Katkovnik, V., Egiazarian, K.: Image denoising by sparse 3-d transform-domain collaborative filtering. IEEE Trans. Image Process. 16(8), 2080–2095 (2007)
10. Dong, W., Zhang, L., Shi, G., Li, X.: Nonlocally centralized sparse representation for image restoration. IEEE Trans. Image Process. 22(4), 1620–1630 (2013)
11. Esser, P., Sutter, E., Ommer, B.: A variational u-net for conditional appearance and shape generation. In: Proceedings of the IEEE Conference on Computer Vision and Pattern Recognition, pp. 8857–8866 (2018)
12. Fei, L., Lu, G., Jia, W., Teng, S., Zhang, D.: Feature extraction methods for palmprint recognition: a survey and evaluation. IEEE Trans. Syst. Man Cybern.: Syst. (2018)
13. Girshick, R., Donahue, J., Darrell, T., Malik, J.: Rich feature hierarchies for accurate object detection and semantic segmentation. In: Proceedings of the IEEE Conference on Computer Vision and Pattern Recognition, pp. 580–587 (2014)
14. Gu, S., Zhang, L., Zuo, W., Feng, X.: Weighted nuclear norm minimization with application to image denoising. In: Proceedings of the IEEE Conference on Computer Vision and Pattern Recognition, pp. 2862–2869 (2014)
15. He, K., Zhang, X., Ren, S., Sun, J.: Deep residual learning for image recognition. In: Proceedings of the IEEE Conference on Computer Vision and Pattern Recognition, pp. 770–778 (2016)
16. Hou, Q., Cheng, M.-M., Hu, X., Borji, A., Tu, Z., Torr, P.: Deeply supervised salient object detection with short connections. In: 2017 IEEE Conference on Computer Vision and Pattern Recognition (CVPR), pp. 5300–5309. IEEE (2017)
17. Ioffe, S., Szegedy, C.: Batch normalization: accelerating deep network training by reducing internal covariate shift (2015). arXiv:1502.03167
18. Krizhevsky, A., Sutskever, I., Hinton, G.E.: Imagenet classification with deep convolutional neural networks. In: Advances in Neural Information Processing Systems, pp. 1097–1105 (2012)
19. Lan, X., Roth, S., Huttenlocher, D., Black, M.J.: Efficient belief propagation with learned higher-order Markov random fields. In: European Conference on Computer Vision, pp. 269–282. Springer (2006)
20. Lawrence, S., Giles, C.L., Tsoi, A.C., Back, A.D.: Face recognition: a convolutional neural-network approach. IEEE Trans. Neural Netw. 8(1), 98–113 (1997)
21. LeCun, Y., Boser, B., Denker, J.S., Henderson, D., Howard, R.E., Hubbard, W., Jackel, L.D.: Backpropagation applied to handwritten zip code recognition. Neural Comput. 1(4), 541–551 (1989)

22. Lefkimmiatis, S.: Universal denoising networks: a novel CNN architecture for image denoising. In: Proceedings of the IEEE Conference on Computer Vision and Pattern Recognition, pp. 3204–3213 (2018)
23. Li, S., Li, W., Cook, C., Zhu, C., Gao, Y.: Independently recurrent neural network (indrnn): Building a longer and deeper RNN. In: Proceedings of the IEEE Conference on Computer Vision and Pattern Recognition, pp. 5457–5466 (2018)
24. Litjens, G., Kooi, T., Bejnordi, B.E., Setio, A.A.A., Ciompi, F., Ghafoorian, M., van der Laak, J.A.W.M., Van Ginneken, B., Sánchez, C.I.: A survey on deep learning in medical image analysis. Med. Image Anal. **42**, 60–88 (2017)
25. Qin, Y., Tian, C.: Weighted feature space representation with kernel for image classification. Arab. J. Sci. Eng. 1–13 (2017)
26. Schmidt, U., Roth, S.: Shrinkage fields for effective image restoration. In: Proceedings of the IEEE Conference on Computer Vision and Pattern Recognition, pp. 2774–2781 (2014)
27. Sermanet, P., Eigen, D., Zhang, X., Mathieu, M., Fergus, R., LeCun, Y.: Overfeat: integrated recognition, localization and detection using convolutional networks (2013). arXiv:1312.6229
28. Shi, J., Malik, J.: Normalized cuts and image segmentation. IEEE Trans. Pattern Anal. Mach. Intell. **22**(8), 888–905 (2000)
29. Simonyan, K., Zisserman, A.: Very deep convolutional networks for large-scale image recognition (2014). arXiv:1409.1556
30. Szegedy, C., Liu, W., Jia, Y., Sermanet, P., Reed, S., Anguelov, D., Erhan, D., Vanhoucke, V., Rabinovich, A.: Going deeper with convolutions. In: Proceedings of the IEEE Conference on Computer Vision and Pattern Recognition, pp. 1–9 (2015)
31. Tai, Y., Yang, J., Liu, X., Xu, C.: Memnet: a persistent memory network for image restoration. In: Proceedings of the IEEE Conference on Computer Vision and Pattern Recognition, pp. 4539–4547 (2017)
32. Tian, C., Zhang, Q., Sun, G., Song, Z., Li, S.: Fft consolidated sparse and collaborative representation for image classification. Arab. J. Sci. Eng. **43**(2), 741–758 (2018)
33. Tripathi, S., Lipton, Z.C., Nguyen, T.Q.: Correction by projection: denoising images with generative adversarial networks (2018). arXiv:1803.04477
34. Wang, H., Kläser, A., Schmid, C., Liu, C.-L.: Action recognition by dense trajectories. In: 2011 IEEE Conference on Computer Vision and Pattern Recognition (CVPR), pp. 3169–3176. IEEE (2011)
35. Wang, L., Liu, T., Wang, G., Chan, K.L., Yang, Q.: Video tracking using learned hierarchical features. IEEE Trans. Image Process. **24**(4), 1424–1435 (2015)
36. Wang, T., Sun, M., Hu, K.: Dilated residual network for image denoising (2017). arXiv:1708.05473
37. Wang, G., Wang, G., Pan, Z., Zhang, Z.: Multiplicative noise removal using deep CNN denoiser prior. In: 2017 International Symposium on Intelligent Signal Processing and Communication Systems (ISPACS), pp. 1–6. IEEE (2017)
38. Wen, J., Fang, X., Yong, X., Tian, C., Fei, L.: Low-rank representation with adaptive graph regularization. Neural Netw. **108**, 83–96 (2018)
39. Wu, Y., He, K.: Group normalization (2018). arXiv:1803.08494
40. Xie, S., Girshick, R., Dollár, P., Tu, Z., He, K.: Aggregated residual transformations for deep neural networks. In: 2017 IEEE Conference on Computer Vision and Pattern Recognition (CVPR), pp. 5987–5995. IEEE (2017)
41. Xie, W., Li, Y., Jia, X.: Deep convolutional networks with residual learning for accurate spectral-spatial denoising. Neurocomputing (2018)
42. Xu, J., Zhang, L., Zuo, W., Zhang, D., Feng, X.: Patch group based nonlocal self-similarity prior learning for image denoising. In: Proceedings of the IEEE International Conference on Computer Vision, pp. 244–252 (2015)
43. You, Y., Zhang, Z., Hsieh, C.-J., Demmel, J., Keutzer, K.: 100-epoch imagenet training with alexnet in 24 minutes (2017)
44. Zagoruyko, S., Komodakis, N.: Diracnets: training very deep neural networks without skip-connections (2017). arXiv:1706.00388

45. Zhang, K., Zuo, W., Chen, Y., Meng, D., Zhang, L.: Beyond a gaussian denoiser: residual learning of deep cnn for image denoising. IEEE Trans. Image Process. **26**(7), 3142–3155 (2017)
46. Zhang, L., Zuo, W.: Image restoration: from sparse and low-rank priors to deep priors. IEEE Signal Process. Mag. **34**(5), 172–179 (2017)
47. Zhang, K., Zuo, W., Zhang, L.: FFDNet: toward a fast and flexible solution for CNN based image denoising. IEEE Trans. Image Process. (2018)
48. Zhang, K., Zuo, W., Zhang, L.: Learning a single convolutional super-resolution network for multiple degradations. In: IEEE Conference on Computer Vision and Pattern Recognition, vol. 6 (2018)
49. Zhang, Y., Tian, Y., Kong, Y., Zhong, B., Fu, Y.: Residual dense network for image super-resolution. In: The IEEE Conference on Computer Vision and Pattern Recognition (CVPR) (2018)
50. Zhu, Z., Wu, W., Zou, W., Yan, J.: End-to-end flow correlation tracking with spatial-temporal attention. Illumination **42**, 20 (2017)

Part XIV
Decision Support Systems and Data Security

Decision Support of Intelligent Factory Evaluation Based on RAGA-AHP

Shaoqin Lu

Abstract This paper systematically studies the intelligent factory evaluation model based on RAGA-AHP, provides decision support and references for government's technical and financial support for enterprise intelligent factory projects. Established evaluation indexes system of intelligent factory and its evaluation model based on RAGA-AHP. It analyzed the evaluation indexes of intelligent factory, the weights ratio of evaluation indexes, and the consistency check of evaluation matrix. The research results show that the accuracy of the data based on the RAGA-AHP intelligent factory evaluation model is higher and the result is more reasonable. When the government evaluates the enterprise intelligent factory project, it is recommended to determine the evaluation indexes of the intelligent factory, and then uses the RAGA-AHP model to analyze the weights of the evaluation indexes and tests its consistency, thus, the evaluation results of the enterprise's intelligent factory construction level are obtained.

Keywords Intelligent factory · Evaluation · Real coding based accelerating genetic algorithm (RAGA) · Analytic hierarchy process (AHP)

1 Introduction

Since the last century, with the development of information technology and Internet of things technology, traditional manufacturing has gradually evolved from automation, digitization, and information to intelligent development, more and more enterprises begin to build intelligent factories. The country and governments need to analyze the status and the level of intelligent manufacturing capabilities of enterprises on the basis of evaluating of intelligent factory construction, formulate the development plan of intelligent manufacturing, and give technical and financial support policy in key

S. Lu (✉)
Changzhou College of Information Technology, No. 22 Mingxin Middle Road,
Wujin District, Changzhou, Jiangsu, China
e-mail: sqlu@163.com

© Springer Nature Singapore Pte Ltd. 2019
J.-S. Pan et al. (eds.), *Genetic and Evolutionary Computing*,
Advances in Intelligent Systems and Computing 834,
https://doi.org/10.1007/978-981-13-5841-8_60

projects of intelligent factories, so as to enhance the enterprises' core competitiveness and the level of intelligent manufacturing in the entire country.

AHP is a commonly used method for comprehensive analysis of quantitative evaluation indexes and qualitative evaluation indexes, and the RAGA-AHP intelligent factory evaluation model integrated with RAGA and AHP can obtain more accurate evaluation index weights and ensures the consistency of the judgment matrix. This paper designs the evaluation indexes system from four aspects of digital design, intelligent equipment, production control, and enterprise management, analyzed the weight of each evaluation index through AHP, and corrected the weights of the evaluation indexes by using RAGA to make up for the inaccuracy of the weights calculation accuracy of AHP. Based on RAGA-AHP, intelligent factory evaluation model provides theoretical models and technical methods for government to evaluate the level of intelligent factory construction.

2   Materials and Methods

2.1   Data Sources

The data used in the study was derived from the intelligent factory review work conducted from April to November 2017 in Changzhou, Jiangsu Province, China. In the evaluation of the intelligent factory in Changzhou in 2017, it mainly involved digital design, intelligent equipment, production control, and enterprise management. The enterprises participating in the intelligent factory review mainly come from industrial enterprises in the 5 districts and 1 town of Changzhou City, there were 359 enterprises participating in the intelligent factory review. According to the assessment data of industrial enterprises participating in the evaluation by RAGA-AHP model, the construction level of the enterprise's intelligent factory was analyzed, and finally, 47 enterprises were awarded intelligent factories.

2.2   Research Methods

(1)   RAGA

RAGA uses a real-number-based coding method, and multiple iterations replace the value range of excellent individual with the optimal range, so that the individual seeking for optimization gradually moves closer to the optimal range, the algorithm runs, until the target function reaches a certain value or reaches a certain number of iterations. RAGA overcomes the problems of poor SGA optimization efficiency and local optimization [1, 2].

Assuming that the general nonlinear optimization is the minimization problem, using RAGA to solve the optimal individual has the following eight steps.

$$
\begin{cases}
\min f(x) \\
a(j) \le x(j) \le b(j)
\end{cases}
\tag{1}
$$

In the formula, $X = \{x(j)\}$ is the set of optimization variables, $j = 1, 2, ..., p$, $[a(j), b(j)]$ is variation range of $x(j), f(x)$ is the desired objective function.

Step 1: Coding. Relative to SGA binary encoding, RAGA uses real-coded methods. Step 2: Parent groups initialization and sorting. Generating random number $y(j)$ in the n groups of $[0, 1]$ intervals, and it can get the values of $x(j)$ and $f(x)$. Sorting the values of n groups of objective functions $f(x)$ will get the first generation. Step 3: Individual fitness evaluation. Setting $F(i) = 1/(f(i) \times f(i) + 0.001)$, the greater value of the fitness function $F(i)$, It shows that the higher the fitness. Step 4: Using selection operators to generate progeny groups. Step 5: crossover. The crossover operator uses the reasonable genetic information of two parent individuals to generate two new child individuals. Step 6: mutation. RAGA increases individual diversity through mutation. Step 7: Evolutionary iteration. The algorithm proceeds to step 3 for the next evolution, and regenerating parent groups with selection, sorting, crossover and mutation operators, and loops. Step 8: Accelerating loop. In the evolutionary iteration cycle, when the value of the objective function of the individual is lower than a certain set value or the algorithm reaches a predetermined number of cycles, the operation of the algorithm ends, so using the average of the best individuals as the result of the RAGA calculation.

(2) RAGA-AHP

The RAGA-AHP-based intelligent factory evaluation model introduces RAGA in AHP, solves the problems of the lack of accuracy and consistency of the evaluation index weight calculation. RAGA is used in AHP, converts the consistency problem of the judgment matrix into the optimization problem of nonlinear index weights. Taking engineering project selection as an example, using the RAGA-AHP model to find the optimal solution includes the following five steps.

Step 1: Creating a hierarchical model. The evaluation model of the project plan is divided into target layer A, index layer B, and programs layer C. Step 2: The index layer B and the program layer C, will have a pairwise comparison according to the upper goal of each evaluation index, and uses 1–9 and its reciprocal as a scale to describe the degree of importance among various indexes and obtains a judgment matrix $B = \{b_{ij} | i, j = 1-n\}$. Step 3: Single rank index weighting and judgment matrix consistency test. The weight of each evaluation index in the judgment matrix B is $\{w_k | k = 1-n\}, w_k > 0\}$, and transforms the consistency check problem of the judgment matrix into the RAGA process of finding the optimal solution of the indexes weights, the objective function for building RAGA is $\min CIF(n) = \frac{\sum_{i=1}^{n} |\sum_{k=1}^{n} b_{ik} w_k - n w_k|}{n}$, $w_k > 0$, $k = 1-n$, $\sum_{k=1}^{n} w_k = 1$. When the value of $\min CIF(n)$ is lower than a certain standard, the weight w_k of each evaluation index obtained is also credible. In the same way, the weights w_i^k of the program layer C based on the evaluation index k

can be obtained [3–6]. Step 4: Calculating the total weight and consistency of the program layer *C*. According to the single layer weights of the index layer *B* and the program layer *C*, the weight of the program layer *C* for the total target layer *A* is calculated. Step 5: According to the total ranking weight of the program layer C, the sorting is calculated and the best solution is obtained.

3 Results and Discussion

3.1 Intelligent Factory Evaluation Index System Construction

The intelligent factory's evaluation index system should be oriented to the characteristics of intelligence, determines whether it has intelligent elements, and analyses the level reached by the intelligent elements [7]. So, this paper constructs four first-level evaluation indexes and 14 second-level evaluation indexes. The first-level indexes mainly include four aspects of digital design, intelligent equipment, production control, and business management. The specific evaluation indexes are shown in Table 1.

Table 1 Intelligent factory evaluation index system

First-level indexes	Second-level indexes	Index description	Type
Digital design	Workshop design	0, 1, 2 Three-scale method	Positive
	Product design	0, 1, 2 Three-scale method	Positive
Intelligent equipment	CNC production equipment	Percentage value	Positive
	Sensing and control devices	Percentage value	Positive
Production control	Workshop production plan management	Fuzzy evaluation	Positive
	Production process management	Fuzzy evaluation	Positive
	Production material management	Fuzzy evaluation	Positive
	Quality control	Fuzzy evaluation	Positive
	Production equipment management	Fuzzy evaluation	Positive
	Data collection	Fuzzy evaluation	Positive
Business management	Information security	Percentage value	Negative
	Economic benefits	Percentage value	Negative
	Business integration	Percentage value	Positive
	Functional management	Percentage value	Positive

3.2 Data Standardization

For the standardization of intelligent factory evaluation indexes data, the fuzzy evaluation method and the three-scale method in AHP are used for dimensionless processing. The specific steps are as follows.

Step 1: Determining the best and worst value of the evaluation index. For the positive index in Table 1, the larger the value, the better, but negative index means that the smaller the value, the better [8, 9]. Step 2: Determining the fuzzy membership function of each evaluation index and building a fuzzy matrix R. Step 3: Building a judgment matrix. According to fuzzy matrix R, Calculates the standard deviation of each evaluation index $s(i)$, the construction of matrix B uses the following formula:

$$b_{ij} = \begin{cases} \frac{s(i)-s(j)}{s_{max}-s_{min}}(b_m - 1) + 1, & s(i) > s(j) \\ \left[\frac{s(j)-s(i)}{s_{max}-s_{min}}(b_m - 1) + 1\right]^{-1}, & s(i) < s(j) \end{cases} \tag{2}$$

3.3 Determining the Weight of Evaluation Indexes and Comprehensive Assessment of Intelligent Factory

The RAGA-AHP model was used to measure the weights of the evaluation indexes of the intelligent factory and the consistency of the judgment matrix. In the process of research, evaluation indexes will be ranked and assigned by experts who are from nine institutions. According to sorting and scoring results, constructs a judgment matrix and uses RAGA to calculate the weight of each evaluation index, as shown in Table 2.

Based on the weights of the evaluation indexes and the standardized values of the participating enterprises, a weighted sum is obtained. Finally, the evaluation results of various enterprises' intelligent factories can be determined, as shown in Table 3. The results are grouped by percentage, giving five-star, four-star, three-star and pending improvement four-grade assessments, among them: The enterprise's rating below 60 points is divided into pending improvement, 60–70 is divided into three-star intelligent factory, 70–80 is divided into four-star intelligent factory, 80 points and above are five-star intelligent factory.

From the results in Table 2, it can be seen that using RAGA-AHP can get the weight and importance of each first-level and second-level indexes. Comparing the relative consistency value of 0.1 with the traditional sum-product method and the square root difference method, using the RAGA-AHP model has a higher consistency. Among the second-level evaluation indexes, sensing and control devices, CNC production equipment, and product design have a large proportion; they are 0.1312, 0.1224, and 0.0791. From the results in Table 3, it can be seen that the comprehensive evaluation scores of the four case enterprises are 67.65, 82.86, 78.45 and 81.35, respectively.

Table 2 Based on RAGA-AHP intelligent factory evaluation index weights

Indexes	Digital design 0.1479	Intelligent equip-ment 0.2535	Production control 0.3576	Business manage-ment 0.2409	Total ranking	*CIF*(*n*)
Workshop design	0.4653				0.0688	0.0003
Product design	0.5347				0.0791	
CNC production equipment		0.4827			0.1224	0.0002
Sensing and control devices		0.5173			0.1312	
Workshop production plan management			0.1704		0.0609	0.0006
Production process management			0.1712		0.0612	
Production material management			0.2150		0.0769	
Quality control			0.2114		0.0756	
Production equipment management			0.0846		0.0303	
Data collection			0.1474		0.0527	
Information security				0.2397	0.0578	0.0001
Economic benefits				0.2470	0.0595	
Business integration				0.2737	0.0659	
Functional management				0.2396	0.0577	

Therefore, it can determine the level of construction of enterprise intelligent factories: Enterprise B > Enterprise D > Enterprise C > Enterprise A.

4   Conclusions

This article starts from the connotation of enterprise intelligent manufacturing, and combines with the project evaluation of intelligent factories, the intelligent factory evaluation indexes system consisting of 4 first-level indexes and 14 second-level

Table 3 Evaluation scores and ratings of intelligent factories for various enterprises

Indexes	Weight	Enterprise A	Enterprise B	Enterprise C	Enterprise D
Workshop design	0.0688	61	89	87	92
Product design	0.0791	63	83	75	78
CNC production equipment	0.1224	63	76	83	91
Sensing and control devices	0.1312	90	85	90	81
Workshop production plan management	0.0609	60	88	91	79
Production process management	0.0612	81	93	77	81
Production material management	0.0769	74	85	77	93
Quality control	0.0756	80	88	78	90
Production equipment management	0.0303	57	86	72	87
Data collection	0.0527	80	93	79	76
Information security	0.0578	73	94	83	88
Economic benefits	0.0577	65	75	84	77
Business integration	0.0659	60	79	85	87
Functional management	0.0595	76	90	79	82
Overall score		67.65	82.86	78.45	81.35
Rating level		Three-star	Five-star	Four-star	Five-star

indexes were constructed. In the intelligent factory project evaluation, an intelligent factory evaluation model based on RAGA-AHP was constructed. Using RAGA to optimize the weight of the evaluation index of AHP will have higher data accuracy. The RAGA-AHP-based intelligent factory evaluation model provides a theoretical model and technical method for government departments to evaluate the level of intelligent factory construction, and formulates intelligent manufacturing related policies for government departments, and also provides targeted decision-making support for enterprises with targeted technology and funding.

Acknowledgements The work is partially supported by Changzhou College of Information Technology Research Platform (CXPT201702R) and Jiangsu University Philosophy and Social Science Key Construction Base (2018ZDJD-B017).

References

1. Weiming, Y., Peiwu, D., Jing, W.: Research on evaluation model of enterprise intelligent manufacturing capacity based on high order tensor analysis. Ind. Technol. Econ. **37**(01), 11–16 (2018)
2. Hao, L., Qingmin, H., Wangjun, Y., Lei, Z.: Research on evaluation criteria and evaluation methods for smart factory in mobile terminal manufacturing industry. Microcomput. Appl. **36**(23), 89–92 (2017)
3. Juliang, J., Xiaohua, Y., Jing, D.: Accelerating genetic algorithm based on real-coded. J. Sichuan Univ. (Engineering Science Edition) (04), 20–24 (2000)
4. Juliang, J., Wei Yiming, F., Qiang, D.J.: Accelerating genetic algorithm for calculating rank weights in analytic hierarchy process. Syst. Eng. Theory Pract. (11), 39–43 (2002)
5. Ministry of Industry and Information Technology National Standardization Management Committee. National Intelligent Manufacturing Standard System Construction Guide (2015 Edition), 2015-12-29
6. Cichocki, A., Mandic, D., De Lathauwer, L., et al.: Tensor decompositions for signal processing applications from two-way to multiway component analysis. Signal Process. Mag. (IEEE) **32**(2), 145–163 (2014)
7. IEC/PAS 63088-2017: Smart manufacturing-reference architecture model industry 4.0 (RAMI 4.0) (2017)
8. Schroth, C.: The Internet of services: global industrialization of information intensive services. In: International Conference on Digital Information Management, vol. 2, pp. 635–642 (2007)
9. Kun, S.: Comprehensive evaluation of intelligent manufacturing capability based on factor analysis method. J. Logist. Sci. **40**(7), 116–120 (2017)

Hot Topic Detection Based on VSM and Improved LDA Hybrid Model

Qiu Liqing, Liu Haiyan, Fan Xin and Jia Wei

Abstract In consideration of the features of microblog content such as short text, sparse feature words, and the huge scale, a hot topic detection method was suggested by this paper based on VSM and improved LDA hybrid model. This method first uses LDA model incorporating the PageRank algorithm, deeply mines the structure of the social network. Then the VSM model and the improved LDA model are used for hybrid modeling and to calculate the similarity of microblog. Finally, the hot topic clustering results are obtained through the single-pass clustering algorithm based on the hybrid model. The experimental results show that the proposed method can effectively improve the microblogging hot topic detection efficiency.

Keywords Vector space model · LDA · Topic discovery · Single-pass clustering

1 Introduction

With the development of computers and the Internet, as an emerging social networking application, microblog which has gradually become an important tool for people to communicate with each other and make a speech. It is of great significance for monitoring and guiding public opinions to obtain the topics which the Internet users are concerning about from the massive microblog texts accurately. While microblog

Q. Liqing (✉) · L. Haiyan · F. Xin · J. Wei
Shandong Province Key Laboratory of Wisdom Mine Information Technology,
College of Computer Science and Engineering, Shandong University of Science
and Technology, Qingdao 266510, China
e-mail: liqingqiu2005@126.com

L. Haiyan
e-mail: 549542683@qq.com

F. Xin
e-mail: 564353949@qq.com

J. Wei
e-mail: 1456514769@qq.com

© Springer Nature Singapore Pte Ltd. 2019   583
J.-S. Pan et al. (eds.), *Genetic and Evolutionary Computing*,
Advances in Intelligent Systems and Computing 834,
https://doi.org/10.1007/978-981-13-5841-8_61

has the characteristics of short content, small and sparse feature words, large scale, and others, so the traditional TDT technology cannot effectively apply to microblog news. In consideration of the features of microblogging content, a method to detect microblogging hot topic was proposed in this paper based on improved LDA and VSM model.

In this paper, combining the advantages and disadvantages of VSM and LDA model, first, VSM model was adopted to conduct directional quantization of microblog; Second, the novel algorithm uses LDA model incorporating PageRank algorithm, deeply mines the structure of social network, enhances the theme distribution of nodes. Then using SVM and improved LDA model to conduct mixed modeling, the potential semantic information is mined out, and the word frequency and semantic information are weighted to obtain the similarity; Finally, the composite similarity is applied to the single-pass clustering algorithm to conduct topic clustering on microblog and extract hot topics.

2 Related Works

In recent years, the research on social network and microblog application has been a hot topic at home and abroad. Many researchers improve the LDA model according to different scenarios. Reference [1] used word clustering cluster as a prior knowledge to influence LDA to carry out the theme-word assignment, so as to extract the subject comment object more consistent with semantic requirements. Reference [2] proposed the DTM (dynamic topic model) topic model, added the discrete time slice information into LDA, so as to acquire text topics that evolve with time. Reference [3] built a dynamic theme model, mined the dynamic theme chain that changes over time, and proposed the calculation method of topic heat. Reference [4] put forward a new social network topic discovery algorithm iMLDA (Importance-Latent Dirichlet Allocation). The algorithm integrated the LDA model with PageRank-based node importance algorithm and improved the accuracy of topic discovery.

In terms of text similarity calculation, the cosine similarity method based on vector space model proposed in Ref. [5] has some defects, and the correlation between words and words is not considered from the semantic perspective; Ref. [6] proposed a calculation method of short text similarity based on semantics and maximum matching degree. Reference [7] combined with microblog native point, proposed a similarity algorithm based on semantic relation, time relation, and social relation, which further improved the accuracy of text similarity measurement and the quality of topic detection.

From the above research, it can be seen that the improvement of similarity can improve the quality of topic detection, while Ref. [8] pointed out the deficiency of traditional text vectorization which only relies on word frequency, and considered using subject model to make up for it. Therefore, this paper introduced PageRank algorithm to improve the LDA model, adopted VSM, and improved LDA model for hybrid modeling and similarity calculation, and applied composite similarity to single-pass clustering algorithm to realize topic discovery.

3 Hot Topic Detection Based on VSM and Improved LDA Hybrid Model

Several important parts of the algorithm are described below.

3.1 Microblog Text Quantification

In this paper, the VSM and improved LDA model are utilized to model the feature items in the microblog text.

3.1.1 The Vector Space Model (VSM)

VSM is a commonly used model for representation of documents in text classification. Let $T = \{t_1, t_2, \ldots, t_n\}$ be the set of features selected from all documents. The vector for a document d_i is represented by $V_i = (w_{i1}, w_{i2}, \ldots, w_{in})$, where w_{ij} represents the weight of the feature t_j in the document d_i. Each element in V_i can be best qualified by TF-IDF. However, because of TF-IDF ignores the semantic information of the word itself, this paper improves the traditional TF-IDF, so the weight w_{ij} of the feature t_{ij} is

$$w_{ij} = \frac{\sqrt{tf_{ij}} \times \log\left(\frac{N}{m_{ij}} + 0.01\right)}{\sqrt{\sum_{j=1}^{M}\left(\sqrt{tf_{ij}} \times \log\left(\frac{N}{m_{ij}} + 0.01\right)\right)^2}} \tag{1}$$

where N is the total number of documents in the collection, m_{ij} is the number of documents that contain at least one occurrence of the term tf_{ij}, and M is the total number of feature items in the document i [9].

3.1.2 LDA Algorithm and Its Improvement. Latent Dirichlet Allocation

LDA (Latent Dirichlet Allocation) is a kind of three-layer tree Bayesian probabilistic generative model, which consists of document layer, theme layer, and word layer.

Gibbs sampling was used in LDA model, the main derivation process is as follows:

- (a) The equation of the joint distribution of the LDA model can be written as follows:

$$p(w, z | \alpha, \beta) = p(w | z, \beta) p(z | \alpha) \tag{2}$$

- (b) According to Bayes Rule and Dirichlet priori, the distribution expectation of Dirichlet are as follows.

$$\varphi_{k,t} = \frac{n_k^{(t)} + \beta_t}{\sum_{t=1}^{V} n_k^{(t)} + \beta_t} \tag{3}$$

$$\theta_{d,k} = \frac{n_d^{(k)} + \alpha_k}{\sum_{k=1}^{K} n_d^{(k)} + \alpha_k} \tag{4}$$

where $\varphi_{k,t}$, $n_k^{(t)}$ respectively denotes the probability and weighted number of the characteristic word t in theme k, $\theta_{d,k}$, $n_d^{(k)}$, respectively, represents the probability and weighted number of topic k in document d, and α and β are super parameters.
- (c) Given the joint distribution of w and z under LDA, we can compute the conditional probability for the Gibbs sampler by

$$p(z_i = k | z_{-i}, w) \propto \theta_{d,k} = \frac{n_d^{(k)} + \alpha_k}{\sum_{k=1}^{K} n_d^{(k)} + \alpha_k} \cdot \varphi_{k,t} = \frac{n_k^{(t)} + \beta_t}{\sum_{t=1}^{V} n_k^{(t)} + \beta_t} \tag{5}$$

The improved LDA model. This paper improves the LDA model by considering the important information of nodes [4]. First, calculate the PageRank value of node i as PR(i) according to the PageRank algorithm, and the higher the PR value, the more important the nodes are. The weight formula for the importance of document d_i is as follows.

$$Importance(d_i) = PR(i) \tag{6}$$

Then the topic document distribution θ_d is multiplied by the weight of importance of the document $Importance(d_i)$, so as to obtain the new topic document distribution of each node, and to strengthen the topic distribution of the node on this account.

3.2 Topic Clustering Algorithm. Similarity Calculation Method Based on SVM and Improved LDA Model

The clustering of hot topics on the microblog first requires the calculation of vector similarity.

- (1) In the SVM model, cosine method combined with semantic similarity is used to calculate the similarity of topic vectors in microblog [10]:

$$sim_{VSM}(d_i, d_j) \frac{\sum_{x=1}^{n} \sum_{y=1}^{n} (w_{ix} w_{iy} S_{ixj})}{\sqrt{\sum_{k=1}^{n} w_{ik}^2} \sqrt{\sum_{k=1}^{n} w_{jk}^2}} \qquad (7)$$

where S_{ixj} denotes the semantic correlation degree between the x characteristic item in the microblog document d_i and the microblog vector d_j, and n is the dimension of VSM.

- (2) In the improved LDA model, the text-topic vector $d_{i-LDA} = (k_1 PR(1), k_2 PR(1), \ldots, k_k PR(1))$ and $d_{j-LDA} = (k_1' PR(1), k_2' PR(1), \ldots, k_K' PR(1))$ are obtained after preprocessing the microblog post, K is the number of topics. Therefore, the cosine similarity based on the improved LDA model is

$$sim_{LDA}(d_i, d_j) = \frac{d_{i-LDA} \times d_{j-LDA}}{|d_{i-LDA}| \times |d_{j-LDA}|} \qquad (8)$$

- (3) The linear weighting of a and b is performed to obtain the calculation formula of mixed similarity as

$$sim(d_i, d_j) = \lambda sim_{VSM}(d_i, d_j) + (1 - \lambda) sim_{LDA}(d_i, d_j) \qquad (9)$$

Single-Pass clustering algorithm. The single-pass clustering method is simple, it belongs to the incremental clustering, and the algorithm is sensitive to the order of input documents, which is suitable for the detection of microblog topics. Preset a similarity threshold δ, the coming new document will be classified into one existing topic category if the maximum value is greater than the threshold δ, otherwise it will be considered as a new topic category. We can get different topic granularity by setting different values for the threshold δ (typically 0.6–0.9). The basic process flow of Single-Pass clustering algorithm is shown in Fig. 1.

4 Experimental Results and Analysis

In this section, we first present the experimental data used in our testbed and then introduce our experimental evaluation criteria, finally we evaluate the experimental results.

Fig. 1 The basic process
flow of single-pass clustering
algorithm

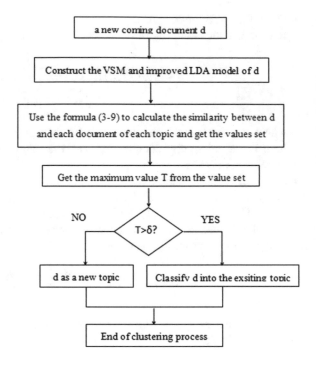

4.1 Experimental Data

The following is a short description of the datasets used in our experiments.

- Sina Microblog. The last one contains microblog content, forwarding relations, and friend relations, obtained in May 2014. The dataset contains more than 27,447 blogs of 3,071 users and 13 topics, with a total of about 580,399 connections.
- Sina Sport Forum. The dataset contains about 26,112 posts of 993 users in Sina Sport forum and the start time of the data is from November 2015 to November 2016.
- Tencent microblog. The paper uses the Tencent microblog provided open API, randomly caught 22,471 users, from April 13, 2013–2013 April 21, totally 9 days, published 331,065 microblogging data.

4.2 Evaluation Criteria

To evaluate the performance of topic detection, we adopt F1-measure, false detection rate (P_{FA}), miss detection rate (P_{Miss}), and cost of wrong detection (C_{Det}). The expression formula is as follows:

$$F1\text{-}measure = \frac{2 * precision * recall}{precision + recall} \tag{10}$$

$$(C_{Det}) = C_{Miss} \cdot P_{Miss} \cdot P_{target} + C_{FA} \cdot P_{FA} \cdot P_{non-target} \tag{11}$$

C_{Miss} is the consideration of coefficients of miss detection rate, C_{FA} is the consideration of coefficients of false, P_{target} is the prior probability of a document belongs to one topic. According to the TDT standards, we set $C_{Miss} = 1.0$, $C_{FA} = 0.1$, $P_{target} = 0.02$.

Because these values vary with the application, C_{Det} will be normalized so that $(C_{Det})_{Norm}$ can be no less than one without extracting information from the source data. This is done as follows [11]:

$$(C_{Det})_{Norm} = \frac{C_{Det}}{\min(C_{Miss} \cdot P_{target}, C_{FA} \cdot P_{non-target})} \tag{12}$$

The $(C_{Det})_{Norm}$ is smaller, the quality of topic detection is better.

4.3   Experimental Results and Analysis

(1) In Experiment 1, Sina Microblog dataset was used to detect the performance of the topic detection method based on the improved LDA model and VSM hybrid modeling algorithm.

In the experiment, LDA model return after 2000 iterations of Gibbs sampling, with K = 50 topics, and Dirichlet hyper-parameters $\beta = 0.01$ and $\alpha = 1/K$. λ is the linear combination parameter of VSM and LDA model, λ choose 0.8. The experimental results are shown in Fig. 2.

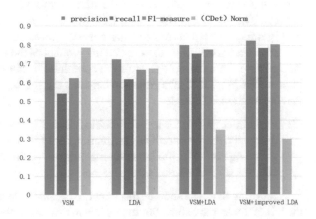

Fig. 2 Algorithm performance comparison diagram

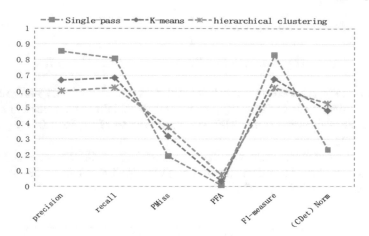

Fig. 3 Comparison of algorithm performance indicators

In Fig. 2, compared with the other three models, the VSM and improved LDA model improved precision, recall and F1-measure value, and $(C_{Det})_{Norm}$ decreased. Therefore, this paper solves the problem of semantic correlation and the deficiency of similarity calculation to a certain extent, thus improving the quality of topic detection.

(2) In order to detect the performance of the single-pass clustering algorithm after similarity improvement, the second experiment was carried out. In this experiment, Sina Sport Forum dataset was adopted, and hierarchical clustering algorithm and k-means clustering algorithm commonly used in topic detection were used for comparison. The average performance of 5 topics was used as the performance index of each algorithm. The experimental results are shown in Fig. 3.

In Fig. 3, compared with the traditional hierarchical clustering and k-means algorithm, the single-pass clustering algorithm in this paper has been improved in all indexes. The novel algorithm uses LDA model incorporating PageRank algorithm, deeply mines the structure of social network, increases the amount of information carried in traditional LDA model, and thus enhances the accuracy of topic discovery; At the same time, multiple models are combined to calculate the similarity, so as to improve the judgment of correlation between microblog. Experimental results show that the method is feasible.

(3) Experiment 3 uses Tencent microblog dataset to detect the performance of topic discovery. About parameter estimation for LDA model, this paper adopts Gibbs sampling methods, the latent theme number is 200, the initial hyper-parameters chooses 0.25. β chooses 0.1. λ chooses 0.8.

According to the method of this paper to conduct cluster, as the results in Table 1, the left of Table 1 are some hot topics in a microblog website at the same time, while the right side are the clustering extracted hot topics using the algorithm in this paper.

Table 1 Comparison of hot topics

Ranking	Online collection of hot topics	Clustering extracted hot topics
1	Earthquake in Ya'an Sichuan province	Sichuan, Ya'an, Earthquake, Lushan, Die
2	Zhigen Zhu injured Yang Sun again	Si'chuan, Ya'an, cheer supplication, safety, love wishes
3	Poisoning case in Fudan University has been solved	Fudan, University, postgraduate, jealous, poisoning, roommate
4	Explosion's suspect on the run in Boston was arrested	Radix isatidis, H7N9, virus, spread, cold, specialist disinfect
5	Seven work days in next week	Boston, explode, malathion, sorry, seven

Table 2 The results of single-pass based on improved LDA and VSM model

δ	0.01	0.02	0.03	0.05	0.08	0.1	0.2	0.3	0.5
P_{Miss}	0.0098	0.0112	0.0256	0.0543	0.0742	0.0957	0.1245	0.2634	0.2967
P_{FA}	0.1498	0.1324	0.1021	0.0789	0.0482	0.0003	0.0000	0.0000	0.0000
$(C_{Det})_{Norm}$	0.0132	0.0102	0.0093	0.0073	0.0047	0.0032	0.0041	0.0049	0.0053

As can be seen from Table 1, clustering extracted hot topics and online collection of hot topics are not totally the same, maybe the difference between this paper and their ranking rules, or due to some hot topic is initiated by their own web site, however it can accurately reflect the hot topics of microblog.

After fixing different similarity threshold δ, the single-pass clustering based on improved LDA and VSM model is executed. The corresponding experimental results are shown in Table 2. Within a certain range, with increasing similarity threshold δ, the missing rate increases gradually, and the fault detection rate decreases gradually, consuming function is down then up.

After fixing different similarity thresholds δ, this experiment compared the results with the results of the single-pass clustering based on VSM model and the single-pass clustering based on LDA [12]. The $(C_{Det})_{Norm}$ of the three algorithms are shown in Fig. 4.

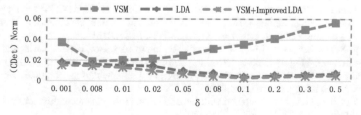

Fig. 4 The $(C_{Det})_{Norm}$ comparison of the three algorithms

As you can see from Fig. 4, with increasing similarity threshold δ, the $(C_{Det})_{Norm}$ of the single-pass clustering based on improved LDA and VSM model is smaller than the other two. Therefore, the single-pass clustering based on improved LDA and VSM model can reduce $(C_{Det})_{Norm}$ to improve the topic detection accuracy.

5   Conclusions

In this paper, we studied how to discover the hot topics in mass microblog messages rapidly and precisely, in this paper, using the hybrid modeling method of VSM and improved LDA model to calculate composite similarity, effectively solve the problem of short text data's sparsity and improved accuracy of results. Experimental results show that the hybrid modeling algorithm proposed in this paper is more accurate than using any model alone, with better clustering effect and better topic discovery effect. However, there are still some deficiencies in this paper, for example, it fails to consider the influence of information such as the number of retweets and comments on the discovery of hot topics. Moreover, this paper only studies the text content. In the future, it can also consider combining the picture, video and other multimedia information to realize the discovery of hot topics.

Acknowledgements This work is supported by the Nature Science Foundation of China (No. 61502281, 71772107).

References

1. Peng, Y., Wan, C., Jiang, T., et al.: An algorithm based on words clustering LDA for product aspects extraction. J. Chin. Comput. Syst. **36**(7), 1458–1463 (2015)
2. Blei, D., Lafferty, J.D.: Dynamic topic models. In: Proceedings of the 23rd International Conference on Machine Learning (ICML), pp. 113–120 (2006)
3. Cao, L., Tang, X.: Trends of BBS topics based on dynamic topic model. J. Manag. Sci. China **17**(11), 109–121 (2014)
4. Qiu, L., Chen, Z., Ding, C., et al.: A novel topic discovery algorithm iMLDA based on modified LDA topic model in social networks. Inf. Sci. **34**(9), 115–188+133 (2016)
5. Zheng, J., Li, Y.: A hot topic detection method for chinese micro-blog based on topic words. In: 2014 2nd International Conference on International Conference on Information Technology and Electronic Commerce, 20–21 Dec 2014, Dalian, China. IEEE Press, New Jersey (2014)
6. Sun, J.W., Lv, X.Q., Zhang, L.H.: Short text classification based on semantics and maximum matching degree. Comput. Eng. Des. **34**(10), 3613–3618 (2013)
7. Huang, X.Y., Chen, H.Y., Liu, Y.T.: Research on micro-blog short text similarity and its application in micro-blog topic detection. Comput. Eng. Des. **36**(11), 3128–3133 (2015)
8. Kai, C.: LDA-Based Theme Evolution Acting and Implementation. National University of Defense Technology, Changsha (2010)
9. Ren, Y., Chen, L., et al.: Combined with semantic feature weight calculation method research. Comput. Eng. Des. **31**(10), 2381–2383 (2010)
10. Zhang, Y., Feng, J.: Discovery method of hot topics on micro-blog based on mixed clustering. J. Hangzhou Dianzi Univ. (Nat. Sci.) **38**(1), 59–64 (2018)

11. Huang, B., Yang, Y., Mahmood, A., Wang, H.J.: Microblog topic detection based on LDA model and single-pass clustering. In: Proceedings of International Conference on Rough Sets and Current Trends in Computing, pp. 166–171 (2012)
12. Huang, B.: Research on Microblog Topic Detection Based on VSM Model and LDA Model. Southwest Jiaotong University, Chengdu (2012)

A Public Auditing Scheme with Data Recovery

Tsu-Yang Wu, Chien-Ming Chen, King-Hang Wang,
Jimmy Ming-Tai Wu and Jeng-Shyang Pan

Abstract In order to solve the integrity verification of outsourced file in cloud, we based on the concept of regenerating code (RC) to propose a new public auditing scheme with data recovery in this paper. With the advantage of RC, our scheme provides a (k, n) recovery functionality, i.e., only encoded data blocks from any k cloud servers, user can recover the original data files, where n denotes the total number of cloud servers and $k \leq n$.

Keywords Public audit · Data recovery · Regenerating code · Bilinear pairing

T.-Y. Wu · J.-S. Pan
Fujian Provincial Key Lab of Big Data Mining and Applications,
Fujian University of Technology, Fuzhou 350118, China
e-mail: wutsuyang@gmail.com

J.-S. Pan
e-mail: jengshyangpan@gmail.com

T.-Y. Wu · J.-S. Pan
National Demonstration Center for Experimental Electronic Information and Electrical
Technology Education, Fujian University of Technology, Fuzhou 350118, China

C.-M. Chen (✉)
Harbin Institute of Technology (Shenzhen), Shenzhen 518055, China
e-mail: chienming.taiwan@gmail.com

K.-H. Wang
Department of Computer Science and Engineering, Hong Kong University of Science
and Technology, Clear Water Bay, Hong Kong
e-mail: kevinw@cse.ust.hk

J. M.-T. Wu
College of Computer Science and Engineering, Shandong University of Science
and Technology, Qingdao 266590, China
e-mail: wmt@wmt35.idv.tw

© Springer Nature Singapore Pte Ltd. 2019   595
J.-S. Pan et al. (eds.), *Genetic and Evolutionary Computing*,
Advances in Intelligent Systems and Computing 834,
https://doi.org/10.1007/978-981-13-5841-8_62

1   Introduction

With the fast development of cloud computing [6], cloud storage services have received much attention by people. Since the outsourced file has lost the physical control by user, it raises several security issues, for example, how to verify the integrity of outsourced file in cloud.

Recently, Ateniese et al. [1] defined a new framework of provable data possession to solve the data integrity auditing problem. Later, numerous literature focusing on audit schemes were proposed in [2, 5, 7–10, 16–18, 20–22]. In this paper, we based on the concept of regenerating code [4] to propose a public auditing scheme with data recovery.

The system model of our scheme is depicted in Fig. 1. There are three entities which are users, cloud servers, and a trusted third-party auditor (TPA) in our scheme. Users have a large number of files outsourced to cloud servers. Cloud servers have a lot of storage resources and powerful computations to provide storage services. TPA responds to auditing the integrity of files. Without loss of generality, we assume that user first encodes the outsourced file with regenerating code and then sends the encoded file to n cloud servers. With the advantage of regenerating code, user only downloads encoded data blocks from any k cloud servers to recover the original data files, where $k \leq n$. In other words, our scheme provides a (k, n) recovery functionality. Finally, we analyze the computational cost and communication cost of cloud server and TPA.

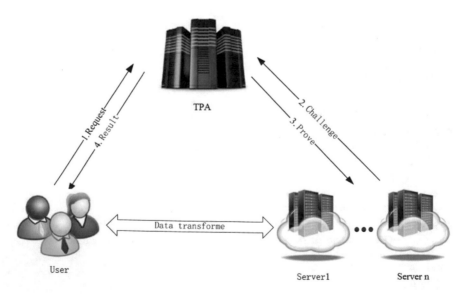

Fig. 1 System model

2 Proposed Scheme

In this section, a public auditing scheme with data recovery is proposed. In our scheme, we assume that a file $\mathcal{F}$ is outsourced to n cloud server and $\mathcal{F}$ can be recovered by any k cloud servers, where $k \leq n$. First, we introduce the underlying mathematical primitive.

2.1 Bilinear Pairings

Let $E(F_p)$ be an elliptic curve over a finite field F_p. We can find that the collection of all points $\{(x, y)|(x, y) \in E(F_p)\}$ with an infinite point O forms an abelian group named $E_{F_p}(x, y)$. We set G_1 ba a multiplicative cyclic group of $E_{F_p}(x, y)$ with a prime order q and G_T be a multiplicative cyclic group of F_p with the same order q.

A bilinear pairing e [3, 11–15, 19] is a map defined by $e : G_1 \times G_1 \to G_T$. We say that e is called admissible, if it satisfies the following three properties:

1. Bilinear. For all g_1 and $g_2 \in G_1$ and $a, b \in Z_p$, $e(g_1^a, g_2^b) = e(g_1, g_2)^{ab}$.
2. Nondegenerate. For any identity $1_{G_1} \in G_1$, $e(1_{G_1}, 1_{G_1})$ is also an identity of G_T.
3. Computable. There exists several algorithms which can be used to compute e.

2.2 Detailed Scheme

An public auditing scheme with data recovery consists of following eight algorithms: system setup ($Setup$), key generation ($KeyGen$), encoding ($Encode$), tag generation ($TagGen$), challenge generation ($ChalGen$), proof generation ($Proof$), verifying ($Verify$), and recovery ($Recover$). The used notations are summarized in Table 1.

1. $Setup(\lambda)$. Inputting a security parameter λ, this algorithm returns public parameters $param = \{G_1, G_T, e, p, g, H\}$, where G_1 and G_T are cyclic groups with same prime order p, $e : G_1 \times G_1 \to G_T$ is a bilinear pairing, g is a generator of group G_1, and H is a collision-resisting cryptographic hash function $H : \{0, 1\}^* \to G_1$.
2. $KeyGen_S(param)$. Some user called U executes this algorithm to generate two secret/public key pairs (S_{PK}, S_{SK}) and (SK, PK), where $SK = (x, S_{SK})$ and $PK = (Y, S_{PK})$. Here, (S_{PK}, S_{SK}) is used to sign and $x \in Z_p$, $Y = g^x$.
3. $Encode(param, \mathcal{F})$. U executes the following steps to encode a file $\mathcal{F}$ as follows:

    (a) $\mathcal{F}$ is divided into n blocks, $\mathcal{F} = \{m_1, m_2, \ldots, m_n\}$.
    (b) To distribute $\{m_1, m_2, \ldots, m_n\}$ into a matrix

$$\mathbf{M} = \begin{bmatrix} \mathbf{S} & \mathbf{U} \\ \mathbf{U}^{\mathrm{T}} & \mathbf{O} \end{bmatrix}, \tag{1}$$

Table 1 Notation

Notations	Meanings
G_1	Multiplicative cyclic group with order p
G_T	Multiplicative cyclic group with order p
e	Admissible bilinear pairing, $e : G_1 \times G_1 \to G_T$
g	Generator of group G_1
H	Collision-resisting cryptographic hash function $H : \{0, 1\}^* \to G_1$
$param$	Public parameters
(S_{SK}, S_{PK})	User's secret/public key pair used to sign
(SK, PK)	User's secret/public key pair
$\mathcal{F}$	A file is divided into n blocks, $\mathcal{F} = \{m_1, m_2, \ldots, m_n\}$
$\mathcal{C}$	An encoded file of $\mathcal{F}$, $\mathcal{C} = \{c_1, c_2, \ldots, c_n\}$ and $c_i = \{c_{i1}, c_{i2}, \ldots, c_{id}\}$.
σ	A signature tuple
$t_{i,j}$	A tag of $c_{i,j}$
TPA	A trusted third party
CS_i	ith cloud server

where $\mathbf{S}$ is a $k \times k$ symmetric matrix, $\mathbf{U}$ is a $k \times (d - k)$ matrix, and $\mathbf{O}$ is a $(d - k) \times (d - k)$ zero matrix.

(c) To select a code matrix

$$\Psi = \begin{bmatrix} \mathbf{I} & \mathbf{O} \\ \Phi & \Lambda \end{bmatrix}, \tag{2}$$

where $\mathbf{I}$ is a $k \times k$ identity matrix, $\mathbf{O}$ is a $k \times (d - k)$ zero matrix, Φ is a $(n - k) \times k$ matrix, and Λ is a $(n - k) \times (n - k)$ matrix.

(d) The encoded file is computed by

$$\mathcal{C} = \begin{bmatrix} \mathbf{S} & \mathbf{U} \\ \Phi \cdot \mathbf{S} + \Lambda \cdot \mathbf{U}^{\mathrm{T}} & \Phi \cdot \mathbf{U} \end{bmatrix}, \tag{3}$$

where $\mathcal{C} = (c_1, c_2, \ldots, c_n)^{\mathrm{T}}$ and $c_i = (c_{i,1}, c_{i,2}, \ldots, c_{i,d})$.

4. $TagGen(param, \mathcal{C}, SK)$. U executes the following steps to generate the corresponding tag:

(a) To select file's identity $F_{ID} \in Z_p$ and an element $u \in G_1$. Then, setting $\alpha = F_{ID}||n||u||d$ and generating a signature tuple σ of α by

$$\sigma = \alpha||Sig_{S_{SK}}(\alpha). \tag{4}$$

(b) For each element $c_{i,j}$ in the encoded file $\mathcal{C}$, to compute the corresponding tag $t_{i,j}$ by

$$t_{i,j} = [H(W_{i,j}) \cdot u^{c_{i,j}}]^x,\tag{5}$$

where $W_{i,j} = F_{ID}||i||j$. Finally, U sends all c_{ij} and its tags $t_{i,j}$ to cloud servers $CS_1, CS_2, \ldots, CS_n$ for $i = 1, 2, \ldots, n$ and $j = 1, 2, \ldots, d$.

5. $ChalGen(\sigma)$. U sends an audit request of σ to a trusted third party (TPA). Upon receiving σ, TPA executes the following steps to generate *challenge*:

   (a) To verify the validity of σ by S_{PK}. If the verification is true, recovering α to obtain F_{ID}, n, u, and d.
   (b) To randomly select a subset $I \subseteq [1, n]$ with l elements as the number of challenged cloud server.
   (c) To randomly select a subset $J \subseteq [1, d]$ and computing $R_i = Y^{r_i}$, where $r_i \in_R Z_p^*$.

Finally, TPA sends

$$chal_i = \{(j, v_{i,j}), R_i\}\tag{6}$$

to each cloud server CS_i for $v_{i,j} \in_R Z_p^*$, $i \in I$, and $j \in J$.

6. $Proof(param, chal_i, t_{i,j})$. Upon receiving $chal_i$, cloud server CS_i computes

$$TP_i = \prod_{j \in J} t_{i,j}^{v_{i,j}}\tag{7}$$

and

$$MP_i = k_i \cdot \sum_{j \in J} v_{i,j} \cdot c_{i,j},\tag{8}$$

where $k_i \in Z_p^*$. Then, CS_i computes

$$DP_i = e(u, R_i)^{MP_i}\tag{9}$$

and sends the corresponding proof

$$proof_i = (TP_i, DP_i, K_{i,1}, K_{i,2})\tag{10}$$

to TPA, where $K_{i,1} = Y^{k_i}$ and $K_{i,2} = g^{k_i}$.
7. $Verify(param, chal_i, proof_i)$. Upon receiving $proof_i$ from each CS_i, TPA verifies all $proof_i$ by

$$DP \cdot e\left(\prod_{i \in I} H_i, K_{i,1}\right) \stackrel{?}{=} e\left(TP, \prod_{i \in I} K_{i,2}^{r_i}\right),\tag{11}$$

where

$$DP = \prod_{i \in I} DP_i, \tag{12}$$

$$TP = \prod_{i \in I} TP_i, \tag{13}$$

and

$$H_i = \prod_{j \in J} H(W_{i,j})^{r_i \cdot v_{i,j}}. \tag{14}$$

If the verification is true, TPA returns "1". Otherwise, TPA verifies

$$DP_i \cdot e(H_i, K_{i,1}) \overset{?}{=} e(TP_i, K_{i,2}^{r_i}) \tag{15}$$

for each $i \in I$ to identify which cloud server's block file is corrupted.

8. *Recover*. In verification phase, TPA can obtain all index of could server for the block file which is corrupted. If the number of index is less than k, then TPA sends the list to U. Then, U downloads other block files $(c_1, c_2, \ldots, c_d)$ from other d cloud servers, where $d \geq k$. Finally, a modified encoded file is computed by

$$C'_d = (c_1, c_2, \ldots, c_d) \cdot \Psi_d^{-1}. \tag{16}$$

3 Performance Analysis

In this section, we analyze the computational cost and the communication cost for cloud server and TPA. The following notations are used in this section summarized in Table 2. The theoretical computational cost of proof generation for cloud server is

$$s \cdot TG_e + (c + s + 2) \cdot TG_{exp} + (c + s) \cdot TG_{mul} + (c \cdot s + s) \cdot T_{mul} + c \cdot s \cdot T_{add} \tag{17}$$

and the theoretical computational cost of verification for TPA is

$$2TG_e + (c + 1) \cdot TG_{exp} + (c + 1) \cdot TG_{mul} + c \cdot T_h. \tag{18}$$

Here, we analyze the computational cost of cloud server and TPA with platform: CPU Intel(R) Core(TM) i3-4160 with 3.60 GHz, memory is 8 GB, the operating system is Win 7 (64 bit), jPairing-Based Cryptography, an open-source library is used. In our experiment, the elliptic curve is the type A curve, security level is 80 bit, $|p| = 160$ bit, $|z_p|$ is 160 bit (20 byte), and $|G_1| = 336$ bit (42 byte), $c = 460$, and $s = 50$ are set. We summarize the computational cost of cloud server and TPA in Table 3.

Table 2 Notations in performance analysis

Notations	Meanings
s	The number of subblocks for each data block
c	The number of data blocks
TG_e	The executing time of one pairing operation
TG_{exp}	The executing time of one exponentiation operation in G_1
TG_{mul}	The executing time of one multiplication operation in G_1
T_{mul}	The executing time of one multiplication operation in Z_p
T_{add}	The executing time of one addition operation in Z_p
T_h	The executing time of one hash function

Table 3 Computational cost of cloud server and TPA

Proof generation (Cloud server)	Verification (TPA)
1420 ms	1655 ms

Then, we analyze the communication cost of cloud server and TPA. In the challenge generation phase of cloud server, the size of $chal_i = \{(j, v_{i,j}), R_i\}$ is $460 \cdot (4 + 20) + 42 = 11082$ byte. In the proof generation phase of TPA, the size of $proof_i = (TP_i, DP_i, K_{i,1}, K_{i,2})$ is $4 \cdot 42 = 168$ byte.

4 Conclusion

In this paper, based on regenerating code (RC) we have proposed a public auditing scheme with (k, n) data recovery functionality. With the advantage of RC, user only downloads encoded data blocks from any k cloud servers to recover the original data files, where n denotes the total number of cloud servers and $k \leq n$. In the future, we will demonstrate the security proofs and comparisons of our proposed scheme.

Acknowledgements The work of Tsu-Yang Wu was supported in part by the Science and Technology Development Center, Ministry of Education, China under Grant no. 2017A13025 and the Natural Science Foundation of Fujian Province under Grant no. 2018J01636. The work of Chien-Ming was supported in part by Shenzhen Technical Project (JCYJ20170307151750788) and in part by Shenzhen Technical Project (KQJSCX20170327161755).

References

1. Ateniese, G., Burns, R., Curtmola, R., Herring, J., Kissner, L., Peterson, Z., Song, D.: Provable data possession at untrusted stores. In: Proceedings of the 14th ACM Conference on Computer and Communications Security, pp. 598–609. ACM (2007)
2. Ateniese, G., Burns, R., Curtmola, R., Herring, J., Khan, O., Kissner, L., Peterson, Z., Song, D.: Remote data checking using provable data possession. ACM Trans. Inf. Syst. Secur. (TISSEC) **14**(1), 12 (2011)

3. Boneh, D., Franklin, M.: Identity-based encryption from the Weil pairing. In: Annual International Cryptology Conference, pp. 213–229. Springer (2001)
4. Dimakis, A.G., Godfrey, P.B., Wu, Y., Wainwright, M.J., Ramchandran, K.: Network coding for distributed storage systems. IEEE Trans. Inf. Theory 56(9), 4539–4551 (2010)
5. Kim, D., Kwon, H., Hahn, C., Hur, J.: Privacy-preserving public auditing for educational multimedia data in cloud computing. Multimed. Tools Appl. 75(21), 13077–13091 (2016)
6. Mell, P., Grance, T., et al.: The NIST definition of cloud computing (2011)
7. Wan, C., Zhang, J., Pei, B., Chen, C.: Efficient privacy-preserving third-party auditing for ambient intelligence systems. J. Ambient Intell. Humaniz. Comput. 7(1), 21–27 (2016)
8. Wang, H.: Identity-based distributed provable data possession in multicloud storage. IEEE Trans. Serv. Comput. 2, 328–340 (2015)
9. Wang, C., Chow, S.S., Wang, Q., Ren, K., Lou, W.: Privacy-preserving public auditing for secure cloud storage. IEEE Trans. Comput. 62(2), 362–375 (2013)
10. Wang, H., Wu, Q., Qin, B., Domingo-Ferrer, J.: Identity-based remote data possession checking in public clouds. IET Inf. Secur. 8(2), 114–121 (2014)
11. Wu, T.Y., Tseng, Y.M.: An ID-based mutual authentication and key exchange protocol for low-power mobile devices. Comput. J. 53(7), 1062–1070 (2010)
12. Wu, T.Y., Tseng, Y.M.: A pairing-based publicly verifiable secret sharing scheme. J. Syst. Sci. Complex. 24(1), 186–194 (2011)
13. Wu, T.Y., Tseng, Y.M.: Publicly verifiable multi-secret sharing scheme from bilinear pairings. IET Inf. Secur. 7(3), 239–246 (2013)
14. Wu, T.Y., Tseng, Y.M., Tsai, T.T.: A revocable ID-based authenticated group key exchange protocol with resistant to malicious participants. Comput. Netw. 56(12), 2994–3006 (2012)
15. Wu, T.Y., Tsai, T.T., Tseng, Y.M.: Efficient searchable id-based encryption with a designated server. Ann. Telecommun.-Ann. Télécommun. 69(7–8), 391–402 (2014)
16. Wu, T.Y., Lin, Y., Wang, K.H., Chen, C.M., Pan, J.S., Wu, M.E.: Comments on a privacy preserving public auditing mechanism for shared cloud data. In: Proceedings of the 4th Multidisciplinary International Social Networks Conference, p. 48. ACM (2017)
17. Wu, T.Y., Lin, Y., Wang, K.H., Chen, C.M., Pan, J.S.: Comments on Yu et al's shared data integrity verification protocol. In: The Euro-China Conference on Intelligent Data Analysis and Applications, pp. 73–78. Springer (2017)
18. Wu, L., Wang, J., Kumar, N., He, D.: Secure public data auditing scheme for cloud storage in smart city. Pers. Ubiquitous Comput. 21(5), 949–962 (2017)
19. Wu, T.Y., Chen, C.M., Wang, K.H., Pan, J.S., Zheng, W., Chu, S.C., Roddick, J.F.: Security analysis of rhee et al.'s public encryption with keyword search schemes: a review. J. Netw. Intell. 3(1), 16–25 (2018)
20. Xu, Z., Wu, L., Khan, M.K., Choo, K.K.R., He, D.: A secure and efficient public auditing scheme using RSA algorithm for cloud storage. J. Supercomput. 73(12), 5285–5309 (2017)
21. Zhang, J., Dong, Q.: Efficient ID-based public auditing for the outsourced data in cloud storage. Inf. Sci. 343, 1–14 (2016)
22. Zhang, J., Zhao, X.: Efficient chameleon hashing-based privacy-preserving auditing in cloud storage. Clust. Comput. 19(1), 47–56 (2016)

Part XV
Data Classification and Clustering

A Novel Approach for Objectness Estimation Based on Saliency Segmentation and Superpixels Clustering

Jie Niu, Yi-Wen Jiang, Liang Huang and Hao-Wen Xue

Abstract Object detection is an important and challenging vision task. It is a critical part in many applications such as image search, robot navigation, and scene understanding. In this paper, we propose a novel objectness measure method, which uses both saliency segmentation and superpixels clustering together. First, we use the single skeleton refinement and fuzzy C-means method to segment the image. Then, the candidate regions are selected by combining the saliency map. At the same time, we used the superpixels clustering and straddling method to filter the windows. The final candidate object windows are obtained based on a fusion of the two results. The experimental results from PASCAL VOC 2007 validate the efficacy of the proposed method, and we get a result of 40.1% on mean average precision which was the best of the tested methods.

Keywords Object detection · Objectness measure · Image segmentation

1 Introduction

In recent years, object detection has become a major research area. However, a variety of methods follow the sliding-window paradigm [1, 2]. In order to accelerate data processing, some researchers try to train an objectness measure has recently become popular [3, 4]. Recently, many deep learning-based approaches also have achieved

J. Niu (✉) · Y.-W. Jiang · L. Huang · H.-W. Xue
School of Electrical and Electronic Engineering, Changzhou College of Information Technology, Changzhou, Jiangsu, China
e-mail: niujie@czcit.edu.cn

Y.-W. Jiang
e-mail: calla_yi@sina.com

L. Huang
e-mail: xwzj1230@163.com

H.-W. Xue
e-mail: jzpfwf@gmail.com

© Springer Nature Singapore Pte Ltd. 2019
J.-S. Pan et al. (eds.), *Genetic and Evolutionary Computing*,
Advances in Intelligent Systems and Computing 834,
https://doi.org/10.1007/978-981-13-5841-8_63

very promising performance at detection, such as Fast R-CNN [5], Faster R-CNN [6], and YOLO [7], etc. However, both the traditional and the CNN-based approaches require manual effort for feature coding and clustering. The difference is since a convolutional neural network contains more layers of simple and complex cells and its learning process is supervised, filter weights can be constantly adjusted according to data and tasks. CNN-based methods have more powerful abilities to characterize images. However, most of the works are using strongly supervised method. As for applications such as robot navigation, since the environment only has a limited amount of data (i.e., a limited number of objects and important landmark information), conventionally embedded robot platforms also do not have the computational resources required for deep learning.

This paper mainly focuses on the problem of objectness measure without deep learning algorithms. We propose an objectness measurement of image windows based on color and superpixel features. The object features' extraction and matching are performed only on the regions suspected to contain objects, which greatly improves computational efficiency. The experimental results show that our new method performs better than other compared methods.

2 Objectness Estimation

2.1 Color Contrast and Saliency Map (CCS)

Color features play an important role in the general object recognition framework, because objects tend to have a different appearance (color distribution) than their surroundings. In the visual attention mechanism, if there is a large difference between the color information of a pixel or area and its surrounding area, then it can generally be considered to be significant.

In this paper, the saliency map of the scene image is obtained by using the central-ambient contrast saliency calculation method [8]. The saliency map contains the location information of an area. The different gray values of the saliency map represent the specificity of the color contrast near an actual scene image. The proposed image segmentation algorithm is then used to combine single skeleton refinement and fuzzy C-means image segmentation. Finally, the prominent area of the image is segmented by combining the previously obtained saliency map. The method flow is shown in Fig. 1.

First, the L0 norm smoothing algorithm [9] is used to smooth the input image, and then obtain the image gradient separately for each color channel of an RGB image. Next, we use the morphological reconstruction technique to mark the foreground objects and the background objects, and the three-channel gradient amplitude image is synthesized so that only the foreground and background marked pixels have local minima. Finally, the image is coarsely segmented using a single refinement of

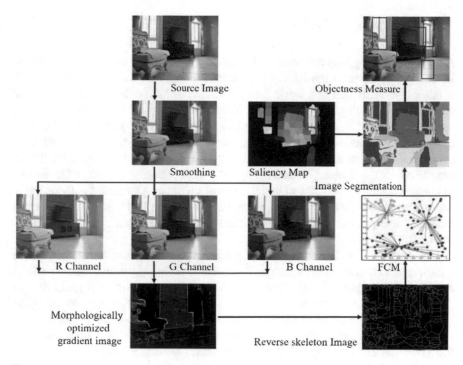

Fig. 1 Outline of the CCS

the skeleton extraction method [10], which can partly overcome shortcomings of traditional methods of over-segmentation and under-segmentation.

After a series of above-mentioned processing steps, a subset of n pixels based on skeleton data can be obtained. These pixels have consistency in the color feature space. Since n is much smaller than the number N of original image pixels, the speed of fuzzy clustering for these subsets will be much faster than clustering N pixels of an original image. Due to the morphologically optimized step, the over-segmentation phenomenon has been significantly restrained. Still, there tend to be some bad segmentation results, and the over-segmented regions still need to be merged to achieve a better effect.

This paper combines the regions with less than 5% of the partition area, then calculates the color features of each region. After that, we use the improved FCM algorithm (as used in [6]) to merge regions and obtain the final segmentation map. Finally, considering the segmentation map results and the previously obtained scene saliency map, we set the threshold T_a to S_u (the average saliency map value, for the entire image), according to the average value S_k of the saliency maps for to the partitioned image regions. When $S\_k > T\_a$, the area is a valid saliency map. Finally, by comparing each segmented image region, a saliency region description based on the color feature is obtained.

2.2   Superpixels Clustering and Straddling (SCS)

A superpixel is a group of connected pixels with similar levels of colors or grays. In this paper, we used the SLIC [11] method to segment the image. Although it is computationally fast and works well, the SLIC method only takes the color and coordinates of each pixel as a feature. When the superpixel segmentation boundaries are not sufficiently clear, the boundary characteristics objects may not be well preserved. Considering these conditions, we used the DBSCAN [12] to optimize the superpixel segmentation results. Compared to other clustering algorithms, DBSCAN has the advantage of being insensitive to noise without having to declare the number of clusters in advance, and it can find clusters of any shape. Combined with the initial candidate split window of the image, the extracted superpixel results are the output from the superpixel straddling-based method. The overall algorithm flow is shown in Fig. 2.

The DBSCAN method can find arbitrarily shaped clusters in noisy data, but its computational requirements are high. This is because the algorithm needs to perform an all-neighbors query for each data sample, which greatly limits its performance. The approach used in this paper improves the algorithm's efficiency, by using the maximum density method to predetermine a limited number of cluster centers.

According to the algorithm flow method, after performing SLIC segmentation on the picture, an improved DBSCAN clustering algorithm is used to optimize the result. It can be seen from the example result in Fig. 3 that, when compared with direct superpixel segmentation, the DBSCAN clustering method can reduce the number of superpixels by more than 50% and also improve the efficiency of subsequent calculations.

Fig. 2 Outline of the SCS

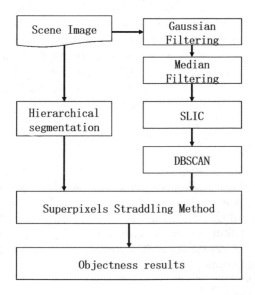

(a) Source Image   (b) SLIC   (c) Improved DBSCAN

Fig. 3 Superpixel image segmentation example input and outputs

---

Algorithm1: Maximum density method

Input: Superpixel Set $S = \{S_1, S_2, \dots, S_n\}$, Search Radius r, Threshold of the number of neighbourhood N_t, Distance threshold D_t.
Output: Cluster Centre Set C
Init $C \leftarrow \emptyset$
For each $S_{i \in [1:n]}$ Do compute the number of superpixel samples N_i in the neighborhood r
END
For each N_i, Select all superpixels S_g with a sample size greater than N_i, Do compute
$D_i \leftarrow \min_{j \in [1:g]} \left(Distance(S_i, S_j) \right)$
End
FOR i from 1: n do
IF $N_i > N_t$ and $D_i > D_t$
       Add S_i to candidate cluster centre set C
       END if
END
Return C

On the other hand, the diversity of objects in the image should be fully considered during the object detection process. It has been shown that object detection at multiple scales can improve accuracy [13]. Inspired by this, we first used graph-based image segmentation algorithm to segment the image into N regions. Then, the similarity between each region and its neighboring regions was calculated, and the two regions corresponding to the maximum similarity value were combined. Since the proportion of the actual object in the image should be within a certain range, a preliminary screening was performed using the combined area of the window region, to eliminate the unlikely candidate windows. The above process was repeated until all areas were merged into the same area. Finally, a series of candidate windows are generated for further evaluation by subsequent algorithms. The overall flow of the algorithm is shown as follows:

Algorithm 2: Hierarchical grouping method

Input: Scene Image R, Object size range $[area_{min}, area_{max}]$
Output: candidate windows set L
Do Segment the image R using the graph cut algorithm
$R = \{r_1, ..., r_n\}$
Initialize similarity set $S \leftarrow \emptyset$
For each Adjacent Area (r_i, r_j) do
        Compute the similarity $s(r_i, r_j)$
        $S \leftarrow S \cup s(r_i, r_j)$
End
While $S \neq \emptyset$ do
      Highest similarity area: $s(r_{i'}, r_{j'}) \leftarrow max\,(S)$
      Merging adjacent areas: $r_t \leftarrow r_{i'} \cup r_{j'}$
      Remove $r_{i'}$ and $r_{j'}$ from S
      If $area(r_t) \in [area_{min}, area_{max}]$
         Save r_t as candidate window $L \leftarrow L \cup r_t$
      End if
      Obtain S_t by computing the similarity between r_t and
its neibor areas
      Update set: $S \leftarrow S \cup S_t$
End
Return L

After evaluating a series of candidate windows L, we use (1) to calculate the score $SS(L_i)$ for whether each window contains an object.

$$SS(L_i) = 1 - \sum_{s_j \in S} \frac{\min(|s_j/L_i|, |s_j \cap L_i|)}{|L_i|} \tag{1}$$

In (1), L_i is a candidate detection window generated by the hierarchical grouping-based method, s_j stands for a superpixel. For each candidate window, we compute the spatial relationship of the window with all superpixels. Among them, $|s_j/L_i|$ means the area of the superpixel outside the window, and $|s_j \cap L_i|$ is the area of the superpixel inside the window. Since an object is usually divided into several superpixels, a target window that might contain an object should contain multiple complete superpixels. Conversely, a window that is unlikely to contain an object will intersect with a superpixel boundary. Therefore, we choose the minimum value between the two to represent the degree of overlap and sort the candidate windows according to the SS scores.

3 Experimental Evaluation

We verified the performance of our method in the published PASCAL VOC 2007 object image dataset. The dataset includes 9,963 images, with a total of 24,640 labeled objects, which can be used for quantitative comparison. Some of the sample images are shown in Fig. 4.

We compare our results with state-of-the-art methods. These were the DNNs-based method [14], CNNs-based method WSL [15], and weakly supervised method WSDDN [16]. The results are compared in Table 1.

According to the results, our method can effectively improve the detection efficiency due to the utility of the objectness measure method. Our method achieved the highest recognition accuracy results for the nine types of objects. The average detec-

Fig. 4 Example images of VOC 2007

Table 1 PASCAL VOC 2007 test detection average precision (%)

Method\Category	Aero	Bike	Bird	Boat	Bottle	Bus	Car	Cat	Chair	Cow
DNNs [14]	28.7	35.2	19.2	14.7	3.7	53.1	50.2	28.1	10.3	35.1
WSL [15]	**54.3**	40.9	31.1	20.8	17.7	48.3	**62.1**	**46.1**	21.8	**53.3**
WSDDN [16]	41.7	**55.3**	32	17.3	10.1	**61.3**	51.2	28.9	3.5	35.9
OUR	52.6	46.9	**35.2**	**23.4**	**27.2**	59.6	52.4	45.6	**23.2**	48.6
Method\Category	Table	Dog	Horse	Motor	Person	Plant	Sheep	Sofa	Train	TV
DNNs [14]	30.1	26.3	45.7	40.7	27.2	10.1	32.2	25.8	38.8	46
WSL [15]	22.1	34.4	46.5	**56.3**	16.2	**29.9**	35.4	15.9	**55.3**	40.3
WSDDN [16]	18.4	30.9	45.8	53.9	10.1	14.8	35.7	**45.1**	50.1	**43.5**
OURS	**33.7**	**36.6**	**51.4**	54.1	**31.2**	25.8	**37.5**	24.5	50.2	41.6

tion accuracy rate reached 40.1%, which was the best of the tested methods. It is worth noting that, compared with the comparison method, we only used the Bag-of-Words (BoW) framework. Therefore, the experimental results fully verify the effectiveness of our objectness measure method on the improvement of the recognition results.

4 Conclusions

In this paper, we propose an objectness measure method based on saliency segmentation and superpixels clustering. The CCS method utilizes the color information and combines with saliency maps to find more accurate windows. However, when the object color contrast is insufficient, the algorithm often fails. On the other hand, the performance of the SCS method mainly depends on the effect of superpixel segmentation. For small target objects, the performance of the method is often unsatisfactory. We use a simple "plus" to output the final objectness windows for the CCS and SCS output results. The experimental results show that our algorithm is effective.

Acknowledgements This work is supported by the Natural Science Foundation of Higher Education Institutions of Jiangsu Province (16KJB520048); Chang Zhou Applied Basic Research Planned Project (CJ20180010); "QingLan" Project of Jiangsu Province; Key Laboratory of Industrial IoT (KYPT201803Z); the Natural Science Foundation of CCIT (CXZK201705Z).

References

1. Felzenszwalb, P.F., Girshick, R.B., McAllester, D., Ramanan, D.: Object detection with discriminatively trained part-based models. IEEE Trans. Pattern Anal. Mach. Intell. 1627–1645 (2010)
2. Harzallah, H., Jurie, F., Schmid, C.: Combining efficient object localization and image classification. In: IEEE 12th International Conference on Computer Vision, pp. 237–244 (2009)
3. Endres, I., Hoiem, D.: Category-independent object proposals with diverse ranking. IEEE Trans. Pattern Anal. Mach. Intell. 222–234 (2014)
4. Uijlings, J.R., Van De Sande, K.E., Gevers, T., Smeulders, A.W.: Selective search for object recognition. Int. J. Comput. Vis. 154–171 (2013)
5. Girshick, R.: Fast R-CNN. In: IEEE International Conference on Computer Vision (ICCV), pp. 1440–1448 (2015)
6. Ren, S., He, K., Girshick, R., Sun, J.: Faster R-CNN: towards real-time object detection with region proposal networks. Adv. Neural Inf. Process. Syst. 91–99 (2015)
7. Redmon, J., Divvala, S., Girshick, R., Farhadi, A.: You only look once: unified, real-time object detection. In: Proceedings of the IEEE Conference on Computer Vision and Pattern Recognition, pp. 779–788 (2016)
8. Niu, J., Bu, X., Qian, K.: Exploiting contrast cues for salient region detection. Multimed. Tools Appl. 10427–10441 (2017)
9. Xu, L., Lu, C., Xu, Y., Jia, J.: Image smoothing via L0 gradient minimization. ACM Trans. Gr. (TOG) 174–186 (2011)
10. Jie, N., Xiongzhu, B., Kun, Q.: Touching corn kernels based on skeleton features information. Trans. Chin. Soc. Agric. Mach. 280–285 (2014)

11. Achanta, R., Shaji, A., Smith, K., Lucchi, A., Fua, P., Süsstrunk, S.: SLIC superpixels compared to state-of-the-art superpixel methods. IEEE Trans. Pattern Anal. Mach. Intell. 2274–2282 (2012)
12. Kriegel, H.P., Kröger, P., Sander, J., Zimek, A.: Density-based clustering. Wiley Interdisciplinary Reviews: Data Mining and Knowledge Discovery, pp. 231–240 (2011)
13. Van de Sande, K.E., Uijlings, J.R., Gevers, T., Smeulders, A.W.: Segmentation as selective search for object recognition. In: IEEE International Conference on Computer Vision (ICCV), pp. 1879–1886 (2011)
14. Szegedy, C., Toshev, A., Erhan, D.: Deep neural networks for object detection. Adv. Neural Inf. Process. Syst. 2553–2561 (2013)
15. Bolei, Z., Khosla, A., Lapedriza, A., Oliva, A., Torralba, A.: Object detectors emerge in deep scene CNNs, pp. 1–12 (2015)
16. Bilen, H., Vedaldi, A.: Weakly supervised deep detection networks. In: Proceedings of the IEEE Conference on Computer Vision and Pattern Recognition, pp. 2846–2854 (2016)

Imbalanced Data Classification Algorithm Based on Integrated Sampling and Ensemble Learning

Yan Han, Mingxiang He and Qixian Lu

Abstract In order to alleviate the impact of imbalanced data on support vector machine (SVM), an integrated hybrid sampling imbalanced data classification method is proposed. First, the imbalance rate of imbalanced data is reduced by the ADASYN-NCL (Adaptive Synthetic Sampling Technique—Domain Cleanup Rule Downsampling Method) hybrid sampling method. Then, the AdaBoost algorithm framework is used to give different weight adjustments to the misclassification of minority and majority classes, and selectively integrate several classifiers to obtain better classification. Finally, use the 10 sets of imbalanced data in the KEEL database as test objects, and F-value and G-mean are used as evaluation indicators to verify the performance of the classification algorithm. The experimental results show that the classification algorithm has certain advantages for the classification effect of imbalanced data sets.

Keywords Mixed sampling · Imbalanced data classification · ADASYN algorithm · NCL algorithm · AdaBoostSVM

1 Introduction

In real life, more and more researches are the imbalance data, and its classification research is also a research hotspot in recent years. For example, abnormal behavior detection, target recognition based on remote sensing images, software defect

Y. Han · M. He (✉)
College of Computer Science and Engineering, Shandong University of Science and Technology, Qingdao 266590, China
e-mail: hmx0708@163.com

Y. Han
e-mail: 13176482397@163.com

Q. Lu
College of International, Beijing University of Posts and Telecommunications, Beijing 100876, China

© Springer Nature Singapore Pte Ltd. 2019
J.-S. Pan et al. (eds.), *Genetic and Evolutionary Computing*,
Advances in Intelligent Systems and Computing 834,
https://doi.org/10.1007/978-981-13-5841-8_64

detection [1], and so on. In these areas, what we really value is the minority class, not a majority, because the misclassification of the minority class is more expensive, making it more important to be correctly identified.

Nowadays, the methods for solving the imbalanced classification problem include the algorithm level, the data level and the combination of the algorithm and the data level. Algorithm level, analyze the problems encountered by existing algorithms in the face of imbalanced data classification, improve existing algorithms or propose a new algorithm, such as single-class learning [2], active learning [3], cost-sensitive learning [4], and so on. At the data level, the main consideration is to convert imbalanced data into balanced data. For example, oversampling and downsampling techniques change the imbalance of the original data set. These methods have achieved good classification results in many data sets, but the downsampling method is easy to cause important information loss while reducing multiple categories. The oversampling method only repeats some small samples, not adding new samples, and it is easy to pay too much attention to rare samples, which leads to overlearning. The adaptive synthetic sampling technique ADASYN algorithm [5] is the most representative algorithm. The combination of the algorithm level and the data level is to sample the data before the model training. The algorithm is mainly combined with the ensemble learning method to obtain a series of different sub-classifiers from the training set, and then use the sub-classifier integration to improve the classification accuracy, for example, SMOTEBoost algorithm [6], AS-AdaBoostSVM [7] algorithm, and so on.

This paper proposes an unbalanced data classification algorithm based on integrated hybrid sampling, the algorithm first uses the combination of oversampling and downsampling (ADASYN-NCL) to solve the data imbalance problem and avoid the defects caused by only one sampling method. Second, the AdaBoost algorithm is used to improve the versatility of the classifier. The algorithm is compared with several existing methods. The test results show the effectiveness of the proposed algorithm.

2 Theoretical Analysis

2.1 Impact of Imbalanced Data on SVM Performance

Support Vector Machine, compared with the traditional learning method, has the advantages of small sample size and strong generalization ability. However, the training results of support vector machines will be affected by the imbalance of sample category distribution, and this effect is negative.

Equation of SVM is

$$\min_{\omega,b}\left(\frac{1}{2}\|\omega\|^2 + C\sum_{i=1}^{l}\xi_i\right)$$

$$y_i\left(\omega^T \phi(x_i) + b\right) \geq 1 - \xi_i, \xi_i \geq 0, i = 1, 2, \ldots, l \tag{1}$$

C is the penalty parameter, $\phi(x_i)$ is Kernel space mapping function, ξ_i is slack variables, b is displacement term, ω is weight vector, y_i is the sample label corresponding to the feature vector x_i.

The purpose of $\frac{1}{2} \|\omega\|^2$ is to make the interval between classes as large as possible. The purpose of $\sum_{i=1}^{l} \xi_i$ is to make the number of misclassified samples as small as possible, and the penalty parameter C is used to reconcile the relationship between the two. Penalty parameter C is used to harmonize the relationship between them. When encountering an imbalanced dataset, even in the class boundary area, there may be more multiclass samples. Thus, the classification hyperplane is offset to rare samples.

The literature [8] points out that the cause of boundary deviation is that the number of minority samples is small. The paper finds that in order to reduce the total number of misclassified samples in SVM learning, the separation hyperplane should be biased toward rare samples. Such an offset will result in an increase in the number of misclassifications of rare samples. When encountered imbalance dataset, this model has reduced predictive performance for rare samples.

The literature [9] points out that another reason for boundary deviation is the imbalanced support vector ratio. The literature finds that as the training data imbalance rate increases, the ratio of positive and negative support vectors also increases. The literature conjecture that due to this imbalance, the neighborhood of the test instance near the boundary is more likely to be dominated by the negative support vector, so the decision function is more likely to divide the boundary point into negative classes.

In summary, when the difference between the training samples of the two types of samples is greater, the difference of the corresponding upper limit of the error rate will be greater, and the effect on the negative influence of the support vector machine is also greater.

2.2 *ADASYN Oversampling*

The main idea of ADASYN oversampling is to make full use of the density distribution information of the sample to determine the frequency of each minority sample used as the main sample, so as to correct the negative impact of the category imbalance distribution as much as possible. The ADASYN algorithm is mainly divided into four steps:

Step 1: Calculate the number of synthetic data examples that need to be generated for the minority class G

$$G = \left(N^- - N^+\right) \times \beta \tag{2}$$

N^- is the number of multiple samples, N^+ is the number of minority samples, $\beta \in (0, 1]$ is a parameter used to specify the desired balance level after generation of the synthetic data.

Step 2: For each of the minority samples, calculate the proportion of synthetic samples to the total number of synthetic sample points G using the K-nearest neighbor algorithm

$$\Gamma_i = \frac{\Delta_i / K}{Z} \tag{3}$$

K is the parameter of the K-nearest neighbor algorithm, Δ_i is the nearest neighbor of the majority sample of the ith minority sample, Z is the normalized parameter, its value is

$$Z = \sum_{i=1}^{N^+} \Delta_i \tag{4}$$

Step 3: Calculate the number of synthetic data examples that need to be generated for each minority example g_i

$$g_i = \Gamma_i \times G \tag{5}$$

Step 4: For each minority sample, from which the latest K samples in a random sample, the synthesis of new minority samples s_j

$$s_j = x_i + (x_{zi} - x_i) \times \lambda \tag{6}$$

x_i is minority sample, x_{zi} is one of the K samples closest to distance x_i, λ is a random number, $\lambda \in (0, 1]$.

2.3   NCL Downsampling

The basic idea of the Neighborhood Cleanup Rule (NCL) method: Sample x in the training sample set, look for the three nearest neighbors of x. If x belongs to a minority class and there are more than two majority classes in the nearest neighbor, the majority of the nearest neighbors are deleted. If x belongs to the majority class and there are more than two minority classes in the nearest neighbor, then x is deleted. The flowchart of the NCL algorithm is shown in Fig. 1.

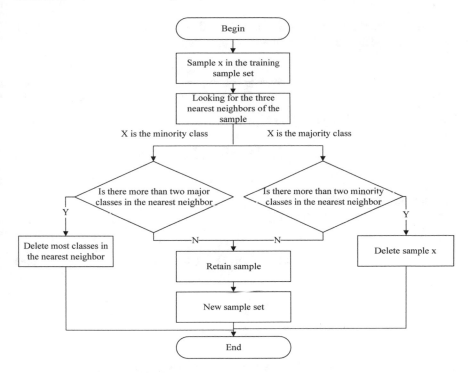

Fig. 1 NCL algorithm flowchart

2.4 AdaBoost Algorithm

The AdaBoost algorithm can effectively improve the generalization ability of the learning model.

The algorithm flow of AdaBoost is as follows:

Input: training data set $S = \{(x_i, y_i), i = 1, 2, \ldots, N, y_i \in \{-1, +1\}\}$.

Step 1: Initialize the weight distribution of the training samples.

$$D_i = \left(\omega^1(1), \omega^1(2), \ldots, \omega^1(i)\right), \omega^1(i) = \frac{1}{N}, i = 1, 2, \ldots, N \qquad (7)$$

Step 2: For $t = 1, 2, \ldots, T$, T is iteration times.
Step 2.1: Learn using a training dataset with a weight distribution D_t, get weak classifier h_t.
Step 2.2: Calculate the classification error rate of the weak classifier h_t on the training data set.

$$\varepsilon_i = \sum_{i=1}^{N} \omega^t(i), y_i \neq h_t(x_i) \qquad (8)$$

$\omega^t(i)$ represents the weight of the sample x_i in the tth iteration.
Step 2.3: Calculate the weight of the weak classifier h_t.

$$\alpha_t = \frac{1}{2} \ln\left(\frac{1 - \varepsilon_t}{\varepsilon_t}\right) \tag{9}$$

Step 2.4: Update training sample weights.

$$\omega^{t+1}(i) = \frac{\omega^t(i) \exp(-\alpha_t \, y_i \, h_t(x_i))}{C_t} \tag{10}$$

C_t is normalization factor, let $\sum_{i=1}^{N} \omega(i + 1) = 1$.
Step 3: Output strong classifier [10].

$$H(x) = sign\left(\sum_{i=1}^{N} \alpha_i \, h_t(x)\right) \tag{11}$$

3   Integrated Hybrid Sampling Algorithm

This paper proposes the basic idea of an integrated hybrid sampling imbalanced data classification method: First, the ADASYN algorithm is used to add a few samples to generate a new data set. Second, the NCL algorithm is used to reduce the majority of the samples in the newly generated data set. Finally, the AdaBoost algorithm is used to integrate multiple weak classifiers. The weak classifier used in this paper is the Radial Basis Kernel Support Vector Machine (RBF-SVM) model. Model parameters include: the penalty coefficient c and the Gaussian width σ determine the SVM classification performance. According to the literature [11], when the C value changed, the classification performance of the classifier changes little. After selecting a roughly suitable c, the performance of the SVM is improved by adjusting σ.

Given training sample set $S = \{(x_i, y_i), i = 1, 2, \ldots, N, y_i \in \{-1, +1\}\}$, $y_i \in \{-1, +1\}$ is the sample label corresponding to the feature vector x_i. RBF-SVM parameters is σ, initial value σ_{ini}, minimum value σ_{min}, and step size σ_{step}, initialize training weights $\omega^1(i) = \frac{1}{N}, i = 1, 2, \ldots, N$.

The algorithm flow in this paper is as follows:

Step 1: All the majority and minority samples are taken from the training sample set S, and the majority training sample set S^- and the minority training sample set S^+ are formed.
Step 2: Set the newly generated sample set S^{new} to be empty.
Step 3: For $i = 1: N^+$, where N^+ is the minority sample number:
Step 3.1: Find the corresponding sample x_i in S^+;
Step 3.2: Find the K-nearest neighbor of x_i in S, and record the majority of its neighbors as N^{maj};

Step 3.3: Calculate its proportional parameter Γ_i using Eq. (3);

Step 3.4: Calculate its main sample frequency g_i using Eq. (5);

Step 3.5: The number of synthetic data samples generated by each sample x_i in the minority samples is as follows:

Step 3.5.1: Randomly take an untaken sample x_i from the minority sample set;

Step 3.5.2: Arbitrarily take one sample x_{zi} from the K neighbors of x_i;

Step 3.5.3: Generating a synthetic data sample $S^{new} = x_i + (x_{zi} - x_i) \times \lambda$;

Repeat steps 3.5.1 through 3.5.2 until the value of i is taken to g_i.

Step 4: Repeat the Step 3 until the number of synthetic minority samples is met.

Step 5: Get the training set $S^{new1} = S \cup S^{new}$.

Step 6: Training the sample x_i in the sample set S^{new1}.

Step 7: Find the three nearest neighbors of x_i:

Step 7.1: $x_i \in S^{new1+}$, and x_i has more than two majority classes in the nearest neighbor, deleting most of the nearest neighbors of x_i, otherwise retaining the samples;

Step 7.2: $x_i \in S^{new1-}$ and x_i has more than two minority classes in the nearest neighbor, and deletes x_i, otherwise, the sample is retained.

Step 8: Retain the sample as the newly generated sample set S^{new2}.

Step 9: When $\sigma > \sigma_{min}, t < T$, For $t = 1, 2, 3 \ldots, T$:

Step 9.1: Training an RBF kernel classifier h_t;

Step 9.2: Calculate the training error ε_t of h_t on the current sample distribution using Eq. (8);

Step 9.3: If $\varepsilon_t > 0.5$, let $\sigma = \sigma - \sigma_{step}, t = 1$, return Step9.1;

Step 9.4: Calculate the weight of the weak classifier h_t using Eq. (9);

Step 9.5: Calculate and update the training sample weights using Eq. (10);

Step 9.6: Let $t = t + 1$.

Step10: Output the final strong classifier $H(x)$.

4 Experiment and Analysis

4.1 Experimental Data

In this paper, the performance of the algorithm is demonstrated by experiments with 10 imbalanced datasets. These data sets are from the Knowledge Extraction based on Evolutionary Learning (KEEL) database. In the experiment, the selected data is shown in Table 1.

4.2 Imbalanced Data Set Performance Evaluation Index

For the problem of unbalanced data, the existing research proposes a variety of evaluation criteria, such as F-value and G-mean. They are built on the confusion matrix as

Table 1 Experimental datasets

Dataset	Sample number	Positive class	Negative class	Imbalance ratio
Pima	768	268	500	1.87
Yeast1	1484	429	1055	2.45
New-thyroid	215	35	180	5.14
Ecoli2	336	52	284	5.46
Glass6	214	29	185	6.38
Ecoli4	336	20	316	15.80
Glass5	214	9	205	22.78
Yeast4	1484	51	1433	28.10
Yeast6	1484	35	1449	41.40
Abalone19	4177	32	4142	129.44

Table 2 Confusion matrix

Classification	Predicted positive	Predicted negative
Positive class	TP	FN
Negative class	FP	TN

shown in Table 2. TP and TN, respectively, represent the number of samples that are originally positive/negative, and are indeed predicted to be positive/negative. FP and FN, respectively, represent negative/positive but are predicted to be positive/negative the number of samples.

The following four evaluation criteria can be calculated from the confusion matrix of Table 2.

(1) Precision

$$precision = \frac{TP}{TP + FP} \tag{12}$$

(2) Recall Ratio: Correct positive class as a percentage of all predicted samples

$$recall = \frac{TP}{TP + FN} \tag{13}$$

(3) F-value is harmonic mean of precision and recall

$$F_{score} = \frac{\left(1 + \beta^2\right) \times recall \times precision}{\beta^2 \, recall + precision} \tag{14}$$

(4) G-mean is the average performance of the algorithm in the correct positive and correct negative classes

$$G\text{-}mean = \sqrt{\frac{TP}{TP + FN} \times \frac{TN}{TN + FP}} \qquad (15)$$

4.3 Experimental Result

The simulation experiments in this paper are based on the Weka platform and implemented in the Eclipse environment. Each dataset was tested using a tenfold cross-validation to prevent random effects. Through comparative experiments, the classification effects of the proposed algorithm and other imbalanced classification algorithms ADASYN, ADASYN-NCL and AS-AdaBoostSVM are compared to verify the effectiveness of the proposed algorithm. The experimental results are shown in Table 3.

We can see that the F-value and G-mean values of the algorithm are the same as those of the ADASYN-NCL algorithm on the Pima and New-thyroid datasets from Table 3, but the F-value and G-mean values on the Yeast4 dataset slightly smaller than the ADASYN-NCL algorithm. The F-value and G-mean values on other datasets are higher than other algorithms, indicating that the reliability of the algorithm for correctly classifying positive classes is improved, and the overall classification performance of the algorithm is improved.

Comparing Table 3, it is found that the hybrid sampling method can solve the data imbalance problem and avoid the defects caused by only one sampling method. The AdaBoost integration algorithm can effectively improve the versatility of the classifier.

Table 3 Comparison of F-value and G-mean values of 4 algorithms on 10 datasets

Dataset	ADASYN		AS-AdaBoostSVM		ADASYN-NCL		Proposed algorithm	
	F-value	G-mean	F-value	G-mean	F-value	G-mean	F-value	G-mean
Pima	0.712	0.712	0.712	0.712	0.858	0.811	0.858	0.811
Yeast1	0.658	0.655	0.663	0.662	0.738	0.568	0.781	0.669
New-thyroid	0.937	0.937	0.941	0.941	0.964	0.946	0.964	0.946
Ecoli2	0.819	0.817	0.828	0.826	0.808	0.809	0.848	0.849
Glass6	0.858	0.856	0.841	0.841	0.891	0.897	0.941	0.934
Ecoli4	0.906	0.903	0.946	0.945	0.936	0.932	0.961	0.959
Glass5	0.951	0.904	0.957	0.978	0.990	0.998	0.997	0.999
Yeast4	0.840	0.839	0.840	0.840	0.885	0.885	0.879	0.866
Yeast6	0.816	0.815	0.815	0.814	0.837	0.838	0.849	0.850
Abalone19	0.776	0.765	0.767	0.767	0.757	0.756	0.793	0.792

5 Conclusion

Support vector machines have higher classification accuracy when classifying data sets with class distribution balance or sample numbers that are roughly equal. However, when the imbalanced data classification is encountered, the classification accuracy of the support vector machine is significantly reduced. For this problem, this paper adopts the hybrid sampling method based on NCL downsampling and ADASYN oversampling to improve the classification performance of SVM and combines AdaBoost algorithm to improve the versatility of classifier.

Acknowledgements This work was supported by the grants of the Shandong Provincial Social Science Planning Project (17CKPJ09).

References

1. Yang, C., Gao, Y., Xiang, J., Liang, L.: Software defect prediction based on conditional random field in imbalance distribution. In: International Symposium on Dependable Computing and Internet of Things, pp. 67–71 (2016)
2. Wang, B.X., Japkowicz, N.: Boosting support vector machines for imbalanced data sets. Knowl. Inf. Syst. **25**(1), 1–20 (2010)
3. Chen, X., Song, E., Ma, G.: An adaptive cost-sensitive classifier. In: International Conference on Computer and Automation Engineering, pp. 699–701(2010)
4. Braytee, A., Liu, W., Kennedy, P.: A cost-sensitive learning strategy for feature extraction from imbalanced data. In: International Conference on Neural Information Processing. Springer International Publishing, pp. 78–86 (2016)
5. He, H., Bai, Y., Garcia, E.A.: ADASYN: adaptive synthetic sampling approach for imbalanced learning. In: IEEE International Joint Conference on Neural Networks, pp. 1322–1328 (2008)
6. Kim, M.J., Kang, D.K., Hong, B.K.: Geometric mean based boosting algorithm with oversampling to resolve data imbalance problem for bankruptcy prediction. Expert Syst. Appl. **42**(3), 1074–1082 (2015)
7. Liu, P.Z., Hong, M., Huang, D.T.: Imbalanced classification algorithm based on combination of ADASYN and AdaBoostSVM. J. Beijing Polytech. Univ. **43**(3), 368–375 (2017)
8. Akbani, R., Kwek, S., Japkowicz, N.: Applying support vector machines to imbalanced datasets. In: European Conference on Machine Learning, pp. 39–50 (2004)
9. Wu, G., Chang, E.Y.: Class-boundary alignment for imbalanced dataset learning. In: ICML Workshop on Learning from Imbalanced Data Sets, pp. 49–56 (2003)
10. Jyoti, D., Udhav, B.: A study of mammogram classification using AdaBoost with decision tree, KNN, SVM and hybrid SVM-KNN as component classifiers. J. Inf. Hiding Multimed. Signal Process 105–109 (2017)
11. Valentini, G., Dietterich, T.G.: Bias-variance analysis of support vector machines for the development of SVM-based ensemble methods. J. Mach. Learn. Res. **5**, 725–775 (2004)

Classification of Snoring Sound-Related Signals Based on MLP

Limin Hou, Huancheng Liu, Xiaoyu Shi and Xinpeng Zhang

Abstract An efficient method to classify snore, breath sound and other noises based on the multilayer perceptron (MLP) was proposed in this paper. The spectral-related feature sets of the sound were extracted and used as the input feature of MLP. The minbatch training was designed to get the effective MLP model in training process. The dropout method was applied to optimize the structure of MLP. The correct rates for distinguishing snoring, breathing sounds, and other noises are 98.88%, 97.36%, and 95.15%, respectively.

Keywords Snoring · Breath · Noise · Multilayer perceptron

1 Introduction

Snoring sounds can be recorded by a noncontact microphone using acoustical property analysis for the auxiliary diagnosis for Sleep Apnea–Hypopnea Syndrome (SAHS) [1, 2]. The classification of the snoring sound-related signals was an important front-end work. The Mel Frequency Cepstral Coefficients (MFCC) of the breath and non-breath sound was used, and the Support Vector Machine (SVM) were employed [3]. The machine learning was applied to detect breathing and snoring episodes during sleep in a wireless system [4]. The pitch, MFCC, and logarithmic energy based on the MLP to distinguish between snoring and non-snoring [5]. For low SNR sound recordings, the Artificial Neural Network (ANN) was applied to

L. Hou (✉) · H. Liu · X. Shi · X. Zhang
Shanghai University, Shanghai 200444, China
e-mail: lmhou@staff.shu.edu.cn

H. Liu
e-mail: lhc521210@163.com

X. Shi
e-mail: xiaoyu1yi@163.com

X. Zhang
e-mail: xzhang@shu.edu.cn

© Springer Nature Singapore Pte Ltd. 2019
J.-S. Pan et al. (eds.), *Genetic and Evolutionary Computing*,
Advances in Intelligent Systems and Computing 834,
https://doi.org/10.1007/978-981-13-5841-8_65

detect snoring and non-snoring sound [6]. Classification of three sound types of snoring, breath and other noises extracted 351-dimensional feature sets [7]. Using a Gaussian mixture model to classify three types of snoring, breath, and other noises [8].

Based on the MLP we designed a classifier for the three type sounds, that is, snoring, breath and other noises. The 43-dimensional feature sets of the sound were extracted and used as the input feature of MLP. The minbatch training and adaptive learning rate were combined to ensure global optimality for the MLP model. A group of optimization parameters of MLP training of models were got by the dropout. The accuracy rates of snoring, breath, and other noises were 98.88%, 97.36%, and 95.15%, respectively.

2 Data and Acoustic Feature

Snore sounds were recorded in the sleep monitoring laboratory in Department of Otolaryngology of Shanghai Jiao Tong University Affiliated Sixth People's Hospital by a noncontact ambient microphone that was fixed on top of the bedhead at a distance of approximately 30 cm from the nose and mouth of the subjects, and simultaneously, polysomnography (PSG) diagnosis was performed. The sound card is Creative Audigy 4 Value, the microphone type is Sony ECM-C10, sampling frequency 16 kHz.

The PSG report contained diagnostic results such as Apnea–Hypopnea Index (AHI). AHI is defined as the average number of apnea–hypopneas events per hour (events/hour). The AHI value corresponds to the severity of SAHS, which is divided into four different levels: AHI < 5 is non-SAHS (N); $5 \leq$ AHI ≤ 15 is Mild SAHS patient (L); $15 <$ AHI ≤ 30 is moderate SAHS patient (M); AHI > 30 is severe SAHS patient (S).

The experimental data were derived from all-night recording of snores at different AHI levels. A total of 42 subjects, including 11 non-SAHS, 11 Mild SAHS patients, 10 moderate SAHS patients, and 10 severe SAHS patients.

The statistical information of subjects with different severity is shown in Table 1. The snoring, breath and other noises in Table 1 are manually cut episode numbers from their overnight recordings. The number of all episodes is nearly 85,000.

The cutting method of snoring is to synchronously compare the snoring of PSG labeled all night recording. The breathing sound was cut after three subjective auditions. Other noises include biological noise (speech, cough, muffled humming, etc.) and nonbiological noise (Music sound, quilt friction, closing door sound, etc.).

There are some differences in the frequency spectrum of the snoring, breathing sound, and other noise. We extracted the spectral energy-related features (SE), the pitch-related features (PR), the MFCC-related features (MF), and spectral cosine similarity-related features (SCS) from the sound. Table 2 is a list of total acoustic parameters.

Table 1 Statistics of experimental data

Severity	N	L	M	S	Total
Age	29.1 ± 16.2	47.8 ± 17.8	44.9 ± 10.2	46.1 ± 14.7	44.4 ± 13.9
AHI	1.5 ± 1.3	9.1 ± 4.0	23.5 ± 3.9	58.6 ± 17.3	22.3 + 22.7
Subjects (M/F)	7/4	8/3	7/3	7/3	29/13
Snoring	9204	12,511	13,147	11,655	46,517
Breath	4974	5983	6257	8143	25,357
Other noises	5215	3965	1779	2007	12,966
Total					84,840

AHI are reported as mean ± SD, M/F male/female. The last line is sum from the shadow

AHI are reported as mean ± SD, M/F male/female. The last line is sum from the shadow

Table 2 List of feature sets

Category	Feature	Dimension
SE	Modified frequency centroid and difference	2
	Sub-band normalized energy and difference	17
	Spectrum entropy	2
	Normalized area of spectral amplitude envelope	1
PR	Pitch (mean, variance, density)	3
MF	$MFCC_0$	1
	Dynamic MFCC distance	1
SCS	FD (Mean, variance, skewness, and entropy)	4
	FO (Mean, variance, skewness, and entropy)	4
	TD (Mean, variance, skewness, and entropy)	4
	TO (Mean, variance, skewness, and entropy)	4

The dimensions of SE, PR, MF, SCS feature set are 22, 3, 2, and 16, respectively. There are a total of 43 dimensions in Table 2. The SCS parameter set consists of four categories: FD represents the frequency-direction spectral cosine similarity of subbands with a width of 31.25 Hz, FO represents the frequency-direction spectral cosine similarity in with sub-bandwidth of 500 Hz. TD represents the spectral cosine similarity in time direction with sub-bandwidth of 31.25 Hz. TO represents the spectral cosine similarity in time-direction with sub-bandwidth of 500 Hz.

3 Classification Design Based on Multilayer Perceptron

Since the perceptron only has the output layer for activation function processing, the learning ability is limited, and only the linear separability problem can be solved. If it needs to solve more complex nonlinear problems, it is necessary to use multilayer functional neuron superposition to add one between the input layer and the output layer. The middle part is called the hidden layer. Each layer of neurons is fully interconnected with the next layer of neurons. Such a neural network model is known as a Multilayer Perception (MLP) [9]. We use multiplayer perceptron for classification of snoring-related signals. The MLP classification flow chart is shown in Fig. 1.

We selected 80% of the episodes in Table 1 as training set, and the remaining 20% episodes as testing set. In training phase, we extracted the features and preprocess them. Then, the minbatch training method was used to train MLP structure, and the dropout method and the adaptive learning rate strategy were applied in the training process. In testing phase, we extracted the same features as the training set. Then, the features were put into the MLP model that had been trained, and the classification result of the test set was obtained.

3.1 Minbatch Training

For optimizing the expected loss, the variance of the gradient estimate using the entire training set is smaller than obtained by any subset. This method is generally called batch training. However, it is less efficient for large-scale data, and it leads to a high degree of reliance on the initial model and do not fully converge to the global best.

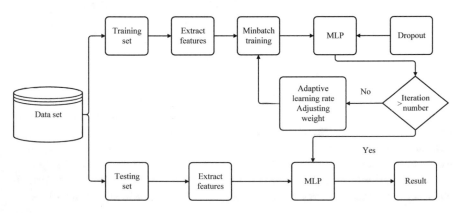

Fig. 1 MLP classification flowchart

Therefore we construct minbatch training and combine stochastic Gradient Descent Algorithm (SGD) [10] in the training phase. The minbatch training refers to selecting small batch data from the training set and estimating the gradient [11].

We adopt the sample randomization method to ensure the independence and distribution of samples. For each iteration of training, the index was reordered after the training set is randomly scrambled, and the sample data were extracted according to the sorted index array. This approach ensured that each sample was used only for training once, so that it did not affect the data distribution and ensured the consistency of the learning output model.

There are many algorithms for adaptive learning rate algorithm. Most algorithms reduce gradually learning rate with the increase in training times. The Adam algorithm [12] was adopted in this paper.

3.2 Parameter Optimization

To ensure the generalization effect of the generated model, the number of repeated training for all training sets is recorded as the number of iterations. The average of the classification error rate of the test set including snore, breath, and other is the test set error rate. Therefore, we can get the relationship between the number of iterations and the average error rate, and use it as the criterion for evaluating the MLP model.

The optimization of parameter adjustment is as follows: first, initialize the parameter model, then control other parameters unchanged, and adjust the number of hidden layers, the number of hidden layer nodes, the learning rate, the minbatch, and the dropout ratio [13], respectively, finally we can get the optimal results corresponding to each parameter.

The number of layers in the hidden layer helps to improve the representation ability of the MLP model. As the number of hidden layers increases, the fitting effect of the training set is better. However, if the number of layers in the hidden layer is too large, it will result in overmatching. And the increase in the number of hidden layers will increase the memory cost during training. Therefore, it is necessary to adjust the number of hidden layers to the appropriate level. Figure 2a shows the test set error rate of different hidden layers during training. Therefore, the number of hidden layers is set to two layers. Similar to this, Fig. 2b shows the test set error rate of different node numbers during training. Therefore, the number of hidden layer nodes is 80.

The parameter optimization adjustment process of learning rate, minbatch, dropout ratio, and iteration number is similar to the above two parameters. The final results of the model optimization parameters are shown in Table 3.

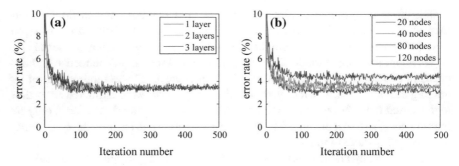

Fig. 2 MLP classification flowchart

Table 3 Optimization parameters of model

Hidden layer	Hidden layer nodes	Learning rate	Minbatch	Dropout ratio	Iteration
2	80	0.3	2048	0.3	400

4 Results and Discussion

Since the training of the neural network model has a certain randomness, the mean and standard deviation of the test results of the three trainings are selected to evaluate the performance of the classification model. The classification results for model derived from the downsampling, upsampling, and the original training set are shown in Table 4.

It can be found from the experimental results that the use of downsampling and upsampling does help the equalization accuracy. The result of using the downsampling method is better than the result of using the upsampling and the original set.

For the downsampling of balances the training set, decreasing training number of the snoring and breathing sound episodes does not cause great damage to the richness of the training set. Because distinguishing feature is significant among the snoring, the breathing sound and the other noise, the MLP module is more effective.

For the upsampling method of balances the training set, it needs to increase the number of the other noise by adding white noise and synthesizing new sound samples. It has little effect on enriching other noise samples in the training set. The other noises include a variety of types such as quilt frictional sound, telephone sound, knocking

Table 4 Classification results (mean ± standard deviation)

	Downsampling	Upsampling	Original set
Snore	98.88% ± 0.09%	99.16% ± 0.09%	99.15% ± 0.08%
Breath	97.36% ± 0.12%	97.68% ± 0.65%	97.56% ± 0.67%
Other	95.15% ± 0.47%	93.49% ± 0.85%	92.62% ± 1.23%

sound, and door opening sound. Therefore, this way of upsampling does not help to improve the classification effect.

For the original set, if the amount of data is larger, the correct rate will be higher for imbalance of three type data in Table 4 last column. How to make better use of unbalanced data is an in-depth study in the future.

The MLP method was widely used in classify between the snoring and the breathing sound [4, 5]. For three types of the snoring, breathing sound, and the other noise, the data set of [8] is equivalent to our paper. A total of more than 80,000 sound segments are classified based on Gaussian Mixture Model (GMM). Compare the GMM result, the MLP method shows classification preponderance from our work.

5 Conclusion

This paper presented a MLP method of classifying snoring, breath sounds and other noises. A scheme of combining minbatch training with gradient descent was designed. Dropout method and adaptive learning rate strategy were adopted. The classifier is trained by downsampling, upsampling and directly using the original training set, respectively, and good classification results are obtained. The model obtained by the downsampling method is optimal. The accuracy rates of snoring, breath and other noises were 98.88%, 97.36%, and 95.15%, respectively. The method used in this paper can effectively classify snoring, breath, and other noises.

Acknowledgements The authors thank the doctors and professors at the Department of Otolaryngology of Shanghai Jiao Tong University Affiliated Sixth People's Hospital for the recording data collection and segmentation. This study was funded by Science and Technology Commission of Shanghai Municipality (No. 13441901600) and National Natural Science Foundation of China (No. 61525203).

Author Contribution Ethical Approval All procedures performed in studies involving human participants were in accordance with the ethical standards of the institutional and/or national research committee and with the 1964 Helsinki declaration and it slater amendments or comparable ethical standards. It has been accepted for approval to the ethics committee of Shanghai sixth people's hospital.

Informed Consent Informed consent was obtained from all individual participants included in the study.

References

1. Dafna, E., Tarasiuk, A., Zigel, Y.: OSA severity assessment based on sleep breathing analysis using ambient microphone. In: 35th Annual International Conference of the IEEE Engineering in Medicine and Biology Society (EMBC), pp. 2044–2047. IEEE, Osaka (2013)
2. Ben-Israel, N., Tarasiuk, A., Zigel, Y.: Obstructive apnea hypopnea index estimation by analysis of nocturnal snoring signals in adults. Sleep 35(9), 1299–1305 (2012)
3. Lei, B., Rahman, S.A., Song, I.: Content-based classification of breath sound with enhanced features. Neurocomputing 141(4), 139–147 (2014)
4. Mlynczak, M., Migacz, E., Migacz, M.: Detecting breathing and snoring episodes using a wireless tracheal sensor-a feasibility study. IEEE J. Biomed. Health Inform. 21(6), 1504–1510 (2016)
5. Swarnkar, V.R., Abeyratne, U.R, Sharan, R.V.: Automatic picking of snore events from overnight breath sound recordings. In: 39th Annual International Conference of the IEEE Engineering in Medicine and Biology Society (EMBC), pp. 2822–2825. IEEE, Seogwipo (2017)
6. Emoto, T., Abeyratne, U.R., Kawano, K.: Detection of sleep breathing sound based on artificial neural network analysis. Biomed. Signal Process. Control 41, 81–89 (2018)
7. Karunajeewa, A.S., Abeyratne, U.R., Hukins, C.: Silence-breathing-snore classification from snore-related sounds. Physiol. Meas. 29(2), 227–243 (2008)
8. Dafna, E., Tarasiuk, A., Zigel, Y.: Automatic detection of snoring events using Gaussian mixture models. In: 7th International Workshop on Models and Analysis of Vocal Emissions for Biomedical Applications, pp. 17–20. Firenze University Press, Florence (2011)
9. Zhang, Z., Lyons, M., Schuster, M., Akamatsu, S.: Comparison between geometry-based and gabor-wavelets-based facial expression recognition using multi-layer perceptron. In: Proceedings Third IEEE International Conference on Automatic Face and Gesture Recognition, pp. 454–459. IEEE, Nara (1998)
10. Bottou, L.: Large-scale machine learning with stochastic gradient descent. In: Proceedings of COMPSTAT'2010, pp. 177–186. Physica-Verlag HD (2010)
11. Li, M., Zhang, T., Chen, Y.: Efficient mini-batch training for stochastic optimization. In: Proceedings of the 20th ACM SIGKDD International Conference on Knowledge Discovery & Data Mining, pp. 661–670. ACM, New York (2014)
12. Kingma, D.P., Ba, J.: Adam: a method for stochastic optimization. Comput. Sci. (2014)
13. Srivastava, N., Hinton, G., Krizhevsky, A.: Dropout: a simple way to prevent neural networks from overfitting. J. Mach. Learn. Res. 15(1), 1929–1958 (2014)

Extension of Sample Dimension and Sparse Representation Based Classification of Low-Dimension Data

Qian Wang, Yongjun Zhang, Ling Xiao and Yuewei Li

Abstract As we know, sparse representation methods can achieve high accuracy for classification of high-dimensional data. However, they usually show poor performance in performing classification of low-dimensional data. In this paper, the increase of the sample dimension for sparse representation is studied and surprising accuracy improvement is obtained. The paper has the following main value. First, the designed method obtains promising results for classification of low-dimensional data and is very useful for widening the applicability of sparse representation. The accuracy of the designed method may be 10% higher than that of sparse representation based on original samples. To our knowledge, no similar work is available. Second, the designed method is simple and has a low computational cost. Extensive experiments are conducted and the experimental results also show that the designed method can be applied to improve other methods too.

Keywords Sparse representation · High-dimensional data · Low-dimensional data · Classification

Q. Wang · Y. Zhang (✉) · L. Xiao · Y. Li
Key Laboratory of Intelligent Medical Image Analysis and Precise Diagnosis of Guizhou Province, College of Computer Science and Technology, Guizhou University, Guiyang, China
e-mail: zyj6667@126.com

Q. Wang
e-mail: wangqyzm@163.com

L. Xiao
e-mail: xl201304@163.com

Y. Li
e-mail: hesjed@163.com

© Springer Nature Singapore Pte Ltd. 2019
J.-S. Pan et al. (eds.), *Genetic and Evolutionary Computing*,
Advances in Intelligent Systems and Computing 834,
https://doi.org/10.1007/978-981-13-5841-8_66

1 Introduction

Sparse representation has been found to be very effective in many pattern classification problems [1–4]. Especially, sparse representation is viewed as a breakthrough of face recognition [5]. Also, sparse representation shows very perfect performance in other images related tasks, such as hyperspectral image classification, image segmentation, and image recovery [6–8]. It is doubtless that sparse representation is highly competent in determining the inherent similarity of different samples [9].

A notable fact is that, when sparse representation performs well in the classification of high-dimension data, its accuracy for low-dimensional data is usually unsatisfactory or even poor. This has been shown in literature. Some attempts have been made to overcome the above issue. For example, kernel sparse representation is partially able to remedy the issue [10, 11]. This is different from the case of conventional dimension extension where samples are very low dimensional and to increase the number of training samples can lead to better recognition of test samples [12]. As we know, the dimension of features in kernel sparse representation is the same as the number of training samples. Thus, if the number of training samples is greater than the dimension of the original samples, kernel sparse representation will lead to higher dimensional features [13–15]. Nevertheless, under the condition that the number of training samples is not large, kernel sparse representation is also not able to obtain very high-dimension "data".

For general pattern classification issues, dedicated algorithms for an increase of the sample dimension have been proposed. Using high-order components of original samples as new features is a widely used way. It has been proved that transforming samples into a high-dimensional space is able to increase the linear separability of samples [12, 16]. Moreover, it turns out that converting low-dimension data into a high-dimensional space may make a non-separable issue to become a linear separable issue. A linear separable issue means that a linear plane or superplane can correctly classify samples into different classes [12]. Figure 1 shows low-dimension data is converted into high-dimensional data and the original non-separable issue becomes a linear separable issue.

It is known that the more used of high-order components will result in more new features and the higher the sample dimension. On the other hand, too many high-order components will cause computational problems and even "dimension disaster".

With this paper, we study the problem which the increase of the sample dimension for sparse representation. The simple design and high computational efficiency result in surprising accuracy improvement in comparison with naïve sparse representation including collaborative representation on original samples. The designed method is also applicable to improving other methods too.

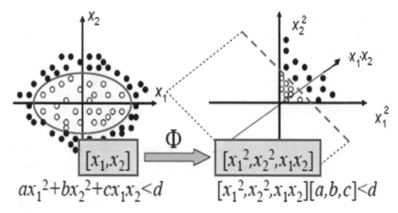

$$ax_1^2 + bx_2^2 + cx_1x_2 < d \qquad [x_1^2, x_2^2, x_1x_2][a, b, c] < d$$

Fig. 1 A case where low-dimension data is converted into high-dimensional data [5]

2 Drawbacks of Sparse Representation and Collaborative Representation

Sparse representation and collaborative representation have non-ignorable drawbacks for low-dimensional samples. First, previous studies and experiments show that sparse representation and collaborative representation are usually very effective for high-dimension data, but may perform badly for low-dimension data. Second, from the procedures of sparse representation and collaborative representation, we are easy to know that discrimination of class residuals of a test sample may be poor in the case of low-dimension data, which implies a high classification error rate. Especially, when the number of training samples is greater than the dimension of the samples, Linear Regression Classification (LRC) is a special sparse representation algorithm, is difficult in determining a class label for the test sample. As for this issue, LRC has been shown as an example in [17]. A regularization parameter has been introduced into collaborative representation and used as a way to overcome the issue [4]. However, for collaborative representation, the use of the regularization parameter values smooths the influence, on the representation of the test samples, of training samples, so it is only a little able to alleviate this negative effect of low-dimension data. Therefore, to determine a proper way to overcome the above problems is very significant, which will make sparse representation and collaborative representation work well for low-dimension samples and can solidly widen their applicable scopes.

Since low-dimensional data may cause a non-separable issue, sparse representation, and cooperative representation have limitations in terms of processing. If it is a non-separable issue, the sparse representation and the cooperative representation will have large errors in classification, and thus the classification accuracy rate will be lowered, which is very unfavorable for the research. In this paper, it is proposed to convert low -dimensional data into high-dimensional data, so as to maximize the non-separable problem into a linearly separable problem and improve the accuracy of classification. Converting non-separable problems into separable ones that have been widely used.

3   The Designed Method

Suppose that there are C classes and each class has q training samples. We define that each sample is a D-dimensional vector. $x_{ij} = \left[x_{ij}^1, x_{ij}^2, \ldots, x_{ij}^D \right]$ denotes the jth training sample of the ith class. x_{ij}^d is the dth component of x_{ij}. Test sample y is denoted by $y = [y_1, y_2, \ldots, y_D]$. The designed method has the following main steps.

Step 1. For sample x_{ij}, new features are generated using formula (1):

$$f_{ij}^k = x_{ij}^m x_{ij}^n. \tag{1}$$

Among them, $m, n = 1, 2, \ldots, D$, $k = 1, 2, \ldots, D\frac{(D-1)}{2}$. The sample x_{ij} is converted into $z_{ij} = \left[x_{ij}^1, x_{ij}^2, \ldots, x_{ij}^D, f_{ij}^1, f_{ij}^2, \ldots, f_{ij}^{D\frac{(D-1)}{2}} \right]$.

Step 2. Each z_{ij} using formula (2) to normalize

$$z_{ij} = z_{ij}/norm(z_{ij}). \tag{2}$$

Step 3. Sparse representation or collaborative representation is applied to normalized test samples and training samples. For normalized test sample y, the objective functions of sparse representation and collaborative representation are usually defined as F_1 and F_2, respectively

$$F_1 = \min \|y - Z\alpha\|_2^2 + \lambda \|\alpha\|_1. \tag{3}$$

$$F_2 = \min \|y - Z\alpha\|_2^2 + \lambda \|\alpha\|_2^2. \tag{4}$$

$Z = \left[z_{11}, \ldots, z_{1q}, \ldots, z_{C1}, \ldots, z_{Cq} \right]$. After the corresponding algorithm is implemented, a solution is obtained.

Step 4. For each test sample, its class residual is calculated using the same formulas as those in sparse representation and collaborative representation. In other words, the class residual of the test samples with respect to the pth class is obtained using

$$S_p = \|y - Z_p\alpha_p\|, \quad Z_p = \left[z_{p1}, \ldots, z_{pq} \right]. \tag{5}$$

Step 5. The test sample is assigned to the rth class if

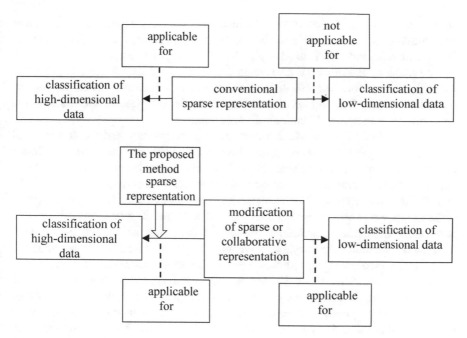

Fig. 2 Illustration of the contribution of the proposed method

$$r = \arg \min_{p} S_p. \tag{6}$$

Figure 2 shows an illustration which the contribution of the proposed method. As shown in this figure, our proposed method allows sparse representation and collaborative representation applicable for classification of high-dimensional data. In other words, conventional sparse representation usually cannot obtain high accuracy for classification of low-dimensional data. However, when we apply our proposed method, the accuracy of sparse representation and collaborative representation of new samples with higher dimension can be greatly improved.

4 Experimental Results

Experiments are mainly conducted for benchmark datasets [18]. Collaborative representation, sparse representation, nearest classification, and the corresponding improved algorithms are tested. Nearest classification works as follows. The nearest neighbor of a test sample is first determined from the set of all training samples. Then the class label of the nearest neighbor is assigned to the test sample. Modification of collaborative representation works as follows. The dimension of the samples is first increased by using our proposed method. Then collaborative representation is applied to the classification of the obtained new samples. Modification of sparse representation is the same as a modification of collaborative representation except that sparse representation is used as the classification procedure.

The mean of classification error rates on all test samples subsets is reported in Table 1. In this table, modification of collaborative representation stands for the procedure to apply the proposed method to collaborative representation. Other terms including "modification of" have similar meanings.

From this Table 1, we see that collaborative representation and sparse representation obtain a greater decrease of the classification error rates than nearest classification. For example, for dataset "Heart", the classification error rates of modification of collaborative representation and collaborative representation are 20.77% and 37.90%, respectively. The classification error rates of modification of sparse representation and sparse representation are 21.81% and 31.52%, respectively. For nearest classification, an increase of the sample dimension may or may not produce a higher classification accuracy, i.e., lower classification error rates. For example, the classification error rates of modification of nearest classification and nearest classification on dataset "Heart" are 24.08% and 22.81%, respectively. The experimental results show that the integration of collaborative representation or sparse representation with the proposed scheme that an increase of the sample dimension is very effective and suitable for real-world classification problems. More importantly, the proposed method greatly enhances the applicability of collaborative representation or sparse representation, allowing it to be competent for classification problems of both high-dimensional and dimensional data.

Table 1 Classification error rates (%) on different benchmark datasets

	Heart	Titanic	Diabetes	Cancer	Thyroid	Banana
Modification of collaborative representation (the proposed method)	20.77	21.87	27.19	26.13	8.36	28.55
Collaborative representation	37.90	29.68	34.80	28.96	21.40	40.08
Modification of nearest classification	24.08	33.01	31.10	32.57	7.35	16.46
Modification of sparse representation (the proposed method)	21.81	22.08	25.83	27.92	5.55	28.90
Sparse representation	31.52	22.53	33.67	31.17	11.65	37.69
Nearest classification	22.81	33.00	30.89	32.06	11.13	33.55

5 Conclusions

The proposed two-order component generation scheme is very effective for increasing the dimension of samples. The combination of the two-order component generation scheme and sparse representation can obtain very promising accuracy improvement. As shown in the experimental results, the accuracy of the designed method samples. In the future, we will explore possible applications of the proposed method such as visual tracking [19] and other problems may be 10% higher than that of sparse representation based on the original.

Acknowledgements This work was supported by the Joint Fund of Department of Science and Technology of Guizhou Province and Guizhou University under grant: LH [2014]7635, Research Foundation for Advanced Talents of Guizhou University under grant: (2016) No. 49, Key Supported Disciplines of Guizhou Province—Computer Application Technology (No. QianXueWei-HeZi ZDXX[2016]20), Specialized Fund for Science and Technology Platform and Talent Team Project of Guizhou Province (No. QianKeHePingTaiRenCai [2016]5609), and the work was also supported by National Natural Science Foundation of China (61462013, 61661010).

References

1. Wright, J., Yang, A.Y., Ganesh, A., et al.: Robust face recognition via sparse representation. IEEE Trans. Pattern Anal. Mach. Intell. **31**(2), 210–227 (2009)
2. Wright, J., Ma, Y., Mairal, J., et al.: Sparse representation for computer vision and pattern recognition. Proc. IEEE **98**(6), 1031–1044 (2010)
3. Xu, Y., Zhang, D., Yang, J., Yang, J.-Y.: A two-phase test sample sparse representation method for use with face recognition. IEEE Trans. Circuits Syst. Video Technol. **21**(9), 1255–1262 (2011)
4. Zhang, L., et al.: Sparse representation or collaborative representation: which helps face recognition. Proc. Int. Congr. Comput. Vision (2011)
5. Kroeker, K.L.: Face recognition breakthrough. Commun. ACM **52**(8), 18–19. https://doi.org/10.1145/1536616.1536623
6. Zhang, E., Zhang, X., Liu, H., Jiao, L.: Fast multifeature joint sparse representation for hyperspectral image classification. IEEE Geosci. Remote Sens. Lett. **12**(7), 1397–1401 (2015)
7. Yang, S., Lv, Y., Ren, Y., Yang, L., Jiao, L.: Unsupervised images segmentation via incremental dictionary learning based sparse representation. Inf. Sci. **269**, 48–59 (2014)
8. Li, X.: Image recovery via hybrid sparse representation: a deterministic annealing approach. IEEE J. Sel. Top. Signal Process. **5**(5), 953–962 (2011). Special issue on adaptive sparse representation
9. Xu, Y., Zhang, B., Zhong, Z.: Multiple representations and sparse representation for image classification. Pattern Recognit. Lett. (2015)
10. Gao, S., Tsang, I.W.-H., Chia, L.-T.: Kernel sparse representation for image classification and face recognition. In: Lecture Notes in Computer Science, vol. 6314, pp. 1–14 (2010)
11. Yin, J., Liu, Z., Jin, Z., Yang, W.: Kernel sparse representation based classification. Neurocomputing **77**(1), 120–128 (2012)
12. Duda, R.O., Hart, P.E., Stork, D.G.: Pattern Classification, 2nd edn. Wiley, New Jersey (2004)
13. Xu, Y., Fan, Z., Zhu, Q.: Feature space-based human face image representation and recognition. Opt. Eng. **51**(1), 017205 (2012)
14. Chen, Z., Zuo, W., Qinghua, H., Lin, L.: Kernel sparse representation for time series classification. Inf. Sci. **292**, 15–26 (2015)

15. Zhang, L., Zhou, W., Li, F.-Z.: Kernel sparse representation-based classifier ensemble for face recognition. Multimed. Tools Appl. **74**(1), 123–137 (2015)
16. Ripley, B.D.: Pattern Recognition and Neural Networks, 1st edn, p. 416. Springer, Cambridge (2007)
17. Xu, Y., Li, X., Yang, J., Lai, Z., Zhang, D.: Integrating conventional and inverse representation for face recognition. IEEE Trans. Cybern. **44**(10), 1738–1746 (2014)
18. Xu, Y., Zhang, D., Jin, Z., Li, M., Yang, J.-Y.: A fast kernel-based nonlinear discriminant analysis for multi-class problems. Pattern Recogn. **39**(6), 1026–1033 (2006)
19. Xu, Y., Zhu, Q., Fan, Z., Zhang, D., Mi, J., Lai, Z.: Using the idea of the sparse representation to perform coarse-to-fine face recognition. Inf. Sci. **238**(20), 138–148 (2013)

Dynamic Texture Classification with Relative Phase Information in the Complex Wavelet Domain

Qiufei Liu, Xiaoyong Men, Yulong Qiao, Bingye Liu, Jiayuan Liu
and Qiuxia Liu

Abstract In recent years, dynamic texture classification has caused widespread concern in the image sequence analysis field. We propose a new method of combining relative phase information of dynamic texture in the complex wavelet domain with probability distribution models for dynamic classification in this paper. Instead of using only real or magnitude information of dynamic texture, relative phase information is an effective complementary measure for dynamic texture classification. Firstly, the finite mixtures of Von Mises distributions (MoVMD) and corresponding parameter estimation method based on expectation-maximization (EM) algorithm are introduced. Subsequently, the dynamic texture features based on MoVMD model for dynamic texture classification are proposed. Besides, the relative phase information of dynamic texture is modeled with MoVMDs after decomposing dynamic texture with the dual-tree complex wavelet transform (DT-CWT). Finally, the variational approximation between different dynamic textures is measured using the Kuller-Leibler divergence (KLD) variational approximation. The effectiveness of the proposed method is verified by experimental evaluation of two popular benchmark texture databases (UCLA and DynTex++).

Keywords Dynamic texture classification · MovMD · Distributions ·
Kuller-Leibler divergence

Q. Liu (✉) · B. Liu · J. Liu
724 Research Institute, Nanjing 211153, China
e-mail: liuqiufei@hrbeu.edu.cn

X. Men · Y. Qiao
Harbin Engineering University, Harbin 150001, China

Q. Liu
Zhoukou Normal University, Zhoukou 466001, China

© Springer Nature Singapore Pte Ltd. 2019
J.-S. Pan et al. (eds.), *Genetic and Evolutionary Computing*,
Advances in Intelligent Systems and Computing 834,
https://doi.org/10.1007/978-981-13-5841-8_67

1 Introduction

Dynamic texture (DT) extends the image texture in time domain, which has certain self-similarity in the time-space domain, such as smoke, sea waves, fountains, and branches that sway with the wind. DT classification is an important issue in computer vision and image understanding [1]. During the past decade, DT classification has caused widespread concern and a lot of different classification methods have been proposed by researchers. Despite efforts, DT classification remains a challenging and interesting area of research.

The extraction of DT features is the basis and key of DT classification. Generally, DT always exhibits two basic properties, appearance and motion. Different methods are proposed in the literature for DT representation and recognition. Existing methods for extracting DT features could be roughly divided into two categories according to properties of DT, motion attributes based methods [2] and the combination of motion and appearance attributes based methods [3]. Optical flow is a kind of typical method among former methods. However, owing to the rapid change in grayscale around a position in DT, estimating real motion with the first category of method is quite difficult. In the meantime, the second category of methods have been further explored in recent years. Transformation-based method is the most widely used method in the second category except for extending image texture analysis to DT analysis methods, such as fractal analysis and local binary mode.

Smith et al. [4] firstly indexed dynamic video content using time-space wavelets among the transformation-based methods. Even though the traditional time-space wavelet transform could perform time-frequency positioning and multi-resolution analysis on DT signals, there are problems such as shift dependence, lack of phase information and direction selectivity. In [5], we successfully realized the DT classification by the combination of magnitude information with probability distribution models. In a way, [5] expands the ideas to realize DT classification. Goncalves et al. [6] successfully achieved the propose of DT classification on the basis of statistic characteristic and energy distribution. Recently, Dubois et al. [7] proposed a new classification approach based on the 2d + t curvelet transform.

In this paper, the finite mixtures of Von Mises distributions (MoVMD) and the corresponding parameter estimation method which bases on EM algorithm are introduced. In view of the dual-tree complex wavelet transform (DT-CWT), a DT feature extraction method based on MoVMD for modeling relative phase information is proposed. In the experimental part, we experimented with two benchmark datasets and the proposed method's classification property was verified.

2 Relative Phase and Its Property

After transforming DT sequences with complex wavelet transforms, we can obtain the corresponding low-frequency as well as the high-frequency complex coefficients. Assuming the expression of complex coefficients is $W(i, j, t) = R(i, j, t) + I(i, j, t)$, where $R(i, j, t)$ and $I(i, j, t)$ denotes real parts and imaginary parts of complex coefficients respectively. We can obtain the following equation according to the definition of phase:

$$P(i, j, t) = \text{atan2}(I(i, j, t), R(i, j, t)) \tag{1}$$

However, we observe that the distribution of phase information in high-frequency sub-bands of DT is approximately uniform as shown in Fig. 1b, which means we can not realize the propose of DT classification with phase information. Therefore, we introduce the definition of relative phase according to [8]. Actually, relative phase means the difference between adjacent phases. And the definition of relative is given as follows:

$$\theta(i, j, t) = P(i, j, t) - P(i + 1, j, t)$$
$$\theta(i, j, t) = P(i, j, t) - P(i, j + 1, t)$$
$$\theta(i, j, t) = P(i, j, t) - P(i, j, t + 1) \tag{2}$$

To eliminate the negative effects, 2π wrap-around, which exists in the application of Eq. (2), we modify the Eq. (2) into the following equation:

$$\theta(i, j, t) = P(i, j, t) \cdot P^*(i + 1, j, t)$$
$$\theta(i, j, t) = P(i, j, t) \cdot P^*(i, j + 1, t)$$
$$\theta(i, j, t) = P(i, j, t) \cdot P^*(i, j, t + 1) \tag{3}$$

(a)   **(b)**   **(c)**

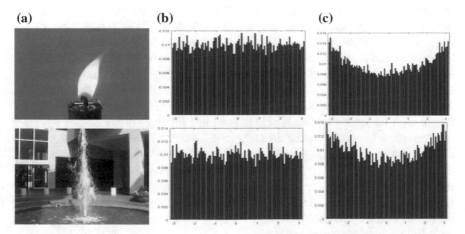

Fig. 1 Distributions of phase information and relative phase information. **a** First frame of DT randomly selected from UCLA dataset. **b** Distribution of phase information. **c** The distribution of relative phase information

where $P^*(i, j, t)$ denotes the conjugate of $P(i, j, t)$. Figure 1b, c denote the corresponding phase information and relative phase information of the same complex wavelet sub-band. According to results in Fig. 1, we can observe that phase information can not provide valuable information for DT classification, while relative information can offer some certain distribution properties. Therefore, we can realize the DT classification on the basis of the modeling analysis of relative phase information. In this paper, we choose the finite mixture of Von Mises distributions to model the relative information in complex wavelet sub-bands of DT.

3 The Finite Mixture of Von Mises Distributions Model

3.1 MoVMD Model

The Von Mises distribution model, also known as the circular normal distribution or Tikhonov distribution, is an approximation of a normal distribution of packages, and the normal distribution of packages is a circular similarity with a normal distribution. When the real and imaginary parts of the first cycle moment are specified, the Von Mises distribution is a maximum entropy distribution of the cyclic data. And the corresponding probability distribution function is usually given as follows:

$$P(x|\mu, k) = \exp(k\cos(x - \mu))/(2\pi I_0(k)) \qquad (4)$$

where the parameter μ is a location measurement (the distribution gathers around μ) and the parameter $k(k > 0)$ is a concentration measurement (a reciprocal measure of dispersion). Besides, $I_0(k)$ denotes the first type of improved Bessel function and the zero-th order which can be defined by $I_0(k) = (1/\pi) \int_0^\pi \exp(v \cos\theta)d\theta$.

The location parameter μ and the concentration parameter k make the Von Mises distribution have the flexibility to fit complex data, which has been shown the potential in the dynamic texture classification. However, it is too hard for a single Von Mises distribution to precisely model complex data, such as relative phase information in high-frequency subbands of DT. The finite mixtures of distributions provide a way to model various random data. Figure 2 shows different classification performance of the single Von Mises distribution (SVMD) and the finite mixtures of Von Mises distributions (MoVMD), from which we can clearly observe that the MoVMD model can catch the real property of relative phase information much more accurately than the SVMD model. Therefore, we introduce the MoVMD to fit relative phase data in this paper. The MoVMD model with N components and the corresponding probability function are defined by:

$$P(x|\Theta) = \sum_{n=1}^{N} \pi_n P(x|\mu_n, k_n) \qquad (5)$$

(a)   (b)

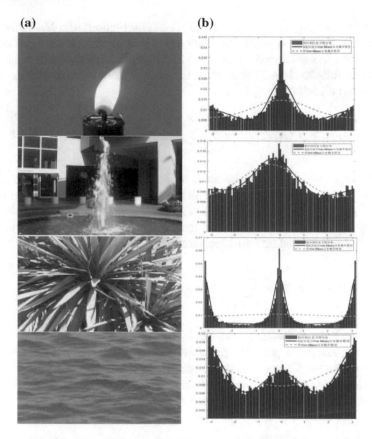

Fig. 2 A histogram fit of the SVMD and MoVMD models on the DT-CWT domain on the UCLA benchmark. **a** Randomly select the first dynamic texture frame from the UCLA benchmark, **b** results matched SVMD and MoVMD mode

where the mixing weight π_n satisfies $0 < \pi_n \leq 1$ and $\sum_{n=1}^{N} \pi_n = 1$, and μ_n and k_n are parameters of the nth component in MoVMD model. The symbol Θ is used to denote the set of MoVMD model parameters $\{\pi_n, \mu_n, k_n | n = 1, \ldots, N\}$.

3.2 Parameter Estimation of MoVMD Model

In order to estimate parameters of MoVMD model, we utilize the expectation-maximization (EM) algorithm in this paper. In the case where the equations cannot be solved directly, the maximum likelihood parameter of the statistical model can be obtained by using the EM algorithm. Often these models contain latent variables except unknown parameters and known data observations.

Suppose $\mathbf{X} = \{x_1, x_2, \ldots, x_M\}$ is a series of observations of a MoVMD model, which will be the relative phase information for dynamic texture classification in the fourth section. When discussing the basic EM algorithm, what we first need to do is to obtain the full-data log-likelihood function of the MoVMD parameters estimation using the EM algorithm.

$$l(\Theta) = \ln(P(\mathbf{X}, \mathbf{Z}|\Theta)) = \sum_{m=1}^{M} \sum_{n=1}^{N} z_{mn}(\ln(\pi_n) + \ln(P(x_m|\mu_n, k_n))) \tag{6}$$

where $\mathbf{Z} = \{\mathbf{z}_1, \mathbf{z}_2, \ldots, \mathbf{z}_M\}$ is the series of latent variables, $\mathbf{z}_m$ is a N-dimensional binary vector, in which a specific element z_{mn} equals 1 and all other elements equal 0 and $P(z_{mn}) = \pi_n$. In fact, z_{mn} denotes that the observation data x_m belongs to the nth component of MoVMD model. In the expectation step, the current parameter value Θ^{t-1} used for finding the expectation of the full-data log-likelihood function,

$$\mathcal{Q}(\Theta; \Theta^{t-1}) = \sum_{m=1}^{M} \sum_{n=1}^{N} \gamma(z_{mn})(\ln(\pi_n) + \ln(P(x_m|\mu_n, k_n))) \tag{7}$$

where $\gamma(z_{mn}) = \pi_n P(x_m|\mu_n^{t-1}, k_n^{t-1}) / \sum_{j=1}^{N} \pi_j P(x_m|\mu_j^{t-1}, k_j^{t-1})$ is the expected value of z_{mn} under the posterior probability $P(\mathbf{Z}|\mathbf{X}, \Theta^{t-1})$. In the maximum step, maximizing the Eq. 7 can obtain the re-estimated parameter set Θ^t. What equations we can get through substituting Eq. 4 into Eq. 7 and then computing the partial derivative of Eq. 7 with respect to μ_n and k_n are as follows,

$$0 = k_n \sum_{m=1}^{M} \gamma(z_{mn})(\sin x_m \cos \mu_n - \cos x_m \sin \mu_n) \tag{8}$$

$$0 = \sum_{m=1}^{M} \gamma(z_{mn})(\cos(x_m - \mu_n) - (I_1(k_n)/I_0(k_n))) \tag{9}$$

According to Eq. 8 and constraint condition $k > 0$, we can obtain the estimated parameter $\hat{\mu}_n$ as follows:

$$\hat{\mu}_n = \arctan 2 \left(\frac{\sum_{m=1}^{M} \gamma(z_{mn}) \sin x_m}{\sum_{m=1}^{M} \gamma(z_{mn}) \cos x_m} \right) \tag{10}$$

Then the parameter $\hat{k}_n$ is estimated with the following formula deriving on the basis of Eqs. 10 and 9,

$$\frac{I_1(\hat{k}_n)}{I_0(\hat{k}_n)} = \frac{\sum_{m=1}^{M} \gamma(z_{mn})\cos(x_m - \hat{\mu}_n)}{\sum_{m=1}^{M} \gamma(z_{mn})} \tag{11}$$

The parameter π_n which estimated as a general EM algorithm is as follows,

$$\hat{\pi}_n = (1/M) \sum_{n=1}^{M} \gamma(z_{mn}). \tag{12}$$

Thus, we have obtained all parameters' estimation equations. However, we can not derive the estimation of concentration parameter $\hat{k}_n$ because of the existence of modified Bessel function. Indeed, we can obtain the solution of Eq. 11 with some numerical algorithms. For convenience, $I_1(\hat{k}_n)/I_0(\hat{k}_n)$ is denoted as $\mathcal{A}$. And then we can obtain the following equation,

$$\hat{k}_n = \begin{cases} 2\mathcal{A} + \mathcal{A}^3 + \frac{5}{6}\mathcal{A}^5 & 0 \le \mathcal{A} < 0.53 \\ -0.4 + 1.39\mathcal{A} + \frac{0.43}{1-\mathcal{A}} & 0.53 \le \mathcal{A} < 0.85 \\ \frac{1}{\mathcal{A}^3 - 4\mathcal{A}^2 + 3\mathcal{A}} & \mathcal{A} \ge 0.85 \end{cases} \tag{13}$$

So far, we have obtained all parameters' estimation equations. In next section, we will introduce how to realize dynamic texture classification with MoVMD model built in this section.

4 Dynamic Texture Classification

4.1 Feature Extraction

Feature extraction is the key to the dynamic texture classification. And the wavelet-based probability distribution model method firstly models the wavelet coefficients of known probability distributions and then uses the model parameters as texture features. Figure 2 shows the performance of MoVMD model. It is easy to see from the figure that the histogram has the characteristics of thick tail and sharp peak, and the outcomes of MoVMD model fitting is better than SVMD model fitting. Therefore, we extract dynamic texture features by modeling the relative phase information of complex wavelet coefficients with MoVMD models in DT-CWT complex wavelet domain.

The MoVMD model inherited from the SVMD model can be used to describe direction data. To describing DT more effectively, decomposing each wavelet sub-band into non-overlapping wavelet sub-bands is a general idea. The relative phase data of these wavelet sub-bands is modeled using the MoVMD model. The proposed dynamic texture feature extraction method of MoVMD model is illustrated in Fig. 3. We decompose each DT with L-level (in our experiments $L = 2$) time-space DT-CWT and then we could get a detailed sub-band of 28L. A complex wavelet detail sub-band of size $M \times N \times T$ is given, which is divided into non-overlapping wavelet blocks of size $2 \times 2 \times T$. The relative phase information of DT-CWT complex wavelet

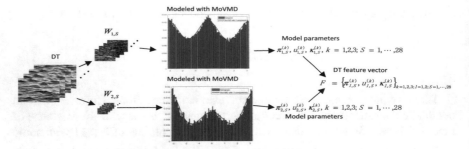

Fig. 3 Description of the proposed feature extraction method. And the symbols $W_{1,S}$ and $W_{2,S}$ are relative information of the detail sub-bands of the first ($l = 1$) and second ($l = 2$) level DT-CWT, and S denotes the Sth sub-band. The parameters of the kth component of the MoVMD model on the sub-band $W_{l,S}$ are $\pi_{l,S}^{(k)}, \alpha_{l,S}^{(k)}, \beta_{l,S}^{(k)}$

coefficients in each wavelet block is calculated. The relative phase data in these sub-bands is then modeled using the MoVMD model, and its parameters are estimated using the EM algorithm. To describe DT, the model parameters of all detail sub-bands are merged to feature vectors in the end.

4.2 Similarity Measurement

In Sect. 4.1, we obtain the feature vector of the dynamic texture. In order to realize the dynamic texture classification, the similarity between the two different dynamic textures (or corresponding feature vectors) needs to be evaluated. The closed-form Kullback–Leibler divergence between the two different SVMD models f and g has been obtained in [8].

$$\text{KLD}(f\|g) = \log\frac{I_0(k_g)}{I_0(k_f)} + \frac{I_1(k_f) - I_1(-k_f)}{2I_0(k_f)}\left(k_f - k_g\cos(\mu_g - \mu_f)\right) \qquad (14)$$

where μ_f and k_f are the position parameter and concentration parameter of the SVMD model f. While the parameters of the SVMD model g are μ_g and k_g. $I_0(\cdot)$ denotes the modified Bessel function of the first kind and the zero-th order, and $I_1(\cdot)$ denotes the modified Bessel function of the first kind and the first order. As to the mixture distribution MoVMD, it is impossible to derive a closed-form KLD between two models, while the KLD can use different methods for approximation [9]. Based on variational approximation, we use the approximated KLD to measure the similarity between two MoVMD models $X = \sum_{i=1}^{K} \pi_i f_i$ and $Y = \sum_{j=1}^{M} \omega_j g_j$ as follows,

$$D_{KL}(X||Y) \triangleq \sum_i^K \pi_i log \left(\frac{\sum_{m=1}^K \pi_m \exp(-KLD(f_i||f_m))}{\sum_{j=1}^M \omega_j \exp(-KLD(f_i||g_j))} \right) \quad (15)$$

According to Eqs. 14 and 15, we can realize the similarity measurement between MoVMD models with the approximated KLD.

5 Experimental Results

In this section, we use two popular data sets, the UCLA and DynTex++ datasets, to evaluate the proposed method for dynamic texture tasks. The proposed method was evaluated on the UCLA dataset using four cross-folding verification scenarios [10]. In this scenario, three of the four DT sequences of each class are used for training and the rest for testing. And we repeated this experiment using different DT sequences for each test sample for four times. Eventually, the final result is the average of the four recognition rates.

In order to evaluate the application of this method to the DynTex++ dataset, we used the same experimental setup as [10] on the DynTex++ dataset. And we randomly divided the 100 DT sequences of each class into two parts for training and testing. Subsequently, this experiment was repeated ten times. Finally, what we want is the average of ten classification results.

In order to successfully conduct experiments on two benchmark dynamic texture datasets, we need to do some settings. Firstly, the number of mixture components is set to be M = 3 in MoVMD models. In addition, we used a good texture classification reference method 1-NN classifier in all experiments. Table 1 shows the performance comparison between MoVMD and other methods on the 50 UCLA-dataset. At the same time, Table 2 shows the performance comparison of MoVMD and other methods on the DynTex++ dataset.

It can be seen that the proposed MoVMD model performs best in Table 1, with results greater than 91% on the UCLA 50 data set. And the results of VLBP, DFS and other methods are similar to those of MoVMD. Although the proposed MoVMD

Table 1 The average experimental results comparison of 50-class UCLA dataset

Method	Classification rate (%)
Space-time oriented [11]	81.00
SVMD	85.50
3D-OTF [12]	87.10
DFS [13]	89.50
KDT-MD [14]	89.50
VLBP [15]	89.50
MoVMD	92.10

Table 2 The average
experimental results
comparison of DynTex++
dataset

Method	Classification rate (%)
DL-PEGASOS [10]	63.70
Chaoic vector approach [16]	64.22
Chaoic feature vector [17]	69.00
SVMD	81.56
MoVMD	86.57
LBP-TOP [18]	89.50

model is not the best performer on the DynTex++ dataset, similar results have been obtained, demonstrating its effectiveness for DT classification. Additionally, the proposed MoVMD model extracts DT feature using L-level DT-CWT. The dimension of feature vector is $28 \times L \times M \times 3$ (in our experiments, $L = 2$, $M = 3$ and the total number of a feature vector for a DT is 504). Besides, texture feature extraction mainly includes several computational aspects such as model parameter estimation, median calculation of each sub-band block, and DT-CWT decomposition. Without any implementation optimization, the proposed method of extracting all 200 eigenvectors in the UCLA 50 data set takes 510 s. At the same time, our proposed method of extracting 3600 feature vectors of DynTex++ dataset takes 22,910 s.

A finite mixture of Von Mises distribution (MoVMD) and a parameter estimation method on the basis of the expectation maximization (EM) algorithm were introduced in this paper. Meanwhile, the model performance of describing complex data in the context of dynamic texture classification was verified. Generally speaking, the experimental results in Tables 1 and 2 demonstrate the validity of our proposed DT classification method.

6　Conclusions

This paper introduces the finite mixtures of Von Mises distributions (MoVMD) model and derives the parameter estimation method based on EM algorithm. According to modelling the DT-CWT coefficients of the dynamic texture, the characterization performance of the model is verified. For the sake of describing the dynamic texture, the MoVMD model is used to model the relative phase information of the complex wavelet coefficients in the discrete wavelet transform domain. Dynamic texture features are extracted by estimating the MoVMD model parameters for each sub-band. Experimental results of two benchmark texture classification datasets show that the proposed hybrid distribution could characterize the dynamic texture of the DT-CWT domain, and the feature descriptors better than the most advanced classification methods can be obtained in dynamic texture classification.

Acknowledgements This work is supported by National Natural Science Foundation of China under Grant 61371175.

References

1. Tiwari, D., Tyagi, V.: Dynamic texture recognition: a review. Inf. Syst. Des. Intell. Appl. **434**, 365–373 (2016)
2. Wang, C., Lu, Z., Liao, Q.: Local texture based optical flow for complex brightness variations. In: 21th IEEE International Conference on Image Processing (ICIP), pp. 1972–1976 (2014)
3. Tiwari, D., Tyagi, V.: Dynamic texture recognition based on completed volume local binary pattern. Multidim. Syst. Signal Process **27**(2), 563–575 (2016)
4. Smith, J.R., Lin, C.-Y., Naphade, M.: Video texture indexing using spatiotemporal wavelets. In: 9th IEEE International Conference on Image Processing (ICIP), pp. 437–440 (2002)
5. Qiao, Y.L., Song, C.Y., Wang, F.S.: Wavelet-based dynamic texture classification using gumbel distribution. Math. Probl. Eng. **2013**(11), 583–603 (2014)
6. Goncalves, W.N., Machado, B.B., Bruno, O.M.: Spatiotemporal gabor filters: a new method for dynamic texture recognition. SIViP **9**(4), 819–830 (2015)
7. Dubois, S., Peteri, R., Mnard, M.: Characterization and recognition of dynamic textures based on the 2d + t curvelet transform. SIViP **9**(4), 819–830 (2015)
8. Vo, A., Oraintara, S.: A study of relative phase in complex wavelet domain: property, statics and applications in texture image retrieval and segmentation. Sig. Process. Image Commun. **25**(1), 28–46 (2010)
9. Allili, M.S.: Wavelet modeling using finite mixtures of generalized Gaussian distributions: application to texture discrimination and retrieval. IEEE Trans. Image Process. **21**(4), 1452 (2012)
10. Ghanem, B., Ahuja, N.: Maximum margin distance learning for dynamic texture recognition. In: 11th Computer Vision—ECCV 2010, pp. 223–236 (2010)
11. Derpanis, K.G., Wildes, R.P.: Dynamic texture recognition based on distributions of space time oriented structure. In: 23th Computer Vision and Pattern Recognition (CVPR), pp. 191–198 (2010)
12. Xu, Y., Huang, S., Ji, H., Fermuller, C.: Scale-space texture description on SIFT-like textons. Comput. Vis. Image Underst. **116**, 999–1013 (2012)
13. Xu, Y., Quan, Y., Ling, H., Ji, H.: Dynamic texture classification using dynamic fractal analysis. In: 13th IEEE International Conference on Computer Vision (ICCV), pp. 1219–1225 (2011)
14. Chan, A., Vasconcelos, N.: Classifying video with kernel dynamic textures. In: 20th IEEE Conference on Computer Vision & Pattern Recognition, pp. 1–6 (2007)
15. Tiwari, D., Tyagi, V.: A novel scheme based on local binary pattern for dynamic texture recognition. Comput. Vis. Image Underst. **150**, 58–65 (2016)
16. Wang, Y., Hu, S.: Chaoic features for dynamic textures recognition. Soft. Comput. **20**(5), 1977–1989 (2016)
17. Wang, Y., Hu, S.: Exploiting high level feature for dynamic textures recognition. Neurocomputing **154**, 217–224 (2015)
18. Zhao, G., Pietikäinen, M.: Dynamic texture recognition using local binary patterns with an application to facial expressions. IEEE Trans. Pattern Anal. Mach. Intell. **29**(6), 915–928 (2007)

Part XVI
Big Data Analysis and Ontology System

GFSOM: Genetic Feature Selection for Ontology Matching

Hiba Belhadi, Karima Akli-Astouati, Youcef Djenouri, Jerry Chun-Wei Lin and Jimmy Ming-Tai Wu

Abstract This paper studies the ontology matching problem and proposes a genetic feature selection approach for ontology matching (GFSOM), which exploits the feature selection using the genetic approach to select the most appropriate properties for the matching process. Three strategies are further proposed to improve the performance of the designed approach. The genetic algorithm is first performed to select the most relevant properties, and the matching process is then applied to the selected properties instead of exploring all properties of the given ontology. To demonstrate the usefulness and accuracy of the GFSOM framework, several experiments on DBpedia ontology database are conducted. The results show that the ontology matching process benefits from the feature selection and the genetic algorithm, where GFSOM outperforms the state-of-the-art ontology matching approaches in terms of both the execution time and quality of the matching process.

Keywords Semantic web · Ontology matching · Feature selection · Genetic algorithm

H. Belhadi (✉) · K. Akli-Astouati
Computer Science Department, University of Science and Technology
Houari Boumediene (USTHB), Algiers, Bab Ezzouar, Algeria
e-mail: hbelhadi@usthb.dz

K. Akli-Astouati
e-mail: kakli@usthb.dz

Y. Djenouri
Computer and Information Sciences Department, Norwegian University of Science
and Technology (NTNU), Trondheim, Norway
e-mail: youcef.djenouri@ntnu.no

J. C.-W. Lin
Department of Computing, Mathematics, and Physics,
Western Norway University of Applied Sciences, Bergen, Norway
e-mail: jerrylin@ieee.org

J. M.-T. Wu
College of Computer Science and Engineering, Shandong University of Technology,
Shandong, China
e-mail: wmt@wmt35.idv.tw

© Springer Nature Singapore Pte Ltd. 2019   655
J.-S. Pan et al. (eds.), *Genetic and Evolutionary Computing*,
Advances in Intelligent Systems and Computing 834,
https://doi.org/10.1007/978-981-13-5841-8_68

1   Introduction and Related Work

Ontology matching is the process to find the correspondences between different ontologies represented by the set of instances, where each instance is characterized by different properties. It is applied in diverse fields such as biomedical data [1], e-learning [2], and Natural Language Processing [3]. Ontology matching is a polynomial problem in terms of number of instances and number of properties, where the existing trivial algorithms for ontology matching only compare each instance of the first ontology with each instance of the second ontology by taking into account all the properties of both ontologies. However, for some high dimensional data like DBpedia ontology,[1] the runtime of those trivial algorithms thus requires high time consumption. To overcome this drawback, several evolutionary approaches have been developed. The proposed systems presented in [4, 5] aim at finding the similarity between the concepts of two ontologies. The useless of genetic algorithms (GAs) is to achieve an approximation case close to the optimal alignment between two ontologies. Several studies have been presented [6, 7] by proposing different methods to compute the fitness of the ontology matching problem, including maximizing precision, recall, F-measure, and optimizing weights for aggregating more similarities, where the work developed in [8] aims at reducing memory consumption using hybrid genetic algorithm and incremental learning process. These evolutionary-based approaches improved considerably the runtime performance of the ontology matching problem. However, the overall performance of these algorithms still requires improvement while dealing with high dimensional data. Motivated by evolutionary techniques, many approaches were designed and applied for solving several complex problems [9–12]. In this paper, we then propose a feature selection approach called GFSOM that explores the genetic process for solving the ontology matching problem. To the best of our knowledge, this is the first work that explores both feature selection and genetic algorithm as preprocessing step for the ontology matching problem. Intensive experiments have been performed to demonstrate the usefulness of the suggested framework. The results reveal that GFSOM outperforms the state-of-the-art ontology matching algorithms on the well-known DBpedia database.

The rest of this paper is organized as follows. Section 2 presents a detail explanation of the GFSOM framework. The evaluation of the GFSOM performance is provided in Sect. 3. Section 4 gives the conclusions and perspectives for our future work.

2   GFSOM: Genetic Feature Selection for Ontology Matching

In this part, we present the main components of the proposed framework called GFSOM (Genetic Feature Selection for Ontology Matching). The aim of GFSOM is to improve the ontology matching problem by taking instances into account as the

---

[1]http://wiki.dbpedia.org/Datasets.

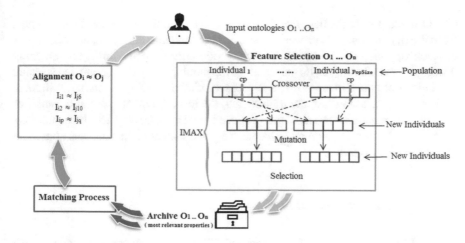

Fig. 1 The architecture of the designed GFSOM

relevant features of the two ontologies. This process allows to boost the matching process for finding the common instances between two ontologies. Moreover, the designed approach aims at improving the quality of the resulted alignment. The developed GFSOM is mainly composed into two steps as: **feature selection** and **matching process**. The flowchart of the designed GFSOM can be described in Fig. 1.

For the feature selection step, it is first performed to the set of attributes for each ontology, which produces an optimal subset of attributes that perfectly represents two ontologies. This step is considered as preprocessing step, and will be only executed once in the designed algorithm. An engineering task is then made to set an archive folder for each ontology. In this step, a genetic algorithm is performed to select the proper features without losing the selection quality. The process starts by generating randomly $PopSize$ individuals from the set of m properties. Each individual is a binary vector of m elements, the ith element is set as 1, if the ith property is selected; otherwise, it is set as 0. The crossover, mutation, and selection operators are then performed in the progress of the GAs. The crossover operator is used to merge two individuals, for instance, x_1, and x_2 of the initial population, which yields two new candidates called x_{11} and x_{12} such as the first cp properties of x_1 are transferred to x_{11}, and the first cp properties of x_2 are transferred to x_{12}. The remaining properties of x_1, and x_2 are transferred to x_{12} and x_{11}, respectively. This is the regular process of the single point crossover operation of GA. Note that cp is the crossover point randomly selected between 1 and m.

After that, the mutation operation is then applied on the new generated individuals by switching randomly a property to 0 if it is presented in the given individual; 1, otherwise. At the end of each iteration, the selection operator is launched to keep the same population size, where all individuals are evaluated using the defined fitness function. It is determined using the information gain value of the selected properties, and the aim is at maximizing the value of fitness function. After that, the designed

GFSOM keeps only the best *PopSize* individuals, and the others are then removed from the population. This process is repeated until the maximal number of iterations is reached, which can be set as the termination criteria. Afterwards, the matching process is applied between the instances of the ontologies by taking only the selected attributes from the above step. The *K*-cross-validation model is used here, where at each pass of the algorithm, the training and the test alignments are performed. For the training matching process, the proposed model is learned to obtain the best parameters. If the alignment rate exceeds the given threshold, then the test alignment is performed and started.

3 Performance Evaluation

To validate the usefulness of the proposed GFSOM framework, extensive experiments were carried out using the well-known DBpedia database.[2] It is a hub data that can be found on Wikipedia. This ontology database contains 4,233,000 instances and 2,795 different properties. All algorithms in the experiments were implemented in Java programming language, and experiments are performed on a desktop machine equipped with an Intel I7 processor and 16 GB memory. The quality of the ontology matching process was evaluated using the *F*-measure. For each reference alignment *R*, and each alignment *A* of the *F*-measure is described as follows:

$$F - measure(A, R) = \frac{2 \times Precision \times Recall}{Precision + Recall},$$ (1)

where $Precision(A, R) = \frac{|R \cap A|}{|A|}$, and $Recall(A, R) = \frac{|R \cap A|}{|R|}$.

The aim of this experiment is to compare the GFSOM with the state-of-the-art EIFPS [13], and the RiMOM [14] algorithms under the DBPedia ontology database. The results are then shown in Fig. 2.

Figure 2 shows the runtime of three approaches considering all instances and properties. When the number of matchings varied from 100 to 1,000,000, the designed GFSOM outperforms the other two approaches. Moreover, the runtime of the GFSOM is stabilized at 105 s, where the other two approaches are highly time consuming, and need more than 900 s for dealing 1,000,000 matchings in the whole DBpedia ontology database for a large number of instances and a large number of matchings. These results were obtained using the preprocessing step, where only the most relevant features were selected using the GA progress.

We then compare the matching quality of three algorithms under the DBpedia ontology database. The results are then shown in Fig. 3.

By varying the percentage of properties from 20% to 100% in Fig. 3, the designed GFSOM outperforms the other two algorithms. Moreover, the results in Fig. 3 show that the quality of the GFSOM was not affected by the increase in the property. Thus,

---

[2]http://wiki.dbpedia.org/Datasets.

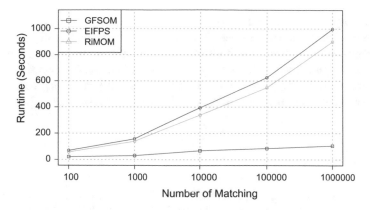

Fig. 2 Comparisons of the runtime performance

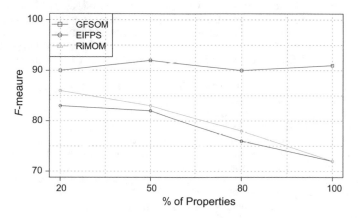

Fig. 3 Comparisons of the F-measure performance

the quality of the developed GFSOM is up to 90%, whereas the qualities of EIFS and RiMOM are under 72%. Thanks to the advantage of the feature selection and the GA progress to select the most relevant properties of the ontologies, the designed GFSOM can thus obtain better performance than the state-of-the-art approaches.

4 Conclusion and Future Work

This paper explores both feature selection and genetic algorithm to improve the ontology matching process. The GFSOM framework is then developed, where the genetic algorithm is first performed to select the most relevant properties, the matching process is then applied to the selected properties instead of exploring all properties of the given ontologies. From the experimental results, it is obvious to see that the designed

GFSOM outperforms the state-of-the-art algorithms on DBpedia ontology database. We also can see that the developed GFSOM benefits from the feature selection and the genetic algorithm in terms of both the execution time and quality of the matching process. In our future works, we will explore other data mining techniques for ontology matching problem. We will further aim at using both clustering, classification, or frequent pattern mining for dealing with big ontology databases.

References

1. Smith, B., Ashburner, M., Rosse, C., Bard, J., Bug, W., Ceusters, W., et al.: The OBO Foundry: coordinated evolution of ontologies to support biomedical data integration. Nat. Biotechnol. **25**(11), 1251 (2007)
2. Cerón-Figueroa, S., López-Yáñez, I., Alhalabi, W., Camacho-Nieto, O., Villuendas-Rey, Y., Aldape-Pérez, M., et al.: Instance-based ontology matching for e-learning material using an associative pattern classifier. Comput. Hum. Behav. **69**, 218–225 (2017)
3. Iwata, T., Kanagawa, M., Hirao, T., Fukumizu, K.: Unsupervised group matching with application to cross-lingual topic matching without alignment information. Data Min. Knowl. Discov. **31**(2), 350–370 (2017)
4. Wang, J., Ding, Z., Jiang, C.: Gaom: genetic algorithm based ontology matching. In: IEEE Asia-Pacific Conference on Services Computing, 2006. APSCC'06, pp. 617–620. IEEE (2006)
5. Acampora, G., Loia, V., Salerno, S., Vitiello, A.: A hybrid evolutionary approach for solving the ontology alignment problem. Int. J. Intell. Syst. **27**(3), 189–216 (2012)
6. Martinez-Gil, J., Alba, E., Aldana-Montes, J.F.: Optimizing ontology alignments by using genetic algorithms. In: Proceedings of the Workshop on Nature Based Reasoning for the Semantic Web. Karlsruhe, Germany (2008)
7. Acampora, G., Loia, V., Vitiello, A.: Enhancing ontology alignment through a memetic aggregation of similarity measures. Inf. Sci. **250**, 1–20 (2013)
8. Xue, X., Chen, J.: Optimizing ontology alignment through hybrid population-based incremental learning algorithm. Memetic Comput. 1–9 2018
9. Djenouri, Y., Belhadi, A., Fournier-Viger, P., Lin, J.C.W.: Fast and effective cluster-based information retrieval using frequent closed itemsets. Inf. Sci. **453**, 154–167 (2018)
10. Djenouri, Y., Djamel, D., Djenoouri, Z.: Data-mining-based decomposition for solving MAXSAT problem: towards a new approach. IEEE Intell. Syst. (2017)
11. Djenouri, Y., Belhadi, A., Fournier-Viger, P., Lin, J.C.W.: An hybrid multi-core/GPU-based mimetic algorithm for big association rule mining. In: International Conference on Genetic and Evolutionary Computing, pp. 59–65. Springer (2017)
12. Lin, J.C.W., Zhang, Y., Fournier-Viger, P., Djenouri, Y., Zhang, J.: A metaheuristic algorithm for hiding sensitive itemsets. In: International Conference on Database and Expert Systems Applications, pp. 492–498. Springer (2018)
13. Niu, X., Rong, S., Wang, H., Yu, Y.: An effective rule miner for instance matching in a web of data. In: Proceedings of the 21st ACM International Conference on Information and Knowledge Management, pp. 1085–1094. ACM (2012)
14. Shao, C., Hu, L.M., Li, J.Z., Wang, Z.C., Chung, T., Xia, J.B.: RiMOM-IM: a novel iterative framework for instance matching. J. Comput. Sci. Technol. **31**(1), 185–197 (2016)

Matching Biomedical Ontologies Through Compact Evolutionary Simulated Annealing Algorithm

Xingsi Xue, Jie Chen, Jianhua Liu and Dongxu Chen

Abstract To overcome biomedical ontology's heterogeneous problem, biomedical ontology matchers are developed to search for bridges of knowledge (or alignment) between heterogeneous biomedical ontologies. Since it is a complex problem, Evolutionary Algorithm (EA) can present a good methodology for matching biomedical ontologies. To improve the efficiency, in this paper, a Compact Evolutionary Simulated Annealing algorithm (CESA) is proposed to match the biomedical ontologies. In particular, CESA utilizes a Probability Vector (PV) to simulate the behavior of population-based EA, and introduces the Simulated Annealing algorithm (SA) as a local search in each generation. The experiment is conducted on the Large Biomed track provided by the Ontology Alignment Evaluation Initiative (OAEI), and the experimental results show the effectiveness of our proposal.

X. Xue (✉) · J. Chen · J. Liu
College of Information Science and Engineering, Fujian University of Technology,
Fuzhou, Fujian 350118, China
e-mail: jack8375@gmail.com

J. Chen
e-mail: jieChen1216@smail.fjut.edu.cn

J. Liu
e-mail: dxchen@fjmu.edu.cn

X. Xue · J. Chen · J. Liu
Intelligent Information Processing Research Center, Fujian University of Technology,
Fuzhou, Fujian 350118, China

X. Xue · J. Liu
Fujian Provincial Key Laboratory of Big Data Mining and Applications,
Fujian University of Technology, Fuzhou, Fujian 350118, China

X. Xue
Fujian Key Lab for Automotive Electronics and Electric Drive,
Fujian University of Technology, Fujian 350118, China

D. Chen
Fujian Medical University Union Hospital, Fuzhou, Fujian 350001, China
e-mail: jhliu@fjnu.edu.cn

© Springer Nature Singapore Pte Ltd. 2019
J.-S. Pan et al. (eds.), *Genetic and Evolutionary Computing*,
Advances in Intelligent Systems and Computing 834,
https://doi.org/10.1007/978-981-13-5841-8_69

Keywords Compact evolutionary algorithm · Simulated annealing algorithm · OAEI

1 Introduction

Biomedical ontology encodes biomedical knowledge in a computationally tractable way to overcome the heterogeneity of biomedical data. Usually, the development of a biomedical ontology is based on the requirements of the research team, which means that even when the encoded domain is identical, the concepts inside may be represented in different ways or with different specificity and granularity. To overcome biomedical ontology's heterogeneous problem, biomedical ontology matchers are developed to search for bridges of knowledge (or alignment) between heterogeneous biomedical ontologies.

Since it is a complex problem (large-scale optimal problem with lots of local optimal solutions), Evolutionary Algorithm (EA) can present a good methodology for matching biomedical ontologies [6]. But drawbacks such as slow convergence speed, premature, and the huge memory consumption limit the application of them. To improve the efficiency, in this paper, a Compact Evolutionary Simulated Annealing algorithm (CESA) is proposed to match the biomedical ontologies. In particular, CESA utilizes a Probability Vector (PV) to simulate the behavior of population-based EA, and introduces the Simulated Annealing algorithm (SA) [4] as a local search in each generation. The main contributions in this paper are as follows: (1) a similarity measure on biomedical concept is proposed to distinguish identical biomedical concepts; (2) an optimal model for biomedical ontology matching is constructed; (3) a problem-specific CESA is presented to efficiently solve the biomedical ontology matching problem.

The rest of the paper is organized as follows: Sect. 2 presents the preliminaries; Sect. 3 describes in details the CESA-based biomedical ontology matching technique; Sect. 4 presents the experimental results; and finally, Sect. 5 draws the conclusion.

2 Preliminaries

2.1 *Biomedical Ontology Matching Problem*

The biomedical alignment quality is evaluated through MatchFmeasure [7], which is a rough evaluation metric to approximate traditional f-measure [5]. Given two biomedical ontologies O_1 and O_2 to match, the optimal model of biomedical ontology matching problem is as follows:

$$\begin{cases} max \quad f(X) \\ s.t. \quad X = (x_1, x_2, \ldots, x_{|O_1|})^T \\ \qquad x_i \in \{0, 1, 2, \ldots, |O_2|\}, i = 1, 2, \ldots, |O_1| \end{cases} \tag{1}$$

where $|O_1|$ and $|O_2|$ respectively represent the cardinalities of source concept set O_1 and target concept set O_2, and $x_i, i = 1, 2, \ldots, |O_1|$ represents the ith concept mapping. In particular, $x_i = 0$ means the ith source class is mapped to none, and the objective function calculates the MatchFmeasure of a solution X's corresponding alignment.

2.2 Similarity Measure on Biomedical Concept

The similarity measure on biomedical concept is the foundation of biomedical ontology matching [3]. In this work, for each sensor concept, a profile is first constructed by collecting the label, comment, and property information from itself and all its direct descendants. Then, the similarity of two biomedical concepts c_1 and c_2 is calculated as follows:

$$sim(c_1, c_2) = \frac{\sum_{i=1}^{f} \max_{j=1\cdots g} (sim'(p_{1i}, p_{2j})) + \sum_{j=1}^{g} \max_{i=1\cdots f} (sim'(p_{1i}, p_{2j}))}{f + g} \tag{2}$$

where p_1 and p_2 are the profiles of c_1 and c_2, respectively.

The function $sim'()$ is calculated by N-gram distance [2] and Unified Medical Language System (UMLS) [1]. To be specific, given two words w_1 and w_2, $sim'(w_1, w_2) = 1$ if w_1 and w_2 are synonymous in UMLS, otherwise $sim'(w_1, w_2) = N - gram(w_1, w_2)$.

3 Compact Evolutionary Simulated Annealing Algorithm

3.1 Probability Vector

The genes are encoded through the binary coding mechanism to represent the correspondences in the alignment. Given the total number of concept in source ontology and target ontology n_1 and n_2, a chromosome (or PV) consists of n_1 genes, and the Binary Code Length (BCL) of each gene is equal to $\lfloor \log_2(n_2) + 0.5 \rfloor$, which ensures each gene can present any target ontology concept's index. Given ith gene bit's value $geneBit_i$, we decode it to obtain a decimal number $\sum_{i=1}^{n} 2^{geneBit_i}$ that represents the index of the target concept.

3.2 Simulated Annealing Algorithm Based Local Search Process

In this work, we utilize the Simulated Annealing algorithm to implement the local search process, which is executed in each generation. As the name suggests, SA simulates the annealing in metallurgy which slowly cools the materials. Such a controlled cooling is implemented in the simulated annealing as the probability to transition to the worse solution, and the probability of a move to a worse solution is proportionate to the temperature. When the temperature is high, it is more likely that the transition to a worse state (based on the fitness function) happens which would help explore the whole search space at the beginning. As the temperature decreases, the odds of moving to a worse solution diminish as well.

SA only operates on one possible solution, called state, and tries to improve it to get a better solution. Such an enhancement is performed by creating a new successor in the neighborhood of the current state, and then probabilistically transition to it. Let $solution_1$ be the current solution and $solution_2$ be the successor (or the neighbor) created based on $solution_1$. The proposed move from $solution_1$ to $solution_2$ happens based on a fitness function, i.e., if the fitness of $solution_2$ is superior to $solution_1$, the transition certainly will happen, and it probably occurs otherwise. The probability of a move when the fitness of the successor is less than the current solution is commensurate with the value of their fitness and the temperature. In more detail, let $f(solution_1)$ and $f(solution_2)$ be the fitness of the current solution and its successor, respectively. If $f(solution_2) > f(solution_1)$, transition to successor will happen. Otherwise, let $\Delta E = f(solution_2) - f(solution_1)$, the probability of moving to the successor $prob$ is calculated according to the following formula:

$$prob = min(e^{\frac{\Delta E}{T}}, 1) \tag{3}$$

where T is the temperature. It is evident that if $f(solution_2) > f(solution_1)$, then $e^{\frac{\Delta E}{T}} > 1$ and $prob = 1$. Thus, the move to $solution_2$ will certainly happen. Otherwise, the transition is reliant on ΔE and T, i.e., the greater ΔE or smaller T, the smaller chance to accept the move to $solution_2$ with lower fitness.

For the sake of clarity, given the maximum iteration number $iter_{max}$ and the initial temperature t, the pseudo-code of SA-based local search process is shown as follows:

```
1. solution_new = solution_elite.copy();
2. iter = 0;
3. while(iter < iter_max)
4.    t_iter = (1 − iter/iter_max) × t
5.    for (i = 0; i < solution_new.length, i = i + 1);
6.       if (rand(0, 1) < p_m^ls)
7.          solution_new[i] = 1 − solution_new[i];
8.       end if
9.    end for
10.   [winner, loser] = compete(solution_elite, solution_new);
11.   if (winner == solution_new)
12.      solution_elite = solution_new;
```

13. $iter = iter + 1$;
14.   else if $(rand(0, 1) < prob$
15. $iter = iter + 1$;
16.   else
17.   break;
18.   end if
19.   end if
20. end while

3.3   Pesudo-code of Compact Hybrid Evolutionary Algorithm

Step (1) Initialization:

1. gen=0;
2. for$(i = 0; i < num; i = i + 1)$
3. $PV[i] = 0.5$;
4. end for
5. generate an individual $solution_{elite}$ through PV

Step (2) Evolutionary Process:

6. $solution_{new} = localSearch(solution_{elite})$;
7. $[winner, loser] = compete(solution_{elite}, solution_{new})$;
8.   if $(winner == solution_{new})$
9. $solution_{elite} = solution_{new}$;
10. end if
11. for$(i = 0; i < num; i = i + 1)$
12.   if $(winner[i]==1)$
13. $PV[i] = PV[i] + updateValue$;
14.   if $(PV[i] > 1)$
15. $PV[i] = 1$;
16.   end if
17.   else
18. $PV[i] = PV[i] - updateValue$;
19.   if $(PV[i] < 0)$
20. $PV[i] = 0$;
21.   end if
22.   end if
23. end for
24. if(all PV bits either > 0.7 or < 0.3)
25.   for$(i = 0; i < num; i = i + 1)$
26.   if$((rand(0, 1) < p_m)$
27. $PV[i] = 1 - PV[i]$;
28.   end if
29.   end for
30. end if

Step (3) Termination:

31.   if (termination condition is met)
32.   stop and output ind_{elite};
33.   else
34.   gen=gen+1;
35.   go to Step 2;
36.   end if

4   Experimental Results

In order to study the effectiveness of CESA, we exploit Large Biomed[1] track from the Ontology Alignment Evaluation Initiative (OAEI 2017),[2] and CESA uses the following parameters which is determined in an empirical way:

- Numerical accuracy $numAccuracy = 0.01$;
- Maximum number of generations $maxGen = 3000$;
- Probability update value $updateValue = 0.02$;
- Initial temperature $t = 1.0$;
- Maximum iteration number in local search $iter_{max} = 10$;
- Mutation probability in local search $p_m^{ls} = 0.5$.

4.1   Results and Analysis

CESA's results in Table 1 are the mean values of thirty time independent executions, and the symbols P, R, and F in the table stand for precision, recall, and f-measure [5], respectively.

As shown in Table 1, CESA's f-measure is the best in task1, task2, and task3 among all the participants in OAEI 2017. In particular, CESA's precision is in generally high, which shows the effectiveness of our similarity measure. To conclude, CESA can effectively match the biomedical ontologies.

[1]http://www.cs.ox.ac.uk/isg/projects/SEALS/oaei/2017/.
[2]http://oaei.ontologymatching.org/2017.

Table 1 Comparison of CESA with the participants in OAEI 2017 on Large Biomed track

Task1: whole FMA and NCI ontologies

Systems	R	P	F
XMap*	0.85	0.88	0.87
AML	0.87	0.84	0.86
YAM-BIO	0.89	0.82	0.85
LogMap	0.81	0.86	0.83
LogMapBio	0.83	0.82	0.83
LogMapLite	0.82	0.67	0.74
Tool1	0.74	0.69	0.71
CESA	0.86	0.94	0.89

Task2: whole FMA and SNOMED ontologies

Systems	R	P	F
XMap*	0.84	0.77	0.81
YAM-BIO	0.73	0.89	0.80
AML	0.69	0.88	0.77
LogMap	0.65	0.84	0.73
LogMapBio	0.65	0.81	0.72
LogMapLite	0.21	0.85	0.34
Tool1	0.13	0.87	0.23
CESA	0.81	0.89	0.84

Task3: whole SNOMED and NCI ontologies

Systems	R	P	F
AML	0.67	0.90	0.77
YAM-BIO	0.70	0.83	0.76
LogMapBio	0.64	0.84	0.73
LogMap	0.60	0.87	0.71
LogMapLite	0.57	0.80	0.66
XMap*	0.55	0.82	0.66
Tool1	0.22	0.81	0.34
CESA	0.72	0.90	0.79

5   Conclusion

To effectively match the biomedical ontologies, a CESA-based biomedical ontology matching technique is proposed to determine the identical biomedical concepts,
which combines compact Evolutionary Algorithm and the Simulated Annealing algorithm to tackle the biomedical ontology matching problem. The experimental results
show that CESA outperforms the state-of-the-art ontology matchers on OAEI 2017's
large biomed track.

Acknowledgements This work is supported by the National Natural Science Foundation of China (No. 61503082), Natural Science Foundation of Fujian Province (No. 2016J05145), Scientific Research Foundation of Fujian University of Technology (Nos. GY-Z17162 and GY-Z15007) and Fujian Province Outstanding Young Scientific Researcher Training Project (No. GY-Z160149).

References

1. Bodenreider, O.: The unified medical language system (UMLS): integrating biomedical terminology. Nucl. Acids Res. **32**(suppl_1), D267–D270 (2004)
2. Kondrak, G.: N-gram similarity and distance. In: International Symposium on String Processing and Information Retrieval, pp. 115–126. Springer (2005)
3. Maedche, A., Staab, S.: Measuring similarity between ontologies. In: Proceedings of the 14th International Conference on Knowledge Engineering and Knowledge Management, pp. 251–263. Ischia Island, Italy (2002)
4. Metropolis, N., Rosenbluth, A.W., Rosenbluth, M.N., Teller, A.H., Teller, E.: Equation of state calculations by fast computing machines. J. Chem. Phys. **21**(6), 1087–1092 (1953)
5. Van Rijsbergen, C.J.: Foundation of evaluation. J. Doc. **30**(4), 365–373 (1974)
6. Xue, X., Pan, J.S.: An overview on evolutionary algorithm based ontology matching. J. Inf. Hiding Multimed. Signal Process. **9**, 75–88 (2018)
7. Xue, X., Wang, Y.: Optimizing ontology alignments through a memetic algorithm using both matchfmeasure and unanimous improvement ratio. Artif. Intell. **223**, 65–81 (2015)

Study on Information and Integrated of MES Big Data and Semiconductor Process Furnace Automation

Kuo-Chi Chang, Jeng-Shyang Pan, Kai-Chun Chu, Der-Juinn Horng and Huang Jing

Abstract The semiconductor process is designed to meet the requirements of photolithography, thin film, etching, cleaning of $12''$ 10 nm advanced commercial semiconductor manufacturing process, and the process integration system obtains the necessary process parameters and product measurement data. The relevant information is transmitted back to the MES system according to the production capacity and the product number of batch. This study first conducted an intelligent review of the $12''$ 10 nm furnace process tools. According to the SEMI specification, the production equipment interface standard for the furnace tube equipment should be discussed. The $12''$ 10 nm furnace module was selected for MES big data analysis. Due to the large number of process data, only the LPCVD process temperature distribution during the TEOS process was selected for distributed computation discussion. However, the yield results of this study can be maintained at 92%, while the equipment utilization rate can reach 97%. It is obvious that good results have been achieved.

Keywords Information integrated · MES · Big data · Furnace · Semiconductor process automation

K.-C. Chang (✉) · J.-S. Pan · H. Jing
Fujian Provincial Key Laboratory of Big Data Mining and Applications, Fujian University of Technology, Fuzhou, China
e-mail: albertchangxuite@gmail.com

J.-S. Pan
e-mail: jengshyangpan@fjut.edu.cn

H. Jing
e-mail: huangj@fjut.edu.cn

K.-C. Chu · D.-J. Horng
Department of Business Administration, Group of Strategic Management of National Central University, No. 300, Zhongda Rd., Zhongli District, 320, Taoyuan City, Taiwan
e-mail: kaykagy@gmail.com

D.-J. Horng
e-mail: horng@cc.ncu.edu.tw

© Springer Nature Singapore Pte Ltd. 2019
J.-S. Pan et al. (eds.), *Genetic and Evolutionary Computing*,
Advances in Intelligent Systems and Computing 834,
https://doi.org/10.1007/978-981-13-5841-8_70

Fig. 1 The proportion of the foundry equipment

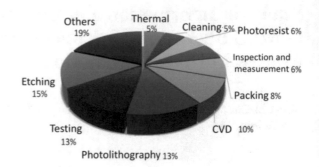

1 Introduction

The global semiconductor output value is expected to reach US$463.4 billion in 2018, which is about 12.4% higher than that in 2017. The semiconductor process is designed to meet the requirements of photolithography, thin film, etching, cleaning of 12″ 10 nm advanced commercial semiconductor manufacturing process, a large number of furnace process tools are used, and accounting for about 15% of the total process, visibility is very important of furnace (Fig. 1) [1].

However, the commercial 12″ 10 nm furnace process tools (Fig. 2), which mainly used in the LPCVD process for vertical type. The current LPCVD process is SiO_2 in the dielectric layer or Si_3N_4 in mask layer, the source materials are SiH_4 and O_2, but methane is very dangerous, so some change to tetraethoxydecane (TEOS). The source materials of Si_3N_4 include SiH_4, N_2, and NH_3. Therefore, the furnace process tool is one of the 12″ 10 nm important process equipment, and the furnace process tool is one of the important equipment's for CVD for the four process modules such as thin film, photolithography, etching, and implantation. Discussing the focus of the integration of manufacturing execution system (MES) information big data and semiconductor process furnace process automation, it will greatly improve the performance of MES, and further apply this study method to other process module applications, which will be integrated into the whole process and fit MES benefits [2].

2 Methodology and Study Procedure

The structure of this study is shown in Fig. 3a. Starting from the intelligent equipment of 12″10 nm LPCVD furnace process tools, since the LPCVD process parameters include temperature, pressure, flow rate, etc., this study reexamines whether the existing equipment can obtain the necessary process parameters, and Transfer to the process integration system, the deposition mold thickness of the process product is sampled by the inspection equipment after the batch product is completed, the relevant data is also transmitted to the process integration system. However, the semi-

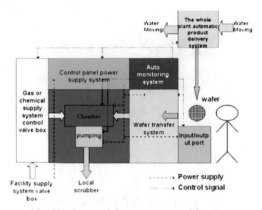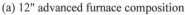

(a) 12" advanced furnace composition　　　　　　(b) Actual status of the machine

Fig. 2　12″ furnace machine integration hardware

conductor process includes photolithography, etching, thin film, and ion implantation in the main processes such as chemical and mechanical polishing; there are thousands of large and small process tools in the clean room of advanced semiconductor manufacturing FAB. Therefore, this study only discusses the LPCVD furnace equipment. The process integration system obtains the necessary process parameters and product measurement data. After that, the relevant information is transmitted back to the MES system according to the production capacity and the product number of batch. The MES then delivers the above information to the data cloud and starts the big data analysis regularly. When the process is biased or the product preparation changes, the system will return the big data and cloud. After the data analysis, the information is returned to the MES and then returned to the furnace process tools for control adjustment, according to which the intelligent production target is achieved. This study and experimental design adopts the waterfall model in the Hsinchu 12″ 10 nm process equipment (Fig. 3b) [3, 4].

This study is mainly used in the 12″ 10 nm LPCVD furnace control process. The process part is for $TEOS(Si(OC_2H_5)_4)$ process, which is applied to the SiO_2, and in order to replace the pyrophoric gas. The extremely dangerous methane to significantly reduce process hazards. TEOS was an organic molecule with high surface mobility, and TEOS-based CVD thin film deposition usually had good step coverage and trench filling capabilities, so TEOS is widely used in the deposition of oxides, including STI, sidewall space layers, PMD, and IMD components (Fig. 4).

However, the industry often uses the furnace process tools to carry out the process; when it is accidentally leaked by the process reaction system, the condensation will not spontaneously ignite and will be dispersed with the clean room air conditioner, and the reaction formula of TEOS is as shown in formula (1).

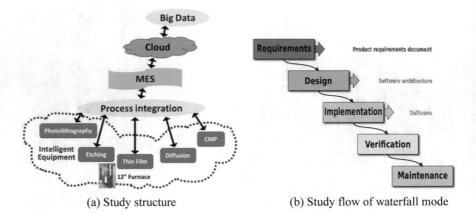

      (a) Study structure   (b) Study flow of waterfall mode

Fig. 3 The research process

Fig. 4 The dielectric and mask layer is in a critical position in CMOS

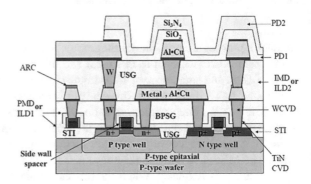

$$Si(C_2H_5O)_{4(g)} \rightarrow SiO_{2(s)} + C_2H_{4(g)} + 2H_2O_{(g)}. \tag{1}$$

For the reaction dynamic mathematical model, we know from (1) and Desu et al. (1989) that the reaction formula of SiO_2 is shown in Eqs. (2), (3), (4) and (5), where (6) Eqs. (7) and (8) are Eq. (5) [5].

1. TEOS decomposes in the gas phase:

$$Si(C_2H_5O)_4 \xrightarrow[k_g]{} I + R. \tag{2}$$

2. Adsorption reaction:

$$I \xrightarrow[k_a]{} I^*. \tag{3}$$

3. Surface chemical reaction:

$$I^* \xrightarrow{K_d} SiO_2 + R +^* . \tag{4}$$

4. The deposition rate dynamic type is

$$r = \frac{k_d k_a (k_g C)^{0.5}}{1 + k_a (k_g C)^{0.5}} \left(\frac{mole}{cm^2 \cdot s}\right), \tag{5}$$

where r is the deposition rate $\left(\frac{mole}{cm^2 \cdot s}\right)$, C is the gas concentration of TEOS $\left(\frac{mole}{cm^2}\right)$, R is the gas constant, and Tr is reaction temperature (K).

Than

$$K_g = 1.38 \times 10^4 \exp(-299240/8.314Tr) \left(\frac{mole}{cm^2}\right). \tag{6}$$

$$K_a = 1.14 \times 10^{10} \exp(-21422/8.314Tr) \left(\frac{cm^3}{mol}\right). \tag{7}$$

$$K_d = 4.74 \times 10^{-8} \exp(-12259/8.314Tr) \left(\frac{mol}{cm^2 \cdot s}\right). \tag{8}$$

For MES big data analysis used distributed computation in study, furnace temperature has the performance of formula (9) at different heating rates. Here p is the sum of all constant thermal powers in the calorimeter, due to stirring, evaporation, joule heating by the resistance thermometer, ε is the energy equivalent of the calorimeter, K is the thermal leakage modulus, and Θj is the effective jacket temperature, Θ approaches with increasing time.

$$g = \frac{d\Theta}{dt} = u + k(\Theta j - \Theta), \text{ where } u = p/\varepsilon. \tag{9}$$

3 12″ 10 nm Furnace Process Tools Intelligent and Communication Interface Analysis

In the Industrial Production Plant of Industry 4.0, the future smart factory has independent and independent capabilities in every production process and all operating equipment, which can automate the production line operation. And each device can communicate with each others, and then monitor the surrounding environment in real time, then find problems at any time to eliminate them, and be more flexible in production process to meet the requirement of different customers. Therefore, this study first conducted an intelligent review of the 12″ 10 nm furnace process tools. According to the SEMI specification, the production equipment interface standard

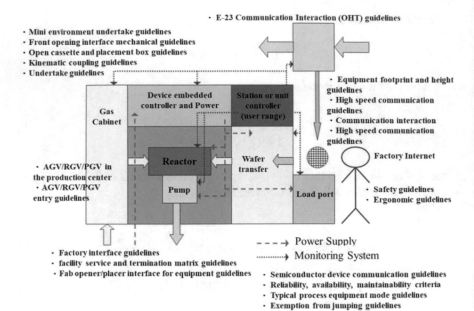

Fig. 5 300 mm wafer production equipment interface standard requirements

for the furnace tube equipment should be discussed. The result is shown in Figs. 5 and 6. These guidelines are very important benchmarks for automation and communication interface for MES and big data in the FAB. Figure 7 is the information operation structure in the FAB [6–10].

4   12″ 10 nm Furnace Process Tools Integration Information and MES Big Data Analysis

In this study, the 12″ 10 nm furnace module was selected for MES big data analysis. Due to the large number of process data, only the LPCVD process temperature distribution during the TEOS process was selected for distributed computation discussion. Figure 8 shows the furnace temperature in the batch process. The five heating zones are differently distributed. Intuitively, the same furnace temperature control should be the same for the same process, and the thin film deposition effect will be the same. Under the big data processing, it can directly master dozens of furnace machines and the specific heating zone heating equipment operation. Therefore, using the big data comparison of each heating zone controller, it will find the difference between different power inputs show in Fig. 9. After the process is completed, the control of wafer cooling is also easy to cause cracks. For PID control, big data can also be used for analysis shown in Fig. 10. However, the yield results of this study can be maintained

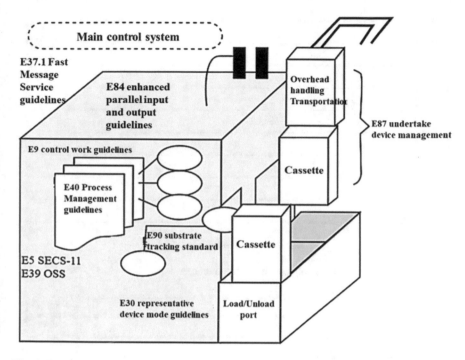

Fig. 6 SEMI device automation software standard function relationship used by major semiconductor process tools

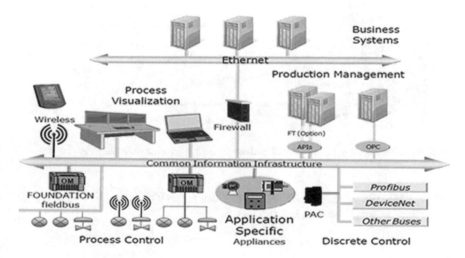

Fig. 7 MES structure of 12″ FAB in this study

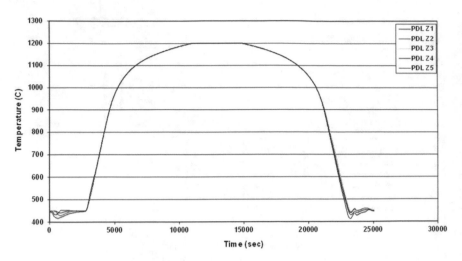

Fig. 8 The furnace temperature big data in the batch process

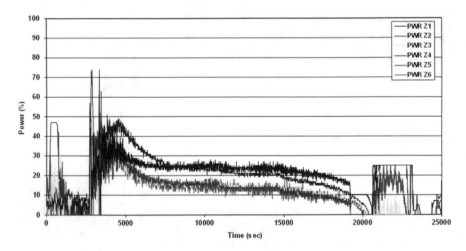

Fig. 9 Each heating zone controller will find the difference between different power inputs

at 92%, while the equipment utilization rate can reach 97%. It is obvious that good results have been achieved [11, 12].

5 Conclusions

The semiconductor process is designed to meet the requirements of photolithography, thin film, etching, cleaning of 12″ 10 nm advanced commercial semiconductor manufacturing process. This study first conducted an intelligent review of the 12″

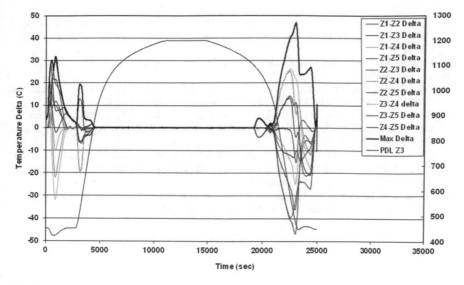

Fig. 10 PID control of big data can also be used for analysis

10 nm furnace process tools. According to the SEMI specification, the production equipment interface standard for the furnace tube equipment should be discussed. The 12″ 10 nm furnace module was selected for MES big data analysis. Due to the large number of process data, only the LPCVD process temperature distribution during the TEOS process was selected for distributed computation discussion. However, the yield results of this study can be maintained at 92%, while the equipment utilization rate can reach 97%. It is obvious that good results have been achieved.

References

1. Lu, C.C., Chang, K.C., Chen, C.Y.: Study of high-tech process furnace using inherently safer design strategies (IV). The advanced thin film manufacturing process design and adjustment. J. Loss. Prevent. Proc. **40**, 378–395 (2016)
2. Chen, C.Y., Chang, K.C., Wang, G.B.: Study of high-tech process furnace using inherently safer design strategies (I) temperature distribution model and process effect. J. Loss. Prevent. Proc. **26**, 1198–1211 (2013)
3. Chen, C.M., Wang, K.H., Wu, T.Y., Pan, J.S., Sun, H.M.: A scalable transitive human-verifiable authentication protocol for mobile devices. IEEE T. Inf. Foren. Sec. **8**(8), 1318–1330 (2013)
4. Desu, S.B.: Decomposition chemistry of tetraethoxysilane. J. Am. Ceram. Soc. **72**(9), 1615 (1989)
5. Chen, C.Y., Chang, K.C., Lu, C.C., Wang, G.B.: Study of high-tech process furnace using inherently safer design strategies (II) deposited film thickness model. J. Loss. Prevent. Proc. **26**, 225–235 (2013)
6. Chen, C.M., Xu, L., Wu, T.Y., Li, C.R.: On the security of a chaotic maps-based three-party authenticated key agreement protocol. J. Netw. Intell. **1**(2), 61–66 (2016)

7. Nguyen, T.T., Horng, M.F., Shieh, C.S., Dao, T.K.: An energy-based cluster head selection algorithm to support long-lifetime in wireless sensor networks. J. Netw. Intell. **1**(1), 23–37 (2016)
8. Boehm, B.: A spiral model for software development and enhancement. Computer **21**(10), 774–778 (1994)
9. Chen, C.H., Chen, T.Y., Wang, D.J., Li, Y.F.: Multipath flatted-hexagon search for block motion estimation. J. Inf. Hiding Multimed. Signal Process. **1**(2), 110–131 (2010)
10. Huang, D.Y., Lin, C.J., Hu, W.C.: Learning-based face detection by adaptive switching of skin color models and AdaBoost under varying illumination. J. Inf. Hiding Multimed. Signal Process. **2**(3), 204–216 (2011)
11. Liu, X., Li, J.B., Hu, C., Pan, J.S.: Deep convolutional neural networks-based age and gender classification with facial images. In: First International Conference on Electronics Instrumentation & Information Systems (EIIS) (2017)
12. Weng, S.W., Pan, J.S., Li, L.D., Zhou, L.Z.: Reversible data hiding based on an adaptive pixel-embedding strategy and two-layer embedding. Inf. Sci. **369**, 144–159 (2016)

An Ontology-Based Recommendation System for ADAS Design

Hsun-Hui Huang, Horng-Chang Yang and Yongjia Yu

Abstract Advanced driver-assistance systems (ADAS) are an important component in a vehicle for these systems to actively improve driving safety. Although technical development for ADAS is already mature, there are still a few areas that can be improved. In particular, the design of newer ADASs are mainly based on the experience and imagination of car designers and their tests are usually based on hypothetical situations and field models. Without user experience data, it is difficult for car makers to refine and improve their ADAS designs effectively and systematically. In order to help designers optimize their designs and shorten design cycles, a framework of collaborative filtering recommendation is proposed, in which domain ontologies, text mining, and machine learning techniques are used to produce multimedia summaries from data repositories for queries of real accidents, or design issues. The recommendation system framework aims to help designers and car makers improve their efficiency and produce safer vehicles. The data and knowledge bases constructed can be used as the basis of tutorial programs for new entrants and the car makers can reduce their managerial cost and manpower needs.

Keywords ADAS · Intelligent recommendation · Ontology

H.-H. Huang (✉)
Tajen University, No. 20, Weixin Rd., Yanpu Township,
Pingtung County 90741, Taiwan, ROC
e-mail: beatrice@tajen.edu.tw

H.-C. Yang
National Taitung University, Taitung City, Taiwan, ROC
e-mail: hcyang@nttu.edu.tw

Y. Yu
School of Software, Changzhou College of Information Technology, Changzhou, China
e-mail: android_yyj@126.com

© Springer Nature Singapore Pte Ltd. 2019   679
J.-S. Pan et al. (eds.), *Genetic and Evolutionary Computing*,
Advances in Intelligent Systems and Computing 834,
https://doi.org/10.1007/978-981-13-5841-8_71

1 Introduction

Driving safety is always essential for car design and manufacture. With recent advances of IoT technologies, an increasing number of modern vehicles have been equipped with advanced driver-assistance systems (ADAS), including 3D Surround View, Lane Departure Warning Systems, Forward Collision Warning Systems, and Parking Assistant Systems, etc. These ADASs rely on information from various data sources such as radar, camera, in-car networking, even vehicle-to-vehicle (V2V) or vehicle-to-infrastructure systems. As ADASs can improve driving safety, car manufacturers have invested in the development of these systems, and usually include a few of these systems in their product cars for appealing to potential customers [1]. Although technical development is already mature, there are still a few areas for improving driving safety. In particular, the design of newer ADASs are mainly based on the experience and imagination of car designers and their tests are usually based on hypothetical situations and field models. For example, a camera-based ADAS may require parameters setting due to conditions such as lighting. But the setting may not cover all the situations. Without feedback, the real situation and problems encountered by the users for long-term use are unknown and optimal user experience is not ensured. Even with questionnaire studies, which usually is not tailored for particular problems, precise information is difficult to obtain. Hence, it is not surprising to see car recalls by manufacturers.

To ensure optimal user experience and continuous driving safety improvement, techniques from machine-to-machine platforms can be applied in the case of an exception. Mainly, the sensor data of the ADAS involved in an exception are recorded and sent for diagnosis and tuning so that the designers can refer to the data for newer design and improvement. In this paper, a framework for assistance to ADAS design is proposed. Domain ontologies, text mining, and machine learning techniques is used to produce multimedia summaries from data repositories for queries of real accidents, or design issues. The recommendation system framework aims to help designers and car makers improve their efficiency and produce safer vehicles. The data and knowledge bases constructed can be used as the basis of tutorial programs for new entrants and the car makers can reduce their managerial cost and manpower needs.

2 Related Work

To design complex systems such as automobiles, associated subsystems are designed and integrated into the target systems. If the subsystems are designed independently, the integrated target systems may not function optimally or not work at all. However, in the beginning of the collaborative design, if the responsible designers know the requirements precisely, the difficulty of system integration will be mitigated [2]. For improving the design of existing products, the user experience is indispensable.

With IoT technologies, the user experience in the form of sensor data is recorded and sent to clouds for analysis. To use the data efficiently, knowledge representation is essential. Common knowledge representations include semantic networks, systems architecture, and frames and ontology [3].

2.1 Ontology-Based Context Modeling

To achieve an intelligent system, it is crucial that information and knowledge are shared among components. An ontology is a representation of knowledge, consisting of formal naming, and definition of the categories, properties, and relations between the concepts, data, and entities that substantiate one, many, or all domains [4, 5]. In the ADAS navigation context, an ontology can be formulated about the vehicle, perceived entities, context (map information), and a description of the entities' interaction [6]. Also, ontology-based computer vision techniques can be used for improving entities recognition [7, 8]. In [9], Town proposed a content-based image retrieval system that allows the users to search image databases using an ontological query language; also, he extended the notion of ontological languages to video event detection. Using deep learning CNN to extract features, Ye et al. developed Event-Net, a video event ontology that organizes events and their concepts into a semantic structure [10]. They suggested that their structure is potential for event retrieval and browsing. Apparently, to achieve context-aware ADAS, ontology is crucial in terms of integration of heterogeneous information sources. Indeed, the past years have seen researchers proposed ontology-based frameworks for such an application [6, 11].

In the context of collaborative design, a well-defined knowledge representation is essential. The approach of integrating ontology and multi-agent techniques to collaborative design is getting attractive. Feng used OWL to represent the knowledge of ship design in a multi-agent environment [12]. Addressing issues in educational project collaboration, Enembreck et al. offer a multi-agent system in which cognitive autonomous agents use ontologies to guide the interactions among the students [13].

2.2 Recommendation Models

According to [14], recommender systems are usually classified into three categories, based on how recommendations are made, (i) Content-based recommendations (ii) Collaborative recommendations, and (iii) Hybrid approaches. A recommendation system employs a utility function $u(c_i, p_j)$, where c_i is a customer, p_j, unseen by c_i, is a product in the product set $\mathcal{P}$. It recommends

$$p_t = \underset{p_j \in \mathcal{P}}{\operatorname{argmax}} u(c_i, p_j)$$

to c_i.

Specifically, in context-based recommendations, the utility function of p_j unseen by c_i is evaluated using the products $\{p_k\} \subset \mathcal{P}$, already rated by the target customer c_i. This approach is based on techniques from information retrieval, in which a product p_j is represented by a vector of content/attribute weights $content(p_j) = (w_1, w_2..., w_n)$. Similarly, $profile(c_i) = (w_1, w_2..., w_n)$, the preferences of customer c_i, is also represented by this content vector aggregated by the rated content vectors of $\{p_k\}$. Thus, the utility function, $u(c_i, p_j)$, is defined by $score(profile(c_i), content(p_j))$. Any measure of similarity of the two vectors such as $cosine$, L_2 distance can be used as the $score$ function. Apparently, accuracy of customer profile is crucial for content-based recommendations. A well-known method for the creation of a customer profile is Rocchio's formula, an adaptation of relevance feedback to text categorization. Also, combining an ontology for improving accuracy of customer profiles was proposed by several researchers [15].

Collaborative filtering recommendation systems evaluate the utility function of a product p_j for customer c_i, $u(c_i, p_j)$, by aggregating the previous ratings of p_j of a set of customers $\{c_k\} \subset C$ who are similar to c_i. Obviously, how to define "similar" customers is crucial for collaborative recommendation. The methods for estimating the similarity between two customers are either based on heuristic or statistics. $cosine$ is a common heuristic similarity measure while correlation and clustering methods are of statistics.

To avoid some limitations of content-based and collaborative recommendation systems, hybrid approaches that combine content-based and collaborative methods are proposed. They can be categorized into four different ways to combine the two categories of methods. For a overview, see [16].

3 An Ontology-Based Recommendation Scheme for ADAS

In order to improve the robustness of an existing ADAS and the efficiency of new ADAS design, an ontology-based recommendation system that collects the user experiences of ADAS and various design knowledge of ADAS is proposed. Figure 1 is the workflow diagram of the system.

The system consists of two subsystems, namely (a) the vehicle situation exception solution recommendation subsystem and (b) the knowledge learning and management subsystem. The operations of the vehicle situation exception solution recommendation subsystem is as follows:

- Real-time sensor record links of the vehicle are sent to this subsystem when an exception occurs or a request by the driver.
- The subsystem processes the multimedia data and searches for existing or collaborative recommendation solutions.
- The multimedia data of the solutions are retrieved, summarized and sent to the responsible users.

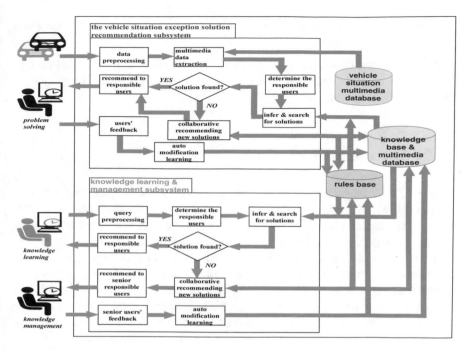

Fig. 1 The proposed collaborative recommendation system work flow diagram

- The users examine the received multimedia reports. They may modify the solutions and provide feedback to the subsystem.
- The subsystem applies machine learning techniques to improve the ontology and the rule/inference bases.

In the vehicle situation exception solution recommendation subsystem, knowledge bases are developed on the Protégé platform of Stanford Medical Informatics. OWL with the XML format is used for the representation of knowledge. Inferring rules are written by using the expert system development language Java Expert System Shell (JESS).

Illustrated by Fig. 1, the knowledge learning and management subsystem operates as follows:

- Newcomer users ask the system for questions about the vehicle situation exception solution.
- The subsystem responds with stored answers for questions already seen before or recommendations/warnings for those unseen before.
- The recommendations are sent to the responsible veteran users of the knowledge management system for verification.
- The veteran users examine the received questions and recommendations. They may modify the recommendations and provide feedback to the subsystem.

- The subsystem applies machine learning techniques to improve the ontology and the rule/inference bases.

Apparently, the subsystem relies on text mining techniques for question semantic analysis and selection. Also, task-technology fit (TTF) is used to evaluate the operations of the system.

4 Discussion and Conclusions

To help ADAS designers optimize their designs and shorten design cycles, a framework of collaborative filtering recommendation is proposed, in which domain ontologies, text mining, and machine learning techniques are used to produce multimedia summaries from data repositories for queries of real accidents, or design issues. The recommendation system framework employs machine learning techniques to refine and furnish its knowledge base from the feedback of veteran designers. However, current methods suffer from the problem of data sparsity. Unfortunately, this can only be solved by the time when data is abundant. However, the data and knowledge bases constructed can be used as the basis of tutorial programs for new entrants and the car makers can reduce their managerial cost and manpower needs.

Acknowledgements This work is supported by Top-notch Academic Programs Project of Jiangsu Higher Education Institutions (TAPP) under Grant No. PPZY2015A090.

References

1. Ziebinski, A., Cupek, R., Erdogan, H., Waechter, S.: A survey of ADAS technologies for the future perspective of sensor fusion. In: Nguyen, N., Iliadis, L., Manolopoulos, Y., Trawiński, B. (eds.) Computational Collective Intelligence. ICCCI 2016. LNCS, vol. 9876. pp. 135–146, Springer, Cham (2016)
2. Klein, M., Sayama, H., Faratin, P., Bar-Yam, Y.: The dynamics of collaborative design: insights from complex systems and negotiation research. Concurr. Eng. **11**(3), 201–209 (2003)
3. Oberle, D., Guarino, N., Staab, S.: What is an ontology? In: Handbook on Ontologies, 2nd edn. Springer (2009)
4. Gruber, T.R.: Toward principles for the design of ontologies used for knowledge sharing. Int. J. Hum.-Comput. Stud. **43**(5–6), 907-928 (1995)
5. Davis, R., Shrobe, H., Szolovits, P.: What is a knowledge representation? AI Mag. **14**(1), 17–33 (1993)
6. Armand, A., Filliat, D., Ibañez-Guzman, J.: Ontology-based context awareness for driving assistance systems. In: 2014 IEEE Intelligent Vehicles Symposium Proceedings, pp. 227–233. Dearborn, MI (2014)
7. Fiorini, S.R., Abel, M.: A Review on Knowledge-Based Computer Vision (2010)
8. Maillot, N., Thonnat, M., Boucher, A.: Towards ontology-based cognitive vision. Mach. Vis. Appl. **16**(1), 33–40 (2004)
9. Town, C.: Ontological inference for image and video analysis. Mach. Vis. Appl. **17**(2), 94–115 (2006)

10. Ye, G., Li, Y., Xu, H., Liu, D., Chang, S-F.: EventNet: a large scale structured concept library for complex event detection in video. In: Proceedings of 23rd ACM International Conference on Multimedia, pp. 471–480 (2015)
11. Zhao, L., Ichise, R., Mita, S., Sasaki, Y.: An ontology-based intelligent speed adaptation system for autonomous cars. In: Semantic Technology, pp. 397–413. Springer (2014)
12. Feng, X.: Semantic web technology applied for description of product data in ship collaborative design. In: Cooperative Design, Visualization, and Engineering, pp. 133–136. Springer, Berlin, Heidelberg (2009)
13. Enembreck, F., Thouvenin, I., Abel, M.H., Barthes, J.P.: An ontology-based multi-agent environment to improve collaborative design. In: 6th International Conference on the Design of Cooperative System, COOP, vol. 4, pp. 81–89 (2004)
14. Balabanovic, M., Shoham, Y.: Fab: content-based, collaborative recommendation. Comm. ACM **40**(3), 66–72 (1997)
15. Lops, P., Gemmis, M., Semeraro, G.: Content-based recommender systems: state of the art and trends. In: Ricci, F., Rokach, L., Shapira, B., Kantor, B. (eds.) Recommender Systems Handbook, pp. 73–105. Springer US (2011)
16. Adomavicius, G., Tuzhilin, A.: Toward the next generation of recommender systems: a survey of the state-of-the-art and possible extensions. IEEE Trans. Knowl. Data Eng. **17**(6), 734–749 (2005)

Application of Big Data Technology in Student Management Decision

Jianjian Luo

Abstract Using big data technology in student affairs management has become the trend of the times. In addition to national policy requirements, the application is not only the need for normative management, but also the call for scientific decision-making. In this application, we should develop standard specifications for big data, build a platform for the application of the big data, and make deep mining from the big data. Based on the application situation, we should continuously strengthen our knowledge of big data, expand the sharing of the big data, and try a variety of big data technologies to improve the work of data collection, storage, processing, and application.

Keywords Big data · Student management · Decision-making · Application

1 Introduction

In recent years, with the rapid development of information technology, using big data technology in the affairs of education management and student management has become the trend of the times. Concepts such as "smart learners" and "digital learners" [1] have also come into being. However, for many college management practitioners, there are still some confusions and puzzles about how to apply big data technology effectively and efficiently in the process of student management and decision-making. Therefore, according to the educational purposes of Changzhou Institute of Information Technology, we try to promote the application of big data technology in the affairs of education management and student management for higher vocational colleges.

J. Luo (✉)
Changzhou College of Information Technology, Changzhou 213164, China
e-mail: 46135651@qq.com

© Springer Nature Singapore Pte Ltd. 2019
J.-S. Pan et al. (eds.), *Genetic and Evolutionary Computing*,
Advances in Intelligent Systems and Computing 834,
https://doi.org/10.1007/978-981-13-5841-8_72

2 Why Use Big Data Technology in Applications

2.1 The Needs for Management

Managing students has long been considered a tedious task. In addition to the complicated content, there are three characteristics in student management for decision-making: the first is the judgment of most of the work is qualitative rather than quantitative; the second is lack of comprehensiveness in the cases of student affairs; the third is that there will be significant differences in opinions and suggestions when different managers evaluate the same work. These existed characteristics have three drawbacks: First, evaluating the effectiveness of management lacks credibility and is easy to cause controversy; second, even if the problem of the individual is found, it is difficult for the whole form an objective judgment; and it is easy to lead to frequent revisions and repeated changes for decision making by empiricism.

2.2 The Call for Scientific Decision-Making

Scientific decision-making means that the decision-making process involves the use of data and logic as opposed to hunch and intuition. Scientific decision-making requires all factors based on facts and data, which comes from precision measurements or organizational assessments. Especially it is needed to collect data extensively to make major decisions in big data. In practical application, the management knowledge accumulated in the concept is sorted and transformed into data knowledge, to form a data center knowledge base by using big data technology. The data is collected, classified, extracted, cleaned, and converted, and the data required for management is extracted and stored in a new combination form in the academic management data warehouse. On the basis of the data warehouse, a general and subject-oriented analysis platform is established to realize statistical analysis from different dimensions, and finally to provide big data support for decision makers at different levels.

2.3 The Requirements of National Policies

The China government has also realized the role of big data in the management of colleges and universities including student work in recent years. The "Guidelines for the Construction of Digital Campuses in Vocational Colleges" [2] clarifies that "the core content of digital campus construction in vocational colleges is the technical system to support the vocational education teaching model and management service system." "The Big data is a clue to support students to study and live in school, and support the student management department to carry out various management tasks

[3]." "Guiding Opinions on Further Promoting the Development of Vocational Education Informatization" [4] proposes: Accelerate the construction and application of management service platform. Encourage vocational colleges to build a big data platform integrating administration, teaching, scientific research, student, and logistics management, and support the school to implement school-enterprise cooperation, project management, post-internship, human resources information, employment information, etc. Education Informatization 2.0 Action Plan is an effective way to accelerate the modernization of education. Without informatization and big data technology, there is no modernization. Education informatization is the basic connotation and distinctive feature of education modernization. It is "educational modernization 2035" key content and important signs.

3 How to Achieve the Application

3.1 Establish a Standard for Big Data Standards

In order to achieve the high integration and sharing of student big data, colleges, and universities must be based on the requirements of the "Higher Education Management Information Standards" promulgated by the Ministry of Education. At the same time, combined with the actual situation of the school, the first is to establish standardized technical information codes and data subsets of college students. Ensure that each valid information of the student can be coded; the second is to scientifically compile the codes of each department, to ensure that the codes of each department are unique, and on this basis, compile each business department's code for the effective information of students.

Changzhou Information Vocational and Technical College broke the information islands. Originally, the entire data scattered in the admissions department, the Academic Affairs Office, the second-level college and other departments, and formed the basic data including more than 30 fields such as student status and family situation. Business data including more than 40 fields including student rewards and punishments, funding status, and quality education. By combining these data with the applications of the Smart Campus, it is convenient for student managers to manage and serve students throughout the process.

In the process of building a university student management big data platform, we must attach great importance to the participation of all employees. From the perspective of the practice process, all students are both big data producers and big data consumers. It is necessary for schools to establish a professional team with informational literacy and data standardization awareness. This team needs both management personnel and technical personnel. It can include school functional department management personnel, secondary school management personnel, counselors, class teachers, and specialized personnel in technical support departments such as information centers. In addition, it is necessary to construct an information management system, clearly define the rights and responsibilities, implement responsibility according to the data category, set up special positions, undertake data collection, data entry, data maintenance, etc., and develop comprehensive big data. Management system, standardizing the collection, storage, processing, and use of student data [5].

3.2 Develop Big Data Platform Applications

It is necessary to use innovation, coordination, green, openness, and sharing as the working concept, to serve teachers and students, demand-driven, innovation-driven, and joint construction and sharing as the work goal, systematically analyze daily key business, and design various student management information applications step by step. Efforts will be made to develop a student service center online platform based on student big data and a student worker management platform to realize the analysis and guidance role of student big data in student self-development decision-making and scientific management decision-making.

Student Service Center Online Service Hall: Based on the online student office hall, the reconstruction and deployment of a one-stop online consultation and service platform. Based on the student's big data, the new online platform is based on the student's perspective system to sort out specific business activities such as apartment affairs, award-winning affairs, academic affairs, and community affairs, and clearly show the business processes and institutional basis for various businesses, to provide students with timely, convenient and diverse learning and living services.

Student Worker Management Platform: Research and development of student worker management platform for the characteristics of student workers' daily affairs and part-time class teachers. Based on the student's big data, the class teacher can easily and conveniently query and count the student information, overall situation, development trend, etc., of the class, and can also query the student business process management, initiate specific affairs, and participate in student management through the management platform.

Other related business applications: Based on the principle of demand-oriented and step-by-step implementation [6], the information center has been developed to develop a number of business applications that are convenient for students, including: new student registration system, student insurance participation, student work-study, online student repair, and student quality. Education, online parent conference, business applications being developed include: student leave, student accommodation resource management, daily check of student dormitory, student apartment access control, etc., and counselor work resource sharing and display platform. The gradual launch of these applications will greatly assist in analyzing the status of student work, improving the effectiveness of student work, and improving the quality of student work.

3.3 Implement Big Data Deep Mining

Through the establishment of big data and the development of the application platform, colleges, and universities can realize the deep mining of user data in the process of student management decision-making. For example, in the work of identifying poor students, data mining technology is used to analyze the students' basic life consumption data and the database of college-impoverished students. The school's basic life consumption data comes from the college students' canteen dining, the data is large, and the data is real. Although these water data are limited to their consumption in the school, this part of consumption water belongs to basic living consumption, which is the main aspect of college students' consumption. To a large extent, it can reflect the consumption behavior of students in school, and can objectively reflect the economic consumption level of students in school. The database of impoverished students comes from the database of poor students in colleges and universities. According to the materials of poor students, the students are surveyed and filled out the questionnaires, including the student's name, student number, student ID number, department, major, class, student source. The total population of the family, the family, the holding of mobile phone brands, and the holding of computer brands. This part of the data is verified by the team and can objectively reflect the student's family economic level. By using the above two parts of data, it can provide data support for poor students' identification work and provide decision-making basis for decision makers.

3.4 Application Management Examples

In the exploration of big data technology applied to student management, Internet companies play an important role, and their products keep up with and even lead the actual needs of higher vocational colleges. "Student Portrait" [7] is a big data product that serves the management and education of college students. It optimizes the traditional management method with "big data + workflow" to provide a total solution for the management of academics and counselors. The application is divided into two levels: basic application and big data application. The basic application solves the information construction, and the big data application supports the management decision. The data generated by each application is transferred through the data center to realize data sharing and interworking. Focusing on the specific problems of students' safety, academics, life, employment, and other aspects during their schooling, of historical loss of events; mining attention to the loss of time period; focus on the status of lost high-frequency students. For students' comprehensive development assessment, it is needed to aggregate students' school data, calculate students' comprehensive development evaluation scores, and provide objective basis for student scholarship evaluation and counselor evaluation. It can flexibly configure the evaluation index system, add, delete and change the evaluation indicators, and flexibly set the calculation formula; it can quickly calculate the evaluation score, automatically fill in the evaluation materials online, match the evaluation scores, and automatically calculate the total score of the comprehensive evaluation; functional analysis of student development assessment score trends, personalized guidance for students' future development.

4 Application Note

4.1 Continue to Strengthen the Concept of Big Data

Despite the remarkable progress in general, compared with developed countries in Europe and the US, China universities still have a lot of room for expansion in the concept of student big data. The big data management of Europe and the US universities mainly relies on the construction of student information management system. Its system construction emphasizes data management and light transaction processing, which means that student big data has become the core of student work information system. For example, the University of California's decision-making system views the campus as a unified, integrated, all-ecological environment that integrates and visualizes existing student big data, as well as mining, analyzing, and analyzing information trends. Provide a basis for scientific decision-making. The North Carolina State University School integrated campus resources and established a virtual computer lab to build a cloud computing-based system big data platform. The Big Data platform effectively integrates all campus resources, monitors and analyzes all business system operations, and supports student management decisions.

4.2 Continue to Expand Big Data Sharing

It must be noted that some colleges and universities have developed information technology in the process of informatization construction. Each department develops or purchases a business management system according to different technology foundations, resulting in difficulties in effectively connecting departments and systems, poor data versatility, and difficulty in resource sharing, resulting in extremely low information utilization. The information in the traditional management mode is too scattered, resulting in an increase in the number of repetitive labor in the information update process, resulting in extremely high errors in the student information and the accuracy and completeness of the data. In the process of data exchange, data quality cannot be effectively controlled, and data loss, data maintenance difficulty, and data quality frequently occur. In addition, in the process of rapid expansion of colleges and universities, the amount of information has increased to geometric multiples, and data updates are faster than ever. Some universities have insufficient capacity in the big data management department and the management model is slow to update. Traditional management concepts and methods are difficult to meet the needs of big data development.

4.3 Continue to Enrich the Means of Big Data

In order to improve the level of student work management decision-making and give full play to the role of big data in student management, we must fully consider network technique in the construction process, and fully utilize the Internet of Things perception technology such as RFID, QR code, video surveillance, etc. The monitoring and comprehensive perception of various types of student data in campus management, the integration, analysis, and processing of perceived big data, and intelligent integration with business processes, and then actively respond. At the same time, we must actively analyze the emerging student data that is often neglected under the traditional management mode, and explore the student management rules behind the massive data to improve the individualized management level of students. Of course, in the process of applying these new technologies, we must pay attention to the safety of students' big data, including ensuring the safety of student management decision-making foundation and process through various technical means such as authority control and network access control.

5 Conclusion Remarks

The extensive use of big data technology is effectively improving the level of student management in China universities, and more scientific management decisions are calling for deeper big data technologies. Big data is not a fashion, but an innovation that is expected by technology push, demand pull, and change in mission philosophy. Faced with the status quo and prospects of big data technology application in the field of student management, it is necessary for Chinese universities to further increase their investment in big data, further understand the value of big data, and further enhance the ability to collect, store, process, and use big data.

Acknowledgements This work is supported by Top-notch Academic Programs Project of Jiangsu Higher Education Institutions (TAPP) under Grant No. PPZY2015A090, and Jiangsu top six talent summit project under Grant No. XXRJ-011.

References

1. "Guidelines for the Construction of Digital Campus in Vocational Colleges" (Vocational Education and Adult Education Correspondence Center [2015] No. 1)
2. "Guiding Opinions on Further Promoting the Development of Vocational Education Informatization" (Vocational education and adult education [2017] No. 4)
3. Education Informatization 2.0 Action Plan (Ministry of Education [2018] No. 6)
4. Wang, Y.: Deepen the education informatization to provide support for education modernization. National Education News. 2014-01-15 (003)
5. Cheng, J., Mei, W.: Thinking and practice of smart campus construction. Inf. Secur. Technol. (05) (2013)
6. Huang, Y., Qi, L.: Technology architecture based on network smart campus and its implementation. Southeast Acad. (06) (2012)
7. Jiang, D.: Let knowledge be more intelligently shared on campus. China Educ. Netw. (11) (2011)

Part XVII
Social Network and Stock Analytics

Detection of Opinion Leaders in Social Networks Using Entropy Weight Method for Multi-attribute Analysis

Qiu Liqing, Gao Wenwen, Fan Xin, Jia Wei and Yu Jinfeng

Abstract Opinion leaders are those users who have great influence, whose emergence has an important effect on social networks. Thus, the recognition of opinion leaders can contribute to a comprehensive understanding of the development trend of information dissemination and other applications. This paper proposes a new opinion leaders detection algorithm, called MAA algorithm, which uses Entropy Weight Method to comprehensively multiple attributes analysis including location attributes, distance attributes, and strength entropy. Specifically, our main contributions are the following two aspects: (1) Multiple attributes analysis is conducted to measure the influence of a user in the dissemination of information; (2) Entropy Weight Method is used to comprehensively evaluate multiple attributes, and calculate the influence weight, respectively. Experimental results illustrate that the proposed algorithm is superior to other traditional algorithms.

Keywords Opinion leaders · Social network · Entropy weight method · Multiple attributes analysis

Q. Liqing (✉) · G. Wenwen · F. Xin · J. Wei · Y. Jinfeng
Shandong Province Key Laboratory of Wisdom Mine Information Technology,
College of Computer Science and Engineering, Shandong University of Science and Technology,
Qingdao 266510, China
e-mail: liqingqiu2005@126.com

G. Wenwen
e-mail: 1141922180@qq.com

F. Xin
e-mail: 564353949@qq.com

J. Wei
e-mail: 1456514769@qq.com

Y. Jinfeng
e-mail: 576476745@qq.com

© Springer Nature Singapore Pte Ltd. 2019
J.-S. Pan et al. (eds.), *Genetic and Evolutionary Computing*,
Advances in Intelligent Systems and Computing 834,
https://doi.org/10.1007/978-981-13-5841-8_73

1 Introduction

Social networks are a relatively stable fabric of society formed by mutual interaction among communitarians, which have complex network structure and information dynamic communication mechanism [1]. With the rapid development of computer technology, social networks have become one of the main channels. Nevertheless, the heterogeneity of social networks' nodes determines the nonreciprocity of node importance. For instance, some nodes have a great influence on the other nodes, which play an important role in information dissemination and other applications, called "Opinion leaders".

Opinion leaders play a crucial role in information propagation. For example, in a malicious attack on the financial scale-free network of national data nodes, attacks on even a small amount of important nodes, which will directly affect the normal operation of the whole system. Consequently, the rank of opinion leaders has great political and commercial values. Therefore, the detection of opinion leader from a large amount of the users becomes an important issue in social networks. The earliest researches on the detection of opinion leaders have focused on the use of opinion leaders' theories and research methods to directly understand Internet leaders, but may fail to achieve ideal results. For instance, these methods proposed by Li et al. [2] overlook the characteristics of objective fact, only rely on logical reasoning, meanwhile, it cannot represent the opinion leaders transmission characteristics on new media of the network. Moreover, some detection methods of opinion leaders based on social network analysis mostly focus on relations in society. These methods often determined opinion leaders based on centrality-based metrics such as methods [3–5]. Other studies proposed PageRank [6] and other modified PageRank algorithms such as LeaderRank [7], which detect opinion leaders based on link structure features. Typically, the importance of nodes in the network does not depend on a single index, but is influenced by multiple indexes at the same time, such as node location attributes and distance attributes. However, most existing methods pay more attention to analyzing individual indexes, ignore how to combine multiple indexes. In addition, the weight of each index is usually based on a subjective method of determination, which can lead to misapprehension when researchers are not familiar with the special filed. Therefore, how to detect the opinion leaders more accurately and effectively by considering multiple attributes such as distance attribute and location attributes, which is still a great challenge for researchers.

This paper proposes a method to address the issue by establishing a multiple-index evaluation model. A novel opinion leaders detection algorithm based on entropy weight method for multi-attribute analysis, named Entropy Weight Method based Multiple Attributes Analysis (MAA). The MAA algorithm consists of four stages: first, the raw data is preprocessed to provide the conditions for subsequent processing. Second, multi-index analysis is performed including distance attributes, location attributes, and strength entropy. After the analysis of the indexes, the weights of different indexes are evaluated based on the entropy weight method. Finally, consider the impact of multifunction measurements by using our MAA algorithm. It is proved

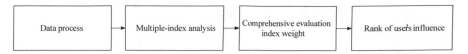

Fig. 1 The framework of the proposed MAA method

to that it has certain reliability and scientificity by the following experiments. The main framework of the proposed method is illustrated as shown in Fig. 1.

2 Related Work

At the current stage, many methods and algorithms have been proposed to detect opinion leaders in social network, but the focus of the studies is different. For example, on social network structures aspect, literature [3–5] detected opinion leaders according to some measurements of users' influence. In other aspects, Larry Page et al. [8] put forward PageRank algorithm, which has been used successfully by Google's search engine. However, PageRank algorithm only considers the link structure of the network, ignoring other additional information. Thus, an improved PageRank method has been proposed to efficiently detect opinion leaders called LeaderRank [9]. A good key nodes identification algorithm was put forward by the literature [10], which took strength entropy into account to measure the importance of nodes throughout the network. Based on the analysis of the above literature, we find that many researches have proposed that opinion leader recognition algorithm to detect opinion leaders through analysis of a single network structure or structure feature, but it seldom combines nodes with multiple attributes detection opinion leaders. Therefore, literature [11] proposed a novel algorithm called MFP algorithm, which detected opinion leaders by constructing social network and based on structural characteristics, behavior characteristics and user sentiment features. It not only considers the topology of the network, but also takes into account the dynamic characteristics and user sentiment towards the topic.

Nevertheless, it is difficult to describe the nonlinear relationship between features accurately by accurate mathematical or mechanical methods. At the same time, the existing multi-feature detection algorithms mainly rely on the subjective way, so the results may be biased when the researchers are unacquainted with the professional areas. To sum up, it is recommended that objective weight determination methods be used to avoid the above-mentioned problem of deviation. Through the analyses above, this study uses the ideas and methods of entropy weight method, which combines the characteristics of the multiple attributes, to further research on the detection of the opinion leaders.

3   Analysis of Multiple Attributes

3.1   Location Attributes

In 2010, Kitsak et al. [10] first put forward that the importance of nodes depends on its location in the entire network. Meanwhile, they found that the most efficient spreaders were those located within the core of the network as identified by the k-shell decomposition analysis. Inspired by this research, we consider the impact of the K-shell on the importance of user's location in the network.

Definition 1 (*Ks*) Given a network graph $G(V, E)$, a subgraph $H(C, E)$, H is referred to $k-$ core of G if and only if the degree of every node in H is at least k, where H is the maximum subgraph. It means that if a node belongs to the k—core but not to the $k + 1$—core, it has a k—core index, denoted by Ks.

However, the Ks value of the smallest k-core node is the same. To get the desired result, it is inappropriate for relying solely on Ks of a node. At the same time, we can consider the degree of node. This mainly because the degree depicts the node from its local connectivity.

Definition 2 (*Degree*) The degree of the node u is equal to the sum of all the edges connected to the node u. It is defined as follows.

$$D_u = \sum_{v \in V} e_{uv} \tag{1}$$

where e_{uv} denotes an edge between nodes u and v. $e_{uv} = 1$, only if there is an edge between nodes u and v; otherwise, $e_{uv} = 0$.

In terms of nodes and edges, degree theory is both theoretical and practical value. However, the method has a serious defect, which has a lower accuracy. For instance, the zombie fans usually have a high degree. To get the desired result, we also can consider betweenness centrality of the node. This because that the betweenness centrality counts the critical path across with global information.

Definition 3 (*Betweenness centrality*) The betweenness centrality of nodes measures the number of shortest paths that will pass a node in the network. The betweenness centrality of the node u is defined as

$$BC_u = \sum_{i=1}^{N} \sum_{j=1}^{N} \frac{g_{ij}(u)}{g_{ij}}, \quad i \neq u \neq v \tag{2}$$

where g_{uv} is the number of shortest paths between nodes u and v, $g_{uv}(i)$ is the number of shortest paths that pass through node i.

Using Euler's formula to calculate the combined effects of three different indicators. The location attribute LA is

$$LA(u) = \sqrt{BC_u^2 + D_u^2 + Ks_u^2} \tag{3}$$

The reasons we choose Euler's formula instead of choosing the coefficient to calculate the combined effects of three different indicators is that the coefficient beyond the scope of this article.

3.2 Distance Attributes

Compared with noncentral nodes, the central node can reach other nodes in the network more quickly. If the average distance from a node to other nodes is shorter, then the node is considered to have a higher centrality and the possibility of becoming an opinion leader is higher. However, how should we measure the distance from a node to other nodes? Here, we investigate the value of the node's location attributes from two different aspects.

Definition 4 (*Harmonic closeness centrality*) Harmonic closeness centrality [12] represents the average distance between a given node and all other nodes. It is defined as follows:

$$HCC_u = \frac{1}{N-1} \sum_{u \neq v} \frac{1}{dist(u, v)} \tag{4}$$

where $dist(u, v)$ represents the distance between nodes u, v. The use of the harmonic mean avoids cases where an infinite distance outweighs the others. Although the harmonic closeness centrality measure is intuitive, its centralized computation in a large network would be challenging in terms of its informational requirements that involve the network topology and the location of type zero stubborn agents. At the same time, we can consider the eccentricity of node.

Definition 5 (*Eccentricity*) The eccentricity is calculated by the shortest distance between nodes. The greater the distance, the weaker the relationship between nodes. Therefore, the distance is used as a reciprocal comparison, so the node with the lowest deviation is the central node. The definition formula is as follows:

$$E_u = \frac{1}{\max_{v \in V} dist(u, v)} \tag{5}$$

As the above analysis, we integrated the two different indicators of the harmonic closeness centrality and the eccentricity to consider the influence of the distance between the nodes of the importance of the nodes. Similar to formula (5), the distance attribute DA is

$$DA(u) = \sqrt{HCC_u^2 + E_u^2} \tag{6}$$

3.3 Strength Entropy

Node strength entropy is a measure of the usefulness of information contained between nodes. Furthermore, strength entropy can take into account the influence of each node on the uncertainty of the whole network. It is defined as follows:

$$Hse_u = - \sum_{v \in N, u \neq v} S_u * \ln S_v \tag{7}$$

The smaller the strength entropy of node u, the greater the possibility that node u will become the opinion leader in a complex network. Where S_u stands for relative point strength, which refers to the relative weight of node u to all nodes strength in the network. Its calculation formula is as follows: $S_u = W_u / \sum_{v=1}^{N} S_v$.

Here, W_u is the sum of weights of the edges that are directly linked to the node u. The calculation formula is as follows: $W_u = \sum_{v=1}^{N} d_{uv}$.

Where d_{uv} refers to the weight of the connection edge between nodes $u, v, d_{uv} = 1$ if the network graph is a weightless graph, that is, the degree of the node.

4 Evaluation of the Weight of Different Attribute Indexes

For each user, this paper proposes three attribute indexes: location attributes, distance attributes, and strength entropy. Therefore, the influence of each user can be described from the above indicators. Then it puts forward a new question: which index can be assigned a larger weight to better reflect the influence of the user? Nevertheless, Entropy Weight Method provides a novel idea to address this issue by empowering method. In the process of determining the weights, information entropy is used to calculate the entropy weight of each index relying on the degree of change of each index, so that the weight of the target index is obtained through the entropy weight.

5 Measurement of User Influence

In Sect. 3, we analyze user's attributes influence from three different aspects including node location attributes influence, distance attributes influence, and strength entropy influence. Then, according to formula (8), entropy weight is applied to weighing the feature data. Specifically, each index is weighed by formula (8). Thus each index data can be represented as

$$F(u) = w_1 * LA(u) + w_2 * DA(u) + w_3 * Hse(u) \tag{8}$$

After defining the influence function $F(u)$, different features of users can be evaluated. And then we rank the users according to the value of the function. The larger the value, the more direct and effective the node's impact on other nodes.

6 Experiments and Results

6.1 Datasets

We conduct our experiments on real networks. The first one, we choose Sinaforum as a main empirical data source, which is one of the most popular forums in China. Meanwhile the network was obtained from January 2015 to December, 2017. The second is conducted on Sina microblog, which has opened up their APIs since 2007. However, in order to protect user privacy, Sina microblog officially restricted public API access since October 2014. Based on this, this article selected part of the May 2014 data as a data set. Each node denotes a user, and the relationship between users is represented by an edge. The basic information of the datasets is shown in Table 1.

6.2 Results of the Weight Evaluation

We define the weight of each characteristic according to Entropy Weight Method, which is allocated to each characteristic with a different value. The results ppear as shown below.

Table 2 illustrates the relative importance of characteristics. It can be seen from Table 2 that the location attribute characteristics play an important role than other characteristics, which have the highest value reached to 0.40217 and 0.52229 in Sinaforum and Sina microblog, respectively.

Table 1 Basic statistics of data sets

Dataset statistics	Sinaforum	Sina microblog
Nodes	1052	9737
Edges	42448	65535
Average path length	4.885	5.573
Average clustering coefficient	0.148	0.008
Number of weakly connected components	2	14
Number of strongly connected components	886	9514

Table 2 The results of the weight evaluation

Dataset	Weight		
	Location attribute	Distance attribute	Strength entropy influence
Sinaforum	0.40271	0.20509	0.3922
Sina microblog	0.52229	0.18603	0.29168

6.3 Results of Attributes Analysis

In order to detect opinion leaders in datasets, we select top-5 opinion leaders with the highest score according to formula (8). We illustrate results analysis including LA, DA, and Hse. The detailed results of the Sinaforum and Sina microblog are shown in Table 3.

6.4 Comparison with Other Algorithm

In order to find the effectiveness of our proposed algorithm, we compare the results of our algorithm and other classical algorithms directly as shown in Tables 4 and 5.

Through experimental analysis, conclusions are listed as follows. Traditional centrality algorithms only depend on single evaluation feature, which has lower accuracy. For example, degree centrality according to the computing the numbers of nodes linked to measuring the influence of nodes. However, the zombie fans usually have

Table 3 The results of attributes analysis

(a) Sina microblog

Rank	LA		DA		Hse	
	ID	Score	ID	Score	ID	Score
1	118959	745998.56	124030	18.00057	161805	1.5892
2	119349	603859.37	121635	18.00053	149685	1.2471
3	119716	602066.89	108581	18.00053	119125	1.162
4	118238	558942.28	123452	18.00051	265627	1.0801
5	123545	443703.35	115309	18.0005	119716	1.0067
(b) Sinaforum						
1	Signin	16368.2	stanley	13.0015	scjyxjb	0.000961
2	scjyxjb	4539.6	sxxj	13.0007	milanc	0.000589
3	Match	3013.3	hhzx	12.005	Tong	0.000184
4	Yining	1939.1	lateman	12.0040	diablo	0.000166
5	pillow	1854.2	Northern	12.0039	Orchid	0.000153

Table 4 Comparison of the results of algorithms of Sinaforum

Rank	CC		BC		DC		PageRank		LeaderRank		MFP		MAA	
	ID	Score	ID	Score	ID	Score	ID	Score	ID	Score	ID	Score	ID	Score
1	Loyal	1	dufeigg	0.0433	scjyxjb	0.2159	milanch	0.0102	s7love	1.1491	Evening	12.5577	Signin	6594.05
2	Pirates	1	milanc	0.0364	milanc	0.1693	Souls	0.0075	Signing	0.5908	Blue	12.1192	scjyxjb	1830.41
3	xhj1802	1	WBW	0.0308	Tong	0.0951	SCT	0.0071	scjyxjb	0.5081	XC	11.8166	Match	1215.74
4	Com	1	BBJ18	0.0248	diablo	0.0903	XC	0.0059	Match	0.3182	Northern	11.7723	Yining	783.14
5	monst	1	Ribless	0.0244	Orchid	0.0865	hero	0.0057	milanch	0.0812	letfoxdash	11.7307	pillow	748.35
6	6009	1	scjyxjb	0.0215	hero	0.0837	Fishing	0.0054	pillow	0.0555	milanc	11.6144	milanc	619.35
7	Milan	1	Kick	0.0194	Signin	0.0818	juv0609	0.0054	Orchid	0.0408	leobettor	11.5518	Orchid	616.27
8	zhangx	1	sisili2513	0.0177	Ribless	0.0818	Favorites	0.0054	Tong	0.0348	LD	11.5476	north	451.83
9	Milan	1	Maple	0.0166	22kaka	0.0713	Sumechel	0.0052	Sheng	0.0314	hero	11.4884	Sheng	439.22
10	Rijkaa	1	Juventus	0.0162	Laughing	0.0704	Orchid	0.0052	diablo	0.0313	joosten	11.4720	hero	276.38

Table 5 Comparison of the results of algorithms of Sina microblog

Rank	CC		BC		DC		PageRank		LeaderRank		MFP		MAA	
	ID	Score	ID	Score	ID	Score	ID	Score	ID	Score	ID	Score	ID	Score
1	119971	1	149685	0.0414	108197	745998	161805	0.0016	104476	14.6299	119104	17.0547	118959	6594.05
2	124298	1	118239	0.0249	117042	603859	119125	0.0014	108919	13.1785	123066	17.0162	119349	1830.41
3	105483	1	119623	0.0169	109397	602066	149685	0.0013	100671	12.6912	109720	16.3111	119716	1215.74
4	106253	1	189380	0.0167	124347	558942	265627	0.0012	122403	12.3581	113552	15.1207	118238	783.14
5	119590	1	119503	0.0146	109249	443703	119716	0.0011	118444	12.2768	123175	14.6007	123545	748.35
6	100056	1	163878	0.0143	124004	403901	280330	0.0010	114363	12.1055	119250	14.2250	113005	619.35
7	100070	1	171392	0.0106	122490	378346	176117	0.0009	103677	12.018	122531	14.1599	121503	616.27
8	100071	1	164439	0.0104	103629	377393	170558	0.0008	117119	11.7104	119321	14.0835	118239	451.83
9	100077	1	189596	0.0103	122815	369484	118238	0.0008	114287	11.5186	123353	13.9164	104202	439.22
10	119736	1	188545	0.0103	119850	361711	162981	0.0008	103308	11.5167	123670	13.8793	103641	276.38

high degree centrality, and thus they can risk being mistaken for influential nodes by degree centrality of node. For instance, the user with ID149685 has high degree centrality, who is likely to be zombie fans. Therefore, relying solely on degree centrality is not a good method to measure opinion leaders.

It can be seen from Table 4 that the sequence of top-10 users is different. There are only 4 users in top-10 users sequence of our proposed algorithm not in the sequence of LeaderRank algorithm. Therefore, the two algorithms have a high consistency from the point of view of the importance of users. Usually, in Sinaforum network, the user with ID milanc is ranked sixth in MAA algorithm, while ranked first in PageRank. It is because that its DA influence is relatively lower depending on distance feature analysis, which decreases the rank in PageRank, but the overall ranking is still in the top 10.

6.5 Correlation Coefficient

Next, we use the Kendall coefficient τ [13] to express the difference between our algorithm and classical algorithms. We can see that the lists of top 100 users ranking by MAA algorithm have a strong correlation with MFP algorithm, while a while a weak correlation with degree centrality. More concretely, on the Sinaforum dataset, the correlation coefficient between the MAA algorithm and LeaderRank, PageRank, DC, and BC is 0.862, 0.416, 0.414, 0.402, 0.489, respectively. From Fig. 2, we can clearly find that there is no correlation coefficient between our algorithm and the CC algorithm. This is because the top 100 users in the CC algorithm have the same value, that is, they are all 1. Therefore, the correlation coefficient is none.

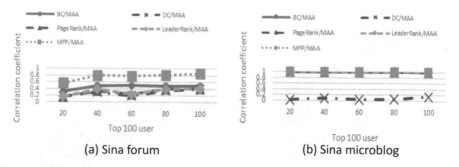

(a) Sina forum   (b) Sina microblog

Fig. 2 Comparison of baseline algorithms about the correlation coefficient

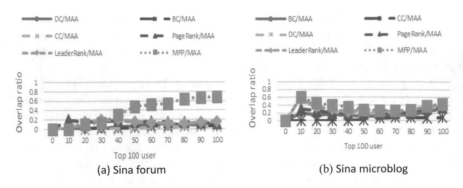

(a) Sina forum   (b) Sina microblog

Fig. 3 Comparison of baseline algorithms about overlap ratio

6.6 Overlap Ratio

We conduct experiments on the two datasets to illustrate results. To simplify the results, the span of opinion leaders are from 1 to 100.

In Fig. 3, we can obtain that with the growth of the number of selected opinion leaders, the overlap ratio is gradually increasing. Our algorithm MAA has a high overlap rate with other algorithms on all datasets. For example, the highest overlap rate of MAA algorithm and MFP algorithm is 68% on Sinaforum dataset, meanwhile, there are 60% overlap rates between MAA algorithm and MFP algorithm on Sina microblog dataset. In addition, the overlap shows that most opinion leaders can be detected. More importantly, most opinion leaders and high concern can be detected while nonoverlapping parts indicate that the MAA algorithm can find nodes where other algorithms are easily overlooked.

7 Conclusion

In this study, we propose a novel algorithm described as MAA algorithm to opinion leaders in the social network. Unlike the existing methods, MAA considers not only different feature analysis but also the weight of different including location feature, distance feature, and strength entropy. In use entropy weight method to comprehensively evaluate multiple With experiment is conducted on two datasets with Sinaforum and The experimental results show that the MAA algorithm is better algorithms including DC, BC, CC, PageRank, LeaderRank, and M

Although we try to provide new directions for the research MAA algorithm still has room for improvement. For instance, some information including location feature, distance feature

However, in many applications, some other information, such as time feature and network behavior are also important.

Acknowledgements This work is supported by the Nature Science Foundation of China (No. 61502281, 71772107).

References

1. Wang, X., Li, X., Chen, G.: Network Science an Introduction. Higher Education Press (HEP) (2012)
2. Li, Y., Ma, S., Zhang, Y., et al.: An improved mix framework for opinion leader identification in online learning communities. Knowl. Based Syst. **43**(2), 43–51 (2013)
3. Kleinberg, J.M.: Authoritative sources in a hyperlinked environment. In: SODA '98: Proceedings of the Ninth ACM-SIAM Symposium on Discrete Algorithms, Philadelphia, PA, USA. Society for Industrial and Applied Mathematics, pp. 604–632 (1998)
4. Brin, S., Page, L.: The anatomy of a large-scale hypertextual Web search engine. In: International Conference on World Wide Web. Elsevier Science Publishers B.V., pp. 107–117 (1998)
5. Kleinberg, J.M.: Hubs, authorities, and communities. ACM Comput. Surv. **31**(4es), 5 (1999)
6. Kandiah, V., Shepelyansky, D.L.: PageRank model of opinion formation on social networks. Physica A **391**(22), 5779–5793 (2012)
7. Yu, X.: Networking groups opinion leader identification algorithms based on sentiment analysis. Comput. Sci. **39**(2), 34–33 (2012)
8. Sergey, B., Lawrence, P.: The anatomy of a large-scale hypertextual Web search engine. Comput. Netw. ISDN Syst. **30**(1–7), 107–117 (1998)
9. Yu, X., Wei, X., Lin, X.: Networking groups opinion leader identification algorithms based on sentiment analysis. Comput. Sci. **39**(2), 34–37 (2012)
10. Kitsak, M., Gallos, L.K., Havlin, S., et al.: Identification of influential spreaders in complex networks. Nat. Phys. **6**(11), 888–893 (2010)
11. Cao, J., Chen, G., Wu, J., et al.: Based on multidimensional analysis of social network opinion leaders, dig. Electron. J. **44**(4), 898–905 (2016)
12. Rochat, Y.: Closeness centrality extended to unconnected graphs: the harmonic centrality index. In: ASNA (2009)
13. Kendall, M.G.: A new measure of rank correlation. Biometrika **30**(1/2), 81–93 (1938)

A K-shell Decomposition Based Heuristic Algorithm for Influence Maximization in Social Networks

Qiu Liqing, Yu Jinfeng, Jia Wei, Fan Xin and Gao Wenwen

Abstract Influence maximization is a widely studied problem that aims to find a set of the most influential nodes to maximize the spread of influence as much as possible. A well-known method K-shell decomposition has a low time complexity, which can be able to quickly find the core nodes in the network who are regarded as the most influential spreaders. Motivated by this, we try to solve the effectiveness and the efficiency problem of influence maximization by utilizing the K-shell decomposition method, and propose a K-shell decomposition based heuristic algorithm called KDBH for finding the most influential nodes. We also design a novel assignment strategy of seeds to effectively avoid producing similar spreading areas by K-shell decomposition method. Furthermore, we take the direct influence and the indirect influence of nodes into account to further optimize the accuracy of seeds selection. At last, we conduct extensive experiments on real-world networks demonstrate that our algorithm performs better than other related algorithms.

Keywords Social networks · Influence maximization · K-shell decomposition · Assignment strategy

Q. Liqing (✉) · Y. Jinfeng · J. Wei · F. Xin · G. Wenwen
Shandong Province Key Laboratory of Wisdom Mine Information Technology,
College of Computer Science and Engineering, Shandong University of Science and Technology,
Qingdao 266510, China
e-mail: liqingqiu2005@126.com

Y. Jinfeng
e-mail: 576476745@qq.com

J. Wei
e-mail: 1456514769@qq.com

F. Xin
e-mail: xhanhaixiangkong@163.com

G. Wenwen
e-mail: 1141922180@qq.com

© Springer Nature Singapore Pte Ltd. 2019
J.-S. Pan et al. (eds.), *Genetic and Evolutionary Computing*,
Advances in Intelligent Systems and Computing 834,
https://doi.org/10.1007/978-981-13-5841-8_74

1   Introduction

With the popularity of social networks, more and more people prefer to share their opinions or ideas on various social platforms such as Twitter and Sina Blog, which has been aroused extensive interests of researches about how to influence or propagate information in social networks. This phenomenon has been broadly applied in viral marketing. For example, a company may select the most influential users, hoping to create a cascade of product adoptions through the word-of-mouth. The influence maximization problem, motivated by the idea of viral marketing, which targets to select a set of initial users that eventually influence other users as widely as possible.

The influence maximization problem was formulated as a discrete issue by [1] with two basic diffusion models, namely Independent Cascade Model (IC) and Linear Threshold Model (LT). Moreover, it was proved NP-hard. Therefore, in order to solve this problem, many approximation algorithms and scalable heuristics have been designed [2–5]. The key problem of these algorithms is how to quickly and accurately find the most influential spreaders. In most cases, nodes with high degree are believed to be the best spreaders. However, if a node with a high degree is located at the periphery of a network, it is likely to be unable to spread continually. In view of this, Kitsak et al. [7] found that the most efficient spreaders were located within the core of the network as identified by the K-shell decomposition analysis. In other words, K-shell decomposition can identify the importance of different nodes' locations which efficiently finds the most influential spreaders. In addition, K-shell decomposition has a low time complexity which can be suitable for large-scale networks.

Inspired by the above discussion, we believe that it is necessary to improve the effectiveness and the efficiency problem of influence maximization by taking advantage of K-shell decomposition method. To account for this problem, we propose a novel K-shell decomposition based heuristic algorithm for influence maximization, called K-shell Decomposition Based Heuristic algorithm (KDBH), to address the influence maximization problem on large networks. In our algorithm KDBH, we first apply K-shell decomposition method to divide network for finding the most influential spreaders. Then, we design a novel assignment strategy of seeds to effectively avoid producing similar spreading areas by K-shell decomposition method. And finally, we take the direct influence and the indirect influence of nodes into account to further optimize the accuracy of seeds selection. Consequently, the proposed method on real-world datasets is superior to other related algorithms in influence spread as well as reasonable running time.

2   Related Work

Since the influence maximization problem is proved to be NP-hard, a lot of researches have been published to obtain approximate solutions. A most effective approach is the greedy algorithm, which can guarantee a good quality of seeds. But it fails to scale to large-scale networks. Therefore, some researchers work on influence maximization to optimize efficiency problem and have been put forward a lot of excellent algorithms [2–6]. Among them, the Degree Discount heuristic [2] roughly estimated influence effect of seeds from discount view which greatly increases the influence spread, while the PMIA [3] was a tree-based scalable algorithm which provides a good result both in influence spread and running time.

Therefore, how to efficiently and accurately find the most influential spreaders is the most important task for all the above algorithms. A recent study [7] found that K-shell decomposition method could identify most efficient spreaders located within the core of the network. Moreover, it could not only identify the most influential spreaders, but also be applied to large-scale networks due to its low time complexity, which completely accords with two most concerned issues on influence maximization, i.e., effectiveness and efficiency. Based on the above analysis, later studies tried to find a good quality of seeds by using the K-shell decomposition method. Cao et al. [8] proposed a core covering algorithm called CCA to cover some redundant nodes by combing the K-shell decomposition method and the distance. Zhao et al. [9] put forward a KDA algorithm to improve the efficiency and accuracy of existing algorithms. Li et al. [10] proposed a K-core filtered algorithm and GIMS algorithm, increasing the influence spread of existing algorithms and reducing running time.

3 Preliminaries

3.1 Problem Definition

In the problem of influence maximization, some nodes are selected to maximize the influence under a certain influence diffusion model. To facilitate the exposition, in this paper, we mainly focus on the Independent Cascade (IC) model, which is applied in all kinds of influence maximization algorithms [2, 3, 6, 8, 10]. In the IC model, suppose a social network is modeled as a directed graph $G(V, E)$, where V is the set of nodes representing users and E is the set of edges representing social relationships (such as friendships or co-authorships) between users. Each edge (v, w) is assigned a probability P_{vw} representing the probability for node w is activated by node v through edge and each individual node has either an active (an adopter of the product) or an inactive state in a network. Note that each node can be allowed to switch from being inactive to being active, but not the reverse direction.

3.2 K-shell Decomposition Method

The goal of influence maximization is to find the most influential spreaders under the diffusion model. Usually, the nodes located at the core of the network are considered as the most influential spreaders. For this purpose, the K-shell decomposition is applied [7–9], which is a well-established method for detecting the core nodes and the

Fig. 1 A layered network by
K-shell decomposition

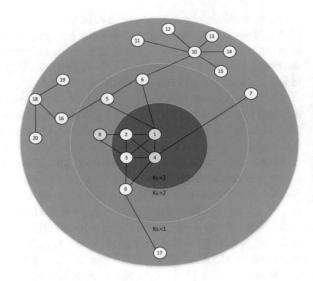

hierarchical structure of a given network. In particular, for a given network $G(V, E)$, the K-shell decomposition method allocates an integer index for each node according to the centrality measurement, denoted as K_s, representing the location of each node in a network. Nodes with a low or high value of K_s correspond to located at the periphery or center of the network, respectively, which directly links to the importance of nodes in influence transmission. In this way, the network is characterized by a layered structure. How to divide a network into a layered structure. The detailed process is similar to peeling onions which can be described as follows (shown as in Fig. 1).

K-shell composition is a constantly pruning process from outside to inside which groups nodes in a network on the basis of the number of connections that nodes have. Initially, some nodes are identified and assigned an integer index $K_s = 1$ if they only have one connection. Once a node is marked, it is removed from the network and search continues until all remaining nodes with one connection have been found, this process corresponds to all nodes, where $K_s = 1$ in Fig. 1. And then the same process is conducted for nodes who have two connections, three connections... and so on, until all nodes of the network have been marked.

4 Influence Maximization Based on K-shell Decomposition

We begin by describing our K-shell decomposition based heuristic algorithm KDBH for influence maximization, which can be efficiently shown as the later experiments. The framework of the proposed approach is shown in Fig. 2. Our proposed KDBH algorithm can be divided into three phases. Firstly, we conduct K-shell decomposition method to partition network, in order to automatically identify the core nodes on networks. Based on this layered network, we next design a new assignment strategy

Fig. 2 The framework of the proposed approach KDBH

of seeds to effectively avoid producing similar spreading areas in the first phase, providing conditions for the next phase. We finally adopted a heuristic strategy to select seeds at each shell layer by taking both the direct influence and the indirect influence of nodes into account, in order to further optimize the accuracy of seeds selection.

4.1 The Analysis of the K-shell Decomposition

As we have introduced in the previous section, the K-shell decomposition method can identify the most influential spreaders, especially, each node is assigned an integer index K_s by the K-shell decomposition method. Moreover, the higher K_s, the closer the node is to the core of the network. Suppose a node is regarded as a user, the user with higher K_s is more influential than the user with lower K_s in the real-world social network. For example, in Fig. 1. If we select a node located at the periphery of the network as a seed such as node 10, it has a little probability (only through node 6) to spread continually. Instead, if we select a node located at the core of the network as a seed such as node 1, it can easily spread out through any of its neighbors. Therefore, the K-shell decomposition method is a good choice to identify the most influential spreaders.

However, a significant drawback of the K-shell composition method for multiple seeds is that nodes with high K_s tend to be clustered close to one another which results in producing similar spreading areas such as the node 2 and node 3 in Fig. 1. We can obviously see that node 2 and node 3 produce an intersection {1, 4, 8} of spreading area between them with a large degree. Furthermore, if these two nodes want to continually spread out only through their common neighbors node 1 and node 4. As a result, comparing to using one node to influence others, using two nodes does not obtain a significantly increased influence spread. This result suggests that if we want to find a better strategy for seeds selection, it may be a good idea to choose nodes with high K_s as well as the requirement that nodes need to be distributed to different shell layers in proportion as much as possible. To achieve the requirement, we design an assignment strategy for seeds to reasonably assign the number of seeds on each shell layer by combining the K-shell index of each node and the size of each shell layer. The specific assignment strategy can be described as follows.

Definition 1 (*Assignment Strategy*) For a layered network $G(V, E)$ obtained by K-shell decomposition method, the assignment strategy of seeds at each layer l can be defined as:

$$N_{seed}(l) = \frac{K_s \cdot S_l}{C} \tag{1}$$

where K_s and S_l denote the K-shell index of each seed and the size of each shell layer, respectively, and C is an adjustment parameter, in order to adjust the number of seeds that should be allocated at each shell layer. It is selected from the set $\{10, 20, \ldots 100, 200, \ldots 1000, 2000, \ldots 10000, 20000, \ldots\}$ according to different types and sizes of datasets. In addition, we use the ceil function to obtain the number of seeds needed. For example, in Fig. 1, we choose 20 for the adjustment parameter C. According to formula (1), we can compute the number of seeds allocated for $K_s = 3$ as follows.

$$N_{seed}(3) = \frac{K_s \cdot S_l}{C} = \frac{3 \cdot 4}{20} \approx 1$$

The calculation results show that we only need to select one node for $K_s = 3$ in Fig. 1, which were consistent with the requirement of a good strategy for seeds selection. Thus, we can see that our assignment strategy is able to effectively assign seeds on different shell layers. Nevertheless, we still cannot distinguish the importance of seeds when seeds located at same shell layer. For this reason, we define a new hybrid degree to further optimize the selection strategy of seeds by considering both the direct influence and the indirect influence of nodes.

Definition 2 (*Hybrid Degree*) For some nodes are located at same shell layer, the hybrid degree of node v can be defined as

$$H_D(v) = Degree(v) + \frac{\sum\limits_{w \in neighbor(v)} Degree(w)}{|neighbor(v)|} \tag{2}$$

where $Degree(v)$ represents the degree of the node v, $neighbor(v)$ is the set of neighbors of node v and $|neighbor(v)|$ is the number of neighbors of node v.

Observe Fig. 1, we select two nodes located at same shell layer randomly, such as node 1 and node 4. We can see that node 1 is more important than node 4 in influence transmission, since node 1 has more and deeper transmission paths than node 4. According to formula (2), the hybrid degree of node 1 is $H_D(1) = 5 + 3 + 3 + 4 + 5 + 5/5 = 9$, while the hybrid degree of node 4 is $H_D(4) = 5 + 4 + 5 + 5 + 1 + 3/5 = 8.6$. Based on the calculation results, we can conclude that node 1 is more important than node 4, which is more consistent with the actual condition of influence transmission. Therefore, we believe that hybrid degree is a good optimization index to select seeds from the same shell layer. However, in the practical situation on influence transmission, once a node is selected as seed, it will influence its neighbor, in which case we do not need to select its neighbors to join in the seed set. Due to this, we apply degree discount idea to discount the hybrid degree of nodes which is shown in pseudocode.

4.2 KDBH Algorithm

Utilizing K-shell decomposition method combined new assignment strategy of seeds and heuristic strategy, we propose KDBH algorithm to optimize the effectiveness and the efficiency problem of influence maximization. Algorithm 1 illustrates the basic idea of KDBH as follows.

Algorithm 1: KDBH algorithm

Input: network $G=(V,E)$ and number k

Output: seed set S

1: $S \leftarrow \phi$ //initialization

// K-shell decomposition method

2: Apply K-shell decomposition method to obtain a $l-layer$ network

// assignment strategy

3: For each shell layer l

4:    Compute $N_{seed}(l)$

//Heuristic strategy

5: For $i \leftarrow \max(K_s)$ to 1 do

6:    For $j \leftarrow 1$ to $N_{seed}(i)$ do

7:       $u \leftarrow \arg\max_{w \in V-S}(H_D(w))$

8:       $S = S + \{u\}$

9:       if $|S| = k$

10:          break

11:       for each $v \in neighbor(u)$ do

12:          update $H_D(v)$ //discount idea

13: Return S

The core idea of the KDBH algorithm is to heuristically find the most influential k nodes from core to periphery of the network. Specifically, we first apply K-shell decomposition method to partition network, which identifies core nodes on networks (line 2). For a layered network obtained by K-shell decomposition method, we then carry out the assignment strategy to effectively avoid producing similar spreading areas by core nodes. As a result, each shell layer is assigned an appropriate number of nodes (lines 3–4). At last, according to the number of nodes assigned on each shell, we iteratively select a new seed that maximizes the incremental changes of H_D into the seed set S from core to periphery of the network until the total number of seeds satisfies k (lines 5–13). Noted that we will constantly update the hybrid degree of seeds' neighbors if a node is selected as seed, in order to discount the influence of seeds' neighbors which avoids the overlapping effect generated by any two adjacent nodes.

5   Experiments

We conduct experiments on four real-world datasets to evaluate our proposed KDBH algorithm, and compare it to other related algorithms except for the greedy algorithm due to its high time complexity.

5.1   Experimental Setup

5.1.1   Dataset Preparation

We use four real-world datasets contained various different networks, the statistics of the datasets are showed in Table 1. NetPHY and DBLP are both collaboration networks from the e-print arXiv and DBLP Computer Science Bibliography respectively. Nodes in these two networks are authors, if two nodes have collaborated, there will be arcs in both directions. P2p is a Gnutella peer-to-peer file sharing network from August 2002. Nodes represent hosts in the Gnutella network topology and edges represent connections between the Gnutella hosts. The last one Oregon, an Autonomous Systems (AS) peering information inferred from Oregon route-views between March 31, 2001 and May 26, 2001.

Table 1 Statistic of network datasets

Datasets	P2p	Oregon	NetPHY	DBLP
#Nodes	10K	10K	37K	3.1M
#Edges	40K	22K	231K	117M
Max. degree	103	2370	286	33,313
Avg. degree	7.355	4.119	12.46	76.17
Avg. CC	0.0062	0.2970	0.899	0.6324

5.1.2   Parameters

Influence activation probability P_{vw} of edge is assigned by the Weighted Cascade model which is an improvement of the Independent Cascade model. P_{vw} is set to be $1/in(w)$, where $in(w)$ is the indegree of node w. In addition, for adjustment parameter C in assignment strategy, we choose 200, 20, 400 and 400 on four datasets P2p, Oregon, NetPHY, and DBLP, respectively.

5.1.3   Algorithms Compared

We evaluate our proposed KDBH algorithm with several other algorithms including Degree Discount, CCA, PMIA, K-shell, and Random, which have discussed in the section of related works. Note that the distance parameter for CCA and the threshold parameter for PMIA are set to 2 and $1/320$, respectively.

To evaluate the quality of seeds returned by different algorithms, we simulate the process 10,000 times for each targeted set and take the average of the influence spread.

5.2   Experimental Results

We investigate the performance of KDBH on influence spread and running time. In addition, we conduct the K-shell decomposition method obtaining the number of shell layers is 7, 17, 66 and 113 on four datasets P2p, Oregon, NetPHY and DBLP respectively.

5.2.1   Influence Spread

We first compare the influence spread of different algorithms on four real-world datasets when the size of the seed set is set to 20, 40, 60, 80, 100, respectively, as shown in Fig. 3. From the results obtained, we can make the following observations. Our algorithm KDBH provides the best performance on all datasets except for the Oregon dataset. For example, on P2p dataset, the KDBH is 5.17%, 224.52%, 47.97%, 198.62% and 323.04% superior comparing to Degree Discount, CCA, PMIA, K-shell, and Random, respectively. In addition, a remarkable trend is that the KDBH algorithm grows rapidly with the number of seeds increasing. This is primarily due to the following reason. The assignment strategy we designed does not rely on the number of seeds, which limits the seeds only located at the high shell layers when the number of seeds is small. As the number of seeds increased, our assignment strategy can be able to spread more shell layers comparing to the based on the K-shell composition algorithms CCA and K-shell, greatly increased the influence spread. It is worth noticing that the performance of the KDBH on Oregon is not well which leads to being far below the PMIA. The reason is probably that Oregon is a relatively sparse dataset and has an unclear network structure which not provides large impact depended on the nodes' location.

5.2.2 Running Time

Next, we illustrate the running time for different algorithms on all datasets when the seed set is set to be 100. From the result, we see that our proposal KDBH has similar speed with the CCA, which can find the 100 most influential nodes on the P2p and the NetPHY datasets within 10 s. Moreover, our algorithm KDBH just slightly slower than the K-shell and Random and run several times faster than the remaining algorithms, CCA, Degree Discount and PMIA on Oregon dataset. The K-shell and Random, though run fast, cannot provide any performance guarantee according to the influence spread shown in Fig. 3. Furthermore, our proposal KDBH only requires 343 s on the largest dataset DBLP which nearly four times faster than the PMIA algorithm showing its extremely high time efficiency (Fig. 4).

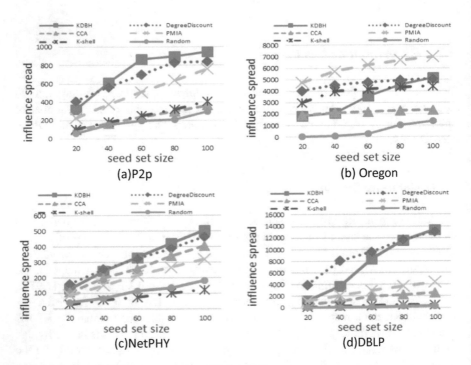

Fig. 3 Illustration of influence spread on different datasets

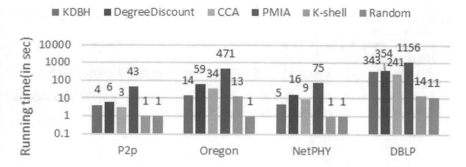

Fig. 4 Illustration of running time on different datasets when the seed set is set to 100

6 Conclusion

In this paper, we propose a K-shell decomposition based heuristic algorithm called KDBH to address the effectiveness and efficiency problem of influence maximization in social networks. In addition, we also design a novel assignment strategy of seeds to effectively avoid producing similar spreading areas by K-shell decomposition method. At last, we take the direct influence and the indirect influence of nodes into account to further optimize the accuracy of seeds selection. The extensive experiments on real-world networks demonstrate the efficiency and effectiveness of our algorithm.

Acknowledgements This work is supported by the Nature Science Foundation of China (No. 61502281, 71772107).

References

1. Kemple, D., Kleinberg, J.M., Tardos, E.: Maximizing the spread of influence through a social network. In: Proceedings of the 9th ACM SIGKDD Conference on Knowledge Discovery and Data Mining, pp. 137–146 (2003)
2. Chen, W., Wang, Y., Yang, S.: Efficient influence maximization in social networks. In: ACM SIGKDD International Conference on Knowledge Discovery and Data Mining, pp. 199–208. ACM (2009)
3. Chen, W., Yuan, Y., Zhang, L.: Scalable influence maximization in social networks under the linear threshold model. In: IEEE, International Conference on Data Mining, pp. 88–97. IEEE (2011)
4. Jung, K., Heo, W., Chen, W.: IRIE: scalable and robust influence maximization in social networks. In: IEEE, International Conference on Data Mining, pp. 918–923. IEEE (2013)
5. Tang, Y., Xiao, X., Shi, Y.: Influence maximization: near-optimal time complexity meets practical efficiency. In: ACM SIGMOD International Conference on Management of Data, pp. 75–86. ACM (2014)
6. Tang, J., Tang, X., Yuan, J.: Influence maximization meets efficiency and effectiveness: a hop-based approach (2017)

7. Kitsak, M., Gallos, L.K., Havlin, S., et al.: Identification of influential spreaders in complex networks. Nat. Phys. **6**(11), 888–893 (2011)
8. Cao, J.X., Dong, D., Xu, S., et al.: A k-core based algorithm for influence maximization in social networks. Chin. J. Comput., 1–7 (2015)
9. Zhao, Q., Lu, H., Gan, Z., et al.: A K-shell decomposition based algorithm for influence maximization (2015)
10. Li, Y.Z., Chu, W.W., Zhong, M., et al.: k-core filtered influence maximization algorithms in social networks. J. Comput. Appl. **38**(2), 464–470 (2018)

Predicting the Co-movement of Stocks in the Hong Kong Stock Market

Chen Chuxun and Jean Lai

Abstract Stock co-movement was examined in Finance research but not in the IT research. Previous studies revealed that the co-movement is usually caused by either the determinants of the stocks' values, habitat movements between stocks, or the change in portfolio composition. Most of the studies used a statistical approach to uncover the co-movement relation between stocks. This paper takes a combination of the statistical approach and the machine learning approach to: (1) prove the existence of stock co-movement; and (2) identify a prediction model that can forecast the stock co-movement. Both supervised and unsupervised methods are used. In this study, the inter-day stock data in the real estate industry were extracted from the Yahoo finance in Hong Kong. After cleaning the data, stocks of the industry were categorized into two groups by its market capitalization. The correlation between the two trading data set is tested. Support Vector Machine (SVM) is used to train the prediction model. The predictive power of the model looks good.

Keywords Stock Co-movement · K-means clustering · Support vector machine

1 Introduction

Stock markets attract investors and analysts. In order to find out the best time to buying and selling stock(s) in the stock market, investors formulate their winning formula using different approaches. Three types of analysis are widely used. They are: Technical analysis, Fundamental analysis, and Behavioral analysis. Trading in the stock market, however, is quite challenging because of its dynamic and complex characteristics. It is difficult to formulate the stock price movement with one sin-

C. Chuxun
Hong Kong Polytechnic University, Hung Hom, Hong Kong
e-mail: cxchen@polyu.edu.hk

J. Lai (✉)
Hong Kong Baptist University, Kowloon Tong, Hong Kong
e-mail: jeanlai@comp.hkbu.edu.hk

© Springer Nature Singapore Pte Ltd. 2019   723
J.-S. Pan et al. (eds.), *Genetic and Evolutionary Computing*,
Advances in Intelligent Systems and Computing 834,
https://doi.org/10.1007/978-981-13-5841-8_75

gle parametric or nonparametric method [1]. Some investors, therefore, use hybrid method as well.

Stock co-movement is defined as the lead and lag changes in stock price of two or more stocks of the same market. Co-movement can be in a positive way or in a negative way. Lo and MacKinlay [2] revealed that the lead–lag effect in returns exists between large capitalization firms and small capitalization firms. The price of the small capitalization firms moves following the price change of large capitalization firms.

In this paper, we use the K-means Clustering to divide the stocks into two groups based on its market capitalization and then train a SVM model to predict the co-movement effect. This paper is organized as follows: Sect. 2 reviews the literature on the machine learning tools that are used in this study; Sect. 3 introduces the research methodologies; Sect. 4 reports the experimental results, and Sect. 5 concludes the study.

2 Literature Review

2.1 Clustering

Clustering algorithms have been widely used for over half of a century. Clustering is considered as an unsupervised method for categorizing data with different attributes into a defined number of groups. Clustering can be performed using observation, and feature vectors extraction [3]. K-means clustering (the most popular clustering algorithm) is still widely used in both academia and industry. K-means clustering is an algorithm, which can intelligently group the datasets into several groups [4]. The procedure of K-means application works in this way: (1) randomly choose K centroids, one per each cluster, (2) calculate the distance between individual points and the nearest centroid; and (3) repeat step 1 and 2 for n iterations or until no more improvement in the result. The selection of centroids causes different result. The better choice is to place the centroids as much as possible far away from each other. The value of K is also a factor affecting the clustering result.

2.2 Pearson Correlation

Correlation is a statistical method for measuring the relation between two variables. The two popular correlation approaches are: Pearson's correlation coefficient and Spearmen's rank correlation coefficient [5]. The correlation of Pearson correlation coefficient is: $R = C_{xy}/(C_{xx}C_{yy})^{1/2}$. Pearson correlation coefficient discusses how similar are the changes of two variables. The highly correlated characteristics existing between stocks is proved and well known in the stock market [6]. Kennett et al.

[7] defined a partial correlation network and reported that the stocks belong to the financial industries are easily affected by the correlated factors in the market.

2.3 Support Vector Machine

Support Vector Machine (SVM) is a kind of classification method for pattern recognition (Vapnik 2013). Al-Radaideh et al. [1] applied SVM to forecast the movement trend of daily stock price in the Korea composite stock price index (KOSPI). Comparing with backpropagation neural network (BPN) and case-based reasoning (CBR), SVM can provide a better output for stock market prediction. Huang et al. [8] used SVM for stock market movement direction predicting, showing that SVM works better than the other supervised algorithms. Kim [9] compared SVM with BPN and CBR in predicting the stock price index in the future. The results showed that SVM outperforms other machine learning algorithms. Lin et al. [10] predicted stock market trend by using SVM. In this paper, the authors used the ranking method for selecting the key factors, showing that SVM performance in hit ratio is better than other traditional methods. Gavrishchaka and Banerjee [11] proposed an SVM-based model for volatility forecasting, the results indicate that SVM model can effectively get information from inputs with numeric time-delay returns.

3 Research Methodology

3.1 Data Collection

The full data sets that are used for this study are from yahoofinance.com. The industry that we examined in this paper is "Real Estate—General" (RSG), there are 96 stocks in this RSG industry in the Hong Kong stock market. Table 1 shows the basic information of the selected stocks.

3.2 Grouping the Stocks

In order to divide the stocks into two groups, we use K-means to cluster the stocks into two groups based on capitalization. The results are as follows:
Set 1 (L set): [hk0016, hk3333, hk0688, hk1109]
Set 2 (S set): [hk0017, hk0020, hk0087, hk0019, hk0083, hk0813, hk3383, hk2777,
hk0014, hk3900, hk0604, hk0127, hk0123, hk0754, hk0272, hk0173, hk0119, hk0588,
hk0059, hk0081, hk1098, hk0035, hk0034, hk0337, hk1176, hk0369, hk0163, hk0147,
hk0996, hk0120, hk0028, hk0715, hk0497, hk0230, hk0190, hk0106, hk1207, hk0124,

Table 1 Basic information of the selected stocks

Id	StockCode	Capitalization	Industry	Id	StockCode	Capitalization	Industry
1	hk0016	370.234B	Real Estate—General	36	hk0715	6.376B	Real Estate—General
2	hk3333	322.866B	Real Estate—General	37	hk0497	5.119B	Real Estate—General
3	hk0688	311.704B	Real Estate—General	38	hk0230	5.018B	Real Estate—General
4	hk1109	215.834B	Real Estate—General	39	hk0190	4.093B	Real Estate—General
5	hk0017	125.411B	Real Estate—General	40	hk0106	3.996B	Real Estate—General
6	hk0020	122.534B	Real Estate—General	41	hk1207	3.763B	Real Estate—General
7	hk0087	116.963B	Real Estate—General	42	hk0124	3.646B	Real Estate—General
8	hk0019	115.444B	Real Estate—General	43	hkll68	3.541B	Real Estate—General
9	hk0083	92.443B	Real Estate—General	44	hk0160	3.494B	Real Estate—General
10	hk0813	84.149B	Real Estate—General	45	hk0050	3.199B	Real Estate—General
11	hk3383	63.77B	Real Estate—General	46	hk0232	3.07B	Real Estate—General
12	hk2777	61.354B	Real Estate—General	47	hk0367	2.944B	Real Estate—General
13	hk0014	48.903B	Real Estate—General	48	hk1278	2.772B	Real Estate—General
14	hk3900	26.749B	Real Estate—General	49	hk0655	2.638B	Real Estate—General
15	hk0604	26.029B	Real Estate—General	50	hk0199	2.607B	Real Estate—General
16	hk0127	22.663B	Real Estate—General	51	hk1838	2.587B	Real Estate—General
17	hk0123	21.826B	Real Estate—General	52	hk0129	2.468B	Real Estate—General
18	hk0754	17.983B	Real Estate—General	53	hk0726	2.432B	Real Estate—General
19	hk0272	17.698B	Real Estate—General	54	hkl222	2.275B	Real Estate—General

(continued)

Table 1 (continued)

Id	StockCode	Capitalization	Industry	Id	StockCode	Capitalization	Industry
20	hk0173	16.187B	Real Estate—General	55	hk0859	2.138B	Real Estate—General
21	hk0119	14.39B	Real Estate—General	56	hk0542	1.967B	Real Estate—General
22	hk0588	14.193B	Real Estate—General	57	hk0231	1.795B	Real Estate—General
23	hk0059	14.047B	Real Estate—General	58	hk0456	1.659B	Real Estate—General
24	hk0081	12.529B	Real Estate—General	59	hk0214	1.614B	Real Estate—General
25	hk1098	11.609B	Real Estate—General	60	hk0112	1.541B	Real Estate—General
26	hk0035	10.68B	Real Estate—General	61	hk1124	1.507B	Real Estate—General
27	hk0034	10.66B	Real Estate—General	62	hk0063	1.367B	Real Estate—General
28	hk0337	10.194B	Real Estate—General	63	hk0298	1.362B	Real Estate—General
29	hk1176	9.572B	Real Estate—General	64	hk0278	1.125B	Real Estate—General
30	hk0369	9.309B	Real Estate—General	65	hk0459	613.795M	Real Estate—General
31	hk0163	8.826B	Real Estate—General	66	hk8155	517.437M	Real Estate—General
32	hk0147	8.229B	Real Estate—General	67	hk2088	401.499M	Real Estate—General
33	hk0996	8.11B	Real Estate—General	68	hk0495	302.564M	Real Estate—General
34	hk0120	7.773B	Real Estate—General	69	hk0616	228.369H	Real Estate—General
35	hk0028	7.353B	Real Estate—General	70	hk0736	204.018M	Real Estate—General
				71	hk1064	178.657H	Real Estate—General

C. Chuxun and J. Lai

hk1168, hk0160, hk0050, hk0232, hk0367, hk1278, hk0655, hk0199, hk1838, hk0129, hk0726, hk1222, hk0859, hk0542, hk0231, hk0456, hk0214, hk0112, hk1124, hk0063, hk0298, hk0278, hk0459, hk8155, hk2088, hk0495, hk0616, hk0736, hk1064]

The capitalization of the stocks in Set 1 (L set) ranges from 215.834B to 372.524B, while the capitalization of the stocks in Set 2 (S set) is lower than 200B. Based on the range of capitalization, we consider the stocks in Set 1 as the "large" stocks and stocks in Set 2 as the "small" stocks. Table 2 shows the grouping result.

3.3 Support Vector Machine

Support Vector Machine (SVM) is a useful learning algorithm for classification and decision-making. Burges [12] shows the whole details of SVM of two classification types. This paper use "One Vs Rest Classifier" SVM method due to three different types of stock movement. A brief introduction of the SVM is as follows.

Assume a giving training data set in features space $T = \{(x1, r1), (x2, r2), ..., (xN, rN)\}$, where $xi \in Rn$, $ri \in \{+1, -1\}$, $i = 1, 2, ..., N$, xi stands for (i) th feature vector.

Giving the linearly separable training set, the generated hyper-plane by learning is $w^* \cdot x + b^* = 0$.

The decision function:

$$f(x) = \text{sign}(w^* \cdot x + b^*)$$

The linear SVM can classify training set into groups with an optimal separating hyper-plane.

We hope to find the w and w_0, for

$$r^t = +1, \ w^T x^t + w_0 >= +1 \tag{1}$$

$$r^t = -1, \ w^T x^t + w_0 <= -1 \tag{2}$$

when $r^t \in \{-1, +1\}$, the combination of Eqs. (1) and (2) is

$$r^t(w^T x^t + w_0)/||w||$$

At least for any ρ, we hope the ρ:

$$r^t(w^T x^t + w_0)/||w|| >= \rho, \forall t$$

In order to get the unique solution: we set $\rho||w|| = 1$. For minimizing the ||w||, the task can be defined as

Table 2 Clustered stock groups

ID	StockCode	Capitalization	Industry	Class	Id	StockCode	Capitalization	Industry	Class
1	hk0016	370.234B	Real Estate—General	1	36	hk0715	6.376B	Real Estate—General	2
2	hk3333	322.866B	Real Estate—General	1	37	hk0497	5.119B	Real Estate—General	2
3	hk0688	311.704B	Real Estate—General	1	38	hk0230	5 018B	Real Estate—General	2
4	hk1109	215.834B	Real Estate—General	1	39	hk0190	4 093B	Real Estate—General	2
5	hk0017	125.41 1B	Real Estate—General	2	40	hk0106	3.996B	Real Estate—General	2
6	hk0020	122.534B	Real Estate—General	2	41	hk1207	3.7638	Real Estate—General	2
7	hk0087	116.963B	Real Estate—General	2	42	hk0124	3.546B	Real Estate—General	2
8	hk0019	115.444B	Real Estate—General	2	43	hk1168	3.541B	Real Estate—General	2
9	hk0083	92.443B	Real Estate—General	2	44	hk0160	3.494B	Real Estate—General	2
10	hk0813	84.149B	Real Estate—General	2	45	hk0050	3.1996	Real Estate—General	2
11	hk3383	63.77B	Real Estate—General	2	46	hk0232	3.07B	Real Estate—General	2
12	hk2777	61.354B	Real Estate—General	2	47	hk0367	2.9448	Real Estate—General	2
13	hk0014	48.903B	Real Estate—General	2	48	hk1278	2.7728	Real Estate—General	2
14	hk3900	26.749B	Real Estate—General	2	49	hk0655	2.6388	Real Estate—General	2
15	hk0604	26.029B	Real Estate—General	2	50	hk0199	2.6078	Real Estate—General	2
16	hk0127	22.663B	Real Estate—General	2	51	hk1838	2.587B	Real Estate—General	2
17	hk0123	21.826B	Real Estate—General	2	52	hk0129	2.4588	Real Estate—General	2
18	hk0754	17.983B	Real Estate—General	2	53	hk0726	2.432B	Real Estate—General	2
19	hk0272	17.698B	Real Estate—General	2	54	hk1222	2.275B	Real Estate—General	2

(continued)

Table 2 (continued)

ID	StockCode	Capitalization	Industry	Class	Id	StockCode	Capitalization	Industry	Class
20	hk0173	16.187B	Real Estate—General	2	55	hk0859	2.1388	Real Estate—General	2
21	hk0119	14.39B	Real Estate—General	2	56	hk0542	1.967B	Real Estate—General	2
22	hk0588	14.193B	Real Estate—General	2	57	hk0231	1.7956	Real Estate—General	2
23	hk0059	14.047B	Real Estate—General	2	58	hk0456	1.6596	Real Estate—General	2
24	hk0081	12.529B	Real Estate—General	2	59	hk0214	1.6148	Real Estate—General	2
25	hk098	11.609B	Real Estate—General	2	60	hk0112	1.541B	Real Estate—General	2
26	hk0035	10.68B	Real Estate—General	2	61	hk1124	1.507B	Real Estate—General	2
27	hk0034	10.66B	Real Estate—General	2	62	hk0063	1.367B	Real Estate—General	2
28	hk0337	10.194B	Real Estate—General	2	63	hk0298	1.3628	Real Estate—General	2
29	hk1176	9.572B	Real Estate—General	2	64	hk0278	1.1256	Real Estate—General	2
30	hk0369	9.309B	Real Estate—General	2	65	hk0459	613.795M	Real Estate—General	2
31	hk0163	8.826B	Real Estate—General	2	66	hk8155	517.437M	Real Estate—General	2
32	hk0147	8.229B	Real Estate—General	2	67	hk2068	401.499M	Real Estate—General	2
33	hk0996	8.11B	Real Estate—General	2	68	hk0495	302.564M	Real Estate—General	2
34	hk0120	7.7738	Real Estate—General	2	69	hk0616	228.369M	Real Estate—General	2
35	hk0028	7.3538	Real Estate—General	2	70	hk0736	204.018M	Real Estate—General	2
					71	hk1064	178.657M	Real Estate—General	2

Fig. 1 Classification sample of SVM [15]

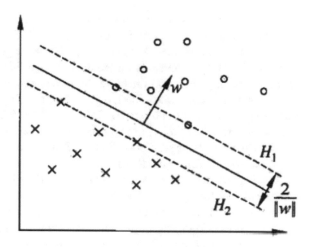

$$\text{Min} \, 1/2 ||w||^2,$$
$$\text{Subject to:} \, r^t(w^T x^t + w_0)/||w|| >= +1$$

Within linearly separable, the sample points in training set that are closest to the hyper-plan are called support vectors, Fig. 1 shows the classification sample of SVM. For positive points, the support vectors in hyper-plane are as follows:

$$H_1 : w \cdot x + b = 1$$

For negative points, the support vectors in hyper-plane are as follows:

$$H_2 : w \cdot x + b = -1$$

H_1 and H_2 are the support vectors.

4 Experiment and Results

4.1 Correlation Measurement and Data Selection

The two variables with a correlation coefficient higher than 0.8 imply they have a strong correlation to each other [13]. To find out the stocks combination between the L set and the S set with high correlation, we use the stock price of 750 trading days for correlation measurements between two sets. The days selected from the S set (t + 1 day) are one day lag than the days selected from L sets (t day). For example, we predict the 751th day's movement of the stocks in the S set with the data of the stocks in the L set (from day 1 to day 750); while we predict the 752th day's movement of the stocks in the S set with the data of the stocks in the L set (from day 2 to day 751). In this way, we use the nearest 750-trade day movement trend of stocks in the L set as training set. This guarantees that the applied information is the most updated. Too many dimensions with relatively small size sample can lead to the curse of dimensionality [14]. We, therefore, limit the number of dimensions and make it smaller than the cube root of the sample size. The sample size that we used is 750-trading day records, the cube root of 750 is: $750^{1/3} = 9.0856$, since each stock has 3 dimensions, which are raise, drop, peace. Hence, we can choose two stocks with $3^2 = 9$ dimensions as our prediction factors. Same as previous mentioned, we update the sample period for correlation measurement, we also use the new updated period for training, which means we use previous 750 days for predicting the 751th day's movement. The movement ratios in price (in terms of percentage) are calculated. In this study, we used a 100-day moving window to examine if there is any lead–lag effect in the individual stocks' movements between large stocks and small stocks.

The basic steps for prediction model are as follows:

1. Pick a small stock from the S set and a large stock from the L set.
2. Calculate the correlation with small stock and large stock.
3. Iterate the Step 1 and 2 for 100 days.
4. Pick the small stock(s) that have strong correlation (which is higher than 0.8) with at least two stocks in the L set every day.
5. Predict the small stocks movement by using SVM, the inputs are the large stocks movement ratio.

4.2 Results

By using the method stated in Sect. 4.1, a small stock coded HK0035 is extracted.

For 100 times prediction, the results are as follows (Fig. 2):

In the back-testing, we used 100-day data dated from December7, 2017 to May 8, 2018, the prediction accuracy is up to 58%, which means the predictions are meaningful.

Stocks	Training date	Testing date	Accuracy
HK0035	16 Nov2014---- 6 Dec 2017	7 Dec 2017 ---- 8 May 2018	58 / 100 (58%)
Two stocks most corre-lated to HK0035	15 Nov2014---- 5 Dec 2017	6 Dec 2017 ---- 7 May 2018	/

Fig. 2 Prediction parameters and results

5 Conclusion and Future Development

This paper uses SVM applying the change of large stock movement in the same stock industry. On average, the accuracy of our prediction is higher than 50%. In the future research, we can further examine the co-movement effect using moving average of stock prices, lead–lag relation between large stocks and small stocks, etc. A larger data set may be helpful for predicting the co-movement effect, as we believe that co-movement between stock prices is quite consistent.

References

1. Al-Radaideh, Q.A., Assaf, A.A., Alnagi, E.: Predicting stock prices using data mining techniques. In: The International Arab Conference on Information Technology (ACIT 2013)
2. MacKinlay, A.C., Lo, A.W.: When are contrarian profits due to stock market over reaction. Rev. Financ. Stud. **3**(2), 175–205 (1989)
3. Jain, A.K., Murty, M.N., Flynn, P.J.: Data clustering: a review. ACM Comput. Surv. (CSUR) **31**(3), 264–323 (1999)
4. MacQueen, J.: Some methods for classification and analysis of multivariate observations. In: Proceedings of the Fifth Berkeley Symposium on Mathematical Statistics and Probability, Volume I: Theory of Statistics (1967)
5. Mukaka, M.M.: Statistics corner: a guide to appropriate use of correlation coefficient in medical research. Malawi Med. J. **24**(3), 69–71 (2012)
6. Bonanno, G., Lillo, F., Mantegna, R.N.: High-frequency cross-correlation in a set of stocks. Quant. Financ. **1** (2001)
7. Kenett, D.Y., Tumminello, M., Madi, A., Gur-Gershgoren, G., Mantegna, R.N., Ben-Jacob, E.: Dominating clasp of the financial sector revealed by partial correlation analysis of the stock market. PloS one **5**(12) (2010)
8. Hung, W., Nakamori, Y., Wang, S.Y.: Forecasting stock market movement direction with support vector machine. Comput. Oper. Res. **32**(10), 2513–2522 (2005)
9. Kim, K.: Financial time series forecasting using support vector machines. Neurocomputing, **55**(1–2), 307–319 (2003)
10. Lin, Y.L., Guo, H.X., Hu, J.L.: An SVM-based approach for stock market trend prediction. In: The 2013 International Joint Conference on Neural Networks (IJCNN)
11. Gavrishchaka, V.V., Banerjee, S.: Support vector machine as an efficient framework for stock market volatility forecasting. CMS **3**(2), 147–160 (2006)
12. Burges, C.J.C.: A tutorial on support vector machines for pattern recognition. Data Min. Knowl. Disc. **2**(2), 121–167 (1998)
13. Zou, K.H., Tuncali. K., Silverman, S.G.: Correlation and simple linear regression. Radiology **227**(3) (2003)
14. Vlachos, M., Domeniconi, C., Gunopulos, D.: Non-linear dimensionality reduction techniques for classification and visualization. In: Proceedings of the Eighth ACM SIGKDD International Conference on Knowledge Discovery and Data Mining, pp. 645–651 (2002)
15. Li, H.: Statistical Learning Method. Tsing Hua University Press (2012)
16. Taylor, R.: Interpretation of the correlation coefficient: a basic review. J. Diagn. Med. Sonogr. **6**(1), 35–39 (1990)
17. Chen, N.F., Chan, K.C.: Structural and return characteristics of small and large firms. J. Financ. **46**, 1467–1484 (1991)
18. Celebi, M.E., Kingravi, H.A., Vela, P.A.: A comparative study of efficient initialization methods for the K-means clustering algorithm. Expert. Syst. Appl. **40**(1), 200–210 (2013)

19. Derby, R., Seo, K.S., Chen, Y.C., Lee, S.H., Kim, B.J.: A factor analysis of lumbar intradiscal electrothermal annuloplasty outcomes. Spine J. **5**(3), 256–261 (2005)
20. Suykens, J., Vandewalle, J.: Least squares support vector machine classifiers. Neural Process. Lett. **9**(3), 293–300 (1999)
21. Vapnik, V.: The nature of statistical learning theory. **3**(2), 147–160 (2006)

A New Multi-criteria Recommendation Algorithm for Groups of Users

Shenghui Guo, Shu-Juan Ji, Chunjin Zhang, Xin Wang and Jianli Zhao

Abstract Group recommendation algorithms have the advantage of helping groups of users find favorite items under less time. And, the multi-criteria recommendation system aims at obtaining the user's preferences over various aspects and make accurate recommendation. This paper presents a new group-oriented multi-criteria recommendation algorithm called GMURec. This algorithm first uses K-means algorithm which generates the groups (i.e., sets of users who have similar interests). Then, it uses the BP neural network to aggregate the groups preferences and learn the implicit relationship between multi-ratings and overall rating of each group. The performance of GMURec algorithm is compared with three baseline algorithms. Experimental results show that: (1) the precision of GMURec is only lower than the individual-targeted personal multi-criteria recommendation algorithm, but is higher than the other two group ones; (2) the recall of GMURec is as good as or better than other algorithms; (3) the run time of GMURec is the least one among the compared algorithms.

Keywords Multi-criteria recommendation · Group recommendation ·
BP neural network · Preferences of group

1 Introduction

With the development of electronic commerce, the transaction volume grows exponentially, which causes an information explosion. It is very difficult to find useful information from such large-scale data. Therefore, the recommendation system

S. Guo · S.-J. Ji (✉) · J. Zhao
Key Laboratory for Wisdom Mine Information Technology of Shandong Province,
Shandong University of Science and Technology, Qingdao, China
e-mail: jane_ji2003@aliyun.com

C. Zhang · X. Wang
Network Information Center (NIC), Shandong University of Science and Technology,
Qingdao 266590, Shandong, China

© Springer Nature Singapore Pte Ltd. 2019
J. S. Pan et al. (eds.), *Genetic and Evolutionary Computing*,
Advances in Intelligent Systems and Computing 834,
https://doi.org/10.1007/978-981-13-5841-8_76

came into being. According to the difference of recommended targets, recommendation algorithms can be divided into individual-oriented ones and group-oriented ones. Some researchers prove that group-oriented recommendation algorithms are more efficient than individual ones [1–4]. The group-oriented recommendation algorithms aim at recommending for an entity with different preferences. In the process of making purchase decisions, the users consider many aspects. For example, when booking a hotel, someone may pay more attention to cleanliness, some people are more concerned about distances, and others may focus more on prices. Therefore, some e-commerce platforms provide multi-criteria ratings for getting users' detailed preferences. Recently, some researchers [5–7] proposed algorithms to improve the accuracy by using multi-criteria ratings. What is more, researchers [8, 9] used linear regression method to learn the relationship between multi-criteria ratings and overall ratings to acquire the preferences of users.

Although researchers developed various kinds of group recommendation algorithms to relieve the data sparseness problem and to improve recommendation efficiency, the recommendation accuracy of group-oriented recommendation algorithms are still quite low. Moreover, the existing multi-criteria recommendation algorithms are still limited by data sparseness problem because there are not enough ratings from individuals. To address these problems, this paper proposes a new multi-criteria recommendation algorithm named GMURec algorithm for users group. The novel characteristics of this paper are as follows. (1) The groups used in this paper are virtual ones that are obtained by clustering of similar users. In the recommendation process, each group can be seen as a unit, which can reduce the impact of data sparseness problem and the consumption of time. (2) The GMURec algorithm first uses BP neural network to aggregate the group's preferences, as well as learn the implicit relationship between multi-ratings and overall ratings of each group. (3) The GMURec algorithm does not need to predict ratings that an individual may give to each item.

The remainder of this paper is organized as follows. Section 2 explains the main idea of our recommendation algorithm. Section 3 describes the experimental results. Finally, we concludes this paper in Sect. 4.

2   The GMURec Algorithm

2.1   *Concepts in Multi-criteria Recommendation System*

In the recommendation algorithm given in this paper, there are three concepts that are frequently used. To specify these concepts explicitly, we define them formally as follows.

Definition 1 (*Rating matrix*) $R_{m \times n} = [\overrightarrow{r}_{ui}]_{m \times n}$ represents the rating matrix that user u rated item i. The set of users and items is denoted as $User = \{u_1, u_2, u_3, \dots, u_m\}$ and $Item = \{i_1, i_2, i_3, \dots, i_n\}$, respectively. The rating matrix can be represented as formula (1).

$$R_{m \times n} = \begin{pmatrix} \overrightarrow{r}_{11} & \cdots & \overrightarrow{r}_{1n} \\ \vdots & \ddots & \vdots \\ \overrightarrow{r}_{m1} & \cdots & \overrightarrow{r}_{mn} \end{pmatrix} \tag{1}$$

where $\overrightarrow{r}_{ui} = [a_1, a_2, \dots, a_k, tr_{ui}]$ is the rating vector that user u rated item i. In addition, a_k is the rating that user u rated item i on aspect k, tr_{ui} is the overall rating that user u rated item i.

Definition 2 (*Users groups*) A users group $g_{num} = [u_1, u_2, \dots, u_p]$ (p is the number of users in the num^{th} group) is a set of users who have the same interests, i.e., who gave similar ratings on various aspects and overall rating of items. The user groups $G = [g_1, g_2, \dots, g_{num}]$ is a set of user groups.

Definition 3 (*User's aggregated ratings over various aspects*) Supposing that there are m users who had given ratings $R_m = [\overrightarrow{r}_1, \overrightarrow{r}_2, \dots, \overrightarrow{r}_m]$ to a type of items over various aspect. The aggregated ratings of user u over the aspects and the overall rating is a vector that are aggregated based on his/her ratings over various aspects, which can be calculated by formula (2), (3), and (4).

$$\overrightarrow{r}_u = [\overline{a}_u^1, \overline{a}_u^2, \dots, \overline{a}_u^k, \overline{tr}_u] \tag{2}$$

$$\overline{a}_k = \frac{1}{n} \sum_{i=1}^{n} a_{ui}^{\ k} \tag{3}$$

$$\overline{tr}_u = \frac{1}{n} \sum_{i=1}^{n} \overline{tr}_{ui} \tag{4}$$

where m is the number of the users, and n is the number of ratings that user u gave in history, $\overrightarrow{r}_u$ is the rating vector that user u rated on the various aspects, $\overline{tr}_u$ is the averaged overall rating that user u gave, k is the number of aspects of the item, $\overline{a}_u^k$ is the averaged rating that user u rated on aspect k, which can be calculated according to formula (3), $a_{ui}^{\ k}$ is the rating that user u rated item i on aspect k.

2.2 Users Group Generation Algorithm

In this section, we use K-means algorithm to cluster similar users into virtual groups. However, the K-means algorithm has a limitation, i.e., the number of groups that we want to get must be given before clustering. To address this limitation, we introduce

Silhouette Coefficient $SC(G)$ [10] (which is a kind of evaluation method of the group G) to the traditional K-means algorithm, which is described in detail in Algorithm 1.

Algorithm 1: Group Generation Algorithm

Input: The data of users $R_m = [\vec{r}_1, \vec{r}_2, ..., \vec{r}_m]$
Output: The optimal groups of users G

1: **for** i in c **do** $//c \in [c_1, c_2]$
2:   $G_i = K - means(R_u, i)$;
3:   Compute $SC(G_i)$;
4: **end for**
5: $G = \arg\max(SC(G_i))$;

2.3 Group-Oriented Multi-criteria Recommendation Algorithm Based on User Groups (GMURec)

After getting the users group, we use a three-layer BP neural network to learn each group's implicit weight over various aspects of the items. In learning process, we take the ratings of users over various aspects of the item as input, and the corresponding overall ratings as output. Moreover, the gradient descent algorithm [11] is adopted to learn each group's weight (denoted as $w_g = \{w_{g_1}, w_{g_2}, ..., w_{g_{num}}\}$) over various aspects and the bias (denoted as $b_g = \{b_{g_1}, b_{g_2}, ..., b_{g_{num}}\}$). Formally, based on the learned w_g and b_g, the trained neural network model can be represented as formula (5).

$$\widehat{tr}_{g_{num}i} = f(w_{g_{num}}{}^T * \vec{r}_{g_{num}i} + b_{g_{num}}) \tag{5}$$

where the $\vec{r}_{g_{num}i}$ is the vector of ratings that the users in num^{th} group rated item i over various aspect, $\widehat{tr}_{g_{num}i}$ is the predicted overall rating the num^{th} group for item i, $w_{g_{num}}$ represents the preferences weight of the num^{th} group for various aspects of the item. $b_{g_{num}}$ is the bias of the num^{th} group on implicit weight learning. $f(\cdot)$ is represents the relationship between the ratings on various aspects and overall rating, which is determined with neural network.

Based on the learned implicit relationship between aspect ratings and overall ratings of each group of users, we can take each group as a unit and select the most suitable items to make a recommendation. To achieve this aim, we must determine that (1) which item may be preferred by a group? (2) the degree that the group like an item. Definition 4 defines the concept of *preference*.

Definition 4 $preference(g_{num}, I_n)$ represents the degree that the item n is preferred by the group num, which is calculated by formulas (6) and (7).

$$preference\,(g_{num}, I_n) = \frac{|\{tr_{I_n} | \widehat{tr}_{I_n} = tr_{I_n}\}|}{|R_{I_n}|} \tag{6}$$

$$\widehat{tr}_{I_n} = f(w_{g_{num}}{}^T * \overrightarrow{r}_{I_n} + b_{g_{num}}) \tag{7}$$

where I_n is the rating that users rated for item n, $\widehat{tr}_{I_n}$ is the predicted overall rating corresponding to $\overrightarrow{r}_{I_n}$, which is calculated by formulas (7), tr_{I_n} is the real overall rating corresponding to $\widehat{tr}_{I_n}$, $w_{g_{num}}$ represents the implicit weight of the num^{th} group over various aspects of the item. $\overrightarrow{r}_{I_n}$ is the vector of ratings over the various aspects for item n, $f(\cdot)$ represents the relationship between the ratings on various aspects and overall rating, which is determined with neural network.

The calculation principle of formulas (6) and (7) is as follows: (1) taking all the ratings that the users rated each item as test samples of the corresponding trained neural network that are represented in formula (5); (2) using the trained neural network to predict the overall rating $\widehat{tr}_{I_n}$ with the dataset R_{I_n} represented in formula (7); (3) check whether the predicted overall rating (i.e., $\widehat{tr}_{I_n}$) is similar to the real one (i.e., tr_{I_n}) that each user rates item n. If so, then it is believed that this item is preferred by the corresponding groups of users; (4) calculate the ratio of the numbers of item n that are preferred by all users to the total number of the samples R_{I_n} (i.e., all the ratings given about the item n) as formula (6).

Based on the concept of *preference*, we present a multi-criteria recommendation algorithm for groups based on users (abbr., GMURec). The detail steps of this algorithm can be summarized as Algorithm 2.

Algorithm 2: GMURec Algorithm

Input: The data set of the num^{th} group and the data set of all item I
Output: The recommendation list of group num

1: Learn the weight $w_{g_{num}}$ using the neural network;
2: **for** I_n in I **do**
3:    Compute $preference_n$ (g_{num}, I_n) according to (6);
4: **end for**
5: $recommendation = \arg\max_{n} \_N(preference_n (g_{num}, I_n))$;

---

3 Experiment

To verify the performance of the recommendation algorithms given in this paper, we select MPRec [8], GBRec[12], PromoRec [13] as baseline algorithm. As we all know, the individual-oriented recommendation algorithms are more targeted than group ones, therefore, it is natural that the precision of individual-oriented recommendation algorithms are higher than the one of group-oriented recommendation algorithms. Hence, this paper chose an individual-oriented algorithm named MPRec as an up line. The GFRec, PromoRec algorithms are chosen as baseline algorithms because they are the recent or typical group-oriented recommendation algorithms.

3.1 Data and Evaluation Criteria

We implement the experiments over Tripadvisor (https://pan.baidu.com/s/1GdUAFf8NP1dleSRFeVvGWA). To decrease the sparseness of experimental data, we filter out the extremely inactive users and unpopular items (i.e., users with less than 10 data and items with less than 100 transactions) and divide the data set into 10 parts in random. 80% of them are used as the training set and the remaining 20% is selected as the test set. Cross-validation is adopted in experiments. The Precision [14], Recall [14], and the run time are selected as the evaluation criteria in this paper.

3.2 Results and Analysis

The comparison result of precisions is shown in Fig. 1. From Fig. 1, we can see the curves of GMURec algorithm and MPRec algorithm maintain smooth with the increase of recommendation list length. Therefore, we can say that these two algorithms are robust. In detail, when the length of the recommendation list is 7, the precision of GMURec algorithm is about 0.777, which is its optimal value. However, this optimal value is still 0.036 smaller than the one of MPRec (whose precision always range about 0.813). That is, because, the MPRec algorithm is a kind of personal multi-criteria recommendation algorithm, which learns individual preferences. Different from MPRec, GMURec algorithm is a group-oriented recommendation algorithm, which uses group preferences instead of individual preferences in recommending. It should be noted that the precision curves of the GBRec algorithm and PromoRec algorithm decrease sharply when the length of the recommendation list increase. That is, because, the GBRec algorithm and the PromoRec algorithm are two kinds of group-oriented algorithms without considering the ratings over multiple aspects of items. Therefore, we can conclude that these two algorithms are not robust ones.

The recall of comparison algorithms is presented in Fig. 2a. From this figure, we can see that the recall are generally low, that is because the processed data set is relatively large (20,443 users, 1755 items, 262,300 ratings, and 99.3% sparsity). In detail, we can see that the recall increases with the length of the recommendation

Fig. 1 The precision of comparison recommendation algorithms

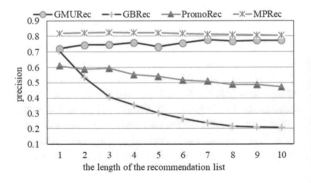

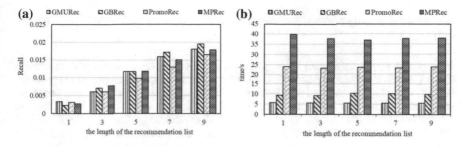

Fig. 2 The recall **a** and run time **b** of different recommendation algorithms

list. It is obvious that the recall of GMURec algorithm is better than the PromoRec algorithm whatever the length of the recommendation list is. That is, because, it considers the implicit relationship between multiple aspects of items and the overall ratings while the PromoRec does not consider. When the length of the recommendation list is 3, 7, or 9, the recall of GBRec algorithm is highest. That is, because, it is group-oriented and aggregate group preference by using fairness strategy. Although the recall of these algorithms is different, the maximum gap of these algorithms on recall is smaller than 0.002, which is almost negligible for large quantities of users.

The results of run time among these algorithms are listed in Fig. 2b. From Fig. 2b, we find that the GMURec algorithm takes the least time (i.e., only 5 s). That is, because, they are group-oriented and need not predict the rating of all users. In contrast, the MPRec algorithm takes the most time. (i.e., almost 40 s). The reason is that the MPRec algorithm is an individual-oriented one. The PromoRec algorithm also takes a long time, i.e., over 20 s. That is, because, the PromoRec algorithm is group-oriented and need to predict the ratings of all users in each group. According to above-experimental results, we can find that our algorithms can reach good precision and simultaneously save much more run time.

4   Conclusion

In this paper, we combine the advantage of multi-criteria recommendation algorithm and the group recommendation algorithm to propose GMURec algorithm. The novelty of the GMURec algorithm is that it can mine the implicit relationship between multi-ratings and overall ratings without predicting users' ratings. Experimental results show that the precision of GMURec is higher than the other two baseline group recommendation algorithms, and the run time of GMURec is greatly less than all the baseline algorithms. Using the GMURec algorithm proposed in this paper, we can learn which aspect of each item should be paid more attention by users. Besides, the group-oriented idea can help solve the problem of data sparsity in collaborative filtering algorithms. In the future work, we can try to integrate text reviews into the prediction of users' on various aspects of the item. That is, because, most e-commerce sites only provide text reviews and overall ratings other than multi-criteria ratings.

Acknowledgements This paper is supported in part by the Natural Science Foundation of China (71772107, 71403151, 61502281, 61433012), Qingdao social science planning project (QDSKL1801138), the National Key R&D Plan (2018YFC0831002), Humanity and Social Science Fund of the Ministry of Education (18YJAZH136), the Key R&D Plan of Shandong Province (2018GGX101045), the Natural Science Foundation of Shandong Province (ZR2018BF013, ZR2013FM023, ZR2014FP011), Shandong Education Quality Improvement Plan for Postgraduate, the Leading talent development program of Shandong University of Science and Technology and Special funding for Taishan scholar construction project.

References

1. Zhang, C., Zhou, J., Xie, W.: A users clustering algorithm for group recommendation. In: Applied Computing and Information Technology. In: International Conference on Computational Science/Intelligence and Applied Informatics. International Conference on Big Data. Cloud Computing. Data Science & Engineering, pp. 352–356. IEEE (2017)
2. Ntoutsi, I., Stefanidis, K., Norvag, K., Kriegel, H.P.: gRecs: a group recommendation system based on user clustering. In: International Conference on Database Systems for Advanced Applications, pp. 299–303 (2012)
3. Ntoutsi, E., Stefanidis, K., Kriegel, H.P.: Fast group recommendations by applying user clustering. In: International Conference on Conceptual Modeling, pp. 126–140 (2012)
4. Garcia, I., Sebastia, L., Onaindia, E., Guzman, C.: A group recommender system for tourist activities. In: International Conference on E-Commerce and Web Technologies, pp. 26–37 (2009)
5. Plantié, M., Montmain, J., Dray, G.: Movies recommenders systems: automation of the information and evaluation phases in a multi-criteria decision-making process. In: Database and Expert Systems Applications, International Conference, Copenhagen, Denmark, August 22–26 (2005)
6. Bilge, A., Kaleli, C.: A multi-criteria item-based collaborative filtering framework. In: International Joint Conference on Computer Science and Software Engineering, pp. 18–22. IEEE (2014)
7. Nilashi, M., Dalviesfahani, M., Roudbaraki, M.Z., Ramayah, T., Ibrahim, O.: A Multi-criteria Collaborative Filtering Recommender System Using Clustering and Regression Techniques. Social Science Electronic Publishing (2016)
8. Majumder, G.S., Dwivedi, P., Kant, V.: Matrix factorization and regression-based approach for multi-criteria recommender system, pp. 103–110 (2017)
9. Zheng, Y.: Criteria chains: a novel multi-criteria recommendation approach. In: International Conference on Intelligent User Interfaces, pp. 29–33. ACM (2017)
10. Sai, L.N., Shreya, M.S., Subudhi, A.A.: Optimal k-means clustering method using silhouette coefficient. **8**(3), 335 (2017)
11. Liang, G.: Neuron adaptive and neural network based on gradient descent searching algorithm for diagonalization of relative gain sensitivity matrix decouple control for MIMO system. In: IEEE International Conference on Networking, Sensing and Control, pp. 368–373. IEEE (2008)
12. Pessemier, T.D., Dooms, S., Martens, L.: Comparison of group recommendation algorithms. Multimed. Tools Appl. **72**(3), 2497–2541 (2014)
13. Zhu, Q., Zhou, M., Liang, J., Yan, T., Wang, S.: Efficient promotion algorithm by exploring group preference in recommendation. In: IEEE International Conference on Web Services, pp. 268–275. IEEE (2016)
14. Xu, X., Liu, J.: Collaborative filtering recommendation algorithm based on multi-level item similarity. Comput. Sci. **34**, 262–265 (2016)

Author Index